Basic Live Sound Reinforcement

Basic Live Sound Reinforcement

A Practical Guide for Starting Live Audio

Raven Biederman and Penny Pattison

NEW YORK AND LONDON

First published 2014
by Focal Press
70 Blanchard Rd Suite 402, Burlington, MA 01803

Simultaneously published in the UK
by Focal Press
2 Park Square, Milton Park, Abingdon, Oxon OX14 4RN

Focal Press is an imprint of the Taylor & Francis Group, an informa business

Library of Congress Cataloging in Publication Data
Biederman, Raven.
Basic live sound reinforcement : a practical guide for starting live audio / by Raven Biederman and Penny Pattison.
pages cm
Includes index.
1. Theaters–Electronic sound control. 2. Auditoriums–Electronic sound control. 3. Sound–Equipment and supplies. 4. Music–Performance. I. Pattison, Penny. II. Title.
TK7881.9.B54 2013
621.382'8–dc23
2013002925

ISBN: 978-0-240-82101-6 (pbk)
ISBN: 978-0-240-82483-3 (ebk)

Typeset in TimesNewRoman
By TNQ Books and Journals, Chennai, India

Contents

Part 4 Getting Ready for the Show

Part 5 Show Time and Beyond

Part 6 Becoming a Live Sound Professional

The companion website can be found at:
www.focalpress.com/CW/biederman

Introduction

We guess, if you're reading this section, that you may be one of our *potential* readers, but want to see what this book and its authors are all about before deciding if this is *the* right live sound book for you. Maybe you have already invested your hard earned dollars in this text, perhaps as part of a school course, or enticed by the prominent placement of the humble word "basic" in the title, and are here now to learn more about our approach. We hope a brief explanation of our aims in writing this book will answer some of the questions that readers of book introductions may be likely to have.

Readers shopping for a basic live audio text who conceive of "the right book" as being one that provides a solid overview of a breadth of topics, without approaching the sort of encyclopedic and comprehensive coverage that can be more confusing and counterproductive than useful for beginning students, may find that our title has something of value to offer them. Readers who interpret "basic" as being intended for someone without much background information on a topic, without assuming that basic students are unable to comprehend complexity or look up a word in the glossary now and then, may also appreciate a text that offers explanations simple enough to remain accessible to beginning learners, but doesn't oversimplify topic discussions and abandon all of the complexity and nuance characteristic of a more advanced analysis.

However, anyone who feels "the right book" should be one that promises to provide "all they need to know" (including all the fundamentals and everything essential) to run a public address (P.A.) system, get a sound company job, or do anything more specific than give readers a good start from which to begin developing their skills and increasing their knowledge of live audio, then we are afraid this is not the text you seek. We aren't saying someone won't be able to do either task with our text as their sole reference, but we don't promise it for all readers, or even think it's a good idea to try. It can be read alone as an introductory text, but was actually designed so that learners who read or reread it in conjunction with other texts on the same subject will get the maximum benefit; it does not matter whether the reader views it as a main text to be supplemented with others or as a supplement to other texts.

Though we feel that there is no single book that can adequately provide comprehensive coverage of a topic as complex as live audio, or even do justice to only the most "essential" information, we understand that many people do, and we wish anyone on this sort of quest the best of luck. We hope no one is offended when we say we are glad something like this is not even possible, since it isn't to gloat over the difficulty of anyone's search, but because we love live sound, and are glad it is such a rich topic that no one source could begin to do it justice. In a world that is always seeking to devalue the human operator and automate as many new tasks as possible, we feel that's a great indicator of future job security for live sound engineers!

We also want to comment on what makes our approach unique enough that there is room for it, considering the marketplace has so many well-written offerings on the subject already, more than one person could read even if, like we do, they feel that learners should always refer to a variety of sources no matter they are studying. Since who we are and what makes our approach unique are related topics, we also touch on that to kill two birds with one stone (a term detested by Raven Biederman to the boundless delight of his coauthor at the keyboard) and satisfy the curiosity of any book shoppers confused by our rather "ordinary Joe" live sound credentials compared with many other live audio authors.

A number of live sound titles boast authors who are respected live sound luminaries with over 40 years in the field. Some of them have written enough on the subjects of audio and live sound engineering to take up a few shelves, including some titles that come as close to being comprehensive as is possible; weighty tomes running to thousands of pages in length and capable of physically injuring someone if thrown at them. Others authors can be argued to be future luminaries; with some names to drop from their list of clients, some name recognition in the industry themselves, and many years of touring and sound-company experience.

This is a pretty big contrast with our text, where neither of us has been on a tour larger than can fit in a V.W. bus, regularly worked a venue that wasn't a bar or large club, or is on a trajectory to ever become a live sound luminary, and neither has any intention to. Raven Biederman teaches audio (both studio and live) at a respected technical college and also DJs, produces, and on occasion still mixes sound at the sorts of venues he loves best, small venues featuring local talent and colloquially referred to as "bars" by the patrons they serve. Penny Pattison is a writer and educator who has long associated with sound engineers, musicians, and DJs as family and friends, and has years of experience being an enthusiastic consumer of live music (also preferring the small-venue, local variety above all other types), but who has no prior technical training, live sound or otherwise, until taking on the challenge of learning live engineering as part of authoring this text. We are the first to admit that our credentials are not nearly as impressive as some of our colleagues and fellow authors, and that they show that we are clearly not among the movers and shakers in the audio industry; however, we must point out that less impressive does not mean less qualified. To the contrary, we feel that our credentials make us uniquely qualified to introduce students to the subject of live audio.

We have read many of the books written by well-known industry players, and some by authors less known but with as much industry experience, and think by and large they are very good books, ones that we have and will continue to recommend to students. We don't think our text is a "better" approach, or should be a replacement for any of them, only that we can add something to the mix that is useful too. As educators, and especially as co-authors in the unique situation where one has just learned everything covered in the text from scratch aided by the other, what we offer is an understanding of the questions students are likely to have about each topic; the concepts that are most likely to be misunderstood or become hurdles to some students; and what in terms of content and format is not found in other texts that would be the most useful to basic students and make the best bridge from what they know already, to help them get the most out of their studies and reach their learning goals.

Our aim was never to provide information that was exactly the same as similar live audio texts or replace them. Our goal was to write a text that could stand on its own as a solid introductory text, but in a way that also would fit well with what was already written. To that end, we covered a number of topics that are less commonly included in live audio texts, or are covered in less detail, to fill in some gaps where we felt more information was needed or could benefit beginning live audio learners. When discussing much of the required technical information found in all audio texts, we tried to repeat information in a way that would provide another unique access point to the topics, and chose to stick with what we know best by focusing on the technical information and live sound issues most common for engineers handling small-venue sound and of most concern to entry-level live sound professionals. In turn, we also felt free to pay less attention to topics and approaches already well covered by other authors.

We have paired each chapter, and the book as a whole, with a link directory (www.focalpress.com/cw/biederman) students can use as a map to reach the greater live sound community that lives online and generously shares a wealth of their audio knowledge and live sound experience with the rest of us by creating and maintaining an ocean of useful online resources. Every chapter has an associated online directory with links to a handful of related resources and supplementary articles for every topic covered, or even just mentioned, in that chapter. A general directory for the book also contains these links, organized for easy access, so students looking for information don't need to remember the chapter they encountered it in. The general directory contains additional links to every kind of resource that we felt might be useful to live sound students, professionals, and educators at all levels. Any topic we discuss in the book can be studied in more depth through these links, and teachers can use them to help tailor our book to their preferred lesson-plan topic list and order, as we link to sources covering most of the relevant topics they might want to cover that we don't include in our text.

We hope that providing educators and students with a directory to the most reliable and useful web resources that are currently available helps them get the most out our book and that they continue to use it as a tool to help them excel at their future studies and work in live sound.

However, we do ask that readers not direct their appreciation for any support and benefit received through use of this tool toward us or the publishers, but only toward those who rightly deserve it, the companies, organizations, and individual men and women who host, maintain, and contribute to these resources, by contacting the sites that are the most helpful and thanking the people responsible for providing the service. We can't accept the thanks due to others, but can speak for the providers of web content in general by assuring you that such simple expressions of gratitude go a long way toward keeping such resources available.

Part 1

Introducing Live Sound

To begin at the beginning we need to answer a few fundamental questions and establish some basic definitions to build upon. To that end, Chapter 1 begins with the question "What is live sound?", and builds a basic answer to this fundamental question by discussing when and how we use live sound reinforcement and, more importantly, why we use it.

Next, in Chapters 2 and 3 we look at the live sound industry, including topics such as the types of businesses and clients that hire live sound employees, the difference between employment status and freelance work, and examples of typical opportunities for entry-level live sound workers. Many live sound beginners start with a limited understanding of the industry, and even those with more background often start with some misconceptions and holes in their knowledge.

Though we revisit our discussion of the live sound industry in the last section to discuss topics aimed at students about to begin work or intern in the field as is most common, we also feel it's important to start with an overview of the live sound industry as soon as as possible. Knowing the different types of opportunities and hurdles they may encounter can help live sound students choose wisely when faced with initial choices such as the kind of training to seek out, what internship to apply for, and even where to move after finishing training, all of which can influence the choices available to them later on. Even those who are making the most basic choices, like whether they really want to begin studying live sound at all, are better equipped to make the best choice for themselves if it is an informed choice.

Chapter 1

What Is Live Sound?

Live sound reinforcement is a process in which an operator uses audio technology to enhance and re-distribute selected sounds. **Natural sound** is the unprocessed sound that comes directly from an instrument or human vocal chords. **Live sound** is what results anytime natural sound is captured and turned into an electric signal by the microphones of a P.A. system, to be processed and enhanced in the system before it is finally converted back into sound at the loudspeakers and re-distributed to the intended audience throughout a location.

Of course this brief description can hardly begin to define live sound reinforcement, when entire books are required to fully explain the process. Rather than try to define live sound reinforcement right away, we will use this chapter to see if we can come to a more complete understanding of what is created by the reinforcement process, the live sound the audience hears. To do this, we can start by examining the P.A. system used to create it, the types of situations where it is used, and the audience it is created for. By looking at the situations we use live sound reinforcement for, we have a better chance to accurately define why we use it, and by extension will be more able to answer the question posed in the title of this chapter; "What is live sound?" While it will take many chapters to explore all the component parts of the P.A. system, for the task at hand we only need to look at some of the broad general categories of P.A. systems.

The P.A. System

The **P.A. system** encompasses all the equipment, starting with the microphones and ending with the loudspeakers, that is used to convert selected natural sounds into the enhanced live sound that is heard by the audience. The P.A. system comes in many sizes and levels of complexity, and the size of any P.A, as well as the amount of work involved in setting it up and running it, will be determined by the amount of power and features required to process sound for the specific event and venue that P.A. is used for. A **venue** is the place where a live sound event occurs, and most venues will have a specific type of event or limited range of events that will typically occur there. There are some "all-purpose" venues, but more commonly the events held at any venue will be limited by logistical limitations imposed by the size, shape, and location of the venue, or simply by custom. Either way these parameters usually also serve to limit the kinds of P.A. used in that venue.

Installed vs. Portable P.A.

Many venues that focus on throwing regular live sound events will invest in an installed P.A. system, or **fixed installation**. These systems are designed and installed by a consultant who tailors the components and placement of the system to the exact needs and capacity of the venue. The advantages to these systems are that precision equipment can be used to analyze the room and calibrate the system in a way only possible with installed sound. Once tuned the ideal settings for components the operator doesn't need to access in order to run the system can be permanently set and the components can be tucked away in storage shelves or even hidden within architectural elements. This keeps delicate equipment and controls safely hidden away from the audience, keeps cords and cables off the floor where they could pose a tripping hazard to all, and has the added benefit of saving space. Some installation companies also offer varying degrees of service; the best may even promise 24/7 on-call service and a guarantee to always have parts on hand. The main disadvantage of fixed installations is the considerably greater cost of these systems, as well as the relative inflexibility that occurs with lack of portability.

While installed systems have some advantages in being tailored to the venue, there are a variety of types and sizes P.A.s can come in, and the flexibility and scalability afforded to users of portable P.A.s means they can tailor the P.A. used to each application it is used for. This makes portable systems the preferred choice of most sound engineers and regardless of size and features, P.A. systems that are not installed are considered to be transportable from one venue location to another and categorized as **portable P.A. systems**. Naturally some systems are more portable than others, and some manufacturers and vendors have P.A. lineups specifically built for extra easy portability, with features like handles and wheels built into all gear in that line. The following three examples of portable P.A.s exemplify the diverse range of size and complexity that can be found in these systems, from the smallest personal P.A.s, to the massive touring systems used for stadium events.

Personal P.A.

The smallest systems are referred to as **personal P.A.s** and range from those that are hardly larger than personal listening devices to those that are just bigger than a brief case or small carry-on travel valet. The tiniest come with clip-on or headset microphones that plug into speakers small enough they can be clipped to the user's belt. These tiny systems are tailored to applications where a system powerful enough to address audiences of more than a dozen people or reach distances beyond 6 to 10 feet would be too powerful and less portable than is needed. An example of a typical application calling for such a system includes a tour guide leading a cluster of people through a museum. In situations like this some amplification is useful since the audience reaches just far enough beyond conversational range that a speaker must project their voice to be heard. This is tiring to do over an eight-hour shift, and difficult to do at a consistent and precise level that is just loud enough to reach the audience, but not so loud as to disturb others beyond that range. In allowing the speaker to easily carry a system than can boost their volume just enough to reach everyone in their audience but maintain a level low enough to avoid disturbing other people in the general area, limited power personal P.A. systems are a perfect tool for this type of use. These systems are also useful for educators, especially for those teaching the hearing impaired.

The larger of these systems are still very compact, containing just the amount of power needed for venues where the user will be speaking or performing for people in a space around the size of an average coffee shop. Such situations include a poetry reading at a coffee shop or an author's book reading and signing in an area about the same size within a larger space such as a large book store or a university commons. Consisting of a microphone, a cord, and a speaker enclosure with a handle at the top for easy carrying, these systems are easy to lift and will fit in the trunk of a small car. Set up is **plug and play** with few controls beyond turning the system on and setting the volume knob; to prepare to use them the operator only needs to connect the two components with a cord and to plug into a power source. These small systems are also the perfect size for applications like Pub Quiz or similar games, being powerful enough to allow everyone participating to hear, but not so powerful that the sound produced will necessarily intrude on patrons in other areas of the bar.

Band P.A.

A band performing an average bar will need a bigger system to be able to capture the input of each performer on stage and to distribute the sound throughout the multiple areas and/or larger spaces typical of these venues. Unlike personal P.A. systems, **band P.A.s** are complete systems, capable of using all the components covered in Part 3 of this book, though each will be configured differently according to the specific requirements of the events and venue they are used at. Band P.A.s can be scaled up or down and provide sound for performances at venues ranging from small to mid-sized to large. Because they are complete systems, scaling them up or down can be achieved merely by adding to or subtracting from the number and/or capabilities of the P.A.'s components, without requiring any qualitative change in the way the system is set up and operated. Flexible enough to process sound for nearly any type of live sound situation, in just about any kind of venue, band P.A.s are the quintessential live reinforcement P.A.s and the benchmark against which all other sound systems can be compared and described according to differences or similarities.

The smallest systems in this category are simplified to make them easier to use, and compact enough to fit in the trunk and passenger seats of the average car. The average band P.A.s, however, are large enough to require some planning and assistance to transport. More than one person will usually be involved with the process, and at least one live sound pro is required to help with setting up and operating the system. Generally, a truck or van will be required to transport the P.A., even those used at small venues, especially if one of the bands is transporting it together with their **backline equipment**, which refers to all the instruments, pedals/effects, and instrument amplifiers that a band uses to create the sounds that the P.A. then captures, processes, and distributes. Back-line equipment is nearly always owned or provided by the band and is not considered part of the live sound reinforcement system itself.

Band P.A.s are not limited to use in small venues and can be scaled up to process sound at all kinds of events and in venues of nearly any size. Mid-sized or large band P.A.s, may need anywhere from one to half a dozen or more large trucks to be transported, requiring proportionally more support personal to set up and run. These powerful P.A.s are more appropriate for events with audiences numbering in the thousands. They are more impressive than their small venue counterparts but not so different as to qualify as a separate class of P.A. altogether, as the next P.A. systems we will look at do.

Touring/Stadium Sound

On the extremely large end of the spectrum, the average P.A. system used for large stadium tours or festival events requires dozens of semi-trucks to transport, and an army of personnel to pack, move, and set up the system. P.A. systems as massive and elaborate as the sound systems used on large stadium tours/world-class festivals are sometimes set apart from run-of-the-mill large P.A.s by being referred to as **tour sound systems** or **stadium sound reinforcement systems**. While these titles sound like they are attempts to set these systems apart from the lowly P.A., these systems are still P.A.s, but on a monstrous scale. On a basic level, scale is essentially all that sets them apart; they contain the same components as the large P.A. systems used at other venues, but these systems are pumped up so that there are more of each type of component and everything is larger.

At the same time, the difference in scale has reached a point where a closer inspection reveals the reality that in order to perform the same function on such a different scale, these systems have evolved beyond simply being the biggest of P.A.s. The logistics of providing live sound on this scale has meant that over time perfecting these systems required them to evolve differently. Despite the many similarities, everything from design parameters for the overall system and component parts, to procedures for setup and operation has diverged slightly from those for any other type of P.A. One might argue that somewhere along the line the scale tipped and tour sound systems have become qualitatively different than other P.A.s, even if they are still related.

To sum things up metaphorically, at some point the differences added up so that instead of being just a bigger breed of P.A. like a Great Dane is a bigger breed of dog, tour systems are now different enough to qualify as a related but entirely unique species, like a wolf is to a coyote or dog. Whether one agrees they have become distinct enough to be the big bad wolf, or thinks they are just the big dogs of the industry, there is no disagreeing that they are the giants of live sound reinforcement. When all set up, top-tier tour sound is beyond impressively large; it is truly awesome and exponentially larger than any other type of sound system on the planet.

What Do We Use Live Sound for?

The initials P.A. stand for "public address," which sums up what the P.A. is used for. No matter how trivial or important the content, we use live sound to convey a message to the audience, and how a particular P.A. is set up and used depends on the type of message we want to communicate and the environments typically

Figure 1.1. Tour systems have come to match the size of the venues we use them in, and easily tower over the rest.

employed for each type of communication. Almost all live sound applications, for all their variety, are examples of three general categories of communication.

Public Announcements

Public announcements are very short messages relayed over a P.A. system, usually one installed for the combined purpose of making announcements on and providing background music. They can occur anywhere that people, in groups or as individuals, are gathered in a public/shared space (such as an airport terminal) or throughout a series of interconnected private and shared public spaces (such as a workplace). Unlike other P.A. applications, the audiences of public announcements are gathered for their own purposes and not specifically to hear announcements. Most of the public hearing an announcement can't see the person speaking and the announcement can be aimed at everyone who can hear the announcement or can name a specific person thought to be among the listeners.

These systems are always installed by a specialist in P.A. installations, and look quite different from other systems. To perform their function these systems typically use many small speakers to cover the intended listening area (such as a system in an office building or school building), or may use large speakers for some areas and many small speakers for others (like ones used in a shopping mall or at a race track or sports park, for example). Because this area often will include every room in an entire building, or even over a large shared outdoor space *and* all the smaller indoor areas of the buildings connecting to this space, these systems are referred to as *distributed systems*. The input for making announcements over the system is typically easy to turn on and off and designed for short-form messages like a citizens' band (CB) input, or a phone handset. Examples of messages sent via distributed systems include:

- *An airport help-desk announcement* aimed at a specific individual and informing them to pick up the green courtesy phone for more information.
- *Mall security* informing a pair of frantically searching (or totally oblivious) parents that their lost kid has been brought to the main help desk and would like to be reunited with them.
- *An office-building security officer* informing people that the alarm is not a mistake and they need to make their way to emergency exits for an efficient exit.
- *A sports-game announcer* reminding attendees it is their yearly "cheap plastic baubles for all" celebration and letting them know that cheap plastic baubles are available at odd numbered exits and are free to all ticket holders.

Direct Public Address

Direct public address is a message directed at everyone in an audience. Even if the message is more directed at a particular subset of the audience, the overall intent is to speak to everyone there. Unlike systems used for announcements, involving simple messages that are usually no more than a minute or two in length and are directed at people not specifically gathered to function as an audience, systems used for direct address are used for complex messages that may take up to an hour or more to relay and are directed at an intentional audience specifically gathered to hear the message. Depending on the application, the P.A. system used for direct address may be installed, or portable. If the message involves a small group of listeners, the P.A. used can be as small as a personal P.A. while if it will be addressed to an entire auditorium of listeners, a bigger band P.A. will be required. Some examples of direct address include:

- *Political debate*: Even though this usually takes the form of two speakers taking turns answering each other, their message is not primarily directed at each other but aimed at convincing the audience they are right.
- *Auctioneering*: This is a business application, and the auctioneer's message involves a preliminary description of the item up for auction, the starting price, and during bidding includes the highly stylized announcement of current and new bids at all times until a time threshold is reached and the auction ends with an announcement to all present of who the winner is and the amount of the winning bid.
- *A sermon*: This is informative or inspirational speech made by a clergy member and directed at church attendees. Delivery can run the gamut from an educational lecture (another form of direct address that the P.A. is used for) to a highly stylized and expressive performance.
- *Demonstration sales pitch:* This is an address made at a large convention center, flea market, or fair-type open marketplace. The salesman making the pitch may use a handheld mic, but if their hands need to be free to complete the demonstration, a hands-free or even wireless headset microphone will be used instead. The message is both about the excellent product, and about your difficult problems that it can solve.

Figure 1.2. Sports stadiums use announcements for informing crowds. Announcements run the gamut from essential information, such as game scores, referee decisions, and emergency safety instructions, to non-essential information, such as birthday announcements and marketing pitches for upcoming stadium events.

Figure 1.3. We use direct public address anytime we want to pass on detailed information to more than a handful of people. *Source: Photograph courtesy of Michal Zacharzewski, SXC.*

Performance

This is what commonly comes to mind when we refer to live sound reinforcement, even though the many business and educational uses such as the examples given above are just as common or even more so. Like all live sound applications, performance carries a message to the audience. The message in any performance will usually reflect the performer's main concern or intent. Performance can be for business and intended to sell a product, or it can be art. Usually, art is meant to make the audience think or feel in some fashion. This may be something specific, with all the details of the art aimed at creating an impression and eliciting a reaction, or this could be just a general desire of an artist to please the audience and make them feel good/have fun. The primary message can even be unintended, or as simple as "look how clever I am" or "admire me"; it is still a message even if the artistic merit of the performance may be debated. Examples of performances include:

- *Song*: A song trying to persuade you to not get caught up in the rat race is obviously art, and a song telling you that "rat race rat traps will never fail to decapitate so buy into the rat race today" could be a sales pitch if there is such a product, or it could be a kind of art called satire. But either way, because it's a song, it counts as performance.

- *Spoken word*: This is sometimes mistaken for a speech application, but it is a performance art. If ever hired to run a P.A. for a speech application and you are told it is for spoken word you will want to treat it as a performance.
- *Live bands/instrumentals*: This is instantly recognizable as performance, and usually as art.
- *Magic shows*: Almost the same sort of setup would work here as with the pitch given by the salesman. Both require P.A. systems that allow for rapid speech and constant movement. Both can make use of sleight of hand, but the magician isn't trying to sell you anything in a commercial sense, just trying to delight you by fooling your senses.
- *Comedy acts*: Technically comedy is just a specific kind of storytelling.
- *Storytelling*: A good example of how it can be important to recognize the kind of use your P.A. will be put to, or get used to asking when you aren't sure, is storytellers. Some storytellers use a conversational voice, and some use props, many of which make sound. Others are even more expressive, including providing multiple sound effects with their voices. For anyone starting up a small business and renting their P.A. services to clients, knowing how performance and business applications differ will allow them to provide a better sound for their clients by ensuring they know what they need to be prepared for the sound needs of each type of communication. If you know how a storyteller is different from a motivational speaker, you won't be too surprised when one begins adding sound effects, and will be ready to adjust the system for the sounds that may occur.

Common Knowledge about Live Sound

By the time most people are young adults, they will have attended at least a few school events where live sound is used as part of the production, and except for very young children, almost anyone can easily tell the difference between natural sound and the enhanced sound at events that include live sound. Live sound is so entrenched a part of our culture that most audience members attending a speech or a performance would be surprised to hear musicians or speakers relying entirely on natural sound, even at smaller venues. We simply take it for granted that sound from the stage will be processed in a P.A. system first, and will flow from loudspeakers placed throughout the venue rather than coming directly from the mouths or instruments of the people on stage.

Because it is so common in our culture for audiences to encounter live sound as a standard aspect of live performance and direct address, it is not surprising that a lot of people take for granted that they know the definition of live sound even if they've never actually thought about it much. Common knowledge holds that the most obvious difference between the sound coming directly from a performer and that same sound after it's gone through a live sound system is so easy to perceive and patently obvious that any fool can tell you the answer to what live sound is in one word—louder.

When prompted to provide a more *precise* definition than this, most people can provide at least a few basic details to flesh out their original definition, for example that microphones and speakers are involved in the process of producing live sound, and the end result of the process is that the audience is (ideally) better able to hear the sound, but none of them will stray too far from their original definition. This is also true of many people with a little more background in audio or technology than the general public; though they may be able to add many more details about how live sound works than most, their basic definition of live sound is effectively the same. Most **lay people** (or those who are not involved in a given profession) define live sound by the common understanding of live sound as **loud sound**.

Common Knowledge: Use as Directed

As with much of common knowledge, the premise that live sound is louder sound is obviously true often enough. However, it's equally true, though less commonly understood, that what's obvious at first glance rarely tells the whole story. This is the main problem with definitions that stem from common knowledge; even when they aren't completely incorrect, because they are based on what is obvious, common knowledge descriptions will usually be incomplete. There is not much under the sun, whether it is an object, a process, a concept, or an area of knowledge, that is so simple or straightforward that it can be fully defined using only the details that are obvious to most observers. This is doubly true when defining something like live sound where most of the process can't be easily observed.

However, those outside the audio field who accept the common knowledge about live sound as their basic definition are using common knowledge as it's commonly supposed to be used—as a general framework for discussing or thinking about topics outside of your own areas of interest or expertise, where the only information you have to work with is what is obvious enough to be picked up from your limited observations or know from common knowledge. Everyone relies on common knowledge to deal with topics that they otherwise would have no information about, and in a pinch it's usually a much better basis for making decisions than relying on a coin toss. For most people the obvious fact that live sound is louder sound will serve perfectly well as a functional definition, even if it's incomplete and imprecise. This basic definition can even help them correctly make decisions such as whether or not an event they are planning should include renting a P.A.

Relying on common knowledge is generally not a problem unless one continues doing so beyond when it is useful or appropriate. As soon as someone begins to study a topic, whether by choice or necessity, then continuing to fall back on the common knowledge about that topic can be counterproductive to learning and hinder a student's ability to reach a more than common understanding.

Students who are beginning to learn about live audio should also note this when they see any educator or professional making a statement that seems to confirm the common understanding and not be tempted to interpret such statements as merely confirming the most limited definition as it's commonly understood. Any time a pro seems to be confirming common knowledge, remember that the professional is doing so with a whole lot of background knowledge and understanding that gives them a specific point of view not commonly shared by those without the same experience and training. Rather than using simplifications as narrow definitions, pros use them as convenient expressions for conveying *one fundamental truth* about a *multifaceted topic*. Any teacher needs to start with simplifications in order to introduce concepts in a way learners can understand and use as a foundation to build on.

They should be taken as important concepts to keep in mind and expand on as you learn more, but not accepted as the *whole story* and used as an excuse not to dig deeper and learn more.

Students who accept simplifications are often looking for the easiest shorthand definition to memorize and check off on their mental list of "known topics." Turning a tool for learning into a shortcut like that only make it take longer to arrive at real understanding. At best the people who do so will eventually need to backtrack to scrub out their mental misconceptions and retrace their steps. At worst are the unlucky learners who end up hopelessly lost from depending on shortcuts, because people too convinced they know the final word on something don't feel the need to think more about that topic, and even pass up on easy chances for learning by ignoring any information that might conflict with what they are certain they already "know."

It's risk inherent in relying too much on any sort of aid, expressed in the quote "A little knowledge is a dangerous thing." If used correctly, there is nothing wrong with using definitions to assist with learning, or starting with common knowledge when the simplest definition is all you have to work with. However, those who don't follow up basic explanations and definitions with further learning, regardless of the topic, know nothing about the subject: *all* they know at best is some of the *terms* used to label the concepts.

In the same vein, beginners in audio do themselves the same disservice if they take the truth "live sound is loud," combine it with their layman's definition of loud, and stamp it on their brain as the idea that best boils down everything about live sound you need to remember before moving on to the cool stuff like handling the equipment. Most who do this only trip themselves up for a while before they eventually figure it out and reframe their understanding, but enough never recognize the error that it's important students know to avoid making this mistake.

Looking for Assumptions

Since it's easy to mistake what's obvious as being the same as what's correct, a great place to start when double checking for errors is to look for the underlying assumptions that help form the foundation of what you "know." Begin by seeking out those conclusions you draw from facts that you can't source, but only know because they seem "obvious" or common knowledge. For most people, the understanding of live sound is louder sound is based both on common knowledge and on personal experience. Most people have been in an audience often enough to hear how different things sound when a microphone cuts out, or that while they are seated about the same distance from the band as from the speakers, the speakers sound louder. This is easily noticeable, and why the common knowledge is taken as a given. It *is* true much of the time.

The problem lies in all the incorrect conclusions about live sound that are drawn from this truth when not taken in balance with other, equally important factors. These conclusions include but may not be limited to ideas such as:

- Live sound is always loud.
- Live sound is used because people always can hear better if it is louder.
- Live sound is loud because the audience always likes louder sound better.
- If an adjustment results in people hearing sound as louder, then that sound *is* louder.
- The way to turn sound into loud sound is cranking up the volume.
- The only goal of live sound is to make sound louder.
- Live sound is never too loud.

Even though some of these statements sound as obvious as the premise they are drawn from, none is true, or at least not that simple. Yet most people accept them without question, and that is the problem with filling in gaps with assumptions: until you question assumptions, they can feel a lot like insight whether they are or aren't.

What the Pros Know that the Rest of Us Don't

You will learn more details about these assumptions as you continue to study live sound, but for now they are only presented as general examples to show how a little background knowledge can make a big difference in how one interprets "obvious" information, and on what kind of conclusions someone will draw (or won't) based on the obvious. Pros occasionally refer to live sound reinforcement as making sound louder, but rather than being the only thing they know about live sound, it's just a small fraction of what they know. Other facts they know about live sound and loudness includes:

- Pros know that even a little too much sound power can make it harder to understand what you hear, instead of easier.
- Pros know the size and design of the venue can determine how loud sound can be and still sound good.
- Pros know that people can hear sound as louder without the sound actually being any louder.
- Pros know there are a number of adjustments an engineer can make to make the P.A. sound louder and volume is not the only one.

For the average audio beginner, the only way to make louder sound is to turn the volume knob on their stereo up. The audio pro doesn't even think of louder in the same terms anymore. The point is, you don't have to dismiss simple truths like live sound is loud sound, but don't turn them into truths so big you don't leave room in your brain for other truths that will follow. More than one audio student has etched the importance of' "loud" so deep into their psyche early on in their studies that later on their instructors could not teach them the rest of what they needed to become proficient. By then, no amount of emphasis on equal truths, including the one that holds that "Less is more," could help them see beyond their self-imposed blinders. If you have ever been to a venue where the sound was so loud it was distorted and uncomfortable to listen to, you may have been experiencing one consequence of this phenomenon.

Exploring the Limits of Loud

The truth is that "What is live sound?" is a bit of a trick question. It encompasses a lot of information and other questions and is simply too large to answer from anything other than a single perspective that represents only one or two aspects of the whole picture. So we need to ask the smaller questions and see if answering those first can get us closer to understanding the nature of live sound, including the live–loud connection.

As we mentioned above, live sound is used for communication situations where the performer or speaker has a message to relay to an audience. There are limitations to natural sound that make it hard to communicate with an audience, particularly when we want to relay a message to a large number of people. Natural sound is clear and easy to understand when the listener is close to it, but it loses power and coherence over longer distances. For a certain distance this loss of power doesn't immediately translate to inaudible sound, but the sound is lackluster and thin, takes more effort on the part of listeners to focus on, and is just not that enjoyable to listen to. If the venue is large enough to allow for significant distance between the sound source and the listener, the sound continues to degrade until it reaches the boundary where "meh" sound becomes "huh?" sound. The threshold where poor sound becomes bad enough to count as incoherent also represents the absolute upper range of the number of audience members that can be addressed as well, though the seats approaching the boundary would be only marginally better than at the threshold and shouldn't be used either. Yet there are often times that people want to communicate with more listeners than

Figure 1.4. Performance is also communication. *Source: Photograph by Tryo Benoît Derrier under a Creative Commons BY-SA 2.0 license.*

will fit between the first row and area that lies at the maximum range they can cover with the sound they produce naturally. What we are primarily concerned with when we engage in live sound reinforcement is overcoming the constraints and limits that hinder people's ability to communicate with their audience. To solve the problem created when an audiences size exceeds the reasonable range of the sounds they came to hear, we use the P.A. to extend the range of those sounds. The effect that live sound engineers are best known for creating while doing this task is loud sound.

So, to enable the message to be heard by the audience requires making it louder, right? Well yes...and no.

Yes, in the sense that most P.A. setups are intended to enhance the sound being reproduced and in fact often create sound output that is louder than the original sound. The scenario above illustrates the limitations on audience size that would occur without use of sound reinforcement to point out the primary communication problem that we solve by using the P.A. to boost the sound level. It's a simple fact that sound for live applications needs to meet a certain loudness level in order for it to be easily heard by the entire audience, and in most live sound situations the P.A. reproduces sound at levels that are louder than could be achieved otherwise.

However, the answer is also "no" in the sense that turning up the P.A. volume is not a one size fits all solution for every audio issue. One function of the P.A. is to amplify sound, but increasing the volume of sound is not the only step on the road to successful live sound reinforcement. Often weak sound is not the issue; instead amplification is used to account for increased distance and excess interference. The optimal sound level required for any situation, and the way to go about providing volume adjustments, will depend on the particular situation. Sound engineers go about producing the right sound level for each situation by choosing from a range of tactics, including many which don't involve increasing the PA "volume" levels.

Providing the best sound for the situation and environment involves more than just making sure it is loud enough to be audible. Good sound engineers know that sound does not need to be low level in order for the output to be incomprehensible; distorted, overly loud output can quickly become impossible to understand. In this case the correct response is to lower the volume, since what we want from live sound is not specifically loud sound, but the perfect amount of loud among other features required for **audible**, comprehensible sound.

Coincidentally Loud

Of all the different combinations of tweaks for making sound better, the vast majority will either involve making the sound louder, or simply making it sound that way, but even though it's always a potential, even probable, part of the process, loud sound is not the main objective of live sound reproduction. It may be the best tool in our toolbox, but it's still only one of the tools we can use to reach our real goal—to make the sound for any application the best it can be.

In truth, making sound audible and comprehensible are only the baseline acceptable features we expect for our sound, but ideally what we are after is **good sound**. In order to make sure the sound arriving at a given area is good, there are a number of strategies that we can use to make that happen once we figure out what the problem is in that area. Increasing the power applied to the signal in the sound system, which serves in effect to turn up the volume of the sound exiting the speakers, is only one of them. The only essential requirement is that we take **control** of the sound and shape it to meet the needs of the particular situation.

A Meditation on Live Sound

- We seek *audible/enjoyable* sound.
- Loud is not the motivator.
- To achieve *good* sound we need *balance.*
- Loud is not the only part of any mix.
- To find *balance* we take *control.*
- Loud is only one path, not the destination.
- With *control* we reach *good* sound.
- *Controlling* the sound is the point.

What is Good Sound?

1. *Good Sound is Tailored to the Application*

Properly reinforced sound is not about loud sound; it is about shaping sound to the optimal level for the venue and application. In many situations, the optimal level will be quite loud. Under some circumstances, such as for very large or outdoor venues, the optimal level will be as loud as you can get it, requiring every ounce of amplification you can wring from your system to do the job. Just as often, however, small venue sound will be loud enough, or almost loud enough, without increasing the P.A. volume. Boosting the signal much in these cases would only result in unpleasant sound far inferior to just leaving the P.A. off and running the backline only. If the P.A. only increased the volume, it would be worthless for half the situations we use it in. Those who focus on only adding power to increase the volume, when what is required is to control the sound power to distribute it more evenly, can be destructive in how they use the P.A. As we will discuss in more depth in a bonus chapter on dedicated industry issues which can be found on the companion website, not only has this approach effectively degraded if not destroyed the reputation of small venue sound, it has been destructive to the hearing of audiences as well.

In addition to considering the venue, we need to remember that live bands play with different purposes in mind. A show designed for the audience to stand and focus all attention toward the stage and band tends to be a bit louder than a live band playing for an event where the main focus is the dance floor. A band playing for a dance tends to be a little louder than a band playing as background for a party, which is louder than a live band playing for an event or venue where the main focus is on dining and drinking.

Finally, *genre* and *band composition* matter as well. Even for shows where the focus is intended to be on the band alone, not all bands want or need you to make it as loud as possible. Nor do all audiences want you to crank it too loud. A lively folk-rock band and their audience will want you to turn sound up a little more than a lone acoustic guitar playing singer/songwriter and their audience. A jazz trio will need a little more help achieving the right sound level, but even if the point is to make things good and loud, you won't need to add much to get there if the band has a ten-piece horn section. And punk bands and their audience will always want it as loud as possible, but classic-rock bands with an audience two to three times older? Likely not as much as the punk audience, even if excesses of their youth were loud enough to leave them with less than perfect hearing.

2. *Good Sound is Clear and Intelligible*

Clear sound is sound that is free from unintended **distortion**. Distortion is the twisting of a sound's form, making it sound harsh and unnatural. It can be desirable in some situations (as in electric guitar sound if that is the artist's aim), but not at all in a live sound P.A. system. Your task is to reproduce electric-guitar sound (distortion included) accurately, but you want to avoid adding any distortion that wasn't already there. Clear sound is also free from noise. If the P.A. with all channels muted sounds like a buzz saw, the sound quality, and the audience, will suffer for it. Clear sound is **intelligible**—its components are all recognizable and the overall sound and words are easily comprehensible.

3. *Good Sound is Balanced and Natural*

Balanced sound is sound where all elements contribute. This doesn't mean all elements must be at the same level or the same at all times. But balanced sound needs to be proportional to ensure one sound doesn't overpower others or get lost in the mix. Guitar so loud you can't hear anyone else in the band clearly, or a snare that obscures the vocals, will always end up disappointing the audience. Natural sound from the P.A. is as true to the sound source as possible. You don't want your singer to sound like she is singing from under a pile of blankets or a cheap megaphone unless she actually is. While small adjustments are necessary, it's important not to go overboard and process the sound so much it is no longer recognizable as belonging to the source—completely removing or excessively boosting any control creates sound that does not sound right or natural. Skillful application of effects is encouraged, but not those so overdone that they make the band sound like they're from another planet (unless that's the band's aim).

It helps to understand your limitations. While a skilled engineer can compensate for some instrument/performer deficiencies, a crappy, badly tuned guitar naturally sounds like a crappy, badly tuned guitar, and there's not much that even the best sound system in the world can do with it. In this same vein, people enjoy some effects, but ones that go overboard to create sound that doesn't sound natural tend to get negative reactions. You can't make bad music sound good by over processing, though you may make good music sound bad. In both cases, however, you will only make it seem like any issues with the quality of the performance is entirely your fault.

4. *Good Sound is Evenly Dispersed*

To achieve good sound, the benefit of your work must not be limited to only one or a few parts of the audience while the rest of the audience sees no benefit, or worse, suffers more sound issues than they would otherwise for the sake of those getting all the care. After all, natural sound does a good job serving the front rows without the P.A., so if only part of the audience is on our minds as we mix sound, we're not doing much to improve on the situation. The reason we resort to the P.A. for all our gatherings is that we want everyone to hear and enjoy it. Much of what needs to be done to achieve this happens when you set up your P.A. speakers, and part of it requires you to keep in mind the differences in sound between front and back, center and sides, and make your choices so that you don't cater the sound to one part of the venue at the expense of another. This is one reason why better sound includes bringing sound closer to the people who need it, because in many cases turning the volume up only makes it painfully loud to the audience closest to the speakers even as it provides great sound to a different area of the room. This is also why so much effort goes into becoming familiar with the science of sound—it allows the engineer to predict the types of issues likely to be problematic at different audience locations. If the sound at any event is going to be able to qualify as good sound, it needs to be good for the whole venue, or as much of it as you can manage, even if that means it's not perfect where you are at.

What Live Sound Isn't

- obvious;
- static;
- your home stereo;
- something you'll master from reading a book;
- rocket science;
- always loud;
- summed up easily in bullets (is or isn't).

Live sound reinforcement supports and strengthens sound, but it isn't simply a process of turning one control up or down. Instead, it's a multi-layered strategy that will be different every time because it's tailored to the situation. Attention to detail is important because the work of live audio engineers is usually about more than just making sound more powerful; it's ultimately about empowering people. What makes live sound so satisfying is that when we exert control and shape sound to provide the cleanest, most audible sound possible, we are bringing people together and helping them hear and be heard by each other.

Chapter 2

Overview of the Live Sound Industry

Live sound engineers are trade professionals skilled in the handling, maintenance, and operation of audio equipment designed for sound enhancement and redistribution. But there is a wide variety of tasks that fall under this umbrella and which ones a live audio pro will be responsible for depends on their title and what industry niche they find themselves working in. One thing is the same across the industry: almost anywhere a P.A. system is used for more than announcements and piped muzak, there will be one or more live audio engineers, on site or on call, who will be responsible for properly handling and running it.

Live Sound Engineer: Crew Positions/Titles

What live sound professionals do each day will depend on where they work, but within any job the focus and types of tasks can be basically divvied up by title. Outside of their coworkers, most can be glad when they run into people who know to refer to them as a sound engineer or sound tech. Most members of the general public don't know much about the fine details of the industry and trade professionals who collectively provide the live sound at events small and large, and will refer to sound professionals as "sound guy/sound lady" or "roadie."

A1, A2, A3

These designations aren't titles, but as indicators of rank or "pecking order" they are just as significant, and usually more so, as your title in determining what your job description will be. The letter refers to audio and the number to order of rank—and three guesses where any new hires, including you, will be. But don't worry; it's actually a good thing. There is a lot to know that can't be gleaned from even the best-written and most complete books. Even if one is the exception and can perform actions expertly with no practice time, the truth is the very act of being in a new environment creates a greater likelihood for error, including for proven experts. Which is why your first job won't be the last time you find yourself at the bottom of the totem pole.

It's not designed to be a punishment, but is actually a precaution meant to safeguard everyone's interests. That's why even those with extensive resumés may have to start new positions as A3s, at least long enough to verify that they mesh well with the new environment and team. If you find yourself lucky enough to find employment as an A3 for a well-established soundco, look over some of the advice in Part 6 to brush up on how to get the most out of the opportunity.

Front of House (FOH) Engineer

Front of house (FOH) engineers are the people who run the board that controls the P.A. sound. They have almost as much influence on the sound the audience hears as the band, but are required to use that power invisibly. They are expected to make their choices in ways that help fulfill the band's artistic vision instead of just their own and, equally important, their sound should always provide the audience with what they need for a good listening experience. The most visible duty of the FOH engineer is to run the main board, though this is probably the task that eats up the least of their overall hours on the job. Whether they work locally for a soundco, tour for a soundco that specializes in large-scale tour sound, or work as dedicated house sound for one or two local venues, a FOH position will often include responsibilities that go well beyond the time they spend mixing shows.

In addition to mixing the shows, the FOH engineer will generally be responsible for the proper handling and maintenance of the P.A. system. To do the job, he/she will need to be there to manage and/or assist with every aspect involving handling gear. Working for touring or local productions with big crews, they will oversee all aspects of the process. FOH engineers

Figure 2.1. Front of house (FOH) engineer. *Source: Photograph by Duncan Underwood under a Creative Commons BY-SA 2.0 license.*

for smaller venues will manage the one or two others helping but also can be found in the thick of the labor, moving gear that needs to be moved and setting up what needs setting up. Regardless of venue type, it will be up to FOH engineers to sound check and tune the system before the show, and to continuously monitor the system and adjust the sound as needed during the show. After the show they will either manage the whole process in reverse, or be assisting with all aspects of packing up and loading sound equipment to be transported.

In many cases they will also be responsible for much of the planning that occurs long before people and gear pack up and converge on a venue to set up for a show. Perhaps their most important function is managing people, as they are the coordinators who keep lines of communication open between all the different parties who must work together to provide sound for a show.

Foldback/Monitor Engineer

The monitor engineer is the engineer who can be found at the side of the stage—if you can see them at all! As often as not they are hidden from view behind P.A. speaker stacks and decorative elements of the production, or tucked behind the stage wing curtains well out of sight. While they don't directly control the sound the audience hears, as the person the band depends on to hear themselves they have almost as much influence on the overall sound quality of a show as the FOH engineer. Their close relationship with the band can sometimes be much like being a band member themselves. Other times the relationship hasn't developed the same way and is more based on the mutual respect of working professionals rather than one grounded in camaraderie. How the monitor engineer and the band they work with relate to each other personally depends on chemistry and how much time they've had to get to know each other, but there is more than one way for them to work well together, as in both of the examples here. Personal interactions aside, in order to do their best work with the band, the monitor engineer still needs to know the music about as well as any band member, along with the particular monitor requirements of each performer according to their preferences and performance styles.

Even though they are less concerned with the overall care of the P.A. than FOH engineers or audio technicians, they will often take on oversight of those parts of the P.A. under their purview, and see to it that stage monitors and mics are set up correctly, that the cables and connections delivering signals between the stage and the monitor mixing board are set up and working, and that wireless monitoring systems are working and all batteries are fully charged. Like any good crew member, if needed they will help out with any set up tasks that remain if they finish their own preparation and set up tasks early enough to be standing around while everyone else is still hustling to finish setup.

The monitor engineer's job at the board is arguably much more difficult than that of the FOH engineer, though both require a lot of skill. However, the average monitor engineer needs to skillfully create many more mixes than the FOH, who will rarely need to produce more than four even if creating mixes intended for broadcast and recording as well as for the house P.A. and one other piped system. Most often, however, the FOH engineer will be responsible for no more than one or two solid mixes. In contrast the monitor engineer will rarely ever find themselves asked to provide anything less than four mixes, and usually can be expected to create anywhere from six to ten distinct mixes, or even a dozen or more depending on the number of performers in the band. These also need to be custom mixes that are tailored to each performer, as the needs of each band member will vary according to their instrument and personal style. The work they do requires both a high level of empathy and strong confidence in their own understanding of how music works in a performance setting and what is best for the artists in their band. Each mix draws from these abilities to create the best balance between what each musician wants and what each musician needs. Empathy helps them tailor each mix to the preferences of the musician receiving it, while their confidence in their own understanding of music allows them to subtly shape the mix, keeping it grounded in what the musician wants, but tempering it with their own judgment of what will bring out the best in each musician.

Usually, the only situations where the mixing duties of the FOH engineer are as intensive as a monitor engineer's is when they are doing the job of both if the band is without a monitor engineer. Except for cases of bad luck securing an engineer for a gig or illness making their current monitor engineer unable to function, it is rare to find the position open for any touring band with a little success to their name. No band who can afford their own monitor mixer will be eager to return to managing without after being spoiled by a good one. Regardless of the ebb and flow of the live sound employment market, a good monitor engineer is so highly prized by musicians that they can rest assured their services will always be in demand.

Finally, their close association with the band, their professional understanding of the challenges and concerns of a house engineer, and the need to coordinate closely with the FOH engineer means that whether they work under the FOH engineer, as is most common, or act as co-manager as happens in some situations, the monitor engineer often functions as the main liaison between the two worlds of artist and audio pro. A monitor engineer adept at facilitating communication between the stage and the FOH can help increase mutual understanding and cooperation between the two, thereby increasing the overall efficient and smooth operation of a production.

Sound Tech/Audio Tech

The specific duties of the **audio technician** will vary based on experience, from one crew to another, and whether working on a local or touring crew. They will typically work under the direction of the FOH and monitor engineers, but their position on the crew can run the gamut from unskilled A3 to an A1 master technician. As an A3 new to the work, a sound tech will be assigned tasks and expected to do them as instructed, while an A1 master sound tech will have a great deal of autonomy to perform their responsibilities as they see fit, consulting with the mix engineers to touch base or troubleshoot issues. Most techs are A2s and A3s falling somewhere between the two extremes.

On tour the audio techs will be responsible for overseeing the loading and unloading of the P.A. from the trucks, making sure the system is set up and dismantled/packed properly, and for generally keeping gear and work areas organized. If the only tech on a small crew, they will take hand in every aspect of handling, setup, and maintenance of the P.A. If on a large crew, their tasks will split along lines of experience and expertise. In these situations the more experienced A1 and A2 techs may each manage the tasks in their area of skill/interest, with one handling the tasks of setting up stage mics and the other flying speaker arrays. A3 techs will be assigned general tasks and can often be found laying and taping down cable and coiling it to put away after the show is over. Other A3s may be occupied mainly with tasks requiring moving messages and gear as instructed if they have no skills at all yet. Additional tasks typically managed by audio technicians include replacing speaker heads, soldering new cable connectors, and cleaning gear.

While audio tech is the position used as the entry-level position for those entering the field, not every audio technician is at entry level; many are skilled and well-respected members of the live sound field. While most mix engineers started as techs, not all techs choose to pursue career paths in audio involving regular mix engineering, and may learn electrical and repair skills instead as a way to increase their marketability and pursue advancement. For those who are most interested in a career focused on technology, there is enough demand for top of the line techs that those who don't feel the call to mix engineering positions can still advance their careers if they focus on expanding and mastering their technical skills.

Sound Designer

This position is a specialist unique to crews working for theatrical venues or productions. Working closely with the director and other designers, sound designers are responsible for designing and implementing all sound elements found in a performance, with the exception of original dialogue and music. Their contribution to any production is to use the incredible ability for sound to evoke feelings and moods in listeners to design the right mix of sound effects to deepen the impact of the overall production. The task requires a good feel for sounds we normally take for granted, and the sound design for each production should

Figure 2.2. View of stage from the monitor mixer position.

greatly increase the emotional response of the audience, but at the same time must fit so naturally into the flow of the performance as to be invisible to the audience.

Sound designers don't decide if the action on stage occurs during a rainstorm or a hot summer day, but they do have a big hand in deciding how to include the sound of the storm so it best furthers the narrative and supports the director's vision. They decide if the rain will come with booms of thunder, howling winds, or both, and are additionally tasked with creating original samples to use, or with finding and licensing existing sounds from sound banks, depending on the production needs and budget constraints.

This audio professional will also work closely with the FOH engineer to plan how the FOH will integrate the designer's vision into the play's action and in the course of planning they will keep track of their choices by recording them on the cue sheets the FOH engineer will depend on during the show. **Cue sheets** are notes telling a mix engineer which events in a production signal an upcoming sound effect or action at the board, as well as exactly what to do when the time arrives. To be able to drop a sound effect into a play you need a cue sheet letting you know the act and scene it belongs in, the stage directions and dialogue leading up the sound, and the exact point when the sound should be played, so you are ready to incorporate it at the right time. In addition the cue sheet lets you know the filename of the sound, the output level to play it at, and how long to play it. Cue sheets are part of the toolbox of any sound engineer mixing a show, but they are especially detailed and important in theatre sound.

The sound designer will also consult with the FOH engineer to best achieve their other major responsibility, which is the design of the P.A. system used for any production, or upgrades to the P.A. in smaller venues. This involves the selection of all components (such as mic number, type, and brand) to be used in any production, and their proper placement. They need to manage this task in a manner that finds the best balance between the productions opposing interests of fulfilling an artistic vision and staying within the allowed budget. For some smaller venues the sound designer will have no need to consult with the FOH engineer for these tasks, because in these cases it is not uncommon for one person to hold both positions.

Who Hires Live Sound Professionals?

As we have seen there are only a few positions that collectively see to the duties of a live sound professional, and the live sound professional will be expected to serve the general functions of one or more of these positions no matter who signs their paycheck. Regarding the people who sign the paychecks in live sound, there are also only a few types of employers who commonly hire for live sound positions. Whether work for them is full time, part time, or by contract, these employers will generally be **P.A. rental companies** (also called **sound companies**, **soundcos**, or **sound providers**), or live performance venue owners/operators with their own installed P.A. systems. In addition to these two types of sound business, freelance employment can be had with the production companies and/or individual performers or performance groups that often are the biggest clients of sound providers and event venues. Many of these prefer to at least provide their own monitor engineers, and sometimes even their own small sound crews for their shows and tours rather than depend solely on the crews working for the soundcos or venues they rent from. Eventually, there is also the option of becoming the owner operator of your own live sound business and signing your own paychecks.

In a nutshell, if one is employed in live sound it will be working as an audio tech, a monitor engineer, or a front of the house engineer, and for one of the two types of small businesses that provide P.A.s for the sound needs of their clients, event venues, or sound companies. Otherwise work is in the form of self-employment either as a small business owner running your own sound company, or as a freelance contract worker for soundcos or their biggest clients, event producers and performers. Judging with these facts alone as a guide, one could easily come to the conclusion that work in live sound is much the same no matter where you end up getting hired. In one sense this is true; no matter the type of production one ends up working on or where it happens, some or all the gear must be hauled and set up, sound checks must be fitted in, and certain basic mixing functions will apply no matter what the situation.

Despite this fundamental similarity, it would be incorrect to assume that the day-to-day reality of working in live sound is necessarily the same across different jobs. Job duties alone are not the only factors contributing to what a live sound pro's workday will be like, and the exact mix of duties for any live sound position will vary depending on where one is employed regardless of title. Straightforward factors like job description and the physical environment/s worked at while employed for a position both contribute to what any live sound job is like, but the unique organizational culture of a particular workplace is just as important in determining the overall experience for the sound professionals that work there.

A group's **organizational culture** is the unique mix of values and practices created collectively by the interaction of all the members of that organization. Even looking at positions that are similar in most ways, one will find that organizational culture will be at least slightly different, and often much different, which means not only will a career working in theater sound be different from a career working with live bands in small venues, but that even two theater sound jobs or small venue positions can be quite different in terms of what it will be like to work in them.

Venues

While venues cover any place where a live sound event may happen, only a fraction of them hire their own live sound professionals. Many venues do not have installed P.A.s and clients renting dates at these venues either will provide their own P.A. or will rent an appropriate P.A. and engineer to run it from a local sound company. Other venues do have installed P.A.s but the type limited in power and intended for playing background music. A few venues have installed P.A.s that use digital auto mixing functions suited for limited types of speech applications rather than a sound board to control sound. Neither of these types of installed P.A. requires the expertise of a live sound professional to perform their limited functions. These venues may rent space to live sound event producers, but will require the renter to arrange for their own P.A.

The positions available at venues that do feature powerful installed or portable P.A.s can involve handling sound for a large variety of productions, though most of them will be at venues with a specialized focus on a limited type of live sound production or limited set of them. Examples of these **house sound** positions might include part-time employment at a small bar with a battered old portable P.A. that hosts local bands on the weekends, and full-time work at venues the same size or larger that feature sophisticated installed systems and host live music shows seven days a week, as can be found in all large cities nationwide. Venues can include the classic midsize halls known for catering to live music fans for decades like The Fillmore in San Francisco and the same classic large auditoriums famed for decades of large events, such as L.A.'s Shrine Auditorium where many Hollywood award shows host their events.

There are commercial venues that are part concert hall, part tourist destination, like the House of Blues chain of music halls, and the smaller music and dining venues at each location. Some small concert venues are focused on one genre, like the well-known jazz venue and restaurant Yoshi's in Oakland, CA. A few midsize venues are also dedicated to one genre, such as the country music venues, including the Grand Old Opry, in cities like Nashville, TN. Other venues providing the in-house P.A. systems required for venues to put on their own live productions include the hotels that provide live entertainment for their customers in special in house nightclubs and bars, and the live music venues at special vacation spots like amusement parks and cruise ships.

Other entertainment venues that may hire live sound professionals full or part time are those that specialize in other types of entertainment than musical performance, including theaters and comedy venues. These are where productions like dance performances, plays, and magic shows can be found, as well as spoken performance events like stand-up comedy, storytelling competitions, and poetry slams. Finally many stadiums hire their own sound professional to assist visiting sound crews with the features specific to that venue, and to handle in house P.A.s at stadiums that have powerful P.A.s used for applications such as half-time shows and the pre-game national anthem, but that are not large enough for full stadium concerts.

Stadiums often hire general support crews who aid with stadium events and also are loaned to help touring crews in the loading and unloading and general labor support of touring productions renting the venue. While work with a local stadium crew lacks the sex appeal of working on a tour crew, it does include the variety of helping for every event that can be found using stadium venues, including ice-skating performances and monster truck shows as well as concert tours. More to the point it allows one the chance to work with crews from a large selection of the largest touring crews as each production comes to their venue, allowing someone interested in working on a touring crew to get a feel for a variety of the biggest tour crew employers nationwide, and make valuable industry connections who can aid them with tips and information or even recommendations, as they pursue a spot on a touring crew.

Alternatively there are positions working for local churches instead of entertainment venues. More churches are starting to pay their sound engineers, at least for bigger weekend services and events; however, this depends on everything from each congregation's budget, to the availability of skilled and dependable free options available to them. Other limited positions may be found working with educational presentation companies who visit schools over a region to give educational presentations or similar small companies that throw presentation or information events in a variety of small venues and may include a sound pro on the team to manage sound.

Note that as with the sound companies below, each venue will provide a different environment and the duties and expectations for employees at each will depend a great deal on the management style and business model of your employers as well as the type of shows at the venue. Some big city club venues may be the classic kind of rockin' employment opportunity that most young students imagine themselves running sound for, but no matter how dedicated to creating a classic rock'n'roll atmosphere a music venue in Disneyland may be, the sound person there will need to abide by the same corporate dress code and hygiene code as the rest of the park employees. It is important to take all factors into account when evaluating what the environment might be like for those working there.

Sound Companies

Collectively, these businesses provide sound equipment and services for clients whenever live sound is desired or required and not provided by the venue or otherwise conveniently accessible. Sound companies manage to fulfill this role with a common focus on providing live sound services for their clients, but the varied businesses created over the years as each new company forges their own path and often their own entirely fresh approach to the basic sound company business model has created the most diverse array of employment environments that can be found in live sound, or any in any other type of business that easily springs to mind, regardless of industry.

While also the source of some confusion for those first looking for sound company employment or services, the flexibility of sound companies to create whichever service products work best to meet the needs of their client base has allowed them to serve the largest volume of clients with the highest efficiency even as their services require more manpower per show than those of venue sound providers. The overall result of this flexibility has been to allow sound companies to adapt to new market demands, and weather market storms throughout their short history, all while providing access to pro audio services for the widest variety of clients and situations of any other kind audio provider, or even media provider, much less any other type of live sound provider. Considering this it's hardly a surprise that sound companies represent the largest employers of live sound professionals among sound businesses, both as the easiest live sound sector where someone looking to enter the live sound field can find internships and entry-level employment, and the easiest sector overall to find steady employment in live sound.

If you crave a variety of work environments, sound companies can see you working at every type of event that might conceivably have reason to rent a P.A. and crew. Sound companies can be classified by size of the sound company, as well as by their particular focus. While determining where you want start looking for employment, starting with sound company classifications can be a way to evaluate what sound companies you think you'd enjoy working with the most, and help narrow the field you want to focus on, but you will still need to investigate the culture of any sound company you are thinking of accepting a job offer from to see if it's a good fit for your needs, or at least to know what you are in for if you work there

Investigating Sound Companies

One way to learn more about sound providers is to look over some examples of the many different types of sound companies on line by looking up all the ones you can find providing sound in your area. Finally, call the ones that interest you the most and ask if they have any additional information about their company for you to look at. Many companies have data sheets with information such as past clients, referrals, case studies, or equipment lists and prices that aren't on their web page, but are made available upon request. Learning as much as you can about a company may provide the right piece of the puzzle, such as the type of clientele they serve, that can help you put other information into context and provide a clearer picture of what they might be like.

Small Sound Companies

Like the client productions they serve, sound companies come in a variety of types and sizes. In terms of number of employees, the smallest sound companies include examples such as the one-man "DJ with P.A. for hire" working backyard parties and doing everything else on their own, the family-run business with one full-time employee and seasonal part-time help to assist with providing personalized karaoke gear and operators for parties and pub events. Small sound companies also include companies with a few small to midsize P.A.s and a dozen or so employees who provide local private and small business clients with a personal touch for their events that larger companies may be less able to provide, or who serve a specialized niche or geographical region overlooked or less saturated by larger sound companies.

Midsize Sound Companies

Midsize sound companies are those with a solid enough selection of sound equipment and an employee pool of a couple dozen or more than twice that many, who are able to serve multiple clients of various sizes over one large metropolitan area, several cities within a larger region, or state wide depending on the company's focus and the size of the market where they are located. Some midsize companies with well-established reputations and contacts may even serve large sound clients regionally or nationwide, as well as serving clients of various size in their home market. The bulk of employment can be found here. Medium companies have many more employees and needs for live sound professionals to help them to serve a larger client base than small companies do, but midsize sound providers are also far more common than large sound companies who, even with a need for more employees, can't provide as many employment opportunities overall as are found amid the numerous midsize sound providers.

Large Sound Companies

Large companies may have a few physical locations they work from, or may maintain local storefronts in most major cities, but in both cases may have up to a hundred or more employees whether across departments in limited locations or in each of numerous locations nationwide. Some large sound providers will work for clients large and small in major markets others may provide sound mostly for small- to medium-sized client events, but from locations in almost every midsized and larger city nationwide. We are limiting our size criteria here mainly to the number of employees at a company, not the size of particular events or specific clients served, but we do include a few who are too world renowned for being dedicated to large-scale sound to classify as anything but large even if they appear smaller in terms of number of company employees. A few may have the same number of employees as a large midsize company but otherwise will provide large-scale sound for the same large concert producers, festivals, or other large-event producers like Red Bull, often for decades, as well as providing sound for massive one-time events, such as the Queen's Diamond Jubilee celebration attended by massive crowds in a number of London locations in 2012. To provide the added numbers of workers on an as needed basis, these companies often will have a pool of employees they have worked with before they can rely on for piecemeal and contract workers, and can also draw on temp workers for unskilled labor.

Sound Company Focus

In addition to coming in a variety of sizes, sound companies vary in their focus and business model. Some may do this as a strategy for competing in a marketplace already served by other sound companies and requiring smaller or newer companies to locate a niche they can compete in; others may do this because they are really only interested in providing one type of service or working with one type of client base and are happy with the work they find there. Still others may discover their specialty purely based on the customers that keep coming back, developing a reputation as skilled providers of specific services over time. These companies may provide services for clientele outside their niche, but by design or accident will find the bulk of their business within one or two specific niches.

- *Some may specialize in serving a particular client base*, such as working with event promoters exclusively serving business clients, or with wedding planners who rely on them regularly to keep even the most savage bridezillas happy.
- *Some companies specialize in meeting the challenges posed by particular types of venues*, such as companies known for providing sound for outdoor events that is robust for the audience area without bleeding unwanted sound into adjacent areas, or companies having a name for delivering stellar sound for events at small to midsize indoor venues that are notoriously difficult to provide top notch sound in.
- *Still other companies will specialize in providing a specific kind of live sound service*, such as DJ mixes, or karaoke systems. Others might specialize in providing line arrays large and small, while some

companies might keep their focus on small sound for house parties.

- *Some sound providers even specialize in being as good as a specialist for every service they offer.* More established sound companies may have grown to become one-stop live sound providers with no client events in their market too large or small for them to work with. These sound companies may retain an area they specialize in, but most specifically make the claim that they have no specialty because they are established and/or large enough to be equally capable of providing sound for all situations.
- *Other companies specialize in rentals that combine live sound services with other non-sound offerings to create a range of full service styles and targeted rental packages.* Some create business models that better serve their clientele. One example is found in those companies offering a one-stop rental experience for business clients, offering sound rentals and services, as well as Wi-Fi boosters, tablet computers, instant feedback modules, and screens/projectors. Other companies provide entertainment bookings, offering clients a selection from bands in a few genres and several types of DJs, and arranging for other performers if their clients prefer. Another example is companies providing stages and dance floors, lighting, and video, as well as P.A. Also of note are the companies targeting performers as clients and offering backline rentals with their P.A.s, some even including instrument repair services as well.

Despite the same overall focus on providing sound gear and sound services for their client, each small sound company will go about it in their own way and each will have a different employment environment depending on everything from the personality quirks of the business owners/managers to the mores of the geographical regions where each is located. Even for companies run by people with similar backgrounds, and within the same region, significant differences can be found according to a company's client base, since in many cases each company will have created their particular business practices to cater to the needs and expectations of their client base. For example a company that specializes in providing DJs and P.A. gear in Napa for the private events of well-to-do locals and for a few winery events will have a different work environment and standard set of practices than a soundco of the same size that has ties to the San Francisco **Electronic Dance Music (EDM)** community and a clientele list featuring local dance-party promoters. The two might seem the same, being the same size, located only an hour apart, and sharing the same specialty of providing DJ P.A. services, but the very different expectations of their two different client bases is going to necessitate they each use practices meeting the expectations of their clients.

Freelance Self-Employment

Even if you don't think you're interested in it now, it is important for any live sound professional to understand a little bit about freelance work. Live sound is a field where so many opportunities are only available as freelance work, you will want to know enough about freelance work to at least have the option to choose or reject opportunities that may help your career, and you may not find enough audio work without freelancing, unless you have other skills you can utilize to get supplemental part-time work. Freelance work is self-employment, but isn't the same as starting your own business. Instead you often will do the same sort of work you might do if employed by a company, but with a few key differences. A job is an agreement to work under an indefinite time frame to continue until either you or your boss gives notice. Freelance work is a contract that defines the time frame the job covers which might even be a couple of years, but usually covers three to six months. After that time frame is up, the contract expires and you need to draw up a new one if you and an employer want to continue working together.

Another difference is that contracts can also include a project with a time range to complete the tasks for the project, but which ends when the project is finished rather than at a set time unless it goes beyond the maximum time allowed, in which case the customer either refuses to pay the large portion reserved for the final part of the project, or just as often draws up a new contract extending the final date for the project to be completed. While extending a project delivery date doesn't seem to apply to a live audio freelancer, who has a set date to mix at a production and is either there or not, there are cases that might apply, such as extending the delivery date of a recording made as part of the contracted services.

Some live sound pros are able to make a career of freelance work (though depending on the market this can require purchase of a P.A. large enough to be adequate for the size of the venues or events they specialize in, if most freelance jobs require the engineer provide the P.A.). Others use freelance work in combination with other positions to round out their hours.

Freelance work can include engineering one-off events, working regular tech/engineering jobs on a show by show basis for a number of dedicated clients with relatively regular needs but not enough to support a full- or part-time dedicated sound pro. Freelance can include longer-term contract work of different lengths and types, from a three-month contract with a sound company needing seasonal help to manage the added summer business, to a 12-month contract with a performer to monitor engineer on their tour covering North and South America.

Jobs pay by the hour or by salary (covering a range of hours) and must conform to laws such as meeting the minimum wage, paying overtime, providing vacation pay, and providing medical benefits. In contrast, the pay rate for freelance work is set by the contract and you negotiate the rate you want or are willing to take. It doesn't need to meet the minimum wage, nor does the law require that a contract include benefits. When not abused, this flexibility can be good for everyone. The problem is by allowing dishonest people to manipulate the system it can mean entire industries find their services devalued almost overnight. The reason it should matter to you is because you need to understand exactly what's going on in the market in order to properly negotiate a good fee. If you have no idea why there are 20 people on every job board offering the same services as you for two dollars an hour, you will be hard pressed to explain why your services are worth ten dollars an hour.

Negotiating contracts is one of the aspects of freelance employment that some people find the most scary. It is not nearly as scary as it seems. If you are going to know enough about freelance work to have the option to choose or reject it, you will need to know how to negotiate, since that is a necessary step in choosing any freelance job. In addition, it is such an essential skill in so many aspects of life it seems foolish even for people who never intend to freelance to avoid learning such a useful skill. After all, you negotiate your paycheck every time you're up for a raise, or you should! Negotiation is not just for freelancers; it's a valuable skill for everyone. Be sure to read the practical advice on how to negotiate a contract in Part 6.

Self-Employment as a Small Business Owner

Unlike freelance work this always involves purchase of P.A. equipment, though the type or amount depends on the type and size of the small business. The most common example of this is the sound company, which involves at least one P.A. for rental, but usually eventually includes at least a few P.A. sizes for rental and a small crew to set up and/or run the P.A.s rented for clients requiring this service. This can also include other related services like backline rentals but doesn't have to. Some sound companies eventually expand to become quite large, and others happily remain small and don't need to expand to successfully serve local clientele needs for many years.

Don't assume you need to go big to successfully start your own small business. Other examples of live sound small businesses include the DJ and P.A. for hire who works local private parties and receptions, and the karaoke DJ who fills their schedule by throwing weekly events at a few local bars and hotels and occasionally at private parties. We know one live sound pro who combined his P.A. and his live sound skills with other skills and now travels throughout his region with a troupe of three actors/educators. His business specializes in providing educational interactive presentations on everything from bullying to safe sex at a variety of schools, summer/day camps, and residential treatment programs. The point is, there are a variety of options for those willing to think and bring their P.A. outside the box.

Your options aren't limited to opening a sound provider business either. There is also the option to open a venue or an event-production company without providing your own rentals. Opening a venue will be the largest amount of work, since if you want to have live performances there, even small ones, you have to have a large enough physical space to do that. An additional complication is the reality that it's hard or even impossible to make many venues profitable without alcohol sales, which requires a licensing procedure and learning a whole set of skills unrelated to audio. The other option, event production, is the easiest of the three. While many audio pros combine event productions and sound provider services, that's not required for a successful event-production company, if buying a P.A. is more than you want to deal with or can afford.

Limitations/Compensations and Rewards and Drawbacks

As with most of life, so it goes in live sound; one sound engineer's drawbacks are another's perks. Like most careers, which is which depends a lot on the individual needs of the person considering work in the field. Ask enough live sound professionals what the pros and cons of working in live sound are, and even if there will be some disagreement as to what the perks and drawbacks are, certain factors will be mentioned pretty consistently. Many of these are included in the lists below. Because the topics touched upon encompass a broad range of issues, doing them all justice here is impractical, even if the chapter's topic requires they be mentioned. Those interested in any of these issues will find them discussed in more depth throughout the text, and/or covered in the recommended articles linked to in the chapter directories.

Compensations

- more potential variety in work environments than other fields or studio audio;
- work with a variety of people;
- work with technology and audio gear;
- work aligned with live music, theatrical performance, or similar performance cultures;
- opportunity for self-employment or self-directed employment;
- non-standard hours;
- potential for touring/travel;
- potential for local employment;
- less industry shake-up than studio sound;
- growing live entertainment industry and more midsize venues present a positive forecast for the future;
- growing church sound industry offers whole new niche and growth opportunities.
- college degree not required for access for those able and motivated enough to self-train and seek alternative training opportunities.

Drawbacks

- local employment sucks—I want to tour;
- touring sucks—I want to work near home;
- non-standard hours;
- noise-induced hearing loss is a risk;
- highly competitive and popular occupation means finding employment can take more hustle than other fields;
- many opportunities offer only part-time or variable hours, while full-time employment may mean working in more than one position;
- the band gets all the glory when things sound great, but you get the flak anytime sound is under par, even if it's because the band has crappy, out-of-tune instruments or the singer has a head cold;
- drunken venue patrons asking you all about your knobs (which may in rare cases feel like a perk);
- non-standardized training means separating the quality training programs from the inadequate ones or the far worse fly-by-night operations can be a frustrating and confusing task;
- live sound engineering is considered a technical field and one of the trades, so like these fields some historical factors combine to make it a little more likely for women to encounter sexist attitudes when working in live sound.

Related Fields

Fields that aren't traditionally considered live sound, but work closely with the industry and/or use similar skills, are always options offering alternatives to employment in live sound, or can be used as a route providing access to a back door to full-time employment in live sound. Taking advantage of opportunities in related fields will keep you working in the same overall music/entertainment industry while you continue to pursue an entry point to a career in live sound. Most will require some supplementary skills in addition to your live sound skills, but the added skills needed are often similar to skills you use for live sound, and gaining experience in some of these fields will add to your marketability as a live sound pro; for example the ability to work with video or lighting as well as audio, or the added professionalism of a resumé listing a few well-written articles on live sound topics in addition to jobs representing experience in the field.

- stagehands and general labor roadies from other crews;
- video and lighting techs/guitar and drum techs/pyrotechnic techs;
- band managers/agents;
- event promoters;
- event production;
- radio/TV/film recording/audio production and post-production;
- studio sound tech/engineer;
- audio media/blogger;
- audio education/training;
- audio installations and acoustic treatments;
- audio equipment sales or repair.

A Closer Look

Now that we've talked about what the aims of live sound are, the positions available to those choosing a career in the field and the general types of employment available in the industry, we can look at the work of live sound in tighter focus, by examining the details that define the actual day-to-day, "boots on the ground" experience of work throughout the field, especially what experiences are most typical at entry level for those beginning to work in live sound.

Chapter 3

Live Sound

Where the Boots Hit the Ground

Who Uses Sound Provider Services and Venue Rentals?

The shortest answer is just about everyone, if not directly, then as a private guest, event attendee, or audience member at an event made possible in part by a sound provider. If you've been to events like the ones below or similar, odds are safe that with only a relatively small number of exceptions, the location, the P.A., or both were rented to the event producer, or just the party hostess:

1. concert at a club and/or larger venue, including any at stadiums;
2. a convention or tradeshow;
3. karaoke bar after it just showed up one Tuesday;
4. a friend or relative's wedding where a mic was floating around for speech givers to use before and after the wedding band played.

If you spend part of your career working for sound companies, or if you try opening shop and being a sound provider yourself, you will at some point be in a position where you will accompany rented P.A.s as part of a crew, to deliver, set up, and run live sound for all kinds of clients and in a variety of venues. These are only a few examples you may find.

Private Individuals

Soundcos are the businesses that provide the general public with access to P.A.s and the professionals required to set up and run them. Venues rent to private citizens as well, allowing receptions to gather in convention halls, or at historic estates and gardens, and restaurants too, but also allowing people to rent a club or bar for an evening or, if they can afford it, an entire amusement park. But whether in a backyard or large venue rental, the mark of a party is in the presence of a P.A. By bringing professional quality sound to the speeches and performances at the private events we cherish most, sound company P.A. and crew rentals have become one of the modern hallmarks that indicate a private celebration is an extra special celebration of one of life's milestones or events of particular value rather than just another party like those we throw and attend many times over the course of each year.

Businesses and Organizations

Businesses and organizations of all types rely on sound companies for their live audio needs as well, from the speech support required by the smallest lone entities such as those who market and book their own speaking tours, to the complex audio support required for huge closed corporate gatherings. Businesses use P.A. rentals at their yearly parties to provide P.A.s for entertainers and bands, and at business gatherings aimed at motivating management and presenting yearly awards and bonuses. Businesses also use P.A. rentals for employee development and training presentations. Business applications also include P.A. rentals used at large gatherings serving much larger communities from entire business sectors, including industry conventions and trade shows. Even tattoo conventions will have a P.A. rented to use for tattoo competitions of the fresh designs tattooed on convention customers.

Sound companies are just as involved in helping non-corporate organizations take care of business. Organizations include nearly all organized groups regardless of size with a method for organizing membership, and a charter stating a shared goal beyond simple monetary profit. Organizations can seek to further social goals, influence political elections or laws, provide charitable services, or fight for rights or justice for a group of people or an animal species. Organizations can advocate for members based on IQ, hobbies, gender, age, regional ties, or ethnicity to name a few.

Religions, government agencies, and your bowling league are not organizations, but non-ecclesiastical groups serving to further the religious goals shared by members, independent groups funded by governmentt grants, and the national group formed to cater to left-handed bowlers are organizations. Organizations use P.A.s to further their day-to-day business, provide training, address and inform members at gatherings, reward members at ceremonies, throw fundraisers, or just at employee holiday parties. Even more than any other group, organizations will use P.A.s and venue rentals for the entire range of reasons and purposes for doing so.

Event Producers and Performers

Finally, sound companies contract with event producers and performers to provide the P.A. systems and logistical support required for one-time local performances and events, as well as touring productions spanning many months. Performers from every corner of the performing-arts world, including dance companies, magicians, musicians, and theater troupes, contract with sound companies to make many of the concerts and performances that audiences enjoy possible.

Where Are Public Address Systems Used?

The following list describes some of the places where live sound may be found. Not all are ones that use a sound professional to run sound, and some are venues that are less commonly used, but most are examples of the kinds of events at which a person may find themselves working if they pursue a career in live sound. The events represented on this list aren't all inclusive, and with more people choosing to use P.A.s for their events, and small powerful systems opening up opportunities for live sound that didn't exist before, the number and type of events where sound professionals are needed is only going to increase.

Stadiums

- concerts;
- sporting events;
- exhibitions.

Stadium concerts are thrown by **production companies** representing touring performers who rent the stadiums for one or more shows. Work here is through hire from a production company, one of the performing acts, the sound company providing the P.A. system, or the stadium venue. Stadium venues hire large crews consisting of a smaller house sound crew who run the house P.A. during games and house-sponsored events, and a large number of stagehands who provide general labor for stadium events and concert tours. For all the rest, working at stadium concerts means working as a member of a touring crew. This is not work that people feel mildly about. People either love it or hate it, and there are lots of great live sound professionals who swear by each. We are of the "local sound rocks, tour crews are crazy" line of thinking, but we hear there are others who feel the exact opposite.

Word on the street about touring is that it is grueling work, but satisfying. In addition to reading more about touring later in this chapter, if you are curious about what it is like to work on a tour, go online and see what the people already working in the trenches have to say. Links in the directory leading to roadie and crew websites and boards are good places to start.

Theaters

- spoken-word performances;
- plays.

Theatrical productions are a bit different than other live sound applications. In addition to practices that apply to any live sound application, theaters have their own additional lingo, a workflow that will be a bit different, and involves working with a whole different cast of characters (pun intended). Similar to most live sound positions, theater sound work will be a slightly different experience depending on whether the position is one working for local theaters, or one working for touring productions. In addition to your further technical studies, you will want to look up the links on theater sound to explore some of the online resources geared toward theater if you want to pursue this avenue of live sound engineering.

If you are interested in eventually looking for live sound work in theaters, start getting to know the ropes now. Begin tracking down your local theaters, or the theater department at your local college, and call to find out who knows the most about theater sound, along with how to get in touch with them. Many cities or regions also have a **civic light opera**. These theaters specialize in musicals and much of the work is done by volunteers. Calling to ask if you can pick the brain of someone on the sound crew, or even to inquire about volunteering some of your time on a production, are also options. See the chapter on networking and social etiquette for more on this.

The best part about theater sound is that theaters provide the opportunity to work your way up to positions where you can help with or be solely responsible for sound effects at every step from choosing which ones to include to incorporating them into the performances. Running theater sound boards requires you be good at keeping track of cue sheets.

Bars/Clubs

- bands;
- DJ/karaoke.

Most people have at least one or two bars and venues that host bands and other live events within driving distance. Bands and **electronic dance music (EDM) DJs** are a staple weekend activity all over the country. Some of these venues hire dedicated sound engineers part or full time, and are small and open enough that they are a great opportunity for any learner who hasn't seen a show mixed close up to go watch what the sound engineer or crew does over the course of a show. As long as you take care not to get in the way (and keep your observation unobtrusive so you come off as curious and not creepy), they won't mind. A number of smaller venues offer employment opportunities for live sound engineers to work locally and to get a chance to mix shows sooner than they are able to at most sound companies.

Another opportunity these venues offer is through the bands that play there. Some of the local bands who play at these venues will be looking to hire their own sound engineers to work at their shows. Most of the bands that end up touring got their start as regular local acts and learned the ropes in the smallest of these venues. Working with local bands is a great way to get experience even if they never end up touring. Some live sound engineers who ended up touring with many bands over their career got their start perfecting their craft and building their rep by working with local bar bands. If you don't know how to take advantage of your local scene, see the chapter on networking/social etiquette for more advice.

Sometimes bars can be a perfect way to open up your own market when no other openings seem likely to open up soon. Many of the bars too small to fit standard live events or without the funds to add the fixtures to bring in bands have owners or managers who are open to ideas for alternative events. Their owners and managers are competing for patrons from the same pool as those venues that offer live sound events, and they would love to be able to offer an alternative to the bar bands on the weekend, or to provide something to draw in new weeknight customers and increase enthusiasm in current customers. Someone with a basic **DJ P.A.** designed for making mixes to dance to with prerecorded music, a **karaoke P.A.** designed to allow audience members to replace the vocals in prerecorded popular music with their own live vocals, or a small personal P.A. for hosting trivia and similar audience-participation games, who can convince the venue owner that an event will increase bar revenues will often be given a chance to prove it. More than one small audio business began with one weekly event arranged in a deal with a bar owner, and then grew as a business from there.

Outdoor Events

- fashion shows;
- fairgrounds;
- auctions.

Outdoor venues are usually seasonal events, and will usually rent sound from a sound rental company. These venue options only are available for use when the weather allows for it, so depending on location are commonly used only four to six months of the year. Some outdoor venues are fairgrounds, race tracks, parks, and large general purpose venues that hold events like flea markets, car shows, dog shows, rodeos, and so on. Other outdoor events that may require sound include yearly hemp fests and Oktoberfests, events thrown by circus and theater troupes, large monthly farmers' markets, and civic events like free concerts in the park events.

Since most people and organizations hosting outdoor events rent their sound from local sound providers, this means many sound companies tend to get more business in the summer season. A number of these sound providers rely on freelancers to fill in for the extra business they get this time of year. If you love the outdoors and want to work live sound at venues like these, look for work at those sound companies that do a lot of outdoor events. Most sound companies will provide sound for some outdoor events, but larger sound companies are the ones most likely to be hired for large outdoor events.

Working at outdoor events that use live sound requires the same sorts of extra preparation you need to remember any time you plan on a day outdoors. Aside from the extra items required to run sound equipment outdoors, if you find yourself working at an outdoor venue there are items you should always remember to bring for your own protection and comfort. Even though equipment requirements mean that there should be shade tent for the gear, there's no guarantee you'll be under it the whole time. Unless you know that these will be supplied for you, you should still remember to bring a bag with at least these items: a cover-up, sunglasses, sun block, chap stick, bug spray, a large water bottle, and rain poncho (depending on location).

Other Venues

- church services;
- university lecture halls;
- convention centers;
- private residences.

Church sound is a niche that is expanding every year. Learners who need more practice and are at ease in the culture of one of the churches looking for volunteers can get some good experience in these positions. There are also more paid positions than ever before. While many churches still rely fully on volunteers, others are recognizing that the skill required for proficiency and the number of hours they are trying to fill for free is making it unrealistic and unfair to expect the professionals in the flock to be available as volunteers without fail. Those pastors who love their live sound too much to part with it are slowly deciding that it's worth it to budget for a modest rate of pay to compensate (and reserve) experienced engineers for large Sunday services and other important events.

Even in church sound, what pay positions are opening up will go to those who've paid their dues first. The good news for live sound learners is that it is unlikely that churches will stop relying heavily on volunteers for all the other events, from choir practice to weekday services, which still require a sound engineer at the board. This means churches will remain a great opportunity for learning and practice for live sound students and new audio pros alike.

Convention centers are a steady source of business for sound companies. A few may have installed P.A.s in the larger halls for playing background music but in general this industry doesn't provide sound for their customers. Instead the customers renting the venue determine what their needs are and arrange with a local sound company to provide sound. Most venues do provide referrals to one or more local sound providers, so if you are interested in working at convention centers look to the sound providers they refer customers to, these will get the most convention clients for that venue. There are notable exceptions to the rule of convention centers relying on sound companies for sound; some do provide in-house audiovisual services and are often billed as "full-service facilities." The kinds of events held at convention centers will depend on the size and location of the particular venue.

Local convention centers are the smallest and generally their bread and butter are locals. They get a lot of employee development and civic meetings when city hall double books, Jaycee events if the local chapter doesn't have a space large enough, and random stuff like driver-training courses if there are no training centers with a classroom locally. They also rent reception-size rooms/small halls for local events, including fundraisers, parties, wedding receptions, and civic celebrations including proms. Depending on location some larger local convention centers will do tradeshows.

Regional and national convention centers get larger conventions than local convention centers and are intended to serve large areas from two to three counties to two to three states. These are places where most of their business comes from clients hosting large events requiring multiple rooms and larger halls. Regionals are near midsize/large cities that aren't destination locations. These will usually be located centrally to whatever area they are intended to serve. This is where regional meetings of civic and professional organizations happen, as well as events like tattoo, gun, antique, and other trade shows.

National convention centers are clustered in the largest cities and considered destination locations, and most are in cities already considered desirable vacation locations. Those in sunshine states always top the lists, others, such as those in Midwestern cities, take turns as hot locations by competing on price. These largest convention centers are where the largest national conventions and shows go every year, including Comicons and other fan conventions, Furcons and other lifestyle conventions, Namm and other industry conventions, Mary Kay and other business conventions, and everything else under the sun.

One way to inquire about how a convention center deals with customers who need sound for an event is to go online or call and ask about their services. Any audio pros or students already considering/choosing to relocate to a bigger market should look at the convention center directory link available in the online section for this chapter to see where these are clustered. For anyone interested in working for a sound company, knowing where these markets are located will enable these to be factored in when comparing different options. All other things being equal, there will usually be more/larger sound companies and opportunities near national convention center destinations.

Private parties will also usually use sound company rentals. These can occur in private residences, but also cover private rentals of all the different venues mentioned above. These include all your standard personal celebrations: anniversaries, birthdays, graduate parties, goodbye parties, and weddings. These also include any events thrown by businesses and organizations that are only open to members and don't include outreach to the general public, such as Christmas parties and employee conventions.

A Look at Live Sound Work: Boots to the Ground

The range of different business models makes it almost impossible to describe a "typical" example of live sound work, not only in general terms, but even when discussing more specific niches like small venues or sound companies. In the live sound field, duties may tend to fall along a small set of position titles and work is usually available from only a few types of employers, but because most live sound positions are with small businesses and organizations as individual and unique as the people found working at them, it is safe to say that no two will be exactly alike, even though they share some or many similarities. Since there are many more descriptions of the work in advanced positions, both in books and online, including many in this chapter's link directory, we will mainly examine the entry-level experience, which is the one most readers will have immediately ahead of them if they pursue live sound employment.

Internships and Entry-level Employment

No reputable live sound professional will tell you there's an easy shortcut to learning everything you need and then to finish mastering your skills. Even those with a training system or product they endorse are only claiming they can offer a more efficient or effective route to climb the mountain ahead. While many programs do an admirable job of teaching students their audio theory, most live students simply don't get enough hands on time working with a P.A., much less mixing. It's not the school's fault; it's the nature of the occupation. The rub of training for a career in live sound is that for a single class to get even a few blocks of real-world experience setting up and running the P.A. during a six-month or one-year course of study, the school or program must have the resources in cash or connections to literally make a big production out of it. While a few schools run their own soundcos, most do not have as many opportunities to provide real-world experiences. And this only refers to students at programs longer than six to eight weeks and attending training at a physical location, there are many students who learn by attending fast-tracked programs or taking classes online, but they often have even less hands-on time. This is why apprenticeships have traditionally been considered a given as part of, or immediately following, all live audio training.

Unfortunately, there is more competition for available internships at the same time as there are fewer official internships to be found. Luckily, there are plenty of unofficial chances for the audio student with an eye open for opportunities. We will discuss how to find and assess official internships, as well as methods for identifying unofficial chances to apprentice in Part 6, but in the meantime, accept that internship is a given, official or not. If you haven't had an internship yet and you nab any sort of live audio position, you can consider your first six months to a year as an internship, anyway. If you are lucky it will be at a sound company where you can benefit from a mentor, but it may be mixing one day a week at bar or monitor mixing for beers for a local band. That's OK; on-the-job learning through trial and error is another live sound tradition. A good sound company is the best place to learn the trade, even if you shouldn't expect to mix much until you learn the many other aspects of the trade that are just as essential. Don't pass up work at a good sound company because of that—there are always exceptions to the rule, and if you work at one that as a rule won't get you time on a mixer anytime soon, volunteer to mix for local bands on your days off and enjoy the best of both worlds.

Entry-level Employment at Sound Companies

Entry-level employment in most sound companies is as an A3 system tech, and employment often will allow for those with the determination and skill to eventually earn advancement to positions of greater responsibility. If not, as many of these hire seasonal workers and don't need any more long-term personnel, it still can be worth working there for the experience, which will help you land and excel at a job down the line where you may find better chances to advance. You will be hired at entry-level again somewhere new, but the knowledge gained working previously will mean you start with basic skills you were lacking in the first time and can do better from the start than you otherwise would have. If you can't get a long-term position right away but are offered seasonal work, take it. A seasonal worker who proves they are a good worker will get a spot on their list and sooner or later the call will come when they get an opening.

No matter the size of the company, at most of them the journey from A3 to A2 will take you at least a year and usually closer to two or three. This shouldn't come as a surprise; while small raises aren't unheard of in the first year at a job, raises will usually be earned before promotions, and most jobs won't promote new hires until the second year at the earliest. There is no reason to hurry anyway, because the route from A3 tech to A1 FOH responsible for handling client accounts, overseeing projects, and mixing shows is a process that takes many years of mastering every other audio tech position within a company, or the equivalent elsewhere. Ultimately, especially for reaching the highest level, there is no standard guaranteed timeline; the process and timetable followed while working up to positions of real responsibility at a sound company depends on both the culture and practices of the particular sound company, and the aptitude and dedication of the individual employee.

Odds are if a company doesn't promote A3s to a higher position for a long time, it will be because their requirements to qualify as A2 are pretty stringent. If a live sound A3 finds the prospect feels like forever, they must not know there will be plenty of milestones to make the time go by quickly. Being an A3 rarely involves being stuck in the warehouse the entire time. A3s go out on crews too, just not right away, and same for mixing opportunities. Ask around if any of the A2s there regret the time it took to get promoted or wish they were trained somewhere a little more lax. We are willing to bet most will say they have no regrets and are glad they stuck it out. Since the road from A2 to A1 is a long one, it evens out in the long run anyway. Often A3s who took the fast track to A2 will just take longer to get to A1.

The Sound Pros Sound Off

Now that you have a general overview of the industry, and the sorts of clients and venues, positions, and tasks a live sound pro may encounter, see what the live sound pros living it have to say. In the link directory are links to profiles of successful sound company owners and their companies, interviews with top FOH engineers, as well as articles, blog and forum posts from the sound pro next door. Each of them is a glimpse into the variety of experiences and thoughts of those living a life with boots on the ground careers in live sound.

For those who do choose to work for a sound company, we advise that when you are feeling impatient at not being promoted/allowed to mix/encouraged to mingle with clients in the office, and it's anytime under at least a year, that you come here and re-read the parts about taking your time, because asking politely will get one pass, and hinting, pestering, complaining will do the opposite of what you hope. There is a great deal of anecdotal evidence that, cable-coiling practices aside, the biggest pet peeve most sound-company owners have about young audio employees is that many start complaining or pestering their managers for more responsibility way before it is reasonable or appropriate.

The Good News

For many audio learners, nothing here will be much of a revelation. For others, especially those who started by reading too many "Become an audio engineer in eight weeks" web advertisements, this news may be more upsetting or seem more dire than it really is. The good news is, look around at all the audio engineers who've been in the industry for five, ten, even twenty years. Well, the process was no less confusing, the mentors not that much more plentiful, than what you will have to deal with, and if you Google "recession," you will be able to double-check that many of them even entered the field during times when the market was tight. Like audio hopefuls today will end up doing, some started by finding a job with a sound company, or a few, and worked their way up. Some started their own companies and worked up, and others took less travelled creative routes to get their foot in the back door if they couldn't find any other way in.

Small Sound Companies

How far you can go in a small sound company usually depends on how small they intend to remain. Lack of advancement opportunities at small companies is usually because they have no plans to expand and are fully staffed with healthy long-term employees. On the other hand, getting in on the ground floor of a well-run small sound company that intends to expand can be a fast track to management for a good employee. Generally advancement will be slower at small companies, but even work at one that's fully staffed and not offering the chance for promotions can be worth applying for because of the learning opportunities. The main advantage of working at small sound companies is access. On site you will be working in closer quarters with managers and owners as well as coworkers and will be able to get to know them more quickly. Closer quarters also means your tasks handling inventory won't keep you a warehouse away from the other aspects of the business, and you have more chances to observe and learn about them, and closer proximity to superiors also means more chances to ask questions. If you are a go-getter, you have a better shot at getting noticed as an asset to the company, because your ideas have a better chance of getting heard.

Because small sound companies usually have more of a family feel and tend to be less hierarchical, they are less intimidating and a more comfortable place to gain experience for those new to the workforce. Some small companies are so small they have no hierarchy besides worker or owner (who is also a worker). Over the long term, as you gain experience and take on more responsibility, raises will increase pay but won't confer a change in title like an official promotion. While there is some stratification, for example financial tasks will almost always be handled by an owner, and sweeping, coiling cables, and cleaning the toilet will go to the newest hire or employee with the least experience, very small companies generally assign tasks based on aptitude, availability, and/or preference. Rather than assigning work based on title, tasks often go to the person who can best complete the task, is available to complete the task, or wants the task/prefers that type of task. This lack of stratification is a benefit to new hires, because even if the only tasks you can do at first is sweep and carry gear, once you start learning you will have chances to work on a wider variety of tasks sooner than at a more stratified company, including a chance to work off site with a (or the) sound crew sooner than is likely at larger companies. If you are observant and not shy about asking questions you can learn a lot working for a company like this. Over time you can learn almost every aspect of running a small business at a small sound company, so they are an especially good place to learn the ropes if you ever intend to start your own sound business

Large and Midsized Sound Companies

Larger companies have a different set of benefits and drawbacks, depending on your point of view. They can offer more opportunity to expand in the long run, but can feel more limited early on because your title and status play a greater role in determining your interactions and tasks. At a large sound company an A3 will work mainly at one location to start, the warehouse where inventory is stored, while much of the company business happens in an office or storefront either on the same property or across town. Either way, they won't be learning as much about that part of the business at first, and may not even get a chance to see the offices again for a while once they are hired. An employee at a large sound company will have more coworkers, but many may work in different departments/areas of the building and there may not be many chances to interact with some of these people early on, so getting to know everyone takes much longer. Unlike smaller sound companies, where everyone works in closer proximity and even the low person on the totem pole is in regular contact with the owner/s, at a large company, entry-level sound techs could easily work under a manager in the warehouse where the inventory is stored, and may not even know who the owner is early on. There are many who enjoy working at larger companies and feel that the larger physical environment to explore and number of people to get to know makes the job more interesting, but it can feel impersonal and be intimidating for some people when they are first hired.

The advantages at such companies are that they may allow for a faster road to long-term employment because they have more positions to keep full. The larger employee pool means working at a large company is a great chance to meet a lot of interesting people, and if/when an employee leaves to work elsewhere, they can take a list with more industry contacts with them. A3 techs at large companies also get work with a wide variety of sound equipment, much of it state of the art. With more experience, when the time comes to work off site on a crew, they may get to work on larger projects than at small companies, though not necessarily right away, or exclusively. Nevertheless, if large events are where you want to work, especially if you want to tour as a sound tech down the line instead of in a nontechnical position, the best way is to get a foothold at one of the large sound providers who handle tour sound.

Mid-level sound companies have a bit of the benefits and drawbacks of both. Not as impersonal as large employment venues, but with less access than small companies, they can represent an even blend of both, the exact middle ground with a hierarchy determining part of tasks, but a number of exceptions that are assigned by ability or expedience, and discrete departments, but areas of the business requiring interaction across them. Usually, they will lean one way or the other more or less. So, for example, the experience at one might be mostly like working at a smaller company, but with a little more structure (or a lot), while another will be very similar to a large company, but offer more access. With such variety there is not the same ease in pinning down a trend that enough of them share to be able to pick three traits they "generally" share as with the small or large companies. But so long as you know they are some mix of the two, a little investigation into any you are interested in, as recommended in Chapter 2, can help provide enough clues to help give you an idea of where they lie between the extremes and which traits they balance.

Entry-level at Venues

Promotion is less of an issue at venue jobs. If you find an entry-level position at a venue, it will either be as the lone sound professional at a small venue or bar, in which case you will find you have arrived at pinnacle of achievement available at that venue, or else employed as part of a crew of two or more at a venue somewhat larger, in which case you will not be promoted until someone above you in rank departs or gives up the ghost. While promotions are rare, these positions can also be great places to learn and practice your craft.

The lone small venue sound pro is the most cut-and-dry experience to describe as there are many commonalities shared across positions. The advantage of working entry level at a small venue for a sound pro is, first and last, having total dominion over the mixing board when other entry level sound pros initially only get to touch one when they are carrying or cleaning it. There are other pluses, such as relative autonomy, but in truth, getting to mix every show right away is the main draw to these positions.

The tasks of the job are easy to describe, as they will be a small-scale version of every task sound pros do as described in this chapter and others. They will do first-year tasks like the ones listed in this chapter, but with one P.A., rather than many, will perform loading and unloading tasks but with the band's backline gear rather than a portable P.A., be responsible for some or all the production tasks discussed in later chapters but for very small productions, perform every mixing duty, both FOH and those the monitor mixer performs but with the set of challenges typical for small venue sound. At first it is easy to mistake it as being an easy job, as all their duties are scaled down to match the size of the venue, and taken alone are the smallest or most simplified version of each task done on a larger scale by other sound pros; however, when you look at all of the tasks together it is clear that to truly master the job requires a lot of focus, organizational skills, and judgment. A certain amount of credit is due those who can't be praised for mastering it, but merely acknowledged for managing it adequately; simply pulling off that lesser achievement involves making strategic choices at regular intervals about what to let slide and how long, and can't be done without putting in a baseline amount of work on what can't be put off.

The venues most commonly worked by lone live sound pros are straightforward as well since there are limits on the ability of a single person to be responsible for every aspect of the live sound for productions. There is no tracking of where live sound pros work with other sound or production staff or work alone, but there is a great deal of anecdotal evidence that the most common venue with this sort of employment is the small venue roughly the size of bar band venues, with audiences generally not larger than 50–150. A few exceptions do exist but, again based on what is humanly possible, any exceptions will be only slightly larger.

There may be cases that seem to represent exceptions, where bigger venues have a single in-house sound pro, but chances are they are on a **production crew**, whose members each focus on different aspects of a production, but whose combined focus and efforts are on the shared goal of making the production run smoothly and so, even if each is the only employee in their specialty, are not really alone. At a venue with a sound pro, a lighting pro, and a stagehand, the sound pro may not be able to ask anyone to take over mixing, but the crew can help each other with heavy lifting, and probably each assist each other running cable, or even just keep an eye out on each other's tool bag when the other has to visit the bathroom. Such a crew will also share experiences of production notable events from similar perspectives, whether it's bossy agents, performers treating the crew like servants, the marimba and conga player melt-down, and so on, the shared experiences and even having someone to listen to your particular gripes and understand from a close-enough perspective create a feeling of teamwork. The sound pro who works with other production staff is on a team working for the same primary goal—a successful production, unlike the bar-venue sound pro, who works where the other employees are focused on serving drinks, door sales, or crowd control and so is the only person employed whose concern is making the production happen for the sake of the show (the other employees' interest in the show is in how it helps or hinders their particular concerns, not the show itself).

The venue will usually present challenges to overcome for good sound, but initially most entry-level hires probably focus on the P.A. The P.A. systems in these venues may range from relatively nice installed systems with soundboards from recognized manufacturers and produced within recent memory to vintage band P.A.s with battered speakers pushed against the walls and everything else creatively jigsawed into a broom closet, including an antique soundboard of unknown make but a design sensibility that brings to mind burnt orange polyester, corduroy wallpaper, and the Doobie Brothers, all fashionable back when it was new. While neither example is an exaggeration, both systems represent outliers and are rare enough that to keep from unfairly raising expectations in any optimists or stirring up fear or despair among any pessimists, it should be noted that the odds are against the average live sound professional encountering either one at the

Figure 3.1. Small venues are good places to gain experience.

small venues they work. The majority of small venue P.A.s will fall between the less extreme extremes of "bare bones but workable" at worst and "nothing fancy, fully functional" at best.

Regardless of what the P.A. looks like, the venue owner will know very little about sound, or know a bit about it and choose maintaining the bottom line instead of aiming for top of the line sound, which can be safely extrapolated based on the fact that they hired and handed control of the sound over to an entry-level sound professional, although this is just a trend, not a rule, and most will have more than one reason for their choices. On the plus side, other than instructing the sound person to turn the volume up or down these types of venue owners will let them do their thing and be satisfied as long as a few simple rules are adhered to, such as "Show up," "Don't break anything," and "Don't drive away customers." Barring any sound mess-ups that annoy customers too badly, or commit the cardinal sin and empty the bar out, both also will be tolerant of the mistakes a new mix engineer will inevitably make as they settle in, giving most first timers the leeway needed to get through the learning curve and to the point they can start producing sound decent enough to keep most of the people around them happy.

On the down side, small-venue owners with a single sound employee have a universally recognized character flaw, at least from the perspective of individuals like the live sound professionals who work for them and share a belief in certain fundamental principles such as "investing in good sound is worth it." While this trait, diplomatically described as "excessive thriftiness" by one live sound pro, does little to detract from venue owners' overall popularity as bosses, it is one of the biggest frustrations for the live sound pros who must deal with it. As a result of this trait, venue owners do have limits to their patience with their entry-level sound pros, and will tend to have no sympathy for those who claim they must have equipment upgrades to function. Upgrade is a word a new live sound employee at a small venue should put out of their vocabulary to begin with; it is an idea that will be shot down with extreme prejudice anyway.

The good news is, despite the fact that thriftiness is indeed an unshakable trait, they do depend on the P.A. for drawing their client base and will replace or repair any broken items that can be proven to be essential to the functioning of the P.A., and may even agree to purchase key items that are missing if the benefit is clearly explained to them. The sound pro seeking some additional investment in the venue P.A. can eventually bring even the flintiest venue owner around enough to meet them halfway or even all the way if they learn to speak their language. But first they must establish some trust, and one of the best ways to earn it is to give it. Sure they should trust the venue owners, but that's secondary to the most important leap of faith they must make. The first step is to show some trust in themselves and their abilities, instead of thinking better gear is needed, and the best way is to roll their sleeves up and make the P.A. work, since the venue owners are right—as long as essential elements are there, there is no reason anything more is required. There are many ways new gear can make a mix engineer's life easier, and a sound pro who can make a basic system work can use extras to make the sound even better, but no equipment upgrade is going to help someone not up to the challenge of making the most with what they've got first.

Sound company workers aren't the only ones who must have discipline as entry-level workers, because what they must sacrifice in paying dues before they get to mix, they gain in the camaraderie of peers and the support and input of mentors. They also get to count on having access to better and more equipment when they do get time at the console. That's the tradeoff made by entry-level sound pros at small venues for the freedom and opportunity to mix right away—sink or swim, they must make the most of minimal resources in less than ideal environments, and they need to figure out solutions on their own or find the help they need on their own. Often they have put in some dues mixing anywhere and everywhere for free well ahead of getting hired, and getting the sound right over the first month or two will count as the rest. Most make it over the hurdle after they accept that this time they are truly on their own.

Those who are hired to work at slightly larger venues with crews of two or more will enjoy greater variability than the lone sound man, working at a wider variety of venues with a wider variety of productions. While teams will work together on all the same tasks as the lone sound pro, the venue and productions are probably larger so being on a team won't mean less work, but it will mean having people to do it with, which always makes tasks more enjoyable. Another difference is that on a crew, monitor mixes will usually be from the monitor mixer, not the FOH, and those hired may have a chance to monitor mix at some point as well. Those hired onto venue crews probably won't have as much immediate access to the main console as the lone sound person at a small venue, but they also may not have to wait as long as most entry-level sound company employees do. The same may be true for the monitor console, or they may have the opportunity to monitor mix sooner. As usual there is no set standard, the policies determining who mixes and when will be different at each venue.

Inventory Tasks Common to Interns and Entry Level

Below is a list of the type of tasks common to A3s and interns at sound companies. Similar tasks are performed on a smaller scale by A3s on large-venue crews and by small-venue sound pros in addition to everything else. These aren't the only tasks that A3s will do, and A3s aren't the only ones doing them, but most of what they do include these or are along the same lines, and they do the bulk of it:

- *Sweeping*: Sweeping the area where gear is stored and worked very thoroughly is done to keep the air free from particulates. Dirt and dust build up on surfaces fast if not cleaned away, and once the buildup is bad enough the slightest air movement can pick up particulates, which can float erratically through the air (and then get into gear) for hours before settling again. Sometimes done once a day, sometimes more, like at the beginning and end of a shift, sweeping isn't done to make the warehouse look good; it's step one in proper gear maintenance.
- *Dusting*: See above.
- *Pulling inventory for upcoming shows*: Whether the gear fits in a large back room or a large warehouse, there will be a system for organizing inventory and every A3 will learn how their sound company does it so they can find and stow gear on command (and should be learning all about the gear as well as how it's organized). In advance of each upcoming show, a list of all required gear is made, and sound techs will refer to a copy when the time comes to retrieve or pull everything needed from inventory. Whatever methods a company follows to determine the right order to pull gear in, all will ensure the gear leaves in travel cases, since aside from speakers which have strong enclosures, all equipment travels in a hard travel case (interchangeably referred to as a **road case**, a **flight case**, or an **ATA case**) that is built to meet stringent airline standards for shipping containers so sensitive equipment can be transported safely. These cases are expensive enough that it's not always cost effective to store inventory in them when only part of inventory gets checked out even on the

busiest days, so stored inventory may be protected by soft cases instead of road cases. It's common for soundcos to make it rule #1 that equipment must be hard cased before moving it, however, so part of pulling inventory may include retrieving the proper case for each piece of equipment first so it can go right into a road case.

- *Staging inventory for transport*: Originally a military term but now used in a variety of contexts, a **staging area** is a temporary location where all equipment, participants, and other assets that will be used for an operation or mission are assembled and prepared for transport or readied for deployment. In a warehouse, the staging areas will be adjacent to the loading docks, and staging equipment for each show involves getting it all packed and stacked here. Packing commonly starts with giving all gear a quick visual inspection as it is packed, to make sure there are no obvious indicators of problems. As already mentioned, electronic gear will go in road cases as it is retrieved from inventory, but other gear will need to be packed up too. Cables, mic and speaker stands, mics, rolls of tape, tools, sound curtains, batteries, CDs for testing the system, headphones and everything else required for a show needs to be stowed in cases or plastic tubs, with each item checked off on a list as it is packed so nothing is left behind. There are different strategies for packing that are used, but one way is to pack items together based on where they will be used on site, and as the lid is closed on each tub, it should be labeled according to where it goes so the crew unloading and setting up gear on site don't have to look inside to know where to put it. Stacking is done so that those loading are able to access the items that go in the truck first without having to move other items out of the way.

Figure 3.2. Flight cases are designed to take the knocks, gear is not.

- *Inventory management*: Even though any business with an inventory has systems to match the virtual inventory that's catalogued and the physical inventory that's warehoused, by tracking each item as it is checked in and out, and removing or adding the corresponding gear listed in the catalogue as items are broken and replaced, over time errors pile up at both ends, and between a clerk's errant keystrokes, a DI (direct injection) box swiped by a wedding-band guitarist unnoticed, and an employee pilferer with a fetish for adapters and gaffer tape, the two inventories diverge. Managing inventory may happen regularly or only once a season, but either way it often means loading the lower-rank employees with scanners, or click counters and inventory lists, and setting them to the task of counting everything in the building and noting each count so the inventory and catalogue can be reconciled and losses assessed. Other inventory management tasks are making sure items are in the correct place, and that beat-up tags on shelves and items are replaced.
- *Loading and unloading equipment*: Wear gloves to protect your hands, and lift from the knees. You will be told where to load items according to proper **load order** by an experienced **loader** using the procedures followed by your company, and will need to follow instructions quickly so you don't hold things up. Load order refers to the best practices for loading trucks that different loaders, such as warehouses, package-delivery companies, and furniture movers use to plan and implement a load in the way that best balances the conflicting requirements of several parties to meet the minimum requirements of all of them while maximizing those priorities of highest value to the loaders. Determining load order is a very complex task involving many strategic decisions and compromises. There is more than one way to plan load order and companies will use the framework that works best for their priorities. Some use software that allows the user to state their priorities and the load type and dimensions and let the computer come up with the best solution and packing instructions. While there may be a loader for large loads, there will also be a set of procedures to guide loading small loads as well. When employed then it's important to stick to the method preferred by the soundco employing you, but some basic load-order priorities

include: following your state's laws regarding truck and axle weight; loading so that unloading is arranged in the ideal order; loading so the load in the truck remains balanced and doesn't shift; loading so customer orders are kept separate from one another; loading large strong objects underneath light fragile ones so nothing gets damaged; and loading to maximize wasted space and avoid incurring extra shipping expense. A comprehensive loading guide is linked to in this chapter's web links, and does a great job of showing the many types of priorities companies balance when planning loads. We recommend anyone who thinks they might need to load a truck in the future take a look at it. You don't need to memorize the tips, but knowing the principles will be helpful.

- *Checking, running and coiling cables*: Cables are inspected for defects before being packed for transport to a show and again after the show, before being coiled and packed for return. Problem cables are separated from the rest and set aside to be fixed. At setup, cables are run, connected to gear, and securely taped down and secured; at teardown they must be gathered, tape removed and coiled using the method preferred at their soundco. One of the complaints voiced most often about entry-level live sound employees is their cluelessness about how to coil cables. There is no faster way to advertise yourself as a rank amateur in front of more experienced live sound professionals than to get caught looping cables over your palm and around your elbow. If you still coil cables this way, there are links in the directory to several online tutorials showing proper cable coiling techniques so you can practice at home and save yourself the eye rolling, teasing, and exasperated coiling lesson you will get if you ever try it on the job. It's worth the time to learn, since so many new hires need to be shown how to coil, being able to do it from the start is an easy way to set yourself apart.
- *Gear cleaning*: Once A3s have learned how to maintain and clean gear this task is also a staple activity. Cables get filthy after being run, and more importantly connectors and inputs get gummed up over time and stop working. Dust and grime builds up in crevices around switches and knobs, making them stick. Finally, no matter how hard you try to prevent it, over time dust creeps inside gear and contributes to overheating. Cleaning gear by blowing out dust and wiping off stuck on grime will increase its lifespan and head off issues caused by unmaintained gear, like overheated amps and blocked signals. These issues can be showstoppers if they happen in the middle of a set, but are easy to prevent with regular gear inspection and maintenance.
- *Restocking inventory*: Checking that all items returning from an event are present, intact, and clean, and returning them to their proper place to be stored.
- *Assigned projects*: Even though some soundcos are more hierarchical than others, there are none that are so completely without structure that a new hire can expect they'll never do grunt work and none that limit tasks by rank so rigidly that they'll leave an important task undone just because there's no one but an A3 around to do it. Assume that random tasks will be part of the job. When they feel like a perk, enjoy it (just not so much you don't do a good job). When they feel like punishment, understand that there is no possible reason to punish you secretly. If you are being punished because you messed up, you will know it. If you think you are being punished because your manager doesn't like you, you are mistaken—people the manager doesn't like get fired; it's perfectly legal. Some tasks just suck. You were there to draw the short stick this time. Smile and get moving; the sooner you start the sooner you'll finish.

Production Tasks: Overview of the Stages of Production

This is a breakdown of steps involved in each show, and after paying their dues as an A3, much of the time a typical live sound pro will be working on the tasks involved in producing a show. Much of this list will occur the day of the event, but the tasks covered in the first point, preproduction, take much longer. We will look at each of these steps in detail in following chapters, but for now this overview should give you an idea of some of the tasks involved in throwing a show:

- *Preproduction*: This can take up to anywhere from a week to a month or more, and basically is everything you do to prepare for a show in advance. Most of these tasks are communication tasks—consulting with people to plan logistics, and passing information on to the people that need it. However, preproduction also includes physical tasks like site visits to inspect the venue and packing. The preproduction tasks will be different depending where you work and much different for a Friday bar gig or bar mitzvah than planning a tour. For a mix monitor hired for a one-show contract by the band then preproduction includes negotiating the contract but also making up a stage plot to send the venue. For an FOH engineer, no matter where they work, listening to the bands recording a few times to learn their music should always be part of it. Sound company preproduction includes consulting with a client and going through your gear list, helping them select what they'll need from your catalog. If you are a tech on a sound crew heading off to tour, getting your passport counts too.

- *Load-in*: Unloading the truck and staging for setup. House sound at a venue with its own P.A. will be helping the band unload and set up the back line, it could take an hour, or much more for a tour production.
- *Setup*: Again setup depends on where you work. It might involve moving a canvas cover back from the console and plugging in mics or it could involve countless hours of humping scaffolding and flight cases.
- *Line check*: This is when you check all your mics, DI boxes, connectors, etc. to make sure everything is plugged in and is working before the band hits the stage for sound check. This also refers to an abbreviated sound check. For example, if you have a gig with four opening bands and a headliner, you probably won't have five sound checks; you'll have four line checks and a sound check.
- *Sound check*: Sound check is when the FOH and monitor mixers dial in their initial mixes and make sure the band has everything they need. It also provides time for the band to adjust to the venue and if necessary for the sound crew to address their worries or just to help them chill out and lose the jitters so that they can perform.
- *Show begins*: Smart engineers take their bathroom breaks ahead of this.
- *Show ends*: Hopefully smiling.
- *Teardown*: This goes fast if the crew have parties to go to.
- *Load out*: Almost home free, so no time to be careless.
- *Post-show (paperwork and follow-up)*: Freelancers need to get paid, ideally before leaving the venue, though this is not always possible. It's always good to spend a few minutes with the band to see if they have any comments or suggestions for next time; same thing for the mix engineers.
- *Following day*: Freelancers need to get paid; the longer they wait the more chance for problems, so make like a detective and track it down in the first 48 hours. While memories are fresh and everyone cares is the best time to evaluate a crew and tell them that they did a good job, and also talk about any issues. Let them know what they did well or not. Also it's important to check in with the client, but they may be tired so it's important not to call too early; be prepared to leave a message, or keep it very short.

Touring: Separating Facts from Fiction

Among any group of people interested in live sound there will be enough who hope to find work on a tour, or are just curious about it, that a bit more needs to be said. What you've seen regarding the variety in type and size of other players in the live sound field holds true for tours as well. There is no one typical tour, and going on a one—bus/five-truck tour of music halls or a six-bus/70-truck tour of stadiums is each going to provide a different experience, just like touring with a rock band, a theatrical production, or a youth orchestra will each have a flavor that makes it unique. Similarly, the problems and perks will be different if you are touring in foreign countries or just in the U.S. As already mentioned, touring work is harder than most people imagine, but other misconceptions abound as well and should be cleared up by any who expects to one day work on a road crew, or work with a touring crew visiting their venue.

One common misconception is that **roadie** is a job title. It's not: rather it describes a group of workers all in different positions, who collectively provide the technical knowhow, services, and general labor to set up and run nearly all aspects of a touring production except for the performance and management/publicity tasks. Anybody who travels with a touring production except management and the band can claim the title roadie, including workers most don't think of as roadies, like costume and makeup artists and catering personnel. Many of them do wear it, and proudly.

Even if you limit the term to only include those jobs referred to as the general public use the term, the variety of jobs included is diverse. There are the highly skilled technicians, usually only one to three per department and all very experienced, such as the sound engineers, *lighting designers*, *pyrotechnics*, *video techs*, *instrument techs*, and *laser techs*. Because of the years of experience required to be in most of these small departments, especially for the largest stadium-type tours, the people staffing these positions are usually older than the average roadie, in their thirties or forties (instrument techs are an exception, though no less skilled). Other roadies are those who build the steel framework of the stage and speaker towers, the **riggers** who hang lights

Figure 3.3. Roadies—they carry stuff (but that's not all there is to know!).

and speaker and speaker arrays, and the **stagehands** who supply general labor and support, and carry the heavy loads. Most of them work for the venue, but some travel with the tour, among them a few specialists, like the *carpenters* and *set builders* and the loaders or packers who oversee the loading of trucks, making sure they are packed in the right load order.

Another common misunderstanding is that roadies hang out with the artists on tour. While some roadies are friendly with band members due to their close association with them as a function of their duties, like monitor and occasionally FOH engineers and instrument techs, non-technical roadies are discouraged from socially interacting with the band and keep interaction polite and brief. Related to this idea is the perception that tours are some sort of 1970s time capsule, and when they are not chatting with rock stars, they are knee deep in groupies and substances of dubious legality. While roadies will try to squeeze in some partying on tour, most young people between the ages of 20 to 25 are known to seek out a party in their free time and some older kids too, even if they have less time for it. Roadies probably manage far less of it, sleeping a dozen to a bus, and living off minimal sleep most days because of work. Roadies of both sexes seek out dates on nights off, but again not to the degree matching the reputation. And their drug use can be no worse than the rest of the nation's; their job is too physically demanding to allow for it. Those thinking they want to tour so they can party hard every night would have more luck enrolling at a state university, and can consult a yearly list to find the ten most known for unrestrained hedonism—
post-1970s, that's where nonstop party animals remain.

Finally, there are those who revile the roadie instead of idealize their imagined hedonism, and assume roadies are uncouth and unwashed sexist brutes. This is an even bigger myth than the others. Roadies are young working-class men, and sometimes women, who engage in physical labor. Most work very hard every day and admittedly get pretty filthy. Then they generally go wash up. Where there are cases that their hygiene is any worse than construction workers or other working-class laborers, it is a result of having to share one bathroom and not having access to on-demand use of a washer and dryer. Even with these limitations they rarely fail to clean up in their off hours.

As to being sexist and uncouth, they probably are similar to most working-class people of their age and sex. They aren't as likely to be as "politically correct" as college track or white-collar workers in their cohort, because an acute awareness of political correctness is mainly taught in colleges and at white-collar workplaces, which they haven't accessed. This doesn't amount to lacking all social graces by any standards nor make them more likely as a group to be rude or insensitive in their interactions, no matter how atrocious the nicknames some of them may have. Their shared behavior is no different than any other young men their age and is more due to hormones and habit than sexism; it is not all that unusual or egregious enough to earn a bad reputation.

Some employees on tour are sexist; and so are some cops and highway workers. Fortunately, most roadies are not, nor are they brutes. To the contrary, odds are high that as a group they have *better* social skills than people of the same demographic employed elsewhere, and they prove it every day by living and working together in very close quarters and without a "time out." Aside from occasional tension and crankiness entirely within normal parameters, they could actually be described as cheerful and easygoing. While one would be foolish to think of any demographic as being perfect or without their fair share of flaws, there are some flaws that are less likely in certain populations.

Surely, here and there a cranky, competitive roadie with a foul disposition surfaces on some tour or another. Just as surely, when that happens they are quickly fired. When it comes to how roadies treat people around them, those readers who feel the call to touring don't need to worry much about that aspect of the work environment, unless it is to assess their own ability to get along with others as well as will be required for them to thrive as part of a road crew.

A few observant readers may have picked up on the sound crew being described as being exclusively highly skilled workers, and to confirm this it's time to correct a misconception held mainly by aspiring live sound engineers who dream of touring as soon as possible, and dream big by imagining themselves working the super-sized tour systems seen at the largest venues. While touring as part of a sound crew is an option they might find open to them much sooner, the tours will be smaller ones scheduled at venues with much lower seating capacities, and if they are persistent they might secure a coveted entry-level road crew position with a large tour, it's fair to say that the entry-level audio tech who hopes to tour with a stadium/large-venue road crew probably won't be doing it as part of the sound crew right away.

The sound crew on top-tier tours need to be experienced enough to handle all of the complexities of their task without fail and so the limited sound crew positions must go to those who not only know how to do every aspect of their duties, but know how to do them under any conceivable combination of adverse circumstances that could possibly occur—blindfolded—and all the while managing the roadies assigned to carry gear and do setup and teardown tasks. There are many chances to learn while on the job in live sound, but not while doing sound at the top tier of live performance tours—those positions only go to engineers at the top of their game, because the stakes are too high to risk canceling even one show.

That shouldn't stop sound techs from pursuing a shot at working a big tour as a stagehand if they get the chance. Stagehands are the entry-level position on road crews, and many highly skilled video techs, sound engineers, and tour managers got their start carrying gear and taking orders as entry-level stagehands. On top of hard work, there is a lot a stagehand can learn about almost every aspect of touring and the experience will look good on the resumé of anyone seeking to enter one of the technical fields involved with event production, so an opportunity to tour as a stagehand shouldn't be passed up over concern it won't further their goals for a career in live sound.

Don't mistake the fact that roadies are on tour to work and don't conform to the pop-culture mythology as meaning they are all work and no play however. Just because they do more working than partying on tour and aren't hobnobbing with the band, doesn't mean that roadies don't get a chance to have fun on tour or meet interesting people. Roadies get days off, and take good advantage of that time to have fun, and traveling the country or the world makes meeting interesting people pretty much a guarantee. One of the perks of being a roadie is the ever-present chance for the unexpected to happen at any time, and every roadie has their share of stories about the adventures of life on the road. If you are curious about the finer details and specific examples of working in a road crew, there are many links in the directory for this chapter where you can read about being a roadie from those who know best—roadies themselves.

Essential Skills and Attributes of a Live Sound Professional

Listed below are the essential skills you will need to attain and maintain work in the live sound field. Depending on the position and or career track an individual is pursuing, these may not all be necessary or necessary at the same time. There are varying degrees for each requirement and the position will determine to what extent an applicant must hold any of these skills and attributes. It takes years to fully develop all the skills that make up the job, so what matters most is that you have the skills to complete the job you are hired or applying for.

An entry-level house sound position may require only basic P.A./backline setup skills, the ability to use an SPL meter to assess sound levels, and the ability to mix the one or two genres that the venue caters to. In contrast, the technical and testing skills for a FOH position overseeing complex festival sound systems from design up would need to be far more advanced, and applicants would need to have management skills and be able to mix a wide range of genres with equal skill as well. Two live sound engineers can be at distinctly different skill levels and still be rightly considered competent professionals, so don't feel you need to be an expert in any or all of these skill sets to seek work in the field. Just make sure to focus on positions that are aligned with your abilities and experience, and keep expanding your skills if you want to eventually move on to more advanced positions.

The general traits and skills needed for most FOH positions, regardless of degree of advancement, are similar. We focus on FOH positions because they encompass all the skills needed for any positions. If you are looking to find entry-level tech work, focus on the social skills involved in being a good team member and don't worry if you don't have management skills. You will develop more advanced skills as you gain experience, and over time you may also choose to put extra work in the skills most relevant to your particular career plan.

The skills and traits essential for live audio pros include:

- *Technical and testing skills.* These include an understanding of audio components, acoustic science, acoustic measurement, and electrical principles.
- *Organizational, planning, and time-management skills.* These are required to coordinate with production teams in advance of shows, to keep track of and transport gear efficiently, and to manage time/prioritize processes, and to manage and delegate work to your crew.
- *Interpersonal skills.* Social skills are essential to negotiate and work with bands, venue and sound company owners, production crews, rental clients, and crews working other departments. You also need interpersonal skills to work well with the roadies and techs in your crew and visiting crews, especially under tour conditions, where sharing close quarters and working on tight schedules can raise everyone's stress levels. Finally, when employed as FOH with a crew under you, in most positions your responsibilities will include acting in a leadership capacity and will need the interpersonal skills to manage other people well.
- *Troubleshooting and problem-solving skills.* The ability to think on your feet and under time constraints is essential. The ability to systematically zero in on problems, evaluate problems to discover root causes, and recognize/choose best options from potential solutions will help you to confidently and efficiently manage sound and technical issues as they arise.
- *Good hearing acuity and an ear for sound/music variation and localization.* The ability to hear the full range of frequencies with equal *acuity* is necessary to mix engineering, although live sound career paths and positions that don't necessarily require mixing are still open to those with less acuity. An ear for music including instrument, and genre is also required for mix engineers. Some training in music theory is helpful.
- *Dependability and professionalism.* While many live sound career tracks are prized by audio professionals for their more casual dress code, variable hours, and laid-back work culture, this laxity doesn't transfer to other areas. No matter how much your boss likes you personally, most tour managers are no more tolerant of habitual tardiness than managers of any business. No matter how sloppy you can look on the job as long as your clothes are clean and cover all the required anatomy, most venue owners still expect their sound crew to be polite to clientele. Professionalism means meeting requirements for respectful communication practices, dependability and timeliness, ability to stay on task, ability to follow instructions with minimal oversight early on and to make decisions autonomously later on. All these are needed by anyone who hopes to thrive in the field, even if owning a tie or high heels may not be required.
- *Flexibility, patience, and a positive attitude.* In a field with as much variety as live sound, it would be impossible not to prefer one type of gig to another, even for someone with the broadest tastes. Live sound pros are bound to find themselves working events they'd rather not, pretty regularly. And this is after they are done paying their dues. While soundcos will try to accommodate everyone's preferences when they can, it's not always possible for everyone to only work their favorite events, not even for the owner/FOH. The audio pro who can smile and have a good day even on the days that don't end up being perfect will be less stressed on the days they do get what they want, and better able to enjoy it. They will also be more pleasant to work with for coworkers and employees, and more able to keep clients happy. Early on, these traits will enable them to stay employed, because rigid, impatient A3s with a bad attitude don't have great career prospects in live audio.
- *Ability to lift and move equipment or equivalent.* Many live sound positions, both at entry level and even some more advanced positions, include load in and load out tasks. This should be expected as a given for able-bodied applicants of most entry-level sound positions. At the same time, this should not necessarily keep people less able to do heavy lifting from

considering work in the live audio field. Depending on the circumstances, accommodations such as task swapping or technological solutions can be found to enable some less physically able applicants to fulfill the requirements of some positions satisfactorily enough to allow them to pull their weight without needing to be able to perform every task or perform the same way as others would. Also, not all positions include this requirement.

Helpful/Supplemental Skills and Attributes

These are not absolutely necessary to entering the field, though they can be helpful. They will also prove helpful or even necessary over the long term in order to thrive in the field, depending on the individual:

- *Motivated, self-starter.* Appeal of the field means there aren't always as many openings as there are applicants depending on time and location, although over time most applicants find placement if they persist. In competitive times/markets, those who tend to actively pursue opportunity through creative or alternative avenues may find positions faster than those who wait for ideal positions to open in their location. Examples include those who are willing to relocate to bigger markets, create demand for their skills locally, or learn supplementary skills, and use positions in related industries as launch pads to attain their preferred positions.
- *Willingness to keep up with industry developments and expand on current skills.* Those willing to add to or expand on current skills and add supplemental skills will naturally enhance their marketability. This can make finding employment and/or achieving advancement easier and likely to occur more quickly. This could be something as simple as adding expertise in multiple genres to your mixing repertoire, or adding basic soldering and instrument repair skills to your technological skill set. Also useful are computer skills, especially mastery of computer software suites in audio production and video-editing suites popular in the live entertainment field. Other related skills that can enhance market value are DJing skills, video/lighting setup and operation, event-production/promotion skills, and band-management skills, to name a few.
- *Networking skills.* Networking skills are helpful for opening avenues of opportunity for learning and employment, and also for increasing career satisfaction through being actively engaged with industry.
- *Passion for the field/contentment in the work.* As with any endeavor, those who love what they do are more likely to put in the effort to thrive in current positions and actively pursue opportunities for advancement/learning in the field. During hard times which happen without warning to everyone now and then, those with passion will be more equipped to overcome obstacles they may encounter.

The More You Know

There may be some who begin their studies in live sound with a good idea of the variety of work falling under the umbrella of live sound, but there are many more who only know part of the story. Some may know about FOH engineering but only learn that monitor engineering is an option as part of their studies. Others may be unaware of sound companies, or unaware that sound companies come in many flavors. Knowing the kinds of work environments and situations that make up the live sound field is important for any learner interested in studying live sound. For those who are curious about live sound but are still deciding whether a life in live sound is best for them, knowing what the opportunities and hurdles are for those working in the field is information that is important to making an informed choice. Others, who already know about a life in live sound need to know about the different avenues in the field so they can choose the ones best for them; you can't plan a career if you don't know all the options that are available to you. While it's not possible to pin down one typical live sound position or experience, it is good to know the broader categories and some examples among the wider array of possibilities.

In the final chapters of the book we will revisit some of the same topics with a focus on information useful to the student actively looking for live sound work or adjusting to it. In the meantime, this overview of the industry covered in Part 1, along with the supplemental links provided online for this section, should give readers a more complete idea of the field. Whether someone is just beginning to consider live sound, beginning a course of training, or with enough training under their belt to begin looking for entry-level jobs in the field, knowledge about the field of live sound will help them make the choices needed to plan their next move wisely.

Part 2

The Art and Science of Live Sound

The chapters in this section will introduce you to some of the science and art useful to a live audio engineer. We hope readers studying to prepare for work in the live audio field will use it as a springboard for continued inquiry into these topics, whether that happens as part of their formal training, or through their own supplementary studies.

Some of the science here any live audio student will be sure to encounter in more depth, no matter where they attend training. Other topics may or may not be covered beyond a basic introduction, or even included in study topics at all, depending on the length of the training programs they choose. Live audio programs run the gamut, from fast-track programs of three to six months that focus only on the essentials, to programs of one or two years that study a broader range of topics, and even four-year degrees that cover a wide range of topics within the audio fields, including live sound. The topics learned in each type of training are different, to account for what can be covered in the time allotted. No approach is intrinsically better or worse; there are successful engineers who started out with each of these types of training

If any readers are attending or will attend fast-track audio training, they shouldn't worry if the bulk of the training is focused on practical techniques and the P.A. system. Technically, audio techs can do very well using a handful of "rules of thumb," without studying most of the principles underlying the science of live sound, and the same strategy works for the art. We think every live sound student, and even professionals, should have a cheat sheet of the essentials to work from regardless of other topics they may have studied; we just don't advise it as a sole strategy and support to hang a career on for the long term. It is easy for students encountering texts and a curriculum geared to one approach, those that teach only the essentials needed for students to begin looking for entry-level jobs, to confuse this practice as proof that essentials are all any live engineer needs to know.

Whether readers just need a list of essential "rules of thumb" to help them learn what they need to prepare for entry-level work, or are looking for what they need to establish a good foundation to branch out from as they continue their studies, we want to encourage learners to approach these topics as just as important as those in the following section about the P.A., and invite them in advance to revisit them even after they have finished with the main portion of this text. These chapters are intended to provide a simple and accessible resource for learners beginning to study live sound. The links included in the online directory for these chapters are arranged to be useful for much more, however, so students should keep them in mind as study resources they can revisit on their own, long after they have mastered the basics, to further develop their understanding of the science and art of live audio.

Chapter **4**

Sound Science

Physics and Perception

Sound in Motion: Anatomy of a Simple Wave

The easiest answer to the question "What is sound?" is that sound is energy that transmits through matter, called the medium, in the form of wave. Sound energy takes the form of **acoustic vibrations** and the medium that transmits most of the sounds we hear is air molecules (although sound can also move quite well through denser matter, like water). Because we can't see sound moving in air, let's compare the process to a process that we can see, and that is familiar to most people: waves moving across water. As we compare both processes for similarities and differences, we can use the process that is easy to visualize as a framework to help us "see" the physical attributes of the sounds we hear.

When you drop a pebble into a smooth pond, what happens? You get ripples that move out in all directions from the location where the pebble hit the water. These ripples are created by the kinetic energy released as the pebble strikes the water's surface. Some ripples will travel smoothly through the water in one direction until their energy becomes diffuse and fades away. Others will react with objects in their way (bounce back, recombine with other ripples), transferring some of their energy each time they react with those objects, until eventually they expend enough energy that they too die out. Over time, all of the **kinetic energy** radiating from the initial source and carried along with the ripples is dissipated and the water's surface returns to its original state, a smooth pond.

Sound interacts with our environment in approximately the same way. From where a sound is initiated by a source (handclap, gunshot, etc.), the ripples of energy (in this case **acoustic energy**) move out in all directions. Like the ripples on a pond, the **sound waves** created travel outward from the **sound source** in all directions. If unimpeded they will travel smoothly, if not they will interact with objects in the way first. (One example of sound interacting with objects in its path is when sound waves bouncing back from a large object produce an echo.) Either way the sound will eventually transmit all its energy until it cannot be heard anymore.

Figure 4.1. Ripples. *Source: Photograph by Roger McLassus under a Creative Commons BY-SA 3.0 license.*

Making Waves, Not Wind

Like water molecules that cycle up and down as a wave passes, but which otherwise remain in their original location, the fluctuating air molecules energized by sound traveling through the air don't travel with it, staying at their original location after the sound has traveled on. Since traveling air is essentially wind, and easily perceived through our skin when it occurs, this fact is easy to verify. To check it out for yourself, stand a couple feet from your TV or stereo at home and turn up the volume for a split second. Unless you left a window open to an airflow that compromises your experiment, you should be able to note that this blast of sound doesn't create even a tiny perceivable change in the air's horizontal movement. Clearly, in spite of a variety of humorous depictions in the popular media, even loud noises will not create so much as a puff of breeze, much less produce enough wind to blow a person's hair back.

Back at the pond, when the pebble releases the energy of its movement, the water itself does not travel. Instead, the water molecules become excited by the expanding energy wave and quickly cycle in place before passing the energy to neighboring water molecules as the wave propagates. The up and down **oscillations** of the water molecules create the wave shape we are all familiar with. At the **crest** of each ripple, all the energy of the wave has pushed the water together, making it rise up. When the energy that created the crest has moved on, the molecules suddenly have extra space between them, creating a **trough** or dip in the water. From this dip the water molecules return to rest at their original positions as the wave energy transfers to the water further out. Depending on how much energy was released by the pebble, for a short while additional waves of energy may follow the first, before all the energy is spent and the pond returns to its original, calm state.

Sound does something quite similar to air molecules as it travels through them as sound waves. From where the sound starts, the acoustic energy spreads out through the air causing the molecules to cycle in place as they pass the energy along to the molecules further out, repeating this cycle until all the energy waves created by the sound source have passed. As in the water wave, where the energy of the traveling wave pushes the water molecules into the crest of the wave, the air molecules excited by sound energy become compacted together into an area of **high pressure**. In audio we call this the **compression** part of the wave cycle.

Then, much like the water drops into a trough as the crest of the wave energy has passed on, as a sound wave's energy moves past, the compressed air molecules no longer have the acoustic energy pushing them together, and snap back out the same relative distance of the compression. In other words the air molecules refract back the same energy that is no longer present, resulting in a brief pulse of **low pressure** roughly equivalent in value to the high-pressure pulse of the compression that preceded it. This low-pressure fluctuation is called the **rarefaction** part of the wave cycle.

Finally, just as the water springs to its original position after dropping into a trough, the air returns from rarefaction to rest once again at its original state, the **ambient pressure** of the atmosphere (also called **atmospheric pressure**). In audio, the level at which the air molecules rested originally, before being compressed or refracted, is called the **zero crossing point**. At this point in the process, after the wave has passed beyond that position, the medium has no more of the compression or rarefaction energy that created the pulse of high- and low-pressure fluctuations, and will remain at rest unless another wave moves through with the energy to repeat the cycle.

The last thing we need to look at before we leave the pond behind are two key qualities that make sound waves different from water waves. The two **waveforms** are so similar that virtually all visual examples of sound waves show the iconic upward swells and downward dips, or the radiating nested circles, of waves on water. This is because water waves are easier to represent in two dimensions. This practice makes it easier to represent and label the components of sound waves in print, but in doing so it disguises a big difference between the two. While their components and the processes that move them work the same, the way they move is not exactly the same. The first key difference between sound and water waves is the direction of their **oscillations**. Water waves have oscillations that are perpendicular to the outward direction the wave energy is traveling and so are called **transverse waves**. Sound waves are **longitudinal waves** because their fluctuating air pressure oscillates in the same direction as the wave energy expanding outwards from the source.

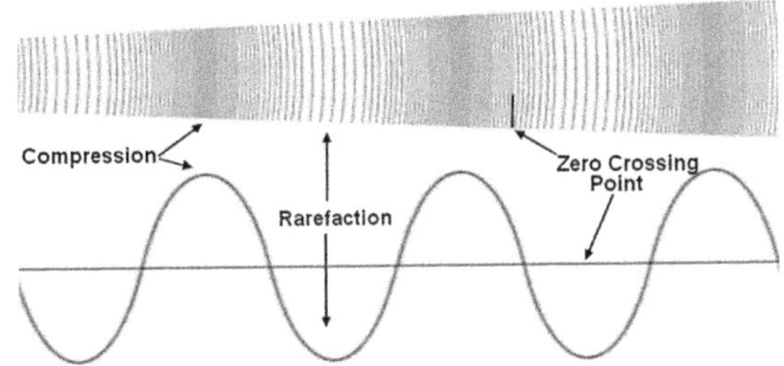

Figure 4.3. In the longitudinal wave at the top, changes in air pressure are represented by the proximity of the bars. Notice that it is much easier to see the zero crossing point in the diagram of a transverse wave below.

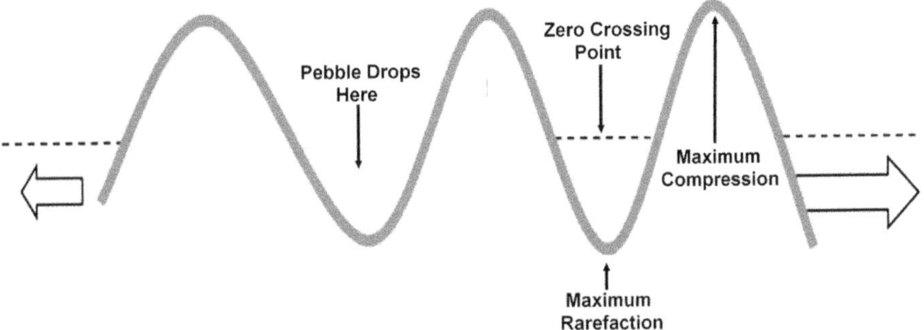

Figure 4.2. Zero crossing point. *Source: Graphic remixed from original released under a Creative Commons BY-SA 3.0 license by John Wetzel, an author at wikipremed.com.*

The next key difference between sound and water waves is the direction they move as they expand outward. Both types of waves expand outward from their source, but water waves are confined to the two-dimensional surface plane of the water by the force of the water's surface tension, radiating in a 360 degree circle, but only along a single plane. Sound energy is not confined to a single plane, so the energy traveling in all directions from the source creates an expanding three-dimensional sphere, similar to a balloon inflating (note that this sphere is distorted when the sound meets objects and barriers). Keeping these two differences in mind will allow you to visualize sound moving outward from its source the way it really behaves, not as undulating up and down waves traveling out from the source 360 degrees on a flat plain, but as pulsating bands of high and low pressure moving out from the central sound source like an expanding bubble.

Attributes of Sound

The science of **physics** describes the physical attributes of all waveforms, including sound. In the section above, we described a fundamental feature of sound, its manner of motion, using the recognizable outline of the traveling water wave as a familiar framework learners can build on as they learn more. It's a good place to start with because from within that familiar outline and moving in easy time with that motion, many of the other physical attributes can be added to the picture with only a few strokes. Yet it would be hard to begin discussing sound without at least a few aspects of the lesson feeling very familiar to students. After all, unlike most other scientific subjects, almost everybody comes to the study of sound with some innate understanding of sound, enabling them to recognize many of the details of sound they already know well, even when cloaked by less familiar academic terms. Similar to a medical student learning human anatomy who can now label all the bones that we use for locomotion, but already knew that tall people have longer legs and strides from experience living among people of all shapes, the audio student learning about sound already comes with personal experiences from a lifetime of hearing sound. Indeed, we have an intimate relationship with sounds, and even as young children we can recognize the qualities in the sounds we hear day to day, and tell one sound from another as easily as we can tell one parent's face from the other.

It is this intimate relationship between people and sounds that makes live sound so essential, and the reason why any scientific inquiry into sound alone is insufficient for those learning live sound. Without ears to hear it, sound is no more than a band of air pressure with no noticeable environmental effect and of no consequence. Because sound only becomes meaningful when there are ears to hear it, any measures of physical sound are valuable to us only if we also understand the relationship between the physical phenomena we can quantify and the sounds we ultimately can hear. It makes sense, then, that as part of our discussion of how the science of physics allows us to measure sound's physical characteristics, we need to also look at how these attributes interact with human physiology and psychology to become the sound we experience. We can do this with the aid of the branch of science called **psychoacoustics.** As the study of how our ears and brains experience sound, psychoacoustics is the science that is born at the crossroads where physics, physiology, and psychology intersect.

The process where sound becomes perception begins the moment the change in cyclical air pressure created by sound touches our ears. As sound is channeled into our ear by our **pinna** (the outside part of the ear), it already is being minutely altered by the unique shape of our ears. The sound pulse then affects our **eardrum**, which is so sensitive it reacts to these tiny variations in the air pressure and vibrates with the air pressure contacting it. Finally, the vibrations of the eardrum are transferred through the bones of the middle ear to finally be picked up by tiny hairs within the **cochlea** deep in the inner ear and sent as electrical impulses to the brain, where these impulses are experienced as the perception of sound.

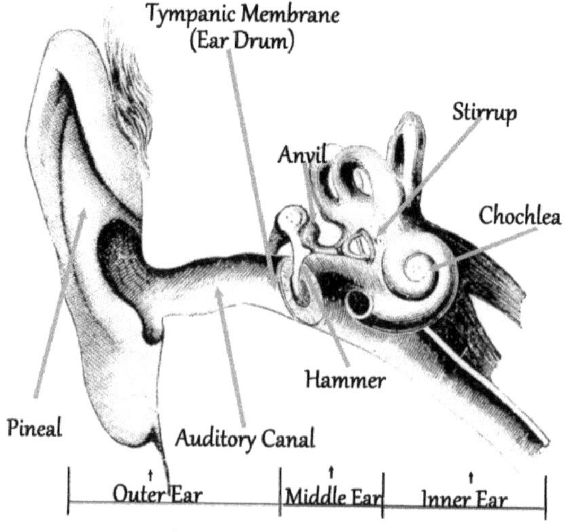

Figure 4.4. Ear diagram.

Our physical form is so attuned to the range of air-pressure waves we know as sound that science has shown our ears and brains are matched perfectly to the unique qualities of sound. The language we use to describe day-to-day sounds, such as referring to them as being stronger and softer or higher and lower, was established long before science discovered and catalogued the physical qualities of sound and learned how they relate to our perception of sound. The fact that these terms we have used for so long to describe how we experience sound actually correctly reflect the way that sound physically behaves according to science is evidence of our human ability to identify intuitively many of sound's parameters that we now can also precisely measure.

Amplitude

In sound waves, **amplitude** is a measure of the maximum distance between the top of a wave's crest and the zero crossing point, and corresponds to the **intensity** (strength) of a wave. The greater the amplitude of a sound wave, the greater the acoustic energy carried by that wave, and the greater the degree of "loudness" that will be perceivable to those listening. While they go hand in hand, amplitude and loudness are not the same thing.

Amplitude is the physical quality of a waveform that can be *objectively* measured by audio engineers, while loudness is subjectively perceived. **Objective measures** are measurements of observable phenomena using methods anyone can use. Objective measurements must be repeatable to qualify as objective. In other words, they can be verified as true because repeated measures of the same phenomenon will produce results that remain stable over time, and vary only within an accepted range of error, no matter how frequently the measurement is repeated, who does the measuring, or which of the agreed-on methods/tools are used to make the measurement. These strict requirements are intended to ensure the highest possible reliability that a specific measurement is a true reflection of physical reality. (This doesn't guarantee it always is; it only provides a framework to help us verify research if we choose to—the work of ensuring our research is reliable by checking for bias and verifying results is still up to us.)

When measuring wave amplitude for live sound, the two main modes of measurement you will encounter are: **peak value** and **root mean square (RMS) value**. Both are used to calculate the change in air pressure from the baseline air pressure, represented by the zero crossing point, to the air pressure at the furthest extremes of each wave cycle's oscillations, represented by the crest and trough of each wave cycle. The peak value of a wave is a measure from the point of no energy (the zero crossing point) to the maximum point of compression at the furthest position of the wave's crest. Rather than measure the peak compression of a sound wave, RMS averages the peaks that occur throughout the wave's cycle. Each method measures the same sound accurately using a different time framework, so neither provides a measurement that is more correct and each is used for different purposes. When choosing which measurement to refer to for general use, RMS is favored because humans tend to hear sound as an average instead of noticing every peak, so a measure of average amplitude is more in line with our perception.

Loudness

Loudness is the subjective perception of how loud a sound is and occurs in our brain. **Subjective measures** are measures that do not meet the criteria for objective measures, but this does not necessarily mean that they are unreliable, only that they incorporate data based on reports of the perception/experience of unobservable inner states, and/or reports that reflect opinion/interpretation of observable physical phenomena into their results, making it more difficult to determine the

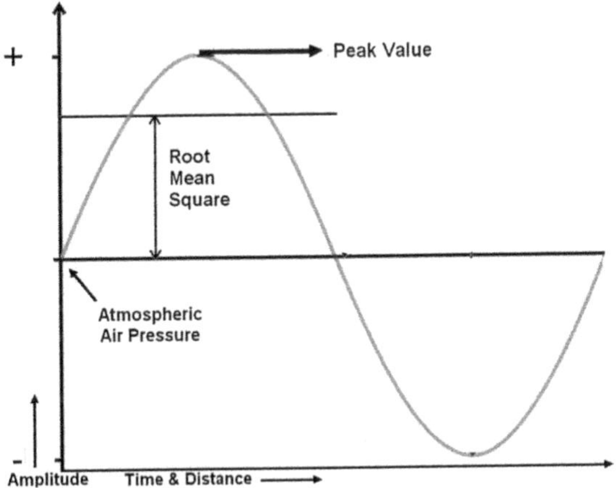

Figure 4.5. Positive and negative peak value are measures of the distance between the zero crossing point and the furthest point of the wave's positive and negative oscillations. RMS calculates the average over the entire wave cycle instead.

accuracy and reliability of the measured results/conclusions. Subjective measure is still an important tool for increasing knowledge, and much of the data we use to base decisions on is subjective, for example the results of opinion polls. To be useful, subjective measures require extra care in crafting the test instrument and procedures to make sure that unintended subjective information (such as experimenter bias) doesn't taint the results, a larger pool of test subjects establishes acceptable reliability of results, and needs to allow for its inherently wider margin of error to be clearly reflected in reported results. Any subjective measure will be an estimate rather than an exact measure, and at best can be correlated with objective measures, but never exactly. Nevertheless, a well-tested, reliable estimate serves nearly as well as an exact measurement in many situations, and that's a good thing, as there are many things that can't be objectively measured, like the loudness of sound, have objective properties which interact with subjective ones and so need to be assessed using both kinds of measurements

Amplitude measurements are essential to being able to discuss the perception of loudness with any precision, and their objective nature ensures an accurate measurement to base our discussions on, but, on their own, objective measures of sound wave amplitude are just numbers that don't fit a pattern we can easily compare to our experience of loudness in any meaningful way. In order to make amplitude values useful, we need to convert them from figures that seem random into ones with a pattern we can easily match with the patterns we hear. The system of measurement we use to convert RMS amplitude measurements into more meaningful values is the **decibel (dB) system**. The decibel is one of the most important tools a sound engineer has for keeping track of sound, because it provides a framework used to measure sound and power in several important ways.

For now, we will look at how we use it to turn amplitude measurements into easy to understand estimates of loudness that closely match the way we hear sound. The subjective nature of the way we experience sound means that any decibel figure will always be an estimate of about how loud a sound wave will feel to average listeners according to where it falls on a scale of the smallest sounds the average listener can hear to the loudest sound the average listener can endure. There will be some variation in how different people experience a specific decibel level because many people's hearing will be a little different than the average, but these estimates are relatively reliable because the data used to create this scale of average loudness perception is derived from decades of scientific studies on loudness perception in test subjects.

The Decibel Ratio

Though we only discuss one decibel measurement in this chapter, the decibel's ability to measure a number of related values useful to live audio professionals is one of its distinctive features. The decibel system so flexible because the dB is a *relative* system of comparison in the form of a ratio, but one where we can fix one of the initial values of the comparison, turning the result into a stable value that can be compared against a set of known quantities that anyone can refer to and that never changes (and so is no longer relative). **Ratios** are simply a method of comparing one quantity against another amount of the same type of thing to discover the difference between the two, and any ratio can be expressed as a fraction where the bottom number, or denominator, represents the original value we are comparing against.

A simple ratio everybody knows well is the one referred to any time we talk about one value being half as much as another, which is just a ratio of 1 to 2 (1:2) and means the same thing in its better-known fraction form ½, which is that there is one unit of what we are comparing for every two units contained in the value we are comparing against. We use it so often because it works with anything and any quantity, but because of this, ½ doesn't mean anything specific except when used in a specific context; how much "value" is contained in ½ changes relative to the situation and the original amount. If I am talking about my dog being half the weight of the 100 lb dog up the street, then the value is 50 pounds (50/100 = ½), but if I'm talking about my car only costing me half of what my brother paid for his $40,000 bling mobile, then the value is $20,000.

All decibels are a kind of ratio that expresses the difference between a quantity of power or pressure against an agreed upon fixed **reference value** of the same thing being measured, serving as the ratio denominator and baseline value for the comparison. The type of pressure or power being used is specified by appending a suffix to the dB notation. For example, to indicate we are using the dB system to measure sound as air pressure, or **sound-pressure level**, we add SPL to the notation dB, to indicate that the denominator is fixed at a specific value of air pressure that is always used when using dB SPL units. This baseline pressure gives us something to compare our amplitude values against, since both peak value and RMS measurements are values of change in air pressure, and enable us to then calculate a value that always will be the same amount; for example 10 dB SPL is always the same amount of pressure. As we will discuss shortly, these fixed values are the ones we match to a scale of known values that corresponds to the way we experience sound and so allows us to understand how we would perceive a 10dBspl sound.

Calculating Decibels: There's an App for That

While it's important to have an idea how decibels work, you typically won't need to know the actual equations needed to convert amplitude measurements into decibels. Whichever testing tool you use to measure amplitude, a sound-level meter, a microphone and laptop with test software, or even an iPhone with the right app will do the calculating for you and deliver a decibel value. Any calculations you may need beyond those that can be solved with simple arithmetic, can be calculated using apps on your smartphone or online.

We still use the decibel as relative measure whenever we use the dB notation with no suffix, as in ±10 dB. Used this way, the overall value is always the difference created by adding or subtracting the dB to the reference value of the decibels that existed prior to the change. Because it is usually used in context, when we refer to ±10 dB, we don't need to explain that this means 10 more or less decibels of whatever power or pressure we are dealing with than there were originally.

Decibel as Logarithmic Scale

Unlike the linear measurement scales we use most often to measure the physical qualities of the world around us, the decibel system is a **logarithmic** system with a base of 10. Logarithms are a tool for dealing with large numbers and are related to exponents, except where exponents quickly make numbers much larger with multiplication; logarithms do the opposite and immediately reduce large numbers by dividing. The relationship between the two can be seen below:

Number	Exponential Expression	Logarithm
1000	10^3	3
100	10^2	2
10	10^1	1
1	10^0	0

The dB is a logarithm with a base 10 ratio (1:10) of power or pressure, meaning it follows the same pattern and brings the number values of pressure or power we would otherwise have to depend on down to a manageable range:

Number	Logarithm	Decibel
10	Log 1 (1 with 1 zero)	10 dB
100	Log 2 (1 with 2 zeros)	20 dB
1000	Log 3 (1 with 3 zeros)	30 dB
10,000	Log 4 (1 with 4 zeros)	40 dB
100,000	Log 5 (1 with 5 zeros)	50 dB
1,000,000	Log 6 (1 with 6 zeros)	60 dB
1,000,000,000,000	Log 12 (1 with 12 zeros)	120 dB

Note that the scale above represents the generic 1 to 10 framework of the decibel ratio for power, and the numbers are provided to show the pattern of exponential increase, as all dB scales increase in terms of a progression of exponential doublings in value. This progression is different than the linear scales most people are more familiar with, where each increase along the whole scale occurs in increments of identical value, rather than increasing exponentially.

The most important thing to understand is that beyond the bottom of the scale, there isn't an identical relationship in terms of value difference from one dB change to another. Each additional dB increment is larger than the one before in terms of the "value" expressed, than the previous ones along the scale, and exponentially more so the further from the baseline of the scale that these increments are measured. So in the general dB chart above, going from 0 dB to 10 dB is only ten times more power, but the same 10 unit increase going from 20 dB to 30 dB is many hundreds of times more power, the increase from 30 dB to 40 dB is many thousands of times more power, and so on.

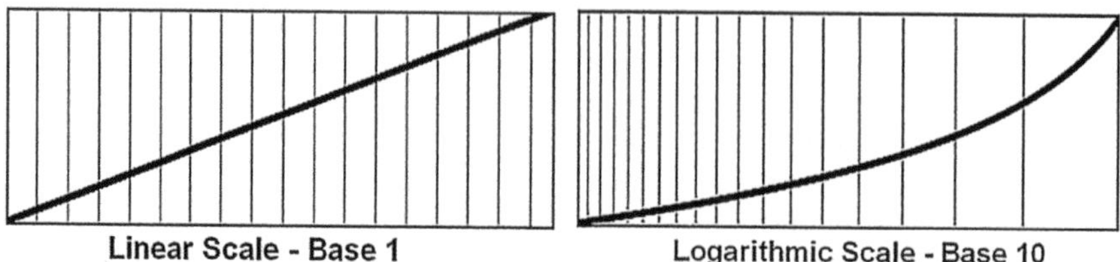

Figure 4.6. A basic visual representation of the general difference in the progression along a typical linear scale vs. a logarithmic-type scale.

Measuring Sound Pressure Levels in dB SPL

For our ears to perceive a change it must be noticeable from what was present before the change. When we calculate dB SPL we set the reference value as 0.00002 **pascal** (Pa), a unit of pressure which references *the smallest air pressure change our ears can detect*. This tiny pressure change becomes the baseline and is set at 0 dB SPL. Zero in this case does not mean the same it usually does; instead of referring to an absence of any value, here 0 dB SPL represents the smallest perceivable sound we can detect, or the **threshold of hearing** and therefore should not be mistaken as meaning "**inaudible**" (i.e., no perceivable sound).

If you recall that we use logarithms as a way to manage large numbers, we can see one of the practical reasons why we save ourselves a lot of headache by converting to the dB system. Our ears are so sensitive they can detect the tiniest changes in pressure, and have a such a large **dynamic range,** or range of sounds difference we can perceive, that is so expansive that the loudest sounds we can comfortably experience are well over a million times greater, in tiny air-pressure changes, than the softest sounds we are able to perceive, or the smallest **audible sounds**. Because this order of magnitude is difficult to even conceive for most of us, the dB system of exponential doublings in value allows us to convert a scale of reference of more than 1 to 1,000,000 into a much easier to use 0 dB to 120 dB. The "value" encompasses the same overall change in air pressure; the only difference is that there are far fewer increments we need to count along the scale making it so much easier to calculate changes along the scale.

Bringing Large Numbers Down to Size with dBs

Without the dB system, a split-second reckoning such as figuring how much to need to lower your output levels in order to move from a very loud 110 dB SPL to a more moderate 95 dB SPL would require you to subtract 5.077728726529775 Pa from 6.324555320336759 Pa to arrive at a figure of 1.1246826503806984 Pa (the equivalent to 15 dB SPL in Pa units). That's time-consuming even using a calculator! Practical use of sound pressure to estimate loudness without the dB would be hard enough; even worse would be having to ask anyone else to turn the level down 1.1246826503806984 shades of difference!

dB SPL and Perceived Volume

The final reason that dB SPL is so convenient is that our hearing happens to work using an almost identical logarithmic scale; the way our brains process loudness matches the decibel scale. This means measuring relative loudness perception with dB SPL is actually quite intuitive and easy to use accurately. When we hear ten successive doublings of sound pressure, it sounds like ten equal increases to our ears. On a logarithmic scale, the sound after ten doublings is around 512 times stronger in terms of pressure than it was to begin with, but because our ear perceives loudness logarithmically it doesn't sound 500 times louder; it sounds ten times louder. So the decibel system is a perfect match with our perceptions, because it reduces the large numbers we would have to deal with otherwise to ones we can perceive and manage more easily, both in our minds and on paper

We have limited our scale to 120 dB SPL, even though some people can comfortably hear a little more without discomfort, and some sources will cut off the scale at 130 dB SPL. We chose our cutoff based on the fact that on average 120 dB SPL is the level where most people begin to feel discomfort, referred to as the **threshold of pain**. Not too far beyond 130 dB SPL most people soon reach a pressure level that causes intense pain. Pressure levels around that part of the scale and just beyond are enough to perforate our eardrum.

The chart below illustrates the corresponding sound-pressure levels in dB SPL of a selection of commonplace sounds covering the range of human hearing and loudness perception. This is the value scale we can compare against if we need to put decibel value into context of how it will generally sound. Notice that each entry indicates the distance from the sound source, because measures of sound in dB SPL only are accurate for the location where the amplitude measures are taken. We can calculate what the measure will be elsewhere but only if we know where the initial measure was taken. If the original measurement location is known, we can calculate for any position using the **inverse square law**, which states that *for every doubling of distance from the sound source, the sound pressure decreases by −6 dB.*

0 dB SPL	The threshold of hearing
10–20 dB SPL	A gentle breeze through the trees
20–30 dB SPL	A soft whisper (at 1 meter/3 feet)
30–40 dB SPL	A quiet auditorium (at 1 meter/3 feet)
40–50 dB SPL	Background music in a cafe, bar, or restaurant
50–60 dB SPL	Average noise of typical residence
60–70 dB SPL	Typical conversation levels (from the listener's position)
70–80 dB SPL	Alarm clock (at 1 meter/3 feet)
80–90 dB SPL	Vacuum cleaner (at 1 meter/3 feet)
90–100 dB SPL	Busy traffic (at 3 meters/9 feet)
100–105 dB SPL	Gas lawn mower (at 1 meter/3 feet)
105–110 dB SPL	Typical dance club
110–115 dB SPL	A loud rock band (front rows of audience)
115–120 dB SPL	Ambulance siren (at 10 meters/30 feet)
120–130 dB SPL	Threshold of pain varies with frequency and by person
130–140 dB SPL	Jet engine (at 3 meters/9 feet)

Some more simple rules to remember when discussing loudness perception are: *each increase or decrease of 6 dB is twice or half the sound pressure level*, and despite some disagreement as to how much of an incremental change in dB SPL best represents a true doubling in the perceived loudness of a sound, the commonly used rule is that it takes *+10 dB to achieve an actual doubling in the perception of loudness. A change of 1 dB is labeled as the smallest perceivable change* we can easily notice, though this actually takes practice and for most people and *+3 dB change is required before the change will be noticeable.*

While pressure and power are physical phenomena we can measure objectively, loudness varies in some really complex ways, independent of its amplitude. Attributes like duration and frequency are also factors. With this in mind, any calculations of loudness based on amplitude alone are never going to be completely accurate. Loudness is a subjective human perception that takes place within the human brain and there are no direct "units" of loudness that can directly measure it. When we use dB SPL to describe the loudness of a sound we use what we can measure (air-pressure changes), convert it to a manageable scale with predictable values (dB SPL) that we can begin to match with common points of reference (familiar sound sources) and experimentally observed physical/perceptual responses (threshold of pain).

Despite the fact that our best way to measure loudness only allows us to make an estimation rather than an exact calculation, it is not uncommon to hear dB SPL referred to as if it were an objective unit of loudness, as in "it needs to be 10 dB louder," or "it was 60 dB SPL loud." Though this may not be perfectly precise, even pros do it for convenience. As long as you understand that this is a figure of speech and clear up any misconceptions if you see others don't, I wouldn't worry too much over referring to volume this way. The best bet is use the lingo that is understood and appropriate in the crew you work with.

Frequency

If it vibrates, oscillates, alternates, or assumes a similar pattern of **periodic motion**, a feature characterized by a series of regularly repeating cyclical movements, its rate of motion can be measured by plotting the **frequency** of its cyclical motions against a fixed unit of time. As the sound waves we have been discussing cycle back and forth between fluctuations of high and low air pressure, they exhibit this pattern of periodic motion and can be measured for their rate of cyclical movement as well. When referring to the frequency of sound, a full **cycle** begins as air is first excited into motion by sound energy, continues as the excited air increases in pressure until it reaches peak compression then decreases in air pressure until the maximum loss in air pressure is reached, ending finally after air pressure increases until the original state of ambient pressure area fluctuates between opposing peaks of high and low pressure, and concludes when the air pressure returns to its original state of ambient pressure. Sound frequency is measured in **hertz (Hz)**, which represents the number of cycles a particular wave will move through in one second. For example, if a tuning fork vibrates at 100 cycles per second when it is struck, then it is said to be vibrating at a frequency of 100 Hz.

In Figure 4.7, the wavy lines represent waves of varying frequency. The horizontal axis represents time, making the waves at the top of the chart of a lower frequency than the waves at the bottom, because the number of repeating cycles that can be counted is clearly lower for the waves at the top of the illustration than for the waves at the bottom.

There is a vast range of frequencies at which energy can vibrate, though the bandwidth of frequencies that fall within the frequency range audible to people is limited to the span of 20 Hz to 20,000 Hz (20 Hz–20 kHz (kilohertz), a range also referred to as the **audio spectrum**. Frequency is related to what the ear perceives as **pitch**, and whether sounds are perceived as having a higher or lower pitch, mostly depends on their frequency though, depending on conditions, other factors may contribute to a lesser degree.

Very generally, frequencies can be are grouped as:

- *low (bass)*: sounds of thunder and gunshots (20 Hz–150 Hz);
- *mid-range*: a guitar or human voice (151 Hz–6 kHz);
- *high (treble)*: small bells and cymbals (6 kHz–20 kHz).

Frequency is perceived as pitch, with waves of higher frequencies perceived as higher-pitched sounds and lower frequency waves perceived as lower-pitched sounds. Pitch is a subjective quality like loudness, and like loudness we quantify it using an associated objective measure, which is frequency. The scale used to measure pitch is the octave, and like the decibel scale, progression along it occurs in a series of doublings. In the *octave scale, each doubling of frequency corresponds to an octave increase in pitch*, no matter what the initial frequency. As with our perception of loudness, our perception of pitch functions logarithmically, making these doublings seem more like a steady and gradual increase to our perception. We will revisit frequency in this chapter and in those to come because it serves as one of the variables of a sound's physical form that has the widest variety of potential ways to impact the way the sound behaves in the environments and how it is perceived by listeners.

Periodicity and Period

A distinctly different measure of periodic motion bears mentioning as it is closely related to frequency but measures periodic motion according to different parameters. Any action that happens repeatedly is referred to as being **periodic,** an example being the repeated cyclic motion in periodic motion, and just as any periodic action can be measured in terms of frequency, it can also be measured by its **period**, which is the time it takes to complete one cycle. *While frequency measures rate* (the number of cycles per second), *period measures time* (the number of seconds per cycle). The value expressed as period is the inverse or reciprocal value of the one described by frequency. The relationship between the two is expressed as

$$\text{Frequency(f)} = 1/\text{Period (T)}$$

and can equally be expressed as

$$\text{Period (T)} = 1/\text{Frequency (f)}$$

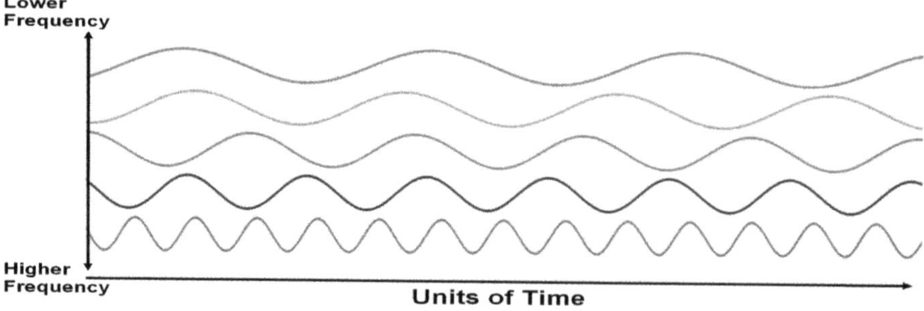

Figure 4.7. High and low frequencies.

With this mathematical relationship it is easy to use the known quantity of either measure to solve for the other. Doing so using the example of a sound that cycles at 100 Hz shows that

$$f = 100 = 1/T$$
$$f = 100 = 1/.01$$

so the period of a wave with a 100 Hz frequency is 1/100 of a second, or 1/.01, which is the same as saying 10 ms (milliseconds).

Though periodicity and period are important concepts, they aren't ones live audio engineers need for their day-to-day tasks. So it is less important to remember the equation than it is to understand the relationship. This equation is not included because you will need to solve for one or the other but to make it clear that *the two values are not the same*. Frequency refers to how often something happens; period refers to how long it takes something to happen.

While the **periodicity** of sounds, the tendency they have toward repeated patterns, is only discussed in reference to certain limited examples and quite often not discussed at all, even in reference to its close relationship to frequency, a basic understanding of the subject will help avoid confusion. It is a topic that has been getting a great deal of attention recently from researchers in related audio fields—a quick Google search will reveal the large number of abstracts generated by recent research and theories about its influence on sound have been given increasingly more weight in related fields such as psychoacoustics, making it a concept anyone in the audio field will encounter more often as they do research or read up on related topics.

Velocity and Wavelength

Velocity and wavelength are attributes related to frequency, so that any two of these values can be used to calculate the third. Since velocity is simple enough to determine and travels at a constant amount, this means as long as either the frequency or wavelength of a sound is known, we can use this known value along with the constant value of velocity to determine the value that is unknown. The **velocity** of a sound is the speed at which sound waves travel. Sound will always travel at the same rate when traveling through the same medium and condition; in other words, the velocity of sound is constant and as long as we know the conditions of the medium sound must travel through, we can calculate the speed sound will travel. At sea level and an air temperature of 68°F, sound travels through air at approximately 343 meters or 1130 feet per second. This works out to 1.3 feet a millisecond.

Velocity, So What?

Velocity of sound in air is a constant, just like the speed of light. It is interesting to note that electrical signals approach the speed of light. So a signal through audio cables is many times faster than the speed of sound. This means that if your system has "back-of-crowd" speakers, it's possible that the back speakers will receive and produce sound before the sound from the main P.A. speakers reaches the same point in space! Obvious if the delay needed for these speakers isn't correct, the sound quality will suffer.

The velocity of sound will vary with altitude, and the temperature and humidity of the air, or when traveling through denser mediums (like water). While the velocity of sound varies with the medium and environmental factors, it can generally be calculated without difficulty, because it varies at a constant (and easily measurable) rate. For example, the velocity of sound increases at a rate of 1.1 feet per second for each 1°F of temperature increase. Velocity also varies from about 1127 ft/s for very dry air to about 1131 ft/s at 100 percent humidity, but because it is such a small variable, this usually doesn't need to be accounted for mathematically.

A sound's **wavelength** is defined as the length (distance) a wave travels as it completes a single cycle from beginning to end. Wavelength can be calculated for any wave where the frequency is known using the formula that defines wavelength: wavelength = velocity/frequency = 1,130 (ft/s) / frequency (cyc/sec).

If you refer again to the figure of multiple waves of varying frequency, you can see that as *a wave's frequency decreases, its wavelength gets longer*. The inverse is also true: as the frequency of a wave increases, its wavelength gets shorter. So the wavelength of a 1000 Hz tone is 1.13 feet, while the wavelength of a lower frequency 100 Hz tone is 11.3 feet.

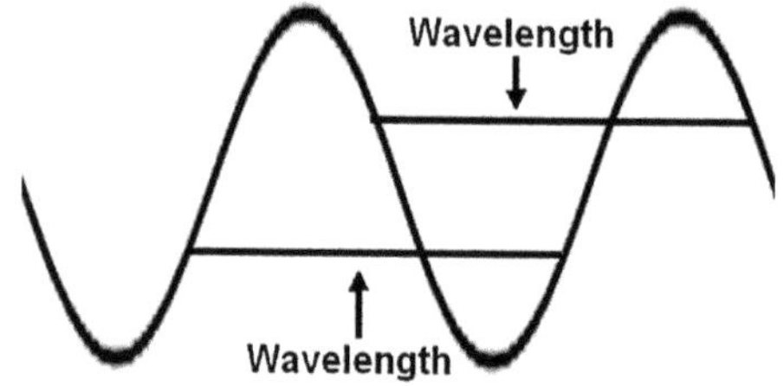

Figure 4.8. Wavelength is derived by dividing the velocity of a wave (which is a constant value) by its frequency. *Source: Image by Mosely under a Creative Commons BY-SA 3.0 license.*

Complex Waves

So far all the graphical diagrams display the smooth even shape of the **sine wave**. This is despite the fact that very few of the sounds we hear are simple sound waves consisting of a single pure tone. Just like the manner in which the transverse sine wave's simplicity makes it preferred for representing and labeling the attributes of sound, they are often used to represent complex waves. The sine wave, however, is simple **waveform** vibrating at one unique frequency. Sine waves do not occur naturally, but they are important for the audio engineer because they're often used as **test tones**. Test tones are a type of artificially created sound that is much simpler and when played through a P.A. can allow the engineer to hear any problems in the system or in the quality of the output much more easily.

Harmonics

All other waveforms comprise multiple sine waves of varying frequency, and these **complex waveforms** can be broken down to their component sine waves using a technique called **Fourier analysis**. This analysis can be represented graphically as multiple superimposed waveforms, or plotted against relative amplitude and frequency, also called a **frequency domain representation**.

The lowest of these frequencies is the **fundamental frequency** and the one which determines the frequency at which the overall complex waveform vibrates and the perceived pitch of the sound. The fundamental frequency is also the lowest resonant frequency. A natural function of an object's physical form, a **resonant frequency** is one of the frequencies an object easily vibrates at (many objects have more than one resonant frequency). The fundamental frequency is also called the **first harmonic**. The higher frequencies, or **harmonics**, are whole-number (integer) multiples of the fundamental, and the number and relative intensity of the upper harmonics is called the **harmonic content**. Resonant frequencies that are not integer (whole-number) multiples of the fundamental frequency are simply referred to as **overtones**.

If these harmonics are simply related to the first harmonic and form a rational number, then that waveform is called a **periodic waveform**. Periodic waveforms are patterned, made up of frequencies that are spaced at regular intervals across their bandwidth, and possess a form and motion that repeats in regular cycles. Periodic waves also are perceived as having pitch. In accordance with our pattern-seeking nature, people find sounds with periodic waveforms more pleasing because they are patterned, and generally the more orderly the pattern of the sound, the more pleasing it will be. This is why we are so enchanted by musical instruments and music: of all the sounds we can produce or hear, music is the most regularly patterned. Many instruments produce harmonics that are exact multiples of the fundamental, usually at octave intervals.

If those harmonics form irrational numbers, then the waveform is non-periodic, also referred to as **aperiodic**. Aperiodic waves do not have a regularly repeating pattern. They either have a non-repeating pattern or no

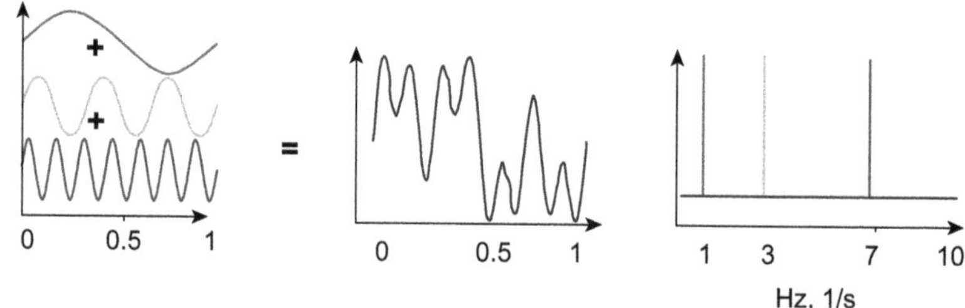

Figure 4.9. A complex waveform analysis revealing the component sine waves as well as a plot of their frequency domain relationship.

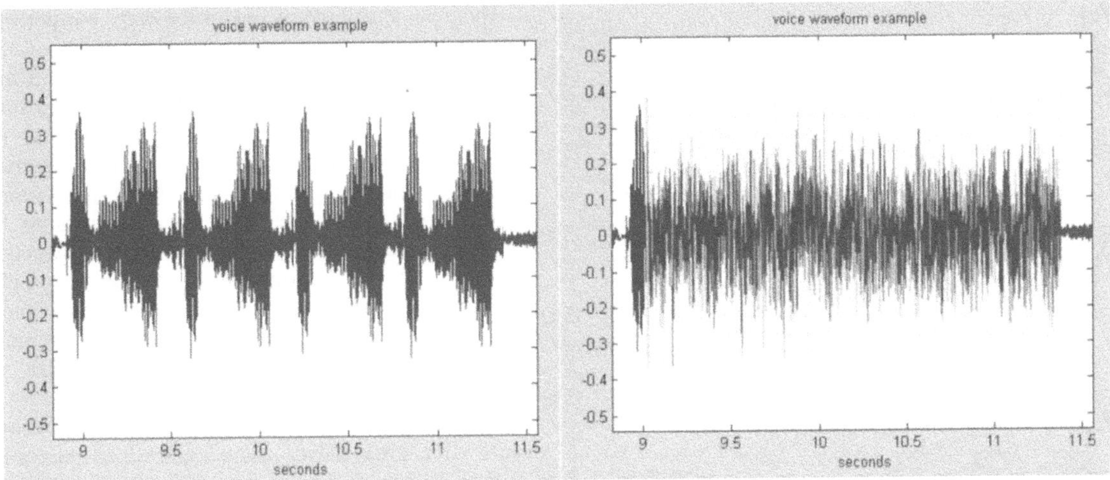

Figure 4.10. Example of a harmonic periodic wave (the vocal sound ahh) and an aperiodic wave (the vocal sound shhh).

pattern at all, in which case their overtones are randomly distributed throughout their bandwidth. Sound composed of random pressure fluctuations is referred to as **noise**. Noise is disorganized information, or a lack of information, and either has content that is distorted or degraded and cannot be understood, or is random and has no useful meaning. While most aperiodic waveforms are considered grating or **dissonant**, this is not always the case; for example, some instruments, especially percussion instruments, do not have pitch, and the sounds they produce are therefore non-periodic.

Other features that distinguish periodic waves from non-periodic waves is their **spectrum** and phase. Periodic waves have overtones, also called **partials**, that are discrete, evenly spaced in proportion with the period of the waveform, and locked in-phase. (This is what keeps them together as one sound instead of a stack of related frequencies that are separate like a musical chord). Because periodic waves have harmonic partials and are **phase locked**, they don't break down and can repeat over and over. Aperiodic waves, in contrast, will have a continuous spectrum, and the relationships between the unrelated frequencies in the partials along this crowded spectrum, though they start in-phase, soon begin to interfere with each other. This interference is what keeps aperiodic waves from repeating.

The Cochlea and Pitch Perception

Much like loudness and amplitude, pitch doesn't always match frequency in an exact fashion. Our ability to break down complex waves has revealed that we can clearly perceive a specific pitch even when the complex wave form is missing the required first harmonic, and therefore vibrates at a completely different frequency from the perceived pitch. This is because of a feature in our cochlea called the **basilar membrane**, which perceives frequency not as randomly individual notes but according to where those notes stimulate the ear, which is mapped to perceive pitch based on widths of 1/3 octave bands. In short we are born with a map of pitch organized as octaves in our ears, so we can identify the mathematical relationship between the overtones and the **missing fundamental** as being the closest possible match, and our brain fills in the correct note. This also means we come wired to listen for and enjoy music, and to perceive certain types of sounds as intrinsically more or less musical based on the mathematical relationship of their harmonics.

Interestingly, these critical bands in our ear create a perceptual quirk that has exactly the opposite effect, where instead of filling in missing sounds so we can hear a pitch that doesn't exist, sounds close enough in frequency to be perceived by the same critical band, with an amplitude difference of 6 dB or greater, will result in the canceling of the softer sound. Whether caused by the function of the critical bands in our ear, or other factors, any time a sound that can normally be heard becomes inaudible in the presence of another sound the result is a form of **acoustic masking**. The (usually louder) sound that erases our perception of the other sound is referred to as the **masking sound**, and the canceled sound is called the **masked sound**.

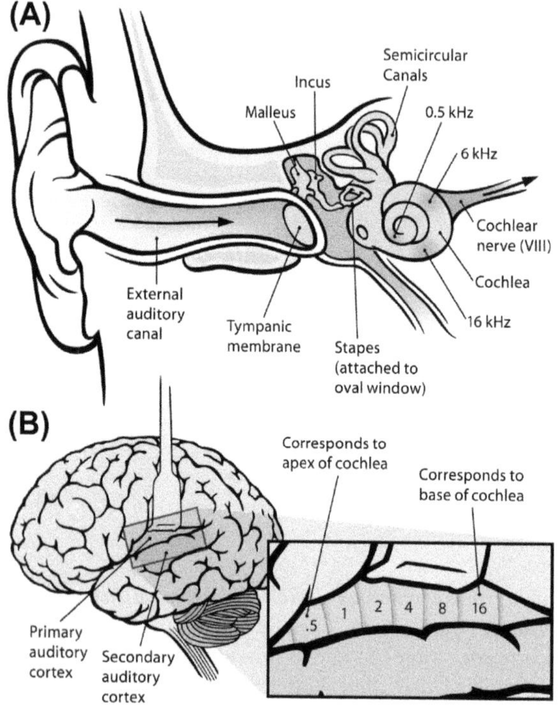

Figure 4.11. Frequency mapping in the cochlea and auditory cortex: our ears and brains are wired to perceive pitch. A-Auditory structure of cochlea, B-Associated brain structure. *Source: image by Chittka L and Brockmann A under a Creative Commons BY-SA 3.0 license.*

This feature influences one of the ways our perception of loudness varies relative to the pitch of the sound. When this occurs, these critical bands in our inner ear react to sounds with **frequency distributions** large enough to stimulate more than one band by adding the amplitude perceived by both bands together, even if the source triggering both bands is a single sound with a single amplitude. This allows sounds to be perceived as significantly louder in certain frequency ranges than can be accounted for by measures of amplitude alone. As a result of both critical band and the interaction of other influences the human ear is much more sensitive, or has a greater **frequency response**, to sounds in some frequency ranges than others.

This is why noises with greater **spectral bandwidth**, like **pink noise**, sound louder than pure sine tones. Like the sine wave pink noise is also used as a test tone, but pink noise is a complex sound made up of many simple sine waves from a broad spread of frequencies making it more like the sounds we hear in the world, while sine waves are tones of a single frequency and aren't representative of sounds we normally hear.

Equal-loudness Curves

Research has shown that testing sound perception with sine waves is not as accurate as tests using complex

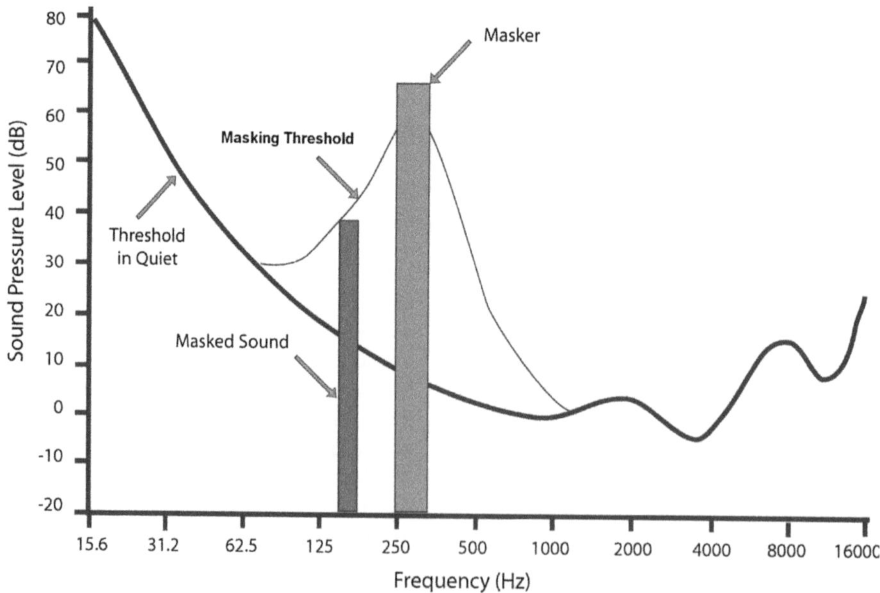

Figure 4.12. Acoustic masking is of consequence to the live pro and can influence how a mix comes together in unexpected ways.

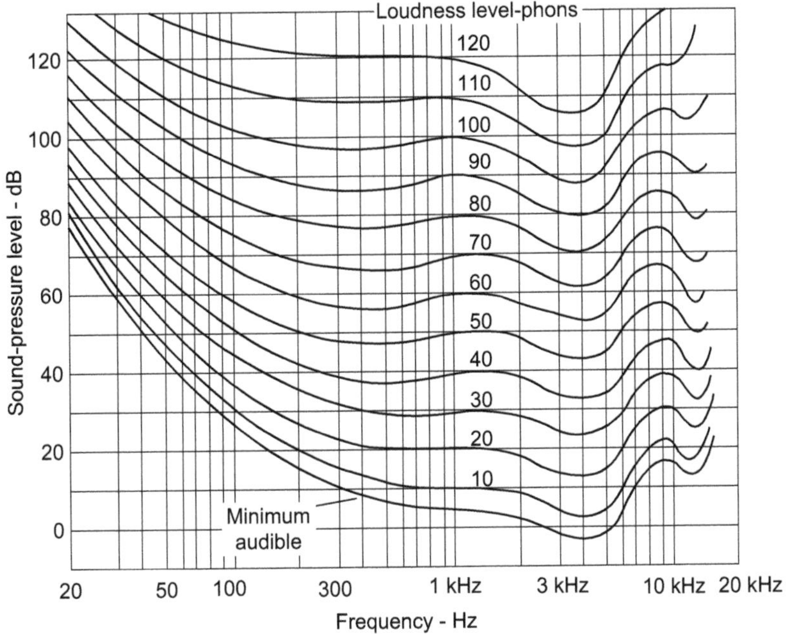

Figure 4.13. Equal loudness curves show that loudness perception is influenced more or less depending on the frequency of the sound.

waves, and therefore recommends that pink noise, made up of the same sort of multiple frequencies over a broader bandwidth as the complex waves we hear daily, should be used instead of tones in measurements of loudness and inquiries into perception. Tests with pink noise reveal an even greater response at some frequencies than we already had discovered in studies using sine waves. Therefore those **equal-loudness curves**, the data maps created in equal-loudness tests, that were derived using sine waves are less accurate than newer ones based on pink noise,

In the 1930s **Fletcher and Munson** originally charted the ear's response to frequencies over the entire audio range and released their results as a set of curves called **Fletcher–Munson curves**, showing the sound pressure levels of pure tones perceived as being equally loud and proving that the ear is not equally attuned to frequencies in the low and high ranges. Equal loudness-level contours are lowest in the range from 1 to 5 kHz, and dip slightly at 4 kHz. This means that the ear is most sensitive to frequencies in this center range. Higher or lower tones must be much stronger in order to create the same impression of loudness as those ranging from 1 to 5 kHz. The phon scale was created to reflect this subjective impression of loudness, since the decibel scale alone refers only to objective sound pressure levels and not how they are perceived.

To account for the ear's frequency response, decibel-testing meters are usually fitted with **filter contours** designed to provide frequency response more like that of the human ear, where the increase in response is most notable at about 1 kHz to 4 kHz (1000 to 4000 vibrations per second) and sensitivity is reduced along the extremes of the low- or high-frequency range. The **weighting scales** (designated A, B, C, D) attempt to do this by reducing response to very low frequencies—quite severely by the **A-scale** which roughly corresponds to the inverse of the 40 dB (at 1 kHz) equal-loudness curve, less by the **B-scale**, and not at all in the **C-scale**. The most easy to use and therefore the most widely used sound-level filter is the A-scale, which gives sound pressures levels in dB(A) or dBA units. The dBA scale gives the most accurate values at low levels. The C-scale is too linear to be suitable for measurements except at very high sound levels, and units on this scale are expressed as dBC.

These scales adjust values based on amplitude measurements to account for the influence of frequency response on perception and weighted dB SPL meters return results that are more accurate representations of

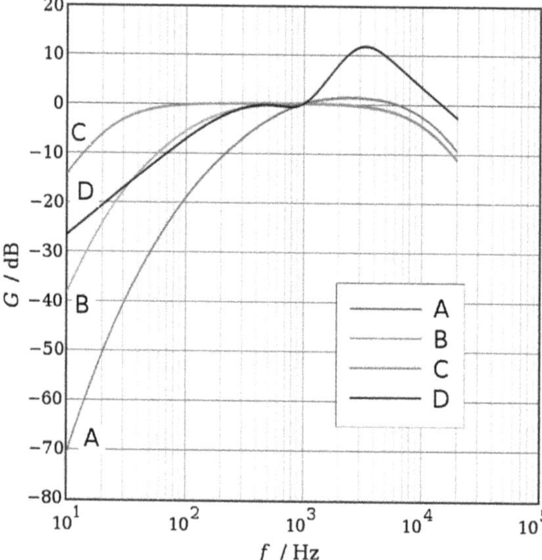

Figure 4.14. Meters for assessing volume levels can have filters to return the most accurate values by using weighted scales.

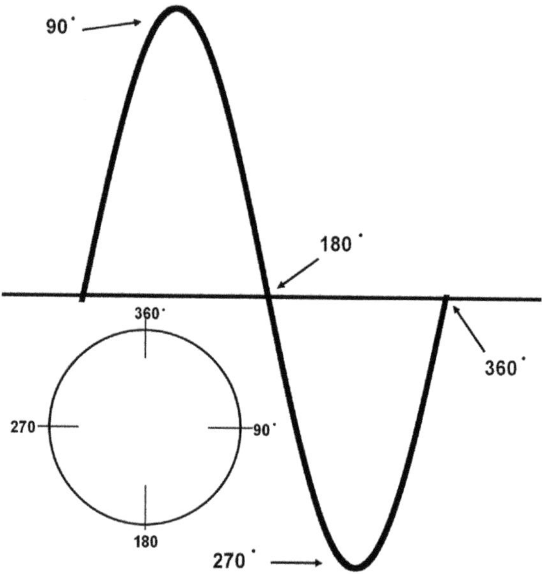

Figure 4.15. Phase.

human loudness perception than tests of amplitude alone.

Phase

Phase is a measure of the relationship between where a sound wave is positioned in space and where it's progress lies along its cycle from beginning to end. Phase relates to the cyclic nature of wave oscillations, because each point along a single wavelength corresponds to a position along one complete rotation of a circle, and can be measured in 360 degrees. This circle can also be visualized by any who have looked down the "tube" of a water wave which best displays that waveforms are cylinders.

When two waves of the same frequency meet and overlap at the same point in space, their identical position in space can be used as a reference point to compare where each wave's progression in their cycle rests in relation to the other. When their position in their cycles is identical to each other at point they overlap, they are **in-phase**. When their positions are offset, they are **out-of-phase**. The amount of **phase difference** between waves that are out-of-phase is described in terms of degrees of offset; waves that are completely offset by 180 degrees are **antiphase**.

For some, it may help to think of each cycle as a clock, instead of degrees of a circle.

When two waveforms of the same frequency overlap completely, they are at the exact same space at the same time, but like dancers in a revue if they are truly in time their position will be the same when they meet. The hands of both will be pointing to the same time. If one waveform is just beginning a new cycle and positioned at 12 o'clock, but the other is at 3 o'clock, then the latter is "ahead" of the other. (Like dancers are in step or out of step, and clocks are in time or out of time, phase can be discussed in the same terms.) Convert the hours back to degrees, and the waves are 90 degrees out-of-phase.

Phase is of consequence because when sounds of the same frequency from different sources (like two speakers) come into contact at the same point in the environment, their phases relative to each other determines how they will interact with each other. When

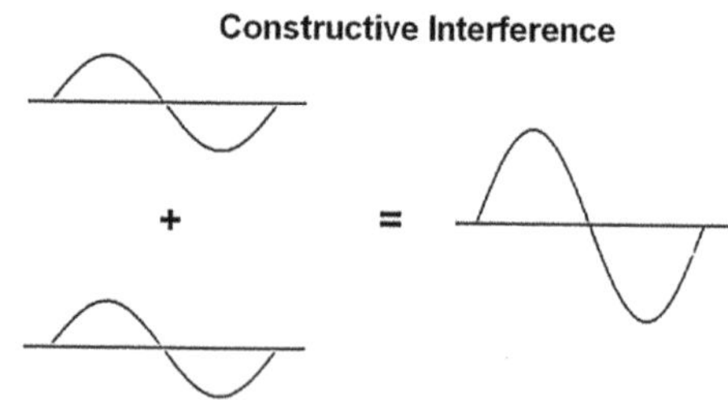

Figure 4.16. Waves that are in-phase will interfere constructively.

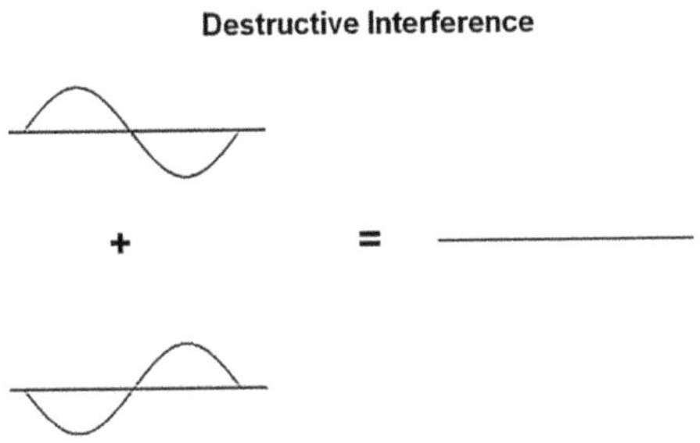

Destructive Interference

+ =

Figure 4.17. Waves that are 180° out-of-phase will interfere destructively.

waves of the same amplitude and frequency become superimposed and are in-phase, their amplitudes completely reinforce each other and mathematically double in amplitude. When identical antiphase waves superimpose, their amplitudes completely cancel each other out and silence each other. Sound waves of differing amplitudes that are completely in or out-of-phase, or sound waves with varying phase relationships of different degree, produce differing sound effects, either partially reinforcing or partially canceling each other according to the variation in their amplitudes and phase.

When waves partially or completely reinforce each other, this process is called **constructive interference**. The process whereby waves partially or completely cancel each other is called **destructive interference**. Waves of different frequency do not interfere with each other and most waves of the same frequency only partially cancel or partially reinforce each other, but when more than one sound source is used during reinforcement, a certain amount of **phase interference** is inevitable.

Phase issues can arise from speaker position and from variations in the shape of the venue. When interference effects that contribute to variable sound quality at different locations in a venue result, it is a concern for the live sound pro. While it is impossible to completely eliminate the effects of phase, in most cases these effects can be minimized, as we will discuss in subsequent chapters. When phase issues stem from factors out of control of the live engineer, like venue acoustics or poor installations, the extent they can be mitigated is more limited.

Envelope

A sound's **duration** is the time frame it covers, starting when the sound initially becomes perceivable and ending when it loses strength and can no longer be heard. Like its frequency, a sound's duration also has an effect on perceived loudness. This is most evident with sounds lasting around half a second in duration up to those that are 1 second in length. Within this range of duration, the longer a sound is, the louder it is perceived, even when there is no difference in amplitude between the sound perceived as louder and the shorter duration sound perceived as less loud.

While the duration of a sound is a measurable aspect of a sound, except for having a small influence on loudness perception, duration alone doesn't provide us with a very useful quantity. Of the multitude of sounds that can be heard, most fall in the same small range in terms of duration. Knowing that a sound lasts for 2.4 seconds, does not tell us much—any number of sounds could occur in 2.4 seconds. If we want to measure a sound's duration in a way that provides us with a unique picture of that sound, the variation in amplitude that occurs over its duration is just as important as the factor of how long is its duration.

When we plot the amplitude of the sound against time, we end up with the unique fingerprint of that sound's **envelope**, which provides a visual map of the sound's amplitude from the beginning to the end of the sound's lifespan, allowing us to see the variation in amplitude for each sound and outlining the difference between different sounds of the same duration. Envelope is determined by four distinct phases that make up a sound, its attack, decay, sustain, and release, and is widely referred to as an **ADSR curve**.

The four phases that create the map of a sound's envelope are:

- *attack*: the time from the initial start of the sound to its peak;
- *decay*: the time from the peak over the slight drop in amplitude to the sustain;
- *sustain*: the time from the beginning of the plateau to the end;
- *release*: the time from the end of the plateau till the sound dies away.

Timbre

Timbre is the quality or **tonal nature** of a sound that makes it unique: it's how we can tell the difference between the same note of the same pitch and loudness played on two different kinds of instruments. The timbre of a sound source is determined by the attack and release of the sounds, the harmonic content, and the **vibrato,** or periodic pitch variations, present in

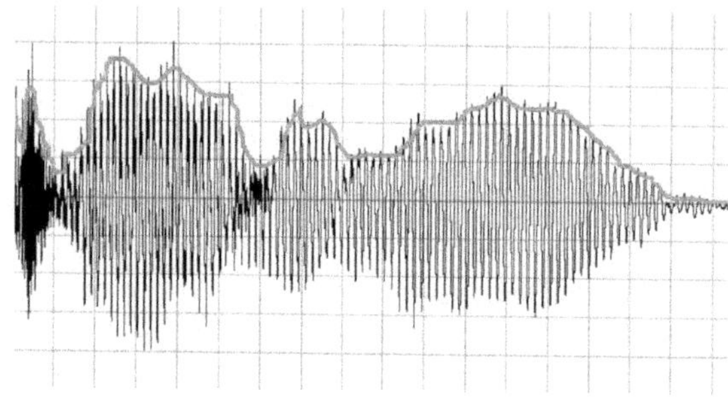

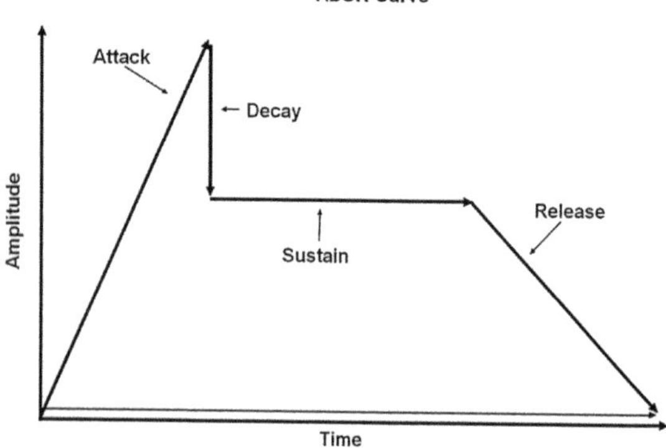

Figures 4.18 and 4.19. The envelope of a particular waveform is also known as its ADSR curve, as it is made up of the four distinct phases of attack, decay, sustain, and release.

a sound. Timbre is the most difficult quality of sound to communicate. Unlike pitch and loudness which can generally be rated by listeners along a scale of "high" to "low," timbre has no subjective rating scale for comparison. When we express a sound's timbre, we can only do so in subjective terms, such as grating, harsh, shrill, and/or mellow, rich, and bright. Since the words are intended to describe the tonal quality of musical sounds, they are usually associated with a specific frequency band, so for each one, you will need to learn the kinds sounds they are associated with or they are effectively useless to you.

Audio engineers need to become accustomed to talking about the tonal quality, or texture, of sound using words like the subjective terms mentioned above, because these types of **timbral descriptors** make up the language we use to convey the overall impression conveyed by certain sounds. This is sometimes a tricky task, because with no uniform standard to refer to, terms so far removed from the thing they are used to describe will be used or interpreted more or less differently depending on who is speaking or hearing them. Some terms, like muddy or bright or tinny, have been proven to be used relatively consistently, and assigned roughly the same meaning by most people no matter where we encounter them used. The meanings of some others seem to be more variable, with different values and shades of meaning attached to the same terms by different parties.

Despite the fact that we may at times need to adjust to the different "dialects" created by such differences, the metaphorical language of timbral descriptors is the best method we have to meaningfully communicate with musicians about their sound.

Psychoacoustics: Sound, Psychology, and Emotion

Sound has a big impact on our emotions and our emotions have a big impact on our perception. Here are some psychoacoustics facts:

- Dave Moulton of Moultan Labs (http://www.moultonlabs.com/more/hearing_the_louds_and_the_softs_of_it/) notes that "The subjective quality 'loudness' has emotional 'meaning' for us. It relates to issues of the emotional intensity of sound, as well as its nearness to us" and further notes that the **dynamics**, or amount of variation in loudness, of a composition conveys much of the emotional intensity in a song and/or a performance.
- Sound has physiological effects that stem from how it stirs our feelings, such as soothing us or making us anxious. As it influences our psychology, by extension it affects our bodies; for example it can de-stress and heal or raise "fight or flight" stress hormones that are not good for overall health.
- Physiologically our emotion centers in our brains are closest to our ears. In the same way our memory centers are closest to our olfactory senses (which is why the smell of someone baking will remind you of past memories of that smell). So, besides all the physical attributes that make sound difficult to quantify, we *feel* sound (in an emotional sense). A song that gets you in the right spot (or wrong one) can make you weep, or become amazingly energized and happy (those two are sometimes linked anyway). When was the last time a random smell made you feel so bad you cried (besides chopping onions of course!)?
- Known as the **cocktail-party effect**, a listener will listen closely to parts of a sound, such as a vocalist, or conversation, while being less aware of other sounds. They will then tend to base their perception of loudness on what they paid attention to, perceiving it as more clear and loud than others. How can we lower

the distracting background dBs in the world just by creating our own silent space in our heads? We don't know. Nevertheless, just by paying attention, we do it all the time.

- Not uncommon at concerts is the fact that the vocals may be hardly audible yet many people are singing along with them, while others are looking around and wondering what they are missing. Expectation can influence what we hear. Similar to the way attention changes our perception by focusing on what we need to hear, expectation changes our perception. But while attention is a tight focus on some part of our reality, expectation can make us perceive things that aren't there.
- Unlike your eyes, you can't shut your ears. Even earplugs cannot completely eliminate your perception of sounds. We depend so much on sound that digital toys and computers that don't have sounds inherently created by different functions, program them in. It was shown early on that people respond better to objects that give aural as well as visual cues.
- Listening to music is a huge computational task, so complex that no computer can process sound in as sophisticated a fashion.

Sound Science and Live Sound

While the P.A. is the tool of the live sound engineer that is most recognizable to the general public, the live sound engineer can't make full use of it unless their choices are guided by a solid understanding of the sound science that pertains to producing sound in live environments. The P.A. can only be used to produce the best sound if the operator knows where to place the P.A., so the sound leaving it will interact in the environment in a way that produces the least interference, and the best mixes only get produced by engineers who are able to justify the adjustments they make to the sound using the P.A., which requires they understand the science first. While early on new engineers can make up for gaps in their knowledge by resorting to a formula for their mixes, inevitably they will reach a point beyond which the formulas will fail them. Just like a carpenter needs to know as much about wood as about hammers, or a soldier needs to understand tactics to get the most out of their weapons, the live sound engineer needs to know the science of sound as much as they need to be familiar with tech of the P.A.

Chapter 5

Acoustics

The Science of Sound in the Environment

As we've seen, the physical attributes in a sound can act as a fingerprint or portrait, distinct even if not completely precise, which we take whenever we measure and catalogue each of the qualities that combine to make the sound unique. As we've developed greater ability to measure the attributes of sound, these sound portraits have in turn allowed us to quantitatively compare sounds against each other and thereby enabled us to study which and how much of the combined qualities of a sound determine how it will be qualitatively perceived. The results of these studies led to the creation of the first systems and scales which have been refined over time into the best ways to measure sounds and sound parameters based on our understanding of the way humans perceive sound. While measuring sound in real time is essential, the greatest benefit all this accumulated knowledge has had for audio engineers is that we can also use these measurements to predict in advance how the sound we work with will be experienced by most people, so we can ensure our P.A. sound is enjoyable before we subject people's ears to it.

Yet if the physics of sound and human biology were the only factors the live audio pro needed to understand or account for to deliver an enjoyable aural experience to the audience, sound engineering would be a relatively simple task. Science has long since worked out much of the problems of measuring sound in air, and sound in ears. If sound and ears were the only variables you had to keep track of, the task of mixing a show to please the ears listening would be very simple. Proper settings could be dialed into the P.A. before the show. At most, mixing would involve measuring dB SBL and double checking that frequencies are all in proportion, and minding that nothing in your mix gets masked by an offending nearby frequency. Then to keep busy and look less useless, you'd probably want to repeat the process every so often. With such simple parameters a perfect mix would be a sure thing for any sound engineer capable of staying awake for a full set. Luckily for any who prefer more complex challenges, this is where we say farewell to the straightforward and predictable sound of the previous chapter. That sound could be managed by a sleepwalker. The sound you will get to know in this chapter will keep you on your toes.

The ways we measure sound revealed in the previous chapter are essential; engineers value them so much because they assist us in tracking the consequences on overall sound arising from an additional factor that throws a building-sized wrench in the works. This factor greatly complicates the stable picture of sound described by the physics of waves undulating neatly through air to be perceived by ears in ways we can clearly define and display on a neat chart. The reality is that sound doesn't just travel from P.A. to ear unmolested; that would only be possible in a world made of nothing but sound and ears. We live in a different world, a world full of stuff of myriad shapes, dimensions, and textures. And as sound encounters stuff in the world, each interaction changes that sound in a different way. Without any way to account for how sound interacts invisibly with the world, there would be nothing to explain the reasons behind the changes in sound, only wonder over the extent, since in a world with an endless variety of objects to interact with, sound can be changed in almost any way except one "easy to predict."

Acoustics

No matter where a P.A. is set up, the sound released from the P.A. will encounter objects and boundaries in the environment, and will be altered to some degree with every interaction. Consider that in a typical venue there are a half dozen discrete objects and groups of objects a sound will ill interact with before even reaching the outer boundaries of the furthest walls, and you can see why a sound pro must know the ways sound interacts with the environment, to be able to optimize sound. Fortunately the area of physics known as **architectural acoustics** can help us to predict the ways sound will interact in the environment, and understand which interactions can cause the biggest problems for our sound. While not all sound issues can be predicted or avoided entirely, understanding which variables you can reasonably manage and which are out of your control can save you a lot of headaches.

Because sound is a kind of energy, the result of any encounter with an object will change it, either by changing its amount, its direction, or both. The variables that determine what will happen are the object's shape and the materials it is made of. Sound can interact with objects by going through the object to be refracted and redirected, by being diffracted around the object, by getting trapped and absorbed, by getting reflected off the object and bouncing back in a different direction, or it can be diffused, and scattered back in all directions. Usually, some combination of two or more of these results will occur at once, such as when sound hits a wall and some of its low frequencies are refracted through, other frequencies are absorbed, and the remaining frequencies are reflected back into the room.

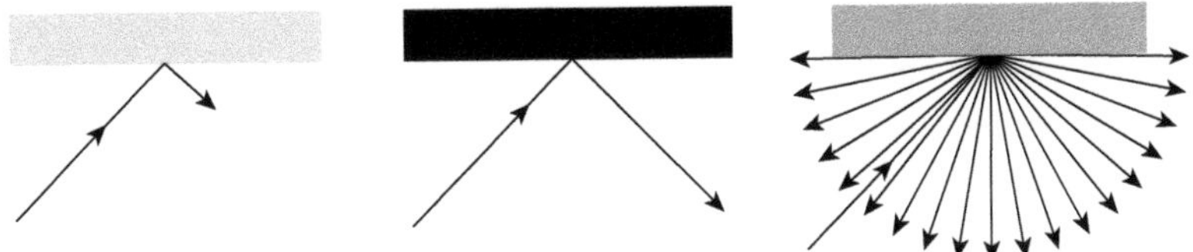

Figure 5.1. Sound can be absorbed, reflected, or diffracted.

Refraction

When crossing boundaries created by media of different densities, a sound wave's speed changes, causing it to bend as it passes from one to the other, such as from the air into a dense object like a wall. Most live audio applications are not particularly affected by sound wave **refraction**, though the amount of sound escaping might create a problem for those with neighbors or roommates. When neighbors are disturbed by sound, it is usually not because the sound is so loud they can hear it clearly, but instead because of the boom and rumble of the bass, low-frequency sound is most likely to transfer into, and beyond, a barrier like walls. If there is worry about disturbing neighbors, turning down the low-frequency bass can help. See the "Outdoor Acoustics" links in the chapter directory for other examples of refraction.

Diffraction

Diffraction refers to the phenomenon that allows sound to travel around corners, over obstacles, and through openings. When a sound wave meets a barrier that is smaller than its wavelength, it bends around it and keeps moving instead of being reflected back like it would with a larger obstruction. The ability to diffract through and around objects is much greater for low-frequency signals than for high frequencies because sound waves cannot diffract around objects with dimensions larger than the sound's wavelength. High-frequency wavelengths are measured in millimeters and centimeters, significantly smaller than most obstructions they're likely to encounter in the environment, while low-frequency wavelengths are measured in yards and can be so large that only objects like walls and buildings are large enough to reflect them. It is because of this that low frequencies travel farther than high frequencies outdoors; the high frequencies in transmitting sounds often get reflected back or absorbed, while the low frequencies move past or through objects. When frequencies are reflected back by an object instead of diffracting the missing energy creates a distinct sound shadow behind the object.

This tendency of low-frequency wavelength sound waves to diffract around corners also accounts for the tendency *for high-frequency sound to be directional*, rather than spreading out, while *low-frequency sounds are omnidirectional*, and move in all directions. This was once exacerbated by speaker design since high-frequency sound can't bend around enclosures and is even more directional when beamed from a speaker, and overcoming this limit has had a great impact on every aspect of speaker design. Our greater understanding of the relationship between wavelength, diffraction, and speaker/enclosure dimensions is at the heart of the look of modern loudspeakers with their parallelogram shape and curved edges so very different from the plain rectangles of the boxes that were the standard not too long ago.

Absorption

Sound that doesn't go around an obstacle can either be reflected or absorbed depending on the frequency of the sound and the material of the obstacle. **Reflective** surfaces tend to be smooth and rigid, while absorptive ones are flexible, soft, and porous, and the less reflective a surface material is, the more **absorption** it possesses. Rather than reflect all the energy that reaches its surface, an absorptive material traps a great deal of it. This energy is then converted to heat as a result of the friction produced as sound passes through the much greater surface area created by the many fibers that make up the bulk of absorptive materials. Commonly found absorptive surfaces include *acoustical tiles, carpets, draperies, fleece, upholstered furniture, and people.*

A material's **absorptive coefficient** is a measure of sonic energy it is able to absorb. Rated along a scale of the 0–1, if an absorptive material were able to absorb 95 percent of sonic energy and let only 5 percent

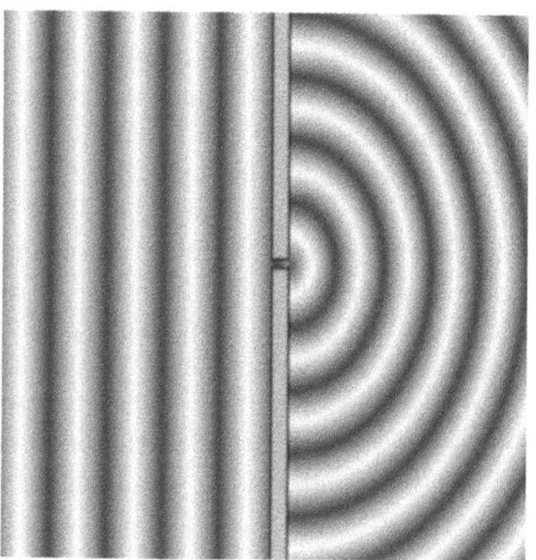

Figure 5.2 and 5.3. Diffraction—note the shadow created in the first image by the inability of shorter wavelengths to diffract around the object. *Source: Graphic user: Mill-eram released under a Creative Commons BY-SA 3.0 license.*

escape, it would have an absorption coefficient of .95. In comparison, reflective materials will have very low absorption coefficients, maybe rating only .05, out of a scale of 0 to 1. Most surfaces are somewhere between totally reflective or absorptive, and some materials allow for sound waves to be diffracted or refracted as well. The ratio is not absolute, however, and many surfaces may absorb the greater part of one frequency while acting equally reflective for sounds at other frequencies. Materials with wooden and metal surfaces display this characteristic to varying degrees, depending on how their surfaces are mounted. This explains why the amount and type of materials in any space will change the character of the sound in that space.

Reflection

Surfaces that are smooth and rigid usually are unable to absorb much sound and reflect it back into the room instead. These types of surface include plaster, glass, concrete, and brick. As we will see, reflected sound is not always a bad thing, and it's part of what makes indoor sound rich and warm. Too much of it can be a problem though, so too many reflective surfaces in a room aren't desirable. The shape of surfaces can also influence how they reflect sound. Sound waves hitting any surface will follow the **law of reflection** which states that the angle made by the incoming wave as it hits the reflective surface (the **angle of incidence**) will be the same angle of its trajectory as it bounces off the surface (the **angle of reflection**).

Diffusion

Sound waves hitting surfaces that curve inward will tend to have their outward trajectory focused on a single spot. This is not desirable because it gathers diffuse sound and sends it in a single direction, lowering the dB in several parts of the room and **beaming** it at one area to raise the sound level there. Sound engineers want to make sure sound is evenly distributed, so concave surfaces are avoided or covered if possible. On the other hand, convex surfaces that bulge outward can be helpful to the live audio pro, because they help scatter sound. Sound hitting convex surfaces from one direction gets diffused in every direction, because the shape of the surface points to all angles instead of angling inward like the concave surface.

Direct Sound and Reflections

In any acoustic environment, the sounds can be classified according to the path they travel and the amount of time they take to reach us. The first sound that reaches the listener is called the **direct sound**, because it is the sound that travels directly from the source to the listener's ear. There is little chance for the data or power within that signal to degrade before reaching the listener, making it the purest form of the signal to occur in any sound event, and the one required for a sound to be perceived clearly.

All successive sounds from the same sound event are reflected sounds that have first contacted surfaces along the path from the source to the listener. As a result these reflected sounds have all undergone some degree of degradation resulting in a smaller amplitude compared

Absorption of Sound in Common Building Materials by Frequency						
Materials	125 Hz	250 Hz	500 Hz	1kHz	2kHz	4kHz
Marble, Tile, Sealed Brick, or Granite	1%	1%	1%	2%	2%	2%
Vinyl Floor (on concrete)	2%	3%	3%	3 %	3 %	3 %
Unglazed Brick	3%	3%	3%	4%	5%	7%
Heavy Carpet (on concrete)	2%	6%	14%	37%	60%	65%
Heavy Carpet (with foam rubber pad)	8%	24%	56%	68%	70%	73%
3/8" Wood Panel (over 2x4's)	28%	22%	19%	9%	10%	10%
3/8" Wood Panel (on brick)	14%	11%	10%	6%	4%	4%
Mid-Weight Curtains (flat)	10%	25%	45%	75%	70%	60%
Heavy-Weight Curtains (pleated)	20%	35%	57%	70%	70%	65%
Heavy Window Panes	18%	5%	4%	3%	3%	2%
Laminated Window Panes	35%	25%	20%	10%	9%	7%
Moderate Density Crowd (Audience)	25%	35%	42%	46%	50%	50%

Figure 5.4. All materials have different absorption coefficients.

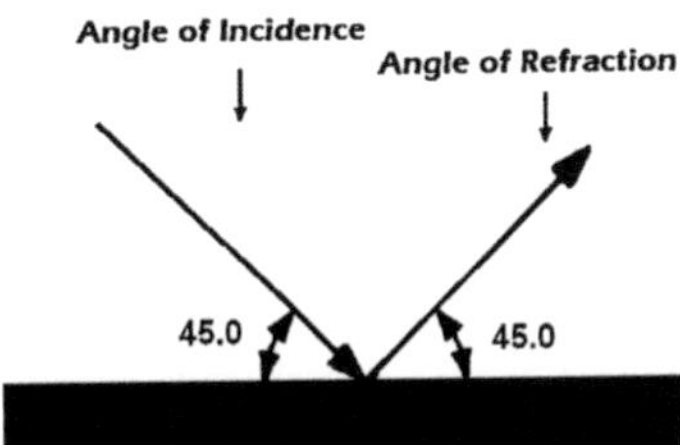

Figure 5.5. The law of reflection.

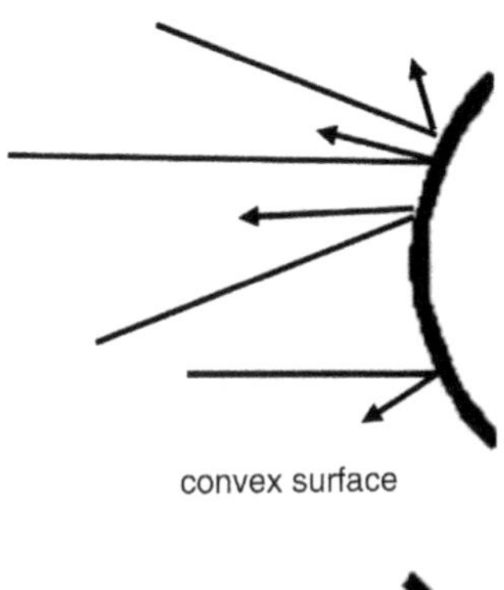

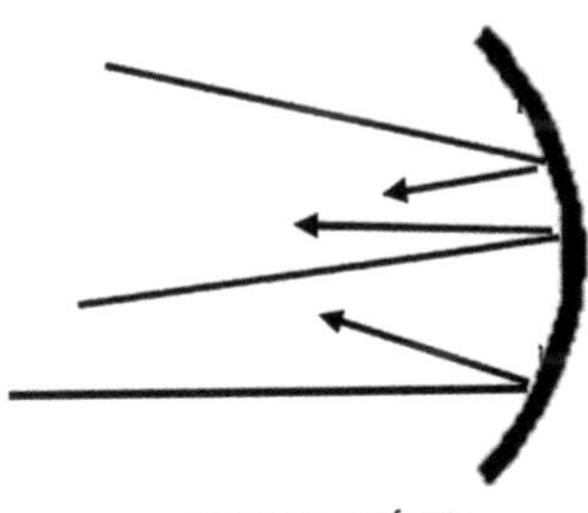

Figures 5.6 and 5.7. Diffusion is the ideal. Concave surfaces should be covered to avoid beaming.

with the original signal. This degradation is due to the energy lost every time a sound contacts a surface and loses a portion of its power before it reflects back. The reflected sounds arriving within approximately one quarter of a second of the direct sound are called early reflections.

Compared with **later reflections** from the same sound event, **early reflections** retain most of the signal's original power, having reflected no more than once before reaching the listener. This class of sound is almost as powerful as the original signal, and because of the tendency of our ears to average all sound qualities in the same range instead of attending to every peak or variation, early reflections are correlated with strong and clear sound perception. Instead of interfering negatively with direct sound, early reflections reinforce the direct sound.

This tendency for the human ear to incorporate and average all sound variants that arrive within 30 ms of a direct sound, into a perception of a single sound is called the **precedence effect**; it is also known as the **Haas effect**, after the researcher who first described the quality. In general, sounds arriving after 30 ms of the direct sound do not benefit from this averaging and begin to interfere with intelligibility. To make use of this information just note this rule of thumb: sounds reaching the listener more than .3 seconds (1/3 of a second) after the direct sound interfere with clarity, so try to keep to short acoustic paths of under 30 feet. Any reflected sound reaching listeners after that is less than ideal.

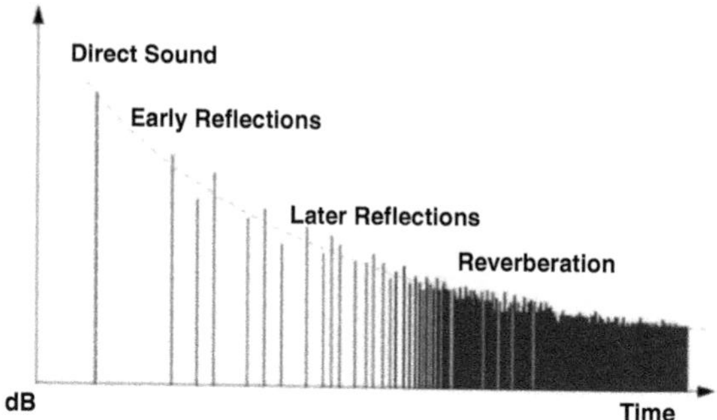

Figure 5.8. Early reflections reinforce sound, but reflections arriving too late make it hard to hear.

Reverberation

The reflected sounds remaining after the source has ceased are referred to as **reverberation**. The effect of reverberation is neither inherently desirable nor automatically problematic. Reverberation is just a basic acoustic property of a room. The influence of reverberation on sound quality is dependent on the sound application, frequency range, the acoustic path the reverb takes, and the distance between the listener and sound source. The same reverberant field that might make a hollow, incomprehensible mess out of 70 percent of a speaker's words, could make an average cello solo sound like Yo Yo Ma. Highly reflective environments increase reverberation because whatever is reflected back into the environment adds to the overall reverberation. The two main measures of reverberation are **critical distance** and **reverberation time**.

Reverberation time (RT60) is the measure of the time reverberation takes to decrease in power by −60 dB. Reverberation time can range from about 0.2 of a second for a typical living room to about 10 seconds for a large empty warehouse or hanger. The boundary between the reverberation time in a room considered **acoustically dry or dead** (no reverberation) and rooms perceived as **acoustically wet or live** (highly reverberant) is a reverberation time of around 1 second (the ideal level for small rooms).

Critical Distance

Critical distance is the distance from the sound source where the reverberant sound energy of a room is equal to the direct energy. Because the direct energy of sound is the only part of the signal that carries the data fully uncorrupted, in order for sounds to be perceived as intelligible by listeners, the amount of direct sound reaching them must exceed the reverberant sound. Therefore critical distance is the boundary representing the distance from the source where reverberant energy begins to sharply interfere with the ability of listeners to understand the message in the direct sound signal. When reverberant sound reaches a level that is +12 dB above the direct sound, the direct sound becomes masked and intelligibility is lost.

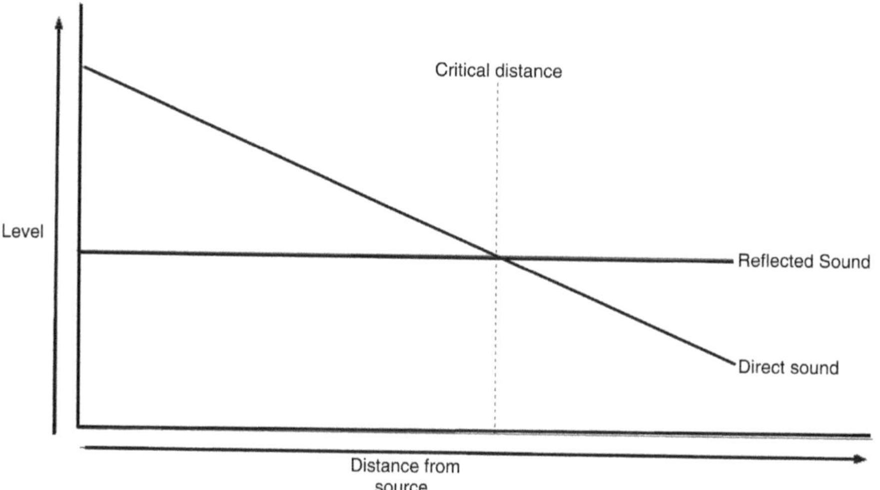

Figure 5.9. Critical distance.

Critical distance is different at all frequencies and is also dependent on the overall balance absorptive material and reverberation time in a room. The relationship between these factors is apparent in the following:

- Critical distance is independent of the dB level of the source because the **attenuation,** or weakening, of the direct sound source over distance as described by the inverse square law can only reach a set distance before the buildup of reverberant energy has had enough time to match it and create a steady state of equilibrium with the direct sound. This distance is determined by the reverberant quality of the room and varies with frequency but is not affected by dB: It remains the same no matter what the dB level of the source.
- Any attempt to increase the distance of intelligibility beyond the boundary set by the critical field by increasing the source dB will only be canceled out by a matching increase in the volume of the reverberant field.
- The more reverberant a room is, the closer the critical distance is to the sound source.
- The more absorbent a room is, the further the critical distance is from the sound source.
- The additional air friction created by humidity, smoke, and other airborne particulates can boost the absorption of high-frequency energy, thereby reducing the level of high frequencies in the **sound field**, as well as the reverberation time and level of the **reverberant field.**

Situations where Positive Reverberant Effects Occur

- the interval between reflections is generally too brief for echoes to be heard;
- the time taken for reverberation to occur is a reasonable match for the size of the venue;
- the listeners are located in the near/direct field (inside the critical distance range);
- the application is a more harmonic music genre where the higher reverberation times characteristic of live rooms is ideal;
- reverberation helps listeners integrate the entire range of distinct sounds from an instrument, including all directional portions into one cohesive tonal sound.

It is important to keep in mind that these positive effects are rarely just a happy accident, but instead are most often found in well-designed halls and venues that were built according to acoustic principles so that the room's reverberation would have a positive rather than a negative effect on the sound.

Situations where the Reverberant Qualities of a Room Can Work Against You

- the reverberant sound's path length is 30 feet longer than that of the original sound, producing interference effects such as echoes which hinder intelligibility;
- the time taken for reverberation to occur is not a perceptual match for the size of the venue, because the visual cues contradict the audio ones;
- the listeners are located in the far/reverberant field (outside the critical distance range of the sound source);
- the application is a percussive music genre or one with a high degree of articulation where the lower reverberation times characteristic of acoustically dry rooms are ideal;
- reverberation causes undesirable changes in the tonal sound of the elements in a music program or reduces the intelligibility of speech for listeners due to the acoustical masking of the direct sound;
- peaks reach a continuous and sustained level 20–30 dB louder than the volume before reverberation, turning a loud noise environment into one that is both uncomfortable and dangerous.

Loudspeakers at Room Boundaries

It is not only distant sound waves that are affected by the tendency of reflective surfaces to influence the sound field. Boundaries are known for their ability to interfere constructively with the low-frequency wavelengths produced by our subs. In essence, *a sub-woofer placed against a room boundary will produce an additional 3 dB of output for each boundary surface.* When placed in a corner where the boundary includes two walls and the floor an increase of 9 dB can be achieved. This effect can be used to get more volume from your system, just remember it only boosts low-frequency sound. This is because low-frequency sound is omnidirectional, radiating out in all directions. When placed against a boundary the energy that would travel in that direction is blocked and reflected back to combine with the sound heading that way. So when you don't want the extra decibels, remember to keep speakers and subs away from walls, and put speakers up on stands as well.

Resonance

Resonance is defined as the tendency for objects to vibrate more intensely at specific frequencies when

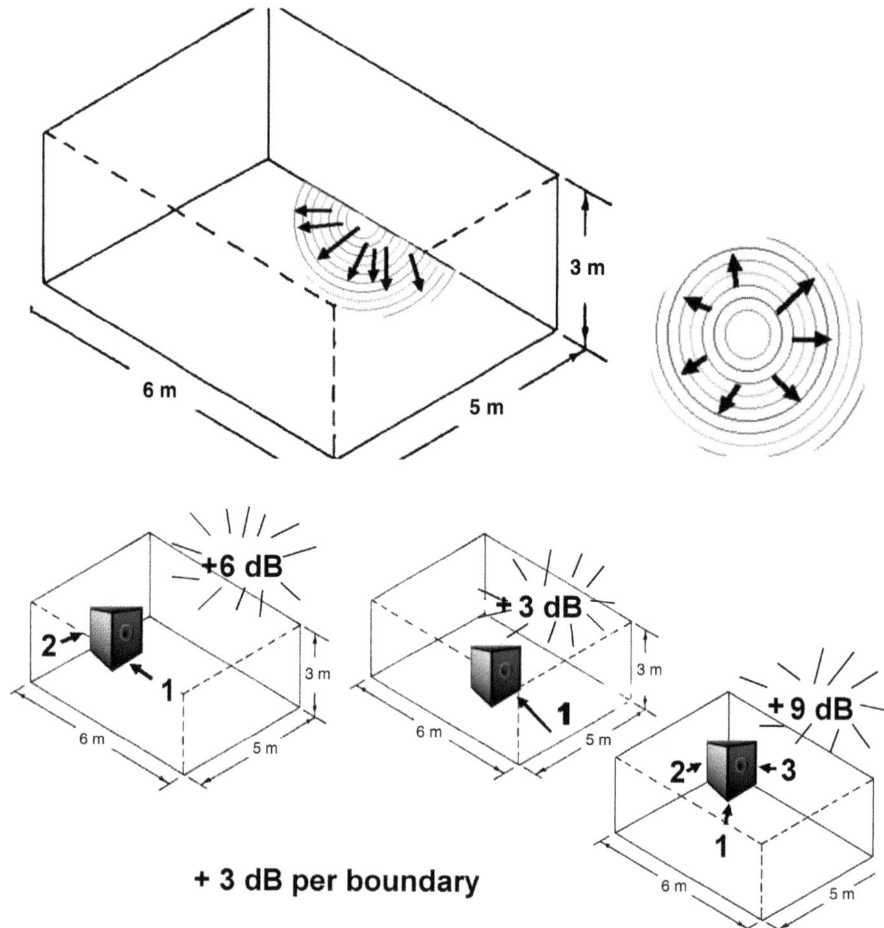

Figures 5.10 and 5.11. A speaker placed against the wall will increase low frequency dB output by 3 to 9dB.

excited by a forcing function. The frequencies at which an object vibrates when excited are that object's **resonant frequencies**. Before discussing resonance in rooms, here are a few key facts to keep in mind about resonance, in general:

- Most objects have more than one resonant frequency, some have many, and everything has at least one.
- Acoustic resonance is a subset of mechanical resonance made up of mechanical vibrations falling in the audible frequency range.
- The forcing function that initiates the vibration and excites the resonant object could be a frequency, mechanical vibration, or another variable force, but in all cases the force needed to initiate and maintain resonance is much less than is normally required to produce a similar energetic response.
- While any object can vibrate in a vast array of frequencies, each object has only one or two where the mathematical relationship between the wavelength of the frequency and the object's form and dimensions is close enough to do so with least resistance and maximum efficiency. This accounts for the bullet point above.
- This energy efficiency extends to stored energy as well, and means that even when no more energy is supplied to keep the resonance going, it will keep going much longer and come to rest more slowly than usual.

Standing Waves

Standing waves occur when sound waves bounce between parallel reflecting surfaces and are the right length to become superimposed. At frequencies above 300 Hz, the wavelengths of sounds are small enough that they aren't affected by the room size as much; when short wavelengths reflect off a room's walls, the distance between the two boundaries will not be related to the wavelength and as the wave bounces back and forth its peaks and troughs will not line up and reinforce. Because of the longer length of low-frequency waves (consider that a 50 Hz wave is 21 feet 8 inches), frequencies below 300 Hz can cause problems, because their wavelengths are large enough to closely match room dimensions. *All rooms have their own low frequencies with wavelengths that are the same dimension (or multiple) as one of the lengths of the room.* When these frequencies bounce back and forth between the walls of that room, they become superimposed and undergo the type of phase interference explained in the previous chapter. One of the results of a standing wave is increased sound pressure near the boundaries (constructive interference), while at the same time sound pressure decreases toward the center of the room (destructive interference).

Room Resonances

In live sound applications, the initiator, or forcing function, that can start a room resonating is a frequency or series of frequencies. These do not have to be the same frequency as the resonant frequency, only be able to initiate that frequency in response, in most cases because one or more of their harmonics is the same frequency or multiple of that frequency. **Room resonances** or **modes** arise from the relationship between a room's dimensions, the size of the resonant frequencies wavelength, and the ratio of surfaces that reflect the resonant frequency and its harmonics more than others. In addition, there may be areas within the room such as the cavities under the stage or balcony that add their own resonant and reflective characteristics to the soup. The types of modes are named according to the number of surfaces in a room resonance standing wave's path. The modes are:

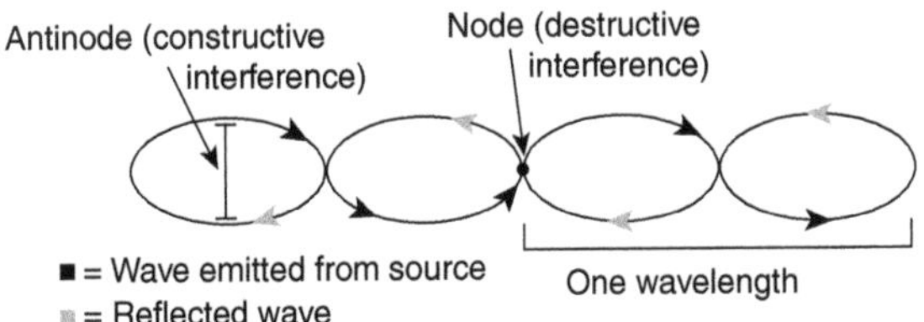

Figure 5.12. Standing waves.

- *Axial modes:* The wave travels a path between two opposing surfaces in the room
- *Tangential modes:* The wave travels a closed path that includes four surfaces. If the room is square, many tangential paths have the same length as the diagonal.
- *Oblique modes:* The wave travels a closed path that includes four surfaces.

Measure and Calculate Room Acoustics

If you have enough information about a room, such as the room's dimensions, the shape and number of objects in it, and the materials these are made of, then you can use some or all of this information to calculate the acoustic parameters of that room, such as its reverberation time, its critical distance at any frequency, and the frequencies it will resonate at. This information can help when making choices about how to set up or the best way to fix sound problems.

Some unavoidable facts about room resonance include:

- All rooms have frequency ranges where modal effects will potentially dominate; all are vulnerable to resonances that can cause variations in the sound.
- Room modes are the main cause of low-frequency acoustic distortion.
- Standing waves can cause peaks and dips in the sound quality of 20 dB SPL or more at some frequencies.
- Nearly all systems experience some negative impact due to modal resonances; even engineers working with high-end equipment and in acoustically treated rooms can encounter standing wave issues under the right circumstances.
- Rooms that have dimensions that are integer multiples of each other are the most likely to present resonance issues, with the worst offenders being cube shaped rooms, square rooms, and rectangular rooms that are exactly twice as long as they are wide.

The variables that contribute to room resonances make them much more difficult to deal with than almost any other acoustic issue most engineers will encounter. It is impossible to fully neutralize room resonance by making frequency adjustments in the P.A., since, as already mentioned, *the frequency at which a system oscillates is largely independent of the forcing frequencies that excite it.* You can identify the frequency a room resonates at and cut it all you want, but since there are still many other frequencies whose harmonics are able to excite the resonance, pinpointing the initial cause and eliminating the offending frequency is extremely difficult. Adjusting frequencies can help a bit, but the best way to mitigate the problem is with speaker placement, especially speakers with a low-frequency output. However, standing waves can occur anywhere the right combination of parallel walls and low frequencies meet, so while this effect can be managed or reduced with careful speaker and mic placement, eliminating the phenomenon altogether isn't possible, simply due to the nature of physics.

One of an engineer's concerns is to make sure the sound they produce has the best frequency response possible. **Frequency response** is the accuracy of the sound reproduction. A **flat frequency response** means that there is no change in volume over the frequency range. Naturally the engineer's job is to try and make sure the frequency response of the room is as accurate as possible, and any kind of interference gets in the way of that. Sound issues that can't be eliminated can be very frustrating, but most engineers learn not to let it drive them crazy. The job description doesn't include miracle working, so when perfect sound isn't possible, and it rarely is, all that can be done is to minimize problems as much as can be done with the tools at hand.

Comb Filter

If you recall from the last chapter that phase can be seen as a clock, then it should make sense that timing issues can cause phase issues as well. Sound from two sources that reach listeners at different times will be out-of-phase when it converges and produce interference. This is why evenly spaced speakers placed on both sides of a stage produce few problems along the centerline, where the sound waves converging are from equidistant speakers and arrive together, but may have more pronounced interference effects to the left or right where one of the speakers is farther than the other, causing one part of the signal to arrive after the other. The term used to refer to the phenomena of *time- or phase-based interference is comb filtering.* While time delays also can't be eliminated in all cases, fortunately they are much easier to deal with than room resonance. As with most acoustical issues, the solution is to avoid the problem by taking care to set up your speakers and mics to avoid timing issues. In Chapter 21 where we discuss setting up the system we will look at the strategies that can reduce delay issues and minimize negative sound effects like comb filters that delay issues can produce.

The Larsen Effect

Larsen effect is a form of positive feedback created when the signal from an audio system output, such as a loudspeaker or monitor, is picked up by a microphone in the system and re-amplified, producing a high-frequency whine or screech and beginning an escalating cycle whereby each output becomes louder and more easily picked up by the mic for another cycle. Originally named after the scientist who first discovered and described it, Søren Absalon Larsen, it is now simply called a **feedback loop**, and the unpleasant sound it makes is **feedback**. The frequency of feedback is determined by the resonant frequencies of the combined interaction of the elements involved in the initiation and escalation of the effect, namely: the venue acoustics, mic, amp, and loudspeakers. Once it starts, feedback is a horrible din that will escalate until eardrums and gear alike fry, unless quickly attended to by a knowledgeable sound engineer. Although again, the best solution begins with preventative measures; setting up the system to minimize the sorts of environmental noises and resonating

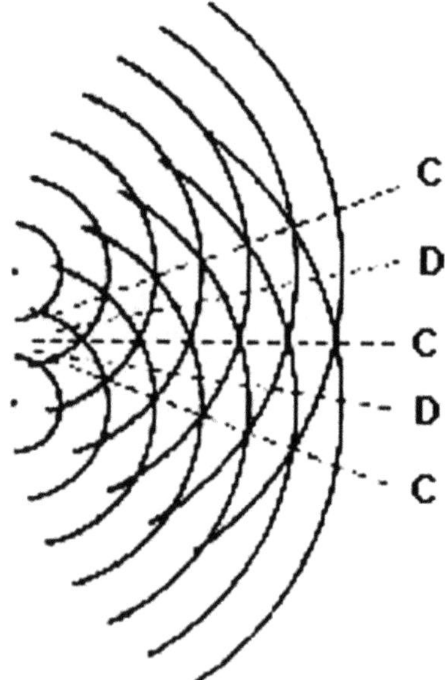

Figure 5.13. Interference pattern of out-of-phase or delayed sound waves as they overlap—note the C and D markers stand for constructive and destructive interference.

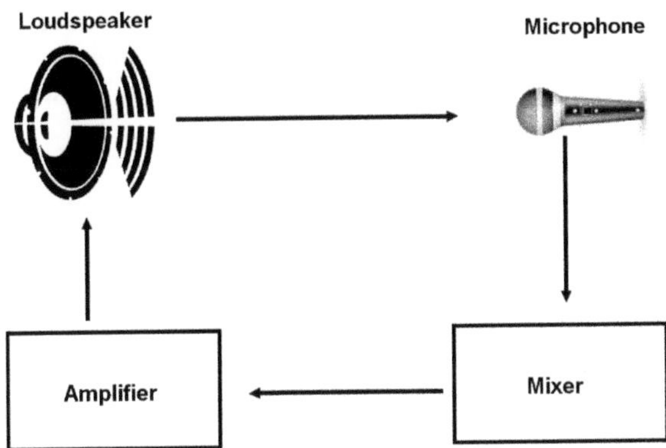

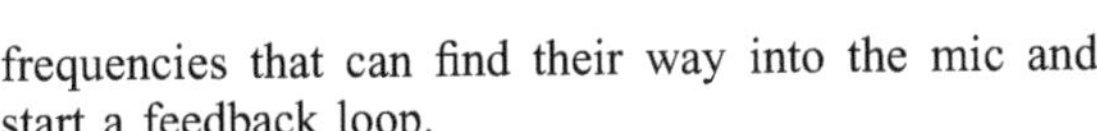

Figure 5.14. Feedback loop. *Source: Image by Borb released under a Creative Commons BY-SA 3.0 license.*

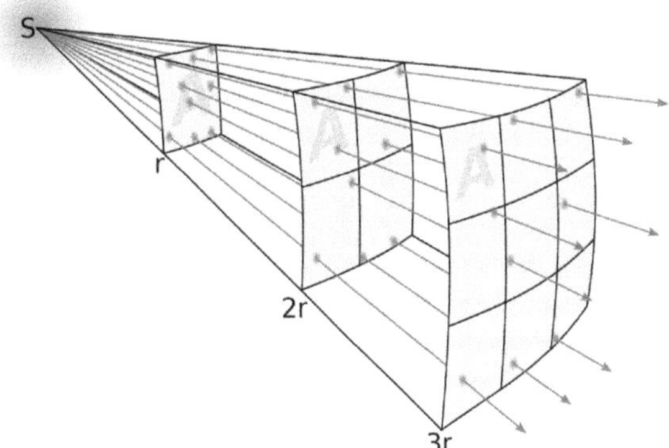

Figure 5.15. Geometric spreading.

frequencies that can find their way into the mic and start a feedback loop.

Outdoor Acoustics

Outdoor venues provide a different mix of sound problems. Phase- or delay-based problems may occur, and problems due to reflection may not be an issue if there are no vertical features nearby, but can occur off buildings, hedges, or woods if there are any nearby. So outdoor venues can present all the same issues one might encounter indoors but also may also present an engineer with problems that stem from an entire set of issues unique to the larger spaces of outdoor venues. The larger area of outdoor venues make attenuation issues a bigger concern, both due to geometric spreading as well as ground-level surface effects. **Atmospheric effects** resulting from humidity, temperature, and wind are more pronounced as well.

Attenuation is the loss of sound intensity. Attenuation can be the result of an intentional action, like turning down the volume knob on a boom box, or it can happen as it interacts with the environment, weakening over time and distance. If you have ever noticed that music from the same portable player that sounds great in your house sounds much thinner outside, you perceived the effects of attenuation due to geometric spreading. **Geometric spreading** refers to the attenuation of sound over distance due to the same amount of power being spread out over a larger area. It is more noticeable outdoors or in stadiums without a ceiling because the sound doesn't benefit from the dB boost sound gets indoors due to reflected sound from the wall and ceiling. We've already mentioned one type of geometric spreading—the inverse square law—where 6 dB are lost each time the distance from the source is doubled. The sound source being referred to in the inverse square law is called a **point source**, and produces its sound power from a single area, and the transmitted sound spreads in all directions.

There are additional sound losses occurring due to **surface effects**, which vary by environment. Outdoor venues that are unpaved will absorb more sound over distance, as dirt and grass are less reflective than concrete. The amount of sound loss will depend on the surface. Concrete surfaces will produce little absorption; while thick grass can reduce sound levels as much as 10 dB every 100 meters at 2000 Hz. In addition outdoor venues tend to have more variable terrain than indoor venues where the ground is perfectly leveled, so obstacles in the terrain like small hills can contribute to sound losses as well.

Because of the often greater distances in the more spread out areas of outdoor venues, **high-frequency attenuation,** the loss of more high frequencies than low frequencies over a distance, is more pronounced in outdoor venues as well, and both types of surface features will create a greater proportion of high-frequency loss than any other loss. High-frequency losses are more noticeable because outdoor venues don't afford the boost sound gets from being reflected off walls and ceilings and becoming diffuse throughout the venue. Unless these issues are accounted for, this can mean less sound overall reaches the audience that is farthest out, as well as that sound being distorted because it's missing much of the original high-frequency content.

The next features that influence outdoor sound in particular are **wind** and **temperature gradients**. Especially common during the morning and evening, temperature gradients occur when the air at ground level and the air at higher altitude become different temperatures. Under normal conditions the atmosphere is warm at the ground with the air above cooler, the different air pressures occurring between cooler air above and the warm ground-level air causes the transmitting sound to be bent upwards. Because the sound is bent upwards attenuation is much faster at ground level and this can be a problem for any audience members expecting to hear sound beyond that point. In the case of the **temperature inversion**, where the opposite pattern holds sway and the air on the ground is cooler than the air above, the sound waves will be refracted downward. When sound is refracted downward, it can be heard over greater distances than usual, which can lead to noise complaints from local residents.

Wind is an issue no matter which way it blows, though the effects will differ depending on if it travels parallel to the direction of the sound or is blowing cross-wind. Wind blowing along the same path the sound travels will cause a sound shadow at one end, and an excess of sound on the other because of the gradient that occurs due to the air at ground level usually remaining still. This gradient causes the same sorts of upward and downward bend of sound occurring with temperature gradients. It makes sense that the louder sound will be in the direction the wind is blowing, and the shadow is what the wind leaves behind; this is no matter which way the sound is traveling.

Cross-winds and gusts also can literally blow phrases and words away. Remember that sound waves travel in air, when a cross-wind blows, that air is suddenly 10 feet to the side. The information contained in that portion of the air is going to be much fainter. In addition, at the same time portions of the sound carried away on the wind are quieter, cross-winds create the most noise (cross-winds are the winds that produce the famous groans and whistles of a windy day) increasing the environmental background noise enough to compete with and mask the sound from the P.A.

Because there is no way to minimize geometric spreading, short of adding four walls and a ceiling, which defeats the purpose of choosing to use an outdoor venue, the main solution to outdoor sound issues is simply to use larger P.A.s that can run more speakers and put out more power. The other solutions to all these issues are for the most part the same as the solutions to acoustic problems indoor, setting up the system so that issues are minimized. We will examine some of these solutions when we look at setting up the P.A.

Chapter 6

Sound Power, Power, and More Science You Need to Know

The Decibel: Power and Pressure

When we discussed measuring sound amplitude we were calculating with the dB SPL to help determine loudness. As we discussed, after a sound has left a source and is moving freely in the air, we can measure the air at any point to get a peak or RMS amplitude measurement of the air pressure fluctuation it creates, which we then convert into decibels to derive a sound pressure level value for that location. (We indicate the location because sound-pressure level can only reliably describe the sound pressure at the distance the amplitude measurement was taken: the same sound measured at a different location will yield a different sound pressure value.) In Chapter 4 we mentioned that decibels could measure anything that could be expressed in power or pressure, but discussed only sound-pressure decibels in detail, because those are the only decibels directly relating to the topics of wave characteristics and sound perception, and represent the only amplitude measurement we can directly take from sound in air. However, the decibel can calculate anything that can be expressed in terms of power or pressure, and all energy, whether it's acoustic waves in air or electricity in power lines, contains power/exerts pressure.

Sound Pressure

To refresh readers on how the decibel system works, the particular kind of power/pressure being measured is specified by appending a suffix to the dB notation, and the agreed-upon reference value for that type of decibel "fixes" the denominator of the fraction and becomes a baseline (in power/pressure), that can be used to compare our amplitude values against and calculate a value that falls along a scale of known values and tells us "how much" power or pressure is contained in our initial amplitude values. You should recall that to indicate we are using the dB to measure sound as air pressure, or sound-pressure level, we add the suffix SPL to the notation dB, which has a specific reference value of air pressure that is always used for dB SPL units.

Sound Power and Sound Intensity

We figure sound-pressure levels by measuring the amplitude of the air-pressure changes produced by a sound wave as it moves. By converting between decibels, we can then use the sound-pressure-level decibel we get to indirectly measure other values that we can't measure directly. We cannot measure a wave's overall power directly, because before being perceived by our ears (at which point measurement is impossible) sound waves are invisible to us. The only evidence we have that sound waves exist in the world is the effect they have on the air, in the form of air-pressure fluctuations. When we measure the amplitude of a wave for pressure variation, we are measuring the amount of visible environmental change created by that wave, which is proportional to the amount of power it carries; the greater the wave power the greater the amplitude of that wave and the larger pressure variation measurement we will derive from it.

The total energy carried in the sound emitted by a source and radiating out in all directions over a specified unit of time is called the **sound power** or **sound intensity** (the two are always mathematically equal and the values can be expressed in decibel units). These two measurements are not as commonly used in live sound applications, but they are related to decibel measurements that are used regularly, and are terms often (incorrectly) used interchangeably with the decibel that describes volume, dB SPL, so you need to know about them.

Decibels in the PA

Before it leaves our speakers as amplified acoustic energy, sound exists as a signal in the P.A. where it is reinforced with electrical power before finally exiting the system. Conveniently the decibel can not only measure the sound when it is in the form of acoustic energy, but decibels are also used to measure sound when it is in the form of an **electrical signal** in analog cables and equipment (with reference values in **voltage**) as well as measure the amount of **current** used to drive the speakers (with reference values in **watts**) and which determines the energy (sound intensity) carried by the sound leaving the speakers. These are the other dB units live engineers use most, along with dB SPL, and the digital dB which we mention later in this chapter.

- *dBV* represents the level compared to 1 Volt RMS. 0 dBV = 1 V. There is no reference to impedance.
- *dBu* represents the level compared to 0.775 Volts RMS with an unloaded, open circuit, source (u = unloaded).
- *dBm* represents the power level compared to 1 mWatt. This is a level compared to 0.775 Volts RMS across 600 ohm load impedance. Note that this is a measurement of power, *not* a measurement of voltage.

Measurement	Decibel Unit	Reference Value
Air Pressure	dB-SPL	20μPa
Voltage Pressure	dBV	1 Volt
Voltage Pressure	dBu, dBV	0.775 Volts
Electric Power	dBm	.001 Watt

Figure 6.1. The reference value for measuring power in your system is watts; the reference value for measuring pressure in your system is volts.

Converting Decibel Values

In other words, the sound-reinforcement process can be described in terms of a series of energy transformations along a timeline of cause and effect. No matter where we look along the flow from sound, to electrical signal that we boost with current, to "loud" sound released from speakers, we can measure the different forms of energy in terms of power (power) or pressure (field quantity). We do this by starting with the RMS or peak amplitude values revealed in the fluctuating air pressure of sound and then calculating with the decibel system to arrive at any related value we want to know. We can start at the beginning of the chain as well since, as we will explain, electricity can take the form of a sine wave that is identical to sound, and when taking that form can be measured for amplitude.

We can't measure power directly; instead we measure **power's observable effect on the environment** in the form of the **pressure it exerts** (amplitude fluctuations,) and then convert those pressure values to power values. Because all the dB values are mathematically related, an increase in one means an increase in all the others, as well as in the amplitude values we can measure. It's important to remember that the relationship between any of them is not 1 to 1, so if you double the power level in the P.A., all the related values along the line, including "loudness," will increase as well, but not in identical proportions. Even then, the proportional relationship that does exist will be influenced by a variety of factors from friction and heat in cords and gear, to the influence of frequency on our perception of sound. Despite not always being perfectly precise, the decibel is essential because it allows us to measure and convert values between every transformation of sound energy as it goes through the process of becoming louder sound.

What we can calculate in one direction (when we measure system power (watts/volts) by converting the values from air-pressure measurements), we can calculate in the other direction as well. We can also convert from power in our system to related pressure values (SPL) of the sound leaving the system and therefore can roughly estimate the power required to achieve a specific dB pressure output. *A commonly used rule of thumb that's easy to remember is that it requires double the power to increase pressure by 3dB. Power is related to the square of pressure, so to double the pressure requires four times the power.*

dB Change	Voltage	Power	Loudness
3	1.4X	2X	1.23X
6	2.0	4.0	1.52
10	3.16	10	2
20	10	100	4
40	100	1,000	16

Figure 6.2a. Starting at the left, we can see that an increase of 3dB results in a voltage increase 1.4 times the original, a doubling of power, and a subjective increase in loudness only 1.23 times the original.

For example, if you are starting at 100 watts of power output, and want to double the pressure (which requires +6 dB), the power applied will need to increase to 400 watts in order to achieve that. You can reference Figure 6.2b to see how to convert to other values.

A few more dB facts and figures you might find useful are:

- If you're dealing with voltage measurements, convert from dBV to dBu: 1 dBV equals +2.2 dBu.
- +4 dBu equals 1.23 Volts RMS.
- The reference level of −10 dBV is the equivalent to a level of −7.8 dBu.
- +4 dBu and −10 dBV systems have a level difference of 11.8 dB and not 14 dB. This is almost a voltage ratio of 4:1 (Don't forget the difference between dBu and dBV.)

To Increase Pressure by	1dB	2dB	3dB	4dB	5dB	6dB	7dB	8dB	9dB	10dB
Multiply Power by	1.25	1.6	2.0	2.5	3.15	4.0	5.0	6.3	8.0	10.0

Figure 6.2b. Double the power to increase pressure by 3 dB.

Dangerous Forces

Anyone entering the audio field must learn about both of the powerful forces that are part of the process they use to serve up the best aural experience. Of the two, electricity is the most dangerous by far. While sound is pretty powerful (one mistake at a crucial moment can literally deafen those present, permanently), it's a lot easier to control than electricity and far less likely to escape the parameters set by the sound engineer. When it does, it isn't fatal.

Electricity is a trickier force to reckon with; it takes less electricity to kill you than it does to run even a single piece of audio gear, much less the entire backline and PA, and when it's where it shouldn't be, it's usually invisible until test equipment reveals there may be an issue, or even worse, until someone discovers it by closing the circuit and receiving a painful or fatal shock. When it's discovered by accident, in the form of a band or crew member receiving an **electric shock**, it can be an extremely painful and upsetting revelation, and that's if you are lucky. A painful shock is the best-case scenario when accidentally finding current flowing where it's not expected; in other, worse-case scenarios someone gets severely injured or even dies. So even though electricity is almost as important a tool as sound itself, learning the science of electricity is a slower and more careful process.

There are things you simply need to learn in person, even if it's a process that seems straightforward in a book. In fact, the need for the sound engineer to be able to safely assess a venue's power supply and tie power to distribution or dimmer racks is part of the reason why any sound tech is expected to apprentice before seeking employment, regardless of education level. It is also one of the many reasons that those new to the field start as A3s, and only advance to the duties of an A2 audio tech and A1 engineer after their time as an A3/A2 has provided them a chance to learn any commonly performed electrical procedures on the job if their training did not include them. Many employers will want to double check such skills first hand before allowing any new techs or engineers to perform them on the job, even if those hired know how either through training or job experience.

Figure 6.3. Electricity is dangerous.

In this chapter we will discuss a few of the basic concepts and terms about electricity that you should know as a sound tech. However, though we may mention situations where the live sound engineer may be required to perform or oversee procedures such as tying power into distribution racks or dimmer racks, we are of the school of thought that when it comes to electricity, a general text such as this one can never provide enough instruction to allow an unqualified reader to safely perform any procedures beyond those commonplace actions, plugging into outlets and flipping switches, that don't require much instruction.

One thing general texts *can* do, if the authors are not careful, is to provide just enough information about procedures to give readers who misunderstand the intent a false sense of security. Such details, unless accompanied with hands-on instruction by a qualified teacher, don't do much to further required skills, but can potentially result in a situation where someone who would not think to do such a task without any information instead has just enough information they feel it's a fine idea to muck about where they shouldn't be, but not enough knowledge to do it safely. Therefore we don't provide more than background and safety information, and leave electrical training to more qualified instructors. While we do link to sources that provide more information about electricity, we ask that none of the sites be used as resources to help you meddle with powered equipment or power distribution boxes until you've had some electrical instruction and/or training.

Many terms used to talk about electricity in general are actually misused (for example using terms like power, energy, and electricity interchangeably is a common practice, though they are all different) and sometimes analogies used to show how electricity works give less than precise impressions about the nature of certain terms but are not corrected to allow the analogy to fit. If you are curious about any of the terms used here that may be used differently than you understood them, or if you have any other questions about electricity, use the links to some of the articles and resources in the chapter directory, or explore the ones in the book directory to find more science information, including information about electricity and electrical systems, than you could read in a year.

Electrical Energy 101

Basically, **electricity** begins with a moving flow of **negatively charged** particles called **electrons**. When electrons are at rest, or **static**, they are referred to by a variety of names (one of these is **charge**) in order to distinguish them from the moving sort. These electrons at rest, or charge, are measured in **coulombs,** but all you really need to know about measuring charge is that one coulomb is a whole lot of electrons. In audio, what matters are not electrons sitting around being useless, but electrons moving around doing work. Moving electrons is a flow of charge and is called **current**. Current is measured in **amperes** or **amps**, which describe the amount of electrons according to the speed of the current's flow (1 amp is equal to a flow of 1 coulomb flowing through a wire per second).

It's the Current that Kills

An interesting bit of electrical trivia is that while volts will make your arm tingle and make you shout "Owww!", amps are the component in electricity that will actually kill you if you don't take care. This is why stun guns can pump a person full of thousands of volts and make them drop and flop from loss of muscle control, yet will usually cause no further damage, because the shock these devices deliver is high voltage but has very low amps. Consider that a typical residential circuit is rated at 15 or 20 amps, but it takes only 50–250 milliamps to kill you. Even a wall socket capable of pulling only 3–4 amps has many thousands of times the amps required to kill you. Don't misunderstand this and think high voltages are safe though; in standard electrical applications, the higher the volts, the more current is present.

Current contacting and flowing through any matter will almost always encounter some amount of **resistance**, which we measure with units called ohms. Ultimately it is the amount of resistance (or lack of it) inherent in any type of matter is the factor that determines how easy or hard it is for the current to enter and flow through it, but before that can happen the current needs to get going. Electrons don't flow just for the heck of it; they flow in response to a force that pushes or pulls them in a given direction. This force is called **voltage**, measured in **volts,** and is the feature that jump starts the process and provides the push required to move current. Finally, as current flows through resistance, a certain amount of energy in the form of heat is transferred or released as **work**, this work is known as **power**, measured in **watts,** and how much is energy is released depends on the resistance. Before we explain the mathematical relationships, here is a small mnemonic device which may help to keep track of what quantity function and measurement go together:

- *Voltage* starts by giving charge a push—a jolt of **volts** to get it on the go.
- *Amps* now can measure newly moving **current**—dancing electrons in a flow.
- *Ohms* will offer its best **resistance** to **impede** the dancers and make them slow.
- "*Watt*" work gets done is so much fun—**power** to the P.A, as we mix live audio.

Now that the basic process has been explained, let's look at some aspects of each quantity and other variables.

The Law of Ohms and Watts

Resistance and the measure we use to determine resistance, ohms, will be the quantity that is the most important to our day-to-day work as audio engineers. This is because without resistance, we don't get much "work" from our current, since moving against resistance is

Quantity	Symbol	Unit of Measurement	Unit Abbreviation
Current	I	Amp	A
Voltage	E or V	Volt	V
Resistance	R	Ohm	Ω
Power	P	Watt	W

Figure 6.4. The quantities that make it happen.

what generates and transfers heat. Resistance acts on electrical current much like friction does on kinetic energy; more resistance makes electron movement more difficult and slows the current down unless more power (voltage) is applied, and less of it makes electron movement easier and requires less power (power) to maintain. Now that you have a tool to remember how each quantity functions, you can start memorizing the most essential of the formulas that describe the relationships between each of these four qualities: they all change in relationship to one another such that if you know any two you can calculate the other two. In the chapter directory are links to easy-to-use calculators that allow you to do this online.

The relationship between these parameters are described mathematically in the two electrical formulas you will encounter most working with audio. The first shows how we calculate resistance and is called **Ohm's Law**. This formula states $R = E/I$, which translates into *1 volt is the force required to push 1 amp through 1 ohm of resistance*. But this law is important because it lets us figure out so many things depending on how we use the equation. So *Ohm's Law is* $R = E/I$, and $E = IR$, and $I = E/R$. Translated, this means:

- With resistance steady, current follows voltage (an increase in voltage means an increase in current and vice versa).
- With voltage steady, changes in current and resistance are opposite (an increase in current means a decrease in resistance, and vice versa).
- With current steady, voltage follows resistance (an increase in resistance means an increase in voltage).

Another equation useful to audio engineers is known as **West VA/Watt's Law**. This law allows one to calculate the power gain or loss at any point and states that *one watt equals one amp multiplied by one volt* (W = VA). A few more facts should be noted before we begin discussing circuits, which is how the building and each piece of gear uses the quantities we've been looking at.

Conductivity v. Resistance

Resistance, is the measure of how much friction a component presents to the electrons flowing through it. While resistance expresses how difficult it is for electrons to flow, then the word expressing how easy it is for electrons to flow is **conductivity**. Almost all materials will have some conductivity and some resistance, except those few that are either 100 percent conductive or 100 percent resistant. As far as electricity is concerned, these two qualities are all that matters, and when it comes to matter, there are only three varieties, where each is defined by how **conductive** or **resistant** it is:

- **Conductors**: Conductors have low resistance and allow electrons to flow through easily.
- **Semiconductors**: Semiconductors are made of a perfect insulator, such as silicon which, when a small amount of conductive impurity is added, can act as either an insulator or a conductor depending on other factors; and can be used to build other essential electronics components.
- **Insulators**: Insulators are materials with high resistance where electrons can only flow with great difficulty.

Conductivity or resistance in any material is determined by the chemical make-up of the material, or rather by how the atoms are arranged in accordance with the material's make-up, by the size of the pathway for current to flow provided by the material, and even by variable features such as the amount the temperature of the material surface is above or below ambient temperature. It's important to remember that conductivity refers to a range of values, materials are not all equally conductive, not even when they are the same type, as variable conditions will add to or reduce an object's resistance/conductivity.

Alternating and Direct Current

There are only two forms of electricity we can generate, direct current and alternating current. **Direct current (DC)** is the type of electricity contained in a battery.

Conductors:	Insulators:
All metal objects	Glass
Salt	Rubber
Dirty Water	Fiberglass
Graphite	Ceramic
Marble/Granite	(Dry) Cotton
Concrete/Brick	(Dry)Wood
Sand/Clay	Plastic

Figure 6.5. Examples of conductors and insulators.

In a battery one terminal is positively charged, the other negatively charged, and electricity always flows from one to the other in the same direction. DC is simple and works well for batteries, but except for very specific applications, it does not travel well over transmission lines because it loses too much of its energy to resistance over most distances. However, much of our electronic equipment needs to convert AC current to DC in order to run. Many of these do so via plug-in **adapters**, less than affectionately referred to as **wall warts**. These contain **transformers** enabling this conversion to occur between the outlet and device.

Alternating current (AC) is the choice for power distribution lines as it can travel well over long distances and is more cost effective to transform to and from higher and lower power levels. In alternating current the **polarity** of the (positive and negative) charge alternates many times per second and with it the direction of flow as well. In the United States, electricity alternates polarity 120 times per second (60 full **cycles per second (cps)** or 60 Hz). Alternating current is the form we use to reproduce the signal in the P.A.; it is a convenient way to reproduce sound in the form of a signal because as it moves back and forth, its shape is analogous to a sine wave. This form gives AC one extra quality that DC lacks—AC has phase as well as magnitude. Therefore the force which is called resistance when discussing DC also gains an extra dimension of phase with AC and is called **impedance**.

Simple Circuits

The electricity we use is contained in **circuits**, and all **electronics** has one or more circuits, many of which are very complicated and contain many components but all share a few basic properties A circuit is a closed **path** (typically using insulated copper wire) for current to flow along. We use circuits to contain and route electricity so we can manipulate and control it safely. The simplest type of circuit contains terminal points enabling the circuit to be connected to a **source** of voltage and electric current, like a battery or power outlet, with a conductive **wire** running from the **hot** (positive) terminal to the **load** (the circuit to be powered, inside the appliance plugged in), and a wire that is grounded and labeled **neutral**, returning from the load back to the negative terminal of the power source to create a **closed loop,** completing the circuit (a requirement for circuits to work). There is also a *switch* designed to *open* or *close* the circuit's path, allowing the free flow of current when closed, and creating a break in the path of the flow when open. The load or appliance we wish to supply power to along that circuit will function only when the circuit is closed and complete.

Going with the Flow

Another important thing to note in relation to circuits is flow direction, because there is more than one accepted method for assessing this. One way that voltage is defined is as a carrier of **potential energy**, because voltage is *always expressed as a difference between two charges, one positive and one negative.* If you recall that according to physics the outer charge of the negatively charged atoms in current would only flow toward a positive charge of the voltage, then you can easily picture the actual direction of this flow as portrayed by **electron flow theory**, where the direction of electrons flows *from the negative charge toward the positive.* On the other hand, if you recall that the dB

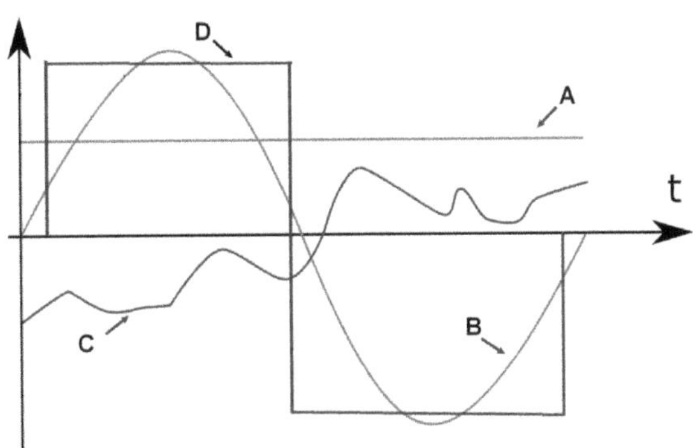

Figure 6.6. Different kinds of current: (A) direct, (B) alternating, (C) random, (D) square.

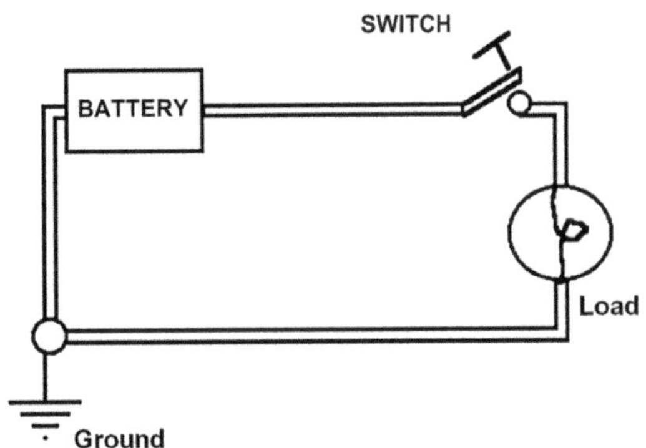

Figure 6.7. A simple circuit.

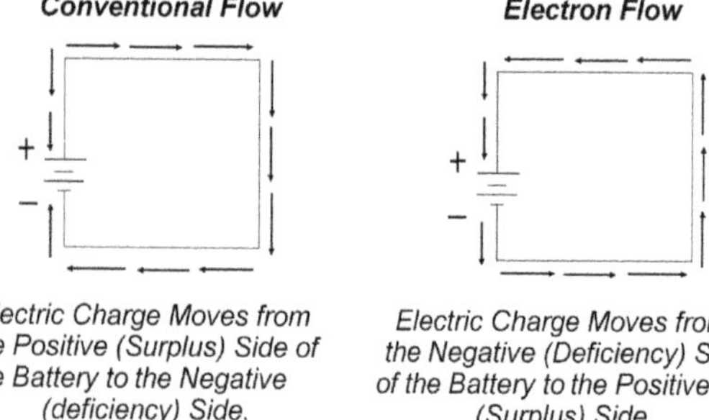

Figure 6.8. Two theories of flow.

for volt is a measure of pressure, it's easy to think of the potential energy as areas of pressure differential instead. In this case, pressure in a contained system is static but as soon as an area of higher pressure is exposed to an area of lesser pressure the gradient will move the matter along a path moving from the area of higher pressure to lower pressure, meaning the different voltages will start moving the current along a path from the positive charge toward the negative, as portrayed by the *conventional flow* theory. Both flows are used and in any specific schematic which one used has no effect and either one is valid, so long as that choice is consistently used. However, electron flow is most commonly used in science texts, while engineering schematics tend to use conventional flow.

Circuit Load Wiring: In Series

Most circuits are more complex, allowing more than one load to be connected and powered at a time. These loads will generally be wired in one of two ways: in series or in parallel, with each method having certain advantages and disadvantages not shared by the other.

In-series circuits are wired so that loads are connected to the circuit along the main path of the circuit requiring the current to pass through the first load to reach the second one and so on, all the way along the series before it can make its return trip to the terminal. This type of wiring is characterized by a few recognizable features:

- In an in-series circuit, all components are connected end to end, forming a single path for electrons to flow along.
- Components in a series circuit share the same current.
- Total resistance in a series circuit is equal to the sum of the individual resistor resistances.

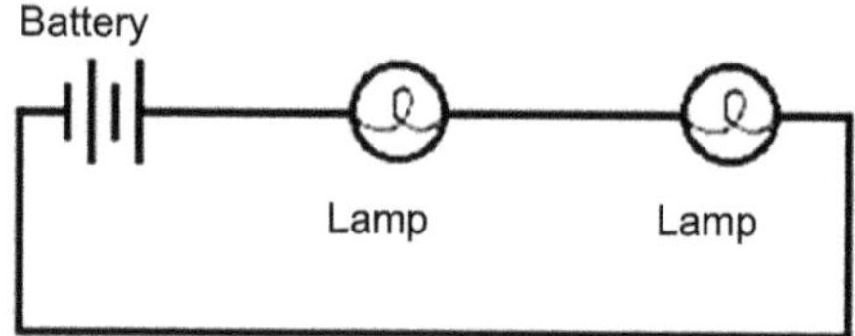

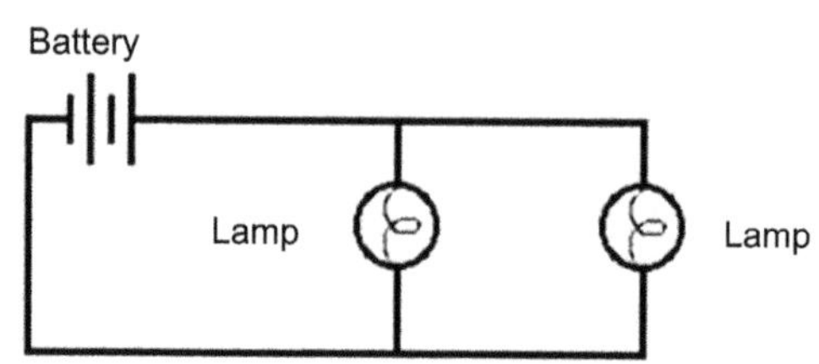

Figure 6.9. In-parallel and in-series circuit wiring.

- Total voltage in a series circuit is equal to the sum of the individual voltage drops (which will equal the voltage applied to the circuit to push current).
- Because current must pass through all the loads in the series in order to complete the circuit, if one load anywhere along the series fails to function and goes out, it will prevent the current from completing the circuit, and the circuit will cease functioning.

Circuit Load Wiring: In Parallel

In-parallel circuits are wired so that each load is connected to the source along the center of the circuit loop in a number of branches that each carry one load and run parallel to each other so the current can access all paths simultaneously:

- Each load has access to the full voltage of the circuit independent of the others such that if one goes out, the rest are not affected.
- Components in a parallel circuit share the same voltage.
- Total resistance in a parallel circuit is less than any of the individual resistances. (Same amount of flow with more places to go, less resistance at each compared with trying to move whole flow through.)
- Total current in a parallel circuit is equal to the sum of the individual branch currents.

Unlike voltage, current, and resistance, the result for total power versus individual power is additive (total power will equal the sum of power applied to each load) and that remains the same for any configuration of circuit: series, parallel, series/parallel, or otherwise.

Power is a measure of rate of work, so following the Law of Conservation of Energy, in a closed system, the power dissipated must equal the total power applied by the source, and circuit configuration won't change power calculations. Most circuits are a variation of these two types of circuit, or a combination that includes features from both.

Made to Be Broken

Circuit breakers and fuses are safety devices designed to prevent current flow from reaching the rest of the circuit when conditions exceed acceptable parameters. Circuit breakers of the kind typically found in buildings and located in the main electrical distribution sit along the circuit paths between the power service supplying the current and the circuit's outlets and/or access points. Fuses and small circuit breakers of the type typically found in sound equipment are *between the plug and the load* internally.

Circuit breakers and **fuses** are safety precautions included in every quality circuit to protect against overload or short-circuit situation. A circuit breaker is a switch that protects the circuit and electronics plugged into it by interrupting the flow of electrons as soon as it detects a fault in the system. Fuses do much the same thing, except they can only blow once and then need to be replaced, while circuit breakers trip a switch to break the circuit continuity and only need to be reset. Both are designed to blow or trip whenever they detect either too much current or too much heat.

Some other basic information about electricity and electrical safety will be touched upon in other chapters, as it's a topic too broad to limit to one chapter. In the meantime check the links for everything electric on the web, including the link to the excellent electrical terms glossary if you want to brush up on basic terms.

Digital Audio

Another area of science the audio engineer will find increasingly important to learn is some basic computer science, at least the principals required to understand *digital technology* and how it works. *Analog technology* will continue to be prominent for the foreseeable future, and will always play an important role in live audio; however, digital audio equipment will be equally common before long, and more people and processes are going digital with each passing season. In order to understand digital technology, one first needs to understand the fundamental *difference between digital and analog technology is in how they reproduce the data,* which in our case is sound, in their respective systems. Both technologies are named for their particular way of handling the data. Analogue audio makes an *analog copy* that reproduces the original in a representative form. While digital audio *encodes* information from the original and stores this data as a *symbolic representation* that looks nothing like the original sound.

In analog equipment the sound is reproduced as an electric signal that is an *analog version* that is almost an exact copy of the original sound. In its original form sound is *organic*, and when it is reproduced as a signal in our P.A., this signal is the sound's analog, it is not exactly the same, because the *fluctuating nature of AC current makes the signal analogous to the original sine wave, with air-pressure fluctuations represented by fluctuating voltage*, but aside from this difference, the signal's form reproduces the original sound exactly. Features common to all analog copies of sound include:

- they are *continuous*, like the original;
- they are measurable for value throughout this continuity, and at each point small changes in waveform will have coinciding changes in values that may be small or large;
- there are effectively an *infinite number of possible air-pressure values* for any signal.

Digital technology now allows us to convert sound to digital data that is much smaller than analog data. When the information to be digitally reproduced already comprises discrete units of symbolic information (like alphanumeric text), the task is merely one of encoding, the act of converting information into a symbolic form for transmission and storage. When the information is analog like sound the information must undergo a more extensive process known as *analog-to digital conversion,* breaking up the continuous form into chunks of data so they can be stored as discrete packets of information instead of a long, continuous waveform:

- In digital technologies all information can be represented in the digital language of binary which *encodes* all data using only 1s and 0s.
- Computers represent sounds as sequences of discrete data, setting pressure values and points in time for each little bit, and changes in the signal form won't change the data.
- In this form there is a finite number of pressure values and a finite number of points in time; to digitize is to set a limit on the potential for infinitely smaller divisions by turning the continuous into a set number of slices, albeit an immense number of very small ones.

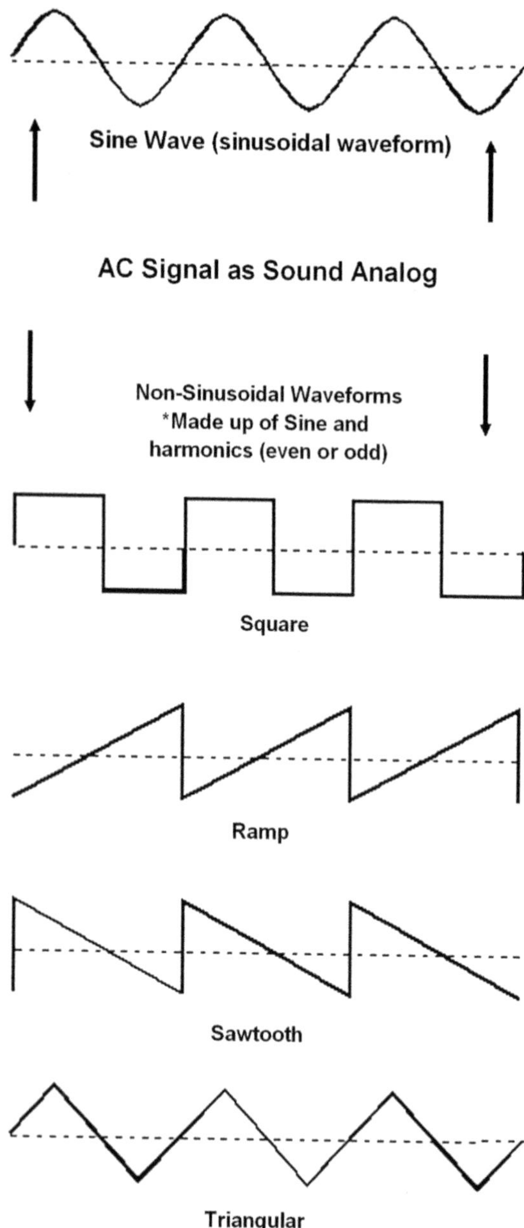

Figure 6.10. AC analogue wave can take a number of forms while reproducing the original soundwave.

Binary Bits

To understand digital audio, you need to know a little about how binary numbers work first. To do that let's start by looking at the number system already most familiar to most of us, decimal numbers. Our decimal system is called *base 10*, meaning that our base number "place" can express 10 values (0–9), and as we count up, each new number "place" we add to enable us to keep counting, increases the value of our count by a power of 10. For example, in a count from 0 to 100, to count higher than 9, we need an additional "place," which lets us increase 9 x 10 and count as far as 99 before we need to add another, and this new number place takes us tenfold again, from 99 x 10 all the way up to 999. Each "place" of a base-10 number (1s, 10s, and 100s) represents a power of 10; by adding one "place" to make room for another series of digits, the value is 10 times the last place.

Binary numbers are modeled after circuits, which we just learned function by means of switches that are either *on* or off. Instead of switches, we use the numbers 1 and 0 to represent each of these two states. Similar to our base-10 system being evident in the very name decimal (from the Greek word for 10, "deca"), it shouldn't take more than two guesses to predict the kind of numerical system binary uses: a base-2 system, with each "place" representing the power of 2 (as opposed to the decimal system's power of 10). The "places" in binary counting then are 2 (to 0), 2 (to 1st), 2 (to 2nd), 2 (to 3rd), 2 (to 4th), or (1s 2s, 4s, 8s, 16s, 32s). However, note that the above example is written for ease of reading, binary counts "up" *from right to left*, as we will show in the next example

Direct your attention to the example of binary numbers and decimal equivalents in Figure 6.11a, particularly the number "9." If you look at the same binary number in Figure 6.11b, and the process for turning it into the decimal again you can see how we can translate the binary back to decimal by simply following their number places from right to left and inserting the decimal equivalent for the exponent value of that "place" in the row below when a 1 shows us "On." Add those up and, as you can see, it is 9 just as the chart said it was.

Binary Bytes—that's a Lot of Bits!

The single *binary digit, called a bit,* is short for "**bi**nary digi**t**," and is represented with the lower case kb. Computers use a standardized number of bits in most circuits, which are usually grouped 8-bit values, or multiples of 8-bit values, called *bytes.* CD audio uses two bytes or 16 bits to encode music. By finding the power of 2 for the number of bits being used, you can find the maximum number of values available, as indicated in Figures 6.12a and 6.12b.

0 = 0	4 = 100	8 = 1000	12 = 1100
1 = 1	5 = 101	9 = 1001	13 = 1101
2 = 10	6 = 110	10 = 1010	14 = 1110
3 = 11	7 = 111	11 = 1011	15 = 1111

Figure 6.11a. Counting in binary.

When dealing with large numbers of bytes, the abbreviations Kilobyte, Megabyte, and Gigabyte are used as bytes increase by an additional power of ten. A couple more binary terms you are likely to run into in working with binary numbers:

- *MSB*: the left-most bit of a binary number is called the *most significant bit* (MSB).
- *LSB*: the right-most bit of a binary number is called the *least significant bit* (LSB).

Analog-to-Digital Conversion

Analog sound can be recorded and digitized using an analog-to-digital converter (ADC). which creates a digital copy by taking data readings along two dimensions:

- X: time (this is called sampling);
- Y: amplitude (this is called quantization).

(Frequencies don't need to be captured as data as they are naturally recreated at playback of the sample amplitudes at the specified rate.)

Digital Dimension: Time

Sampling reduces a continuous signal (wave) to a discrete signal (bits and bytes). This can be done in a couple of ways. Either a digital circuit takes *samples* at regular intervals of the instantaneous amplitudes of sound in real time, or digital synthesis software called a *sampler* selects representative points of data at uniform intervals. In both cases each sample is a measurement of the instantaneous amplitude of the original signal.

Digital Dimension: Amplitude

Samples are then *quantized* or associated with representative amplitudes to produce a representation of an analog signal. The *fidelity* to the original signal (how true to life the copy is) depends on the sampling rate and bit depth. *Sampling rate* is how often samples were taken. *Bit depth* indicates the number of bits (the amount of information) recorded for each sample as it is quantized. In both cases the higher the number the more original data goes into making the copy, and the better the sound of the final recording.

Sampling Rate

Sampling rate (or *sampling frequency*) refers to the frequency at which samples are taken per second; the greater the frequency the greater the number of samples taken, so higher sampling rates use more of the original data to create the copy). The two balancing factors that typically must be weighed when deciding upon a sampling rate have been quality of the recording and the storage space required.

- Higher sampling rate preserves sound quality.
- Lower sampling rate saves disk space (which is no longer much of an issue).

Aside from these two considerations, there is a limit to how low our sampling rate can go in audio because of a phenomenon known as the Nyquist Theorem where the highest-frequency component (including partials and overtones) that can be captured with a given sampling rate is one-half that sampling rate. This frequency threshhold is called the *critical frequency* or

Binary Base 2 "places" ←	2^3	2^2	2^1	2^0	*Binary Counts Right to Left* ←
Equivilant Decimal Value →	*8's*	*4's*	*2's*	*1's*	
4- Bit Binary Number —	*1*	*0*	*0*	*1*	*Solve for On (1)* *Skip Off (0)*
Translate to Decimal →	*8+*	*-0-*	*-0-*	*1+*	*= 9 (Decimal)*

Figure 6.11b. Converting binary to decimal values.

(a)

Bit Depth Value is bits squared (8^2 = 256)	Binary System Byte (8 bits)							
	8	7	6	5	4	3	2	1
Binary Number (= to 77 in Decimal)	0	1	0	0	1	1	0	1
Base 2 Places Powers of 2	2^7	2^6	2^5	2^4	2^3	2^2	2^1	2^0
Bit Weight (In Decimal)	128	64	32	16	8	4	2	1

(b)

Binary Digit	Power of 2	# of Place Values
8 bits (1 byte)	2^8	256
16 bits (2 bytes)	2^{16}	65,536
24 bits (3 bytes)	2^{24}	16,777, 216
32 bits (4 bytes)	2^{32}	4,294, 967, 296

Figures 6.12a and 6.12b. Binary bytes and bit depth.

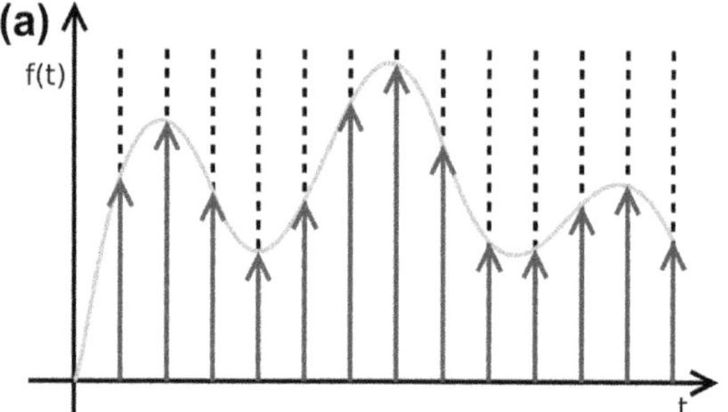

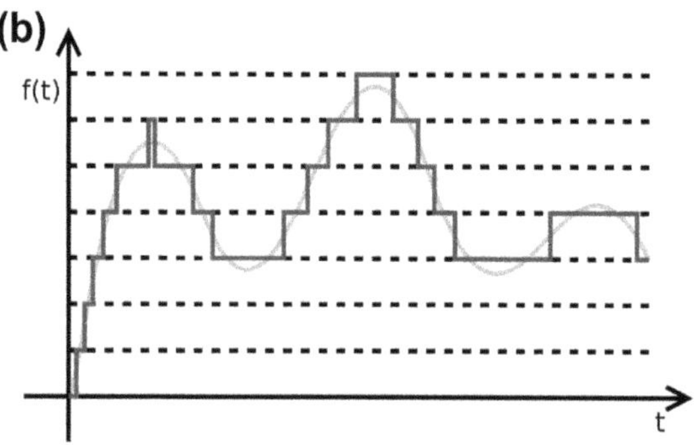

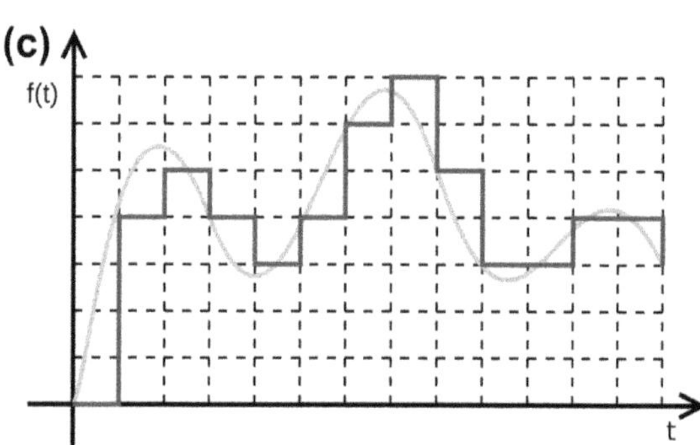

Figures 6.13a, 6.13b, 6.13c. Sampling, quantizing, and the two dimensions together in the form of sampled square wave.

Nyquist frequency. To ensure we capture the entire bandwidth of frequencies the human ear can perceive, an adequate audio sampling rate has to be *at least* twice as much as any frequency components in the signal that you'd like to capture. It is important to note that sample frequency is not the same as sound frequency, even though we must base sample rate on it.

Aliasing

As a result of the Nyquist Theorem a system sampling a waveform at a sampling rate of 20,000 kHz cannot reproduce frequencies above 10,000 Hz. When a frequency above the Nyquist frequency is sampled and played back, the frequencies don't simply disappear. Instead recorded frequencies falling above the Nyquist frequency result in an audio distortion called *aliasing* (also called fold over or biasing). These distorted frequencies show up in the audio at the same distance below the Nyquist frequency as the distance the original frequencies were above it, at the original amplitudes. With a sampling rate of 20,000 kHz a frequency of 12,000 kHz will alias at 8,000 kHz, 2000 kHz below the 10,000 kHz Nyquist frequency. Note that particularly nasty sounds can happen when a sampled frequency is exactly the same as the Nyquist frequency: a zero amplitude signal is produced. So to be sure to capture the sound we want, the sampling rate should be greater than twice the bandwidth of the highest frequency we wish to reproduce.

- To ensure the highest fidelity recording, the sampling rate should be somewhat over twice the critical frequency. Since the optimal audio bandwidth we wish to recreate is 0–20,000 Hz, a good rule of thumb is to record 16 Bit CD quality audio, which is sampled at least at 44100 Hz.

Anti-Aliasing

The two solutions to the problem of aliasing are both implemented by current digital audio systems and are together known as anti-aliasing. Anti-aliasing addresses the distortion caused by frequencies above the Nyquist frequency by:

- band-limiting the frequencies allowed to enter the system (low pass);
- recording at a higher sampling rate, thereby increasing the frequency range;

An *anti-aliasing filter* is placed in front of an analog-to-digital converter to prevent frequencies above the Nyquist frequency from ever being sampled. Since there is no true "brick wall" filter that completely eliminates unwanted frequencies without having any effect on those in the "legal" range, there is still a roll-off that attenuates frequencies closest to the cutoff. How much is attenuated is determined somewhat by the sampling rate. The higher the sampling rate used the less the impact of the roll off has because more of the roll off will fall above the audible range. Advances in computer processing speeds and increased storage speed and size, have made it possible to reach sampling rates higher than 44.1 kHz (96 kHz, 192 kHz).

Quantization and Bit Depth

Every sample has a range of pressure values associated with it determined by the number (0s and 1s) of bits allocated to each sample, also called *bit depth* or *bit resolution*. Each additional bit doubles the number of values available (so multiply 2 by the same power as the number of bits to get bit depth). As with sampling rate, there is a minimum bit depth we can use and maintain fidelity to the original signal. In order to avoid distortion and keep the dynamic range of the reproduction in line with the original there is also a maximum bit depth you don't want to exceed.

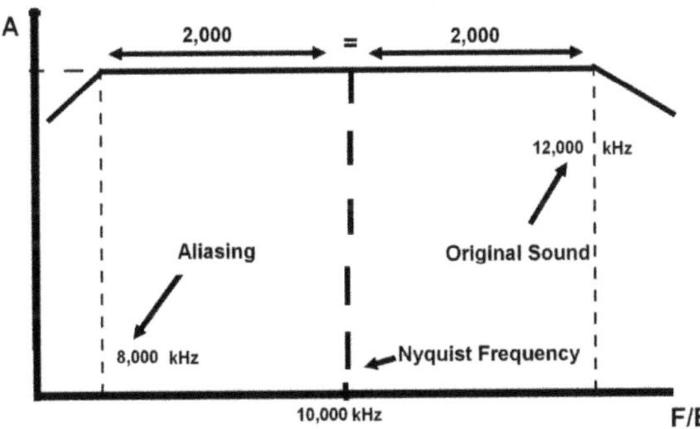

Figure 6.14. Nyquist frequency.

Bit Depth Too Low

Just as sample rate affects frequency response of a digital recording, sample size or bit depth affects dynamic range. When the sample size is too small given the same sampling rate, we lose information and errors like *quantization noise* (a form of distortion) can creep in. When a sample is quantized, the instantaneous snapshot of its analog amplitude has to be rounded off to the nearest available digital value, in a process called *approximation*. The smaller the number of bits used per sample, the greater the difference between the analog value and the digital value. This is called the approximation or *quantizing error*, and the greater the magnitude of approximation errors, the greater the level of digital or quantizing noise produced.

The solution to reducing *digital noise* is to use larger sample word sizes (greater bit depth). Greater bit depth not only reduces the magnitude of data lost through quantizing errors, it also corresponds to the dynamic range of the system, raising the signal-to-noise ratio. Signal-to-noise ratio compares the signal level to the amount of unwanted noise that has crept into the signal and is expressed in decibels. The abbreviation S/N Ratio is commonly used to represent the term, and a larger number means a better ratio. For example, if your P.A. has a signal-to-noise ratio of 100 dB, it produces an audio signal that is 100 dB higher than the level of noise created by the system. Therefore a 100 dB signal-to-noise noise ratio contains less signal noise than a system with only 85 dB S/N ratios. A general rule of thumb is an added 6 dB of dynamic range for every additional bit used per sample.

The Digital Decibel (dBFS—dB Full Scale)

Digital signal levels are measured with dBFS, with dynamic range beginning at a variable noise-floor value (according to bits) and ending in all cases at the top of its range. This full scale ceiling is reached at 0 dBFS, and represents the highest possible level in digital gear. With the top a set boundary, additional dynamics are added at the bottom end, so all other measurements expressed in terms of dBFS will always be less than 0 dB (expressed in negative numbers). 0 dBFS is the same as the binary number for 65,536, which is the maximum bit depth at the maximum sample frequency that fits in 16 bits (and the fact that every bit is used is clear as it's the only 16-bit number with all digits switched on, i.e., all reading 1 —which translates to "not a single bit of byte left"). The sample immediately before running out of room at the top of the scale is 1111 1111 1111 1110, with the last zero left sitting in the spot with the least value, a single bit. On the other end of the scale, the inverse pattern of 1s and 0s shows the lowest possible sample and bit depth for digital audio is 0000 0000 0000 0001, with one single bit being the only one turned on.

The difference between the two puts the floor at −96 dBFS, and as long as the operator doesn't use the final bit, they can make use of most of that dynamic range. This is because digital systems do not need to have the "headroom" of analog systems, which have a gradation of distortion at the top before the sound is unacceptable, digital sound is clean all the way up till the last bit of data capacity is used up. At the same time, it's wisest to leave a few dB between the system volume and that last byte—skate to close to the limit and all it takes is a voltage peak to go a fraction over the limit and immediately hit maximum distortion (which sounds atrocious).

While a 96 dB dynamic range seems small if one tries to compare it 1 to 1 as being the same thing as 96 dB SBL, which would not be loud enough for most music events, it's not as low as it seems; the actual SPL capacity at the top intensity is similar to the top range in analog systems. For example for live applications, the United Kingdom and most European countries specify their meters can be calibrated +18 dBu at 0 dBFS and that's generally the most conservative sound output. In Japan, France, Germany, and some other countries, converters may be calibrated for +22 dBu at 0 dBFS, while in the US most installations will read +24 dBu above 0 dBFS. To clarify in case there is any confusion, because the top of 16-bit digital dynamic range is 96 dB at 0dBFS, analog meters would read at a VU equivalent of at least 96 dB SPL at 0 dBFS, and since most P.A.s have a noise floor starting well above 0, that means the SPL would be more likely to be closer to 100−106 dB SPL. Then the +24 dBu spoken of is in addition to that, making top digital volume equivalent to 120−130 dB SPL, which is as loud as any mix should be.

Though the decibel levels match up fairly well at the loudest levels, one reason why 96 dB dynamic range is enough is that the operator can choose where to shave off to if needed, to account for the full 120 dB range of the audience's ears, making a little room to stretch 96 dB in whichever direction is best make it fit. So if you need it loud, then you mix so that any difference is not shaved off the decibels at the top of the intensity scale but pulled up along with raising the minimum decibel floor, reducing the full range at the bottom instead (a little added system noise at low dB SPL which will already be a little noisy in many cases anyway due to quantization errors, will hardly be noticed anywhere where you'll be mixing over 120 dB SPL). In some cases keeping noise from being heard at the bottom of the range while also trying to get every decibel at the top might be too much of a juggling task to be worth it, and explains why many studio engineers record at 20 or 24 bits, but in most live situations if you need to play loud the ambient noise of the audience will be more than enough to cover a little noise at the bottom, and if you are mixing somewhere you need to keep the sound clean at low decibels you'll probably not be mixing loud enough to have to get every SPL of intensity at the top.

Studio engineers will want to use current CPU capabilities to convert sound into 20-bit or more digital audio, because recording at 16 bits can bring the floor up enough to make it tricky to mix a good sound in a quiet studio environment where system noise could be noticeable at lower volumes. Since computing and storage advances mean engineers can now easily record at 20 bits, which has a much more cushioned 120 dB range, and then use a process called *dithering* to reduce the file to a size closer to the 16 bit we are used to (and we don't need to explain dithering except to say it is also not enough of a strain on modern processing or memory to have any reason to need to avoid using it to reduce a file recorded at a higher bit rate). Choosing this path specially makes sense for studio engineers since recording at higher resolution followed by dithering can help clean up all the other unrelated low-amplitude problems that can occur in the A to D process at the lower bit depth.

For most live sound applications 16-bit conversion is good enough; though 20-bit is just as easy, and which one is chosen doesn't matter as much for live mixing; it's really just a matter of gear capability and personal preference.

Pulse Code Modulation

The method of sampling and quantizing audio for encoding in digital form described above is called pulse code modulation or *PCM* for short. The technique was first developed for transmitting digital signals over analog channels. Variants are based on different mathematical techniques for quantization, including linear, logarithmic, and adaptive. One common variant is the encoding used for audio bit streams of the type that are used by web applications for streaming audio.

Most variants, like Linear PCM, are uncompressed formats, though compressed variants are widely used for low-bandwidth applications. Used for transmission, binary data is transmitted as a series of fluctuations in electrical amplitude (the origin of the name *pulse*). First, an electrical value for 0 and a second for 1 is chosen. The data stream is then timed to a clock so that sequences of the 0s or 1s can be determined.

Digital Audio File Formats

The digital audio files created by digital encoding processes are collections of samples organized in a standard form to be transferred to other computers, shared on the Internet, stored on computer drives, or played back in real time. They are different from audio CD tracks, which mostly contain only the raw sample data which must be *ripped* to an audio file format to be usable by a computer application. A standard 16-bit, 44.1 K stereo file eats up about 10 megs of disk space per minute of sound. Audio files come in a variety of types, and differ in their bit depth, compression scheme, sampling rate, and the amount of non-sample information stored in an area called the header. Three familiar encoded formats are:

- aif or AIFF (Audio Interchange File Format): A PCM variant that travels well between almost all computers and software, and includes header information like file name, sampling rate, and the number of bytes. Also capable of 24-bit and 32-bit resolution.
- .wav or Microsoft WAVE: (A PCM format designed for PCs and Windows, but now compatible with most audio programs, Mac or PC). Similar to AIFF for bit-depth and sample rates.
- *.mp3* (MPEG I-audio layer 3 compression): The most commonly used audio file format, but not the best choice for recording, the MP3 can't be beaten for its ratio of size to perceivable quality of reproduction. With the proper codecs, compression rates of up to 24 times can be achieved with near (but not quite) CD quality. The most convenient aspect of the MP3 is its ability to be downloaded and then loaded into the flash memory of MP3 players. It can also be streamed to MP3 client software, and is recognized by most Web browser audio helper applications. Files are encoded at certain bit-rates for target download speeds, like 128 kbps; however, very good quality can't be attained below160 kbps.

While live audio engineers don't usually work with the same number of files and formats as their counterparts in broadcast and studio sound, with the increasing number of digital systems finding their way into live applications, the recent availability of high quality, affordable handheld digital recording units, and the low cost of storage, it is becoming a lot more common for sound companies to offer recordings as part of their rental package, and more and more bands now make it a practice to record all or most of their shows.

rate	recording method
32K	older DATs, voice quality
44.1K	CD, DAT, digital recording software/hardware
48K	DAT, DVD-Video, digital recording software/hardware
96K	digital recording software/hardware
192K	digital recording software/hardware

Figure 6.15. Rate and recording methods.

The Digital Advantage

As is clear in the brief look at sampling and quantization above, digitalization is not an error-free process; errors can occur, and are most common at the point of analog data conversion (and to a lesser degree errors also occur converting digital data back to analog despite many recent advances in quality of reproduction, accuracy, and speed). There are also still many who argue analog sounds better, and is more user friendly. A number of audio professionals proficient at working with digital will still choose analog when given the option. So with these facts on top of the fact that converting data is not always easy, why do we use digital more and more often?

Because once the work of digitizing is done, the data is much more stable and less prone to degrading in storage, for one. A look at the main benefits of digital over analog shows us that:

- Digital signals do not get corrupted by noise during the process of transmission.
- Digital signals don't degrade over time or with copying.
- Digital signals typically can be compressed without unacceptable loss of perceivable quality, therefore they use less bandwidth, and you can transfer more data at a lesser cost.
- Digital can be encrypted so that only the intended receiver can decode it.
- Digital storage allows us to archive a wide variety of information in easily searchable and retrievable formats, with an ease that is difficult, even impossible, to manage with the same data in analog form.

- Techniques such as parity and checksum error detection allow us to further guard against data corruption in ways that analog does not lend itself to.

As with electricity, this will not be the last word on digital audio topics. While they deserved a chapter of their own to ensure a proper introduction, both are so essential to the live audio learner and pro that they will be relevant topics of discussion over and over again; neither can be kept completely segregated from other topics. However, now a change of gears is in order, as we move on to a chapter with a very different focus, and look at the art of live sound.

Chapter 7

Beyond Ears that Hear

The Art of Live Sound

Before we can look at the art of live sound, we should remind ourselves of the purpose of live audio from Chapter 1: *live sound is about bringing people together and facilitating communication.* There are only two essentials here, people and a message. Though to be more specific—people to convey the message, people to hear it, people to facilitate it, and one or more messages. (If you want to give some credit to the venue owners you could also note that a warm dry place big enough for all the people to come together is also needed, and using every tool available that will help achieve the task.)

Invisible Collaboration: The Art of Facilitation

The art of live sound is what happens when all the live engineer's technical, scientific, and musical knowledge combine with an ear attuned to the nuances, or subtle distinctions and variations of sound, combine together and become something that is greater than the sum of its parts, the art of live sound. This artistry is expressed in the choices that an engineer makes as he begins with the sound sculpted by the performers, and without changing the form of it, uses the mixer to control aspects of the sound such as frequency mix and power level, and with a series of small carefully chosen adjustments, smooths out the rough areas in the sound while adding definition and weight to make it stronger, and finally polishing the whole thing until any dull areas disappear and the piece shines with a balanced, even, natural glow. As implied by one name used for the art, when applied together on the same sound all these small adjustments and improvements reinforce the sound, that is to say they support and make it better without changing the fundamental nature of it. Even though we can name each of the subject areas and skills that contribute to the art, what happens when they combine isn't always easy to put a finger on, and defining what makes the best mixes is a bit like trying to explain magic even for those who practice it.

Part of that is because a great mix is very subtle. The sound engineer doesn't start with a blank canvas or formless lump of clay to produce their art as the performer does. Instead the nature of their art is always to help put the finishing touches and the final stamp on what the performer has produced out of nothing, without getting in the way of the performer's vision. Because the sound engineer's art is by nature collaborative, as much as each engineer's sound will have their signature on it, perceivable to those who have ears who can pick up on the subtlest patterns, their sound is never static because done right it's always tailored to the performer/performance and will reflect the performer's aesthetic and energy above all.

The FOH engineer takes the performer's piece and makes it the best it can be for that performance, which is determined by the room acoustics and particular combination of equipment in use, but also more ephemeral things like the vision of the performers for that show, as well as the energy of the room flowing between the audience and band and the audience members with each other. While it is no less an art, it is different than the performer's art, because it is meant to be invisible. Even those in the audience who logically know the influence the sound engineers have on the sound shouldn't be thinking of the fact because the music shouldn't call attention to this. Ideally what comes out of the speakers should sound natural, convincing all ears that the sound is flowing directly from the performers to the room, and even when less than ideal there should be no sense that the message from performer to audience is being mediated by the mix engineer very intently doing exactly that. The magic part is that for all this subtlety, if you've ever been exposed to the contrast of sound mixed by a practiced sound pro with a good ear, and the same type of sound plugged directly into the amplifiers and played unmixed the following day, or worse mixed badly by a careless or inexperienced engineer, there's no denying that the subtle influence of a sound engineer on the sound output can make all the difference in the world to how enjoyable the show is for all involved.

The monitor engineer's task is even more collaborative, for their contribution isn't at the end of the creation, but takes the form of a recursive loop that starts at the beginning of the process of creation and becomes a continual back and forth flow between the band and the monitor mixer, and for good or ill, influences the sound at every step of the creation process.

The exchange starts as the sound is created by the performers, and this is in turn parsed and layered into separate tracks for each artist by the monitor engineer who takes care to tailor each mix to the preferences and needs of the artists before sending it back to them with the aim of helping them produce the best sound they can at each moment of the unfolding creative process that is a live performance. So the monitor engineer's direct crafting of each mix is not the end or whole of his/her creative contribution, but just the beginning or first part, and crafted so that each mix can support the hands and voices that craft the sound. In short, monitor mixers works with the sound created by the band, and then returns it to them where ideally it supports the artists so they in turn can craft even better sound.

Like the house engineer, the monitor engineer must also be tuned into the energy of the performers, but because they must be even more closely attentive to the moods and energy of each performer, they focus only there, and the crowd's energy becomes only a concern in how it affects the performers from moment to moment. It is not a stretch to say that, because their focus is primarily on affecting the sound to positively influence the performers, their art also requires nearly as great a depth of understanding about people, an intuitive feel for the moods of those around them, and the ability to read the body language and facial expressions of the performers, as it does all the other skills that go into creating great mix.

The best monitor engineers make it a habit to attend to people's reactions so closely that. even with new bands, by the middle of the show they have already unconsciously noted that the guitarist's mouth turns down when he's losing himself in the mix, the singer cocks her head to the left before she's about to ask for a volume adjustment and will be already making the needed adjustments as the musician signals for them. Sometimes, the monitor engineer will be able to anticipate the exact adjustments the artists require based on non-verbal cues and will make them before the performer has a chance to signal at all, as indicated by a sudden smile breaking out on a performer's face that just held a frown or a confused look, and often accompanied by a quick glance of acknowledgment toward the perceptive sound engineer. Just as often the art involves recognizing when they should not try to interpret cues they pick up and should wait and make no noticeable adjustments until the artist asks for them. If it sounds complicated, that's because it is, and while the person and their influence on the sound are even less visible to the audience than the FOH contribution, there are many who agree that, of the two positions, the difficulty of their task and their contribution to the overall sound is the greater.

Deep Truths about the Art of Live Sound

- *The art of live sound is an intentional and voluntary act.* The only requirement of live sound to succeed is that the intended message is communicated, art is an optional component. If the message is from an artist to their audience, the live sound pros can pass the art along in the form of a decent mix without putting any of themselves in; an acceptable or serviceable mix fulfills their duty, as it is more than what the artist and P.A. can do alone, but the art is only in what the artist contributed. This should be a rare practice. If the engineer is mixing from more than a formula and putting themselves in with their intent, whether that looks like passion or just goodwill it is art; it is not required to be high art or fine art (with the understanding that an artist putting themselves into their art practices and does the tasks required before and after performances so that their best is above serviceable).
- *The art of live sound is collaborative.* While the live engineer may engage in other art forms that are solitary, or "one voice," while practicing the art of live sound one needs, at a minimum, an artist to work with and an audience. Many definitions of art contain a requirement for an audience, contending that performance done alone is self-talk and not communication, but it is especially true of live sound, without both an audience and artist there's no reason for, live sound.
- *The art of live sound is more than the sum of its parts.* This is true of many things, and especially the products of creative tasks.

Advancing Your Art is Worth the Work

While part of the knowledge that contributes to the artistry is the science and tech you can learn from a book, you can't learn to mix well from a tutorial or a book like you can learn to hook up a P.A. *As with any skilled craft or art, the only way to hone your mixing abilities is to mix live sound whenever you can.* However, as we discuss in more detail in the final part of the book, one of the hurdles some live sound students must overcome is finding ways to get more practice time at a live mixer. Unlike students of studio sound, who can mix down pre-recorded tracks if they can't find a band, and who for a while now have been able to use computer simulations of studio equipment to practice on daily at home when they can't get studio time, live engineers still need a live performer or band and a live mixer to further their mixing skills, which is usually hard to arrange for oneself on a daily basis.

If the only way to learn to mix is to practice at a mixing board, then the obvious place to get practice in is through your school or training program, and then later your job. If you recall our warning about being misled by the obvious in Chapter 1, however, you'll remember that it's not always safe to count on things that seem obvious to give you an accurate view of the broader picture. Most schools and live audio training centers put a much greater amount of their instructional focus on the science of sound and the technical skills needed to troubleshoot and run the P.A. than on the art of mixing. Scientific and technical knowledge are the focus because they are the prerequisites needed to be able to mix and it is essential for students to have a firm grounding in these skills first. Not because the ability to mix is not important but because it is pointless knowledge until you can get the P.A. properly set up and keep it running without feeding back every five minutes.

While students do begin to learn to mix as they practice at school, it doesn't really matter what school they are attending; with a skill like mixing, there's only time to begin learning. At the best programs, that beginning is a solid one that lays a lot of ground work; at the ones providing the least hands-on instruction that beginning is barely enough to start to get the lay of the land. But neither can hope to teach students more than a fraction of the craft anyway. So whether your school or training can provide ample practice opportunities or only enough time to be able to say they offer some, their job is not to teach you everything you need to know, or provide all your practice before the last day of class, their job is only to give you what you need to be able to continue honing your abilities with further practice if you take the steps to arrange it. *If you take it as understood that mastering the art of the mix is a process and no one person or organization can provide you with all the practice you need, except yourself, you will be less likely to make the common error of wasting time waiting for someone else to do it for you.*

Good Things Come to Those Who Don't Wait

So before you finish training, whether you get time to mix only once a week or as often as four or more, you would be wise to begin to put some effort into seeking more venues to practice in as soon as possible, especially if you are attending a fast-track training program, which will be over before you know it. Learners who are attending more traditional two or four-year programs should get in all the practice they can as well, since at graduation they need to be that much farther ahead in their skills than if they had attended shorter programs. And it goes without saying that students training online and through workshops should be proactively looking as well.

No matter how you are training, odds are you aren't going to get the kind of time at a mixer required to get the same amount of immersive practice that students in many other media sectors take for granted, being unable to practice at home or after class without having to depend on anyone else as they can. Even if you are happy with the amount of practice you are afforded by your training, you will want to be ready to take responsibility for your own practice time once classes end, but to do that you'll need to start looking before then. The right opportunity may not be in the first place you look, and it's well known that new skills are the quickest to be lost if they aren't used regularly and steadily built upon. Even getting a job or internship right away won't a be good enough excuse to slack on this; you could be waiting anywhere from months to well over a year or more before you may get a chance to do much more than carry or clean a mixer as you pay your dues. However, this doesn't mean that you need to stop mixing.

There are those who do chose to wait until they get a chance to mix again on the job. If they do so because of a conscious choice, such as because they prefer to focus on other parts of life and are not in a hurry to advance their skills on any schedule faster than their own pace, or because they aren't that interested in mixing and intend to advance in their careers by giving priority to the skills needed to become a master technician instead, then it is a reasonable choice. However, it's a shame when the same decision is made by people who would like to keep making steady progress learning their art, and who wait anyway. These students are harmed by the sudden brakes they apply to their further development, because they do so not as a reasoned choice, but from the assumption that this is expected or a lack of awareness of the options available to those who want to seek them out.

When you are told that "You'll have to pay your dues before you'll get a chance to mix much" don't forget to add the missing two words left off of the end on the assumption they are obvious, "at work." You shouldn't expect to mix anytime soon **at work**. After work, you are the boss of you; if you got in the game to be a mix engineer, then you don't want to go a month without practice; and you certainly don't want to lose momentum and find yourself waiting for a year, or when you do start practicing again you may find you have lost much of the ground you gained in school or training. So practice your mixing unless you don't particularly like mixing and would rather take amps apart and rebuild them, or your biggest priority now is some other hobby that requires all your focus, like heavy drinking, chainsaw juggling, or running for office.

I also want to be sure to dispel any romantic notions that sometimes get attached to the tradition of "paying dues." Though a thousand pop-culture references may imply there is a deep wisdom the apprentice can only learn from being purified—refraining from further study and performing menial and frustrating tasks until the master deems them purged of bad lessons and a perfect empty vessel to pour teachings into—your boss is not Mr. Miyagi, and you are not the Karate Kid. The only similarity this beloved trope has with the reality of starting out in live sound is that there are more than enough menial and frustrating tasks to keep an apprentice busy between lessons and they will do a lot of them before it's their turn to do the really fun stuff, not because they need purification but because every single person working above them went through the same thing, and who's going to coil that damn mountain of cables if not the new kid?

Opening Access to Practice

So during training and beyond, you will usually want to find more practice opportunities on your own, and you should start looking for them early enough to have something going by the time you will be depending on it to keep moving forward. This is especially true if you are more interested in working small-venue sound than starting at a sound company, since you will likely need to practice mixing to ensure you will be qualified enough to have a shot. So how do new audio pros find venues to practice their live mixing and further their art once their initial training is over and while they are still looking for or paying their dues at work? There are a couple of long-proven ways to begin mixing on your own and the cost of small mixers and portable P.A.s are more affordable than they've ever been, so if you are serious about a career in live sound there is no reason to ever lose ground or to stop progressing all of your skills.

Volunteer to Mix for Local Organizations

There are often opportunities to mix at local institutions that can use the services of a live engineer regularly, but can only afford to pay for one for the most important applications. Sometimes theater groups that can pay for a pro for their major performances can't budget it for other activities like weekly youth workshops, early practices, or minor productions, and are glad to have any who want to volunteer. Similarly churches that hire a pro, or have one who volunteers for large Sunday services, still need to have a few volunteers to run the sound at all the smaller day-to-day events that use live sound, from choral practice to weekday services. Yet sometimes having a system to practice on is at such a premium that vacant spots can't be found even as a volunteer. However, maintain a friendly and patient attitude and keep checking back. Even volunteers get sick and take days off, make sure yours is the friendly face that was there last expressing an interest, and you are more likely to get the call to fill in when a spot opens up. Eventually, turn over or expansion will leave an opening. Don't be one of the people that give up when they can't get what they want immediately, because with enough friendly persistence, you usually will find opportunities opening up to you. Refer to part 6 for tips on how to be persistent without being pushy.

Buy Your Own Mixer or PA and Mix for Local Bands and Groups

Affordable, Portable Sound Gear

One big advantage held by today's up-and-coming sound pro is access to professional quality compact and affordable gear. Nowadays, you can have the same kind of quality system for a few hundred dollars that not too long ago would've cost thousands. For those looking to buy a quality mixer on a tight budget, the *Yamaha MG102c 10-channel compact analog mixer* can be purchased for around \$100, and the *Mackie ProFX16 16-channel 4-bus digital mixer* with on board effects retails from around \$650, but can be found on sale for as low as \$400–\$500. And for those that need more than just a mixer, decent, affordable portable P.A.s can be had at just about any price point between \$500 and \$1,500 depending on your needs.

If you can't find a regular spot mixing using the method above, you may have a lot easier time finding places to practice mixes using the method below. There are mixers and P.A.s available for every budget, so buying either doesn't need to be as brutal a pillaging of your wallet as most people assume. Nor should you think of it as only an investment to help you find a venue to practice in, since once you build up your skills, a small P.A. is all you need to start your own small sound business renting your services for small events. In the meantime, especially if there is a lot of competition for volunteer positions, this is often the easiest way to ensure you will have all the practice time you need, as there are even more groups who would love a chance to utilize the skills of a sound engineer, but don't have a mixer or even a proper system.

Once you decide what to purchase you can decide where to offer your skills. Naturally, with a small P.A. you could go anywhere, while with only a mixer you'd be looking to meet up with groups that have their own speakers, like local bands. Local bands are your most likely resource, and you'll only need to tap into your local music scene to find prospects. See the chapter on networking for ideas on how to locate and meet local bands and other resources in your local music scene.

While the people most mentioned as being the best prospects to go offer mixing services to are local bands, there are other groups you could make arrangements with instead of or in addition to working with a local band and many reasons you might want to. You may find a band that only practices together one night a week, when you want to fill two. You may want the experience of mixing in different rooms, different audiences, and different performers, or you could choose to work with others because they are gigs where you could mix FOH, while you continue to practice monitor mixing with a band. In the case it takes a while to find a band you want to work with, you might check with these other options in the meantime, or instead. Finally, you may enjoy working with organizations where volunteering sound services allows you to give back to your community.

Groups that might not already have access to a mix engineer with P.A., but who might have use for one, or could be shown how one might be useful to them, include small churches, small independent theater groups, small pubs/coffee houses, local charities, community groups, elderly care homes, and even small private schools. Look in your local phone directory for establishments like these and ones like them, and begin reaching out.

Some may know right away how to utilize your offer to volunteer, and others might need you to show them areas you might volunteer usefully (so be ready with some creative ideas and remember to explain that live sound is even for those throwing small events that don't need much added volume). A small school may have a small school band, choir, or debate club who would be glad for sound at practice each week and for performances. A home or center for seniors may have a barbershop quartet, or hold regular dances, and if you do a little selling about how much fun residents would have if the next dance featured a live P.A. performance by the barbershop quartet *and* offered to ask your contacts at a nearby private school if their little jazz ensemble would be willing to play a set, this could be another chance to mix in your pocket. A charity might only throw two fundraisers a year, but you may convince them to hire a band for the next one, if you offer them your services and P.A. for free. Just make sure to only offer what you are capable of doing and to follow through on your promises!

Don't Be Daunted

If you can't find anybody to work with immediately, don't panic. It does not mean that you will have to

Figures 7.1 and 7.2. These Yamaha and Mackie Mixers are examples of how affordable and compact pro live mixers can be.

wait many months for your employer to allow you to mix, nor does it mean that people don't want to work with you. There will be bad days, but there will be as many good days on average. Try again the following week and keep at it. There are seven billion people on this planet, and while a band is preferred, a solo musician, or even a single person willing to speak into the microphone for a while each week (track down actors and poets) will suffice almost as well for a time, and the latter will show you right from the start how tricky spoken applications can be—something that is often not realized by engineers who have never mixed for speech applications. Remember, we are not asking you to find Shangri-La, just one person with something to say into a microphone. Don't give up; we know you'll find them.

We understand that for people who are naturally shy, or merely lack the confidence they will one day have, reaching out to people confidently is going to be harder than for people who are more outgoing or less self-conscious. So some of you may feel pretty nervous about doing what we suggest in this chapter. Calling people you've never met to make a suggestion, offer something, or make a request is called *cold calling*, a common sales technique. Those who choose to go into business for themselves or to be freelance workers will have to do this from time to time no matter what their business. If you want to find work during those times when it doesn't fall into your lap, you will have to get over your fear of this. We're lucky that we haven't had to overcome this hurdle ourselves, but we have seen people that have overcome it. It also is important that you build your self-confidence and learn to be a little bold to work with the variety of characters you will encounter working in live sound. Please see the links in the networking chapter directory that lead to sites with information and support for people who need a little support as they practice being bold. It's a skill you need, but luckily, it's not something you have to be born with to use; it's a skill you can learn and improve with practice like any other skill.

Paint by Numbers Mixing

As important as mixing live sound regularly is to mastering the art, you could spend years mixing rather badly, several times a week, and never get any better at it, if you don't know how to truly listen to the sounds that you alter, and mixes that you produce. There are sound engineers who have made rather lackluster careers out of the practice, unfortunately. The sound waves in the air, the mixer and P.A, the sound qualities of the venue: all these things are tools and materials that you must understand and become accustomed to working with to advance your skills as a mix engineer. But you could learn every fact there is to know about all of them and still not advance beyond producing mediocre mixes. To make the most of your time at the mixer, and keep building your skills, you need to develop your most important tool, and move beyond the "paint by numbers" style of mixing that all live engineers use when learning to mix.

When we refer to *"paint by numbers" mixing*, we are speaking of the mixing that occurs by following a set of memorized "rules of thumb" and mixing tips gleaned from reference material alone rather than mixes based on an understanding of sound and frequency. If you are not already familiar with them, you will run into them often as you study live audio, or any audio topics; they include tips such as, "cut guitar frequencies X MHz to reduce clangy quality in the mix" and "to create an emulated stereo effect, run a copy of the lead vocal signal through reverb and route the signal to the right speakers, while routing the original signal to the left speakers after adding a one millisecond delay." Most are frequency related, but there are rules of thumb about any control or procedure that can influence the sound quality, position, or level.

Such memorized rules of thumb can be used by learners to guide their initial mixes and help them to achieve a baseline acceptable sound quality when they haven't been mixing long enough to know by ear what should be adjusted or to predict before making an adjustment roughly what effect that adjustment will have on the sound. However, for this purpose it is more effective if a limited set identified as useful with the broadest range of sounds and situations is compiled by an experienced sound professional and identified as ideal for live audio learners to use while mixing their first performances. When collected by the live sound learner from a variety of online and text sources, they provide only a little more scaffolding than random experimentation unless they are cautiously applied.

All engineers in training use it to some extent early on to bridge the gap until their mixing skills catch up with the rest of their skills and it's a normal part of the learning process. Without these formulas to guide them through their first mixes early learners would be essentially tasked with mixing in the dark without even a dim LED to see by, as even during some of their earliest practices live audio learners often have the sound needs of others to consider: the performers practicing of putting on a show, and quite often an audience, who even at student run productions will require that sound not fall below a baseline acceptable standard. The pressure to perform as expected without such a cheat sheet would be more stressful than it is fair for learners to have to undergo just to practice.

Like any tool used to support learners, there is nothing inherently wrong with these rules of thumb. We include a page of them ourselves in the chapters on mixing. But early on, the rules should be used as temporary support to ensure their mixes have some stability as learners go through the process of *testing* each rule alone and in various combinations *paying attention* to the results and slowly starting to build a more permanent foundation based on the *understanding*, acquired with experimentation and experience, of exactly when and why one rule is preferable to another. After many, many mixes, the engineer should have taken complete control of the mix, and the rules can be recognized for what they really are: conditional ways to convey actions taken in a particular mix.

Recipes and Formulas: Arrested Development

The problem arises when engineers become too reliant on the rules and recipes, either getting stuck halfway through a mix, or serving up the same mixes again and again (hard-rock mix, jazzy mix, acoustic mix, pop mix, repeat) because they kept to a limited set of mix recipes. Some chronic abusers don't even use two or three recipes but lock themselves into using a one size fits all mix formula to produce mixes ranging from bland to bone-wrenchingly painful on the ears. If producing a great mix were as simple a task as following a set of rules then top-notch engineers would be a dime a dozen and there wouldn't be any bad mixes, and very soon live engineers would be out of work because an algorithm would soon be written allowing the task to be automated. Luckily for the future job prospects of live sound pros mixing is not a one size fits all activity, and a rule of thumb may or may not sound good depending on how similar or different circumstances are to those spawning that particular rule.

Rules of thumb and mixing tips used this way are the best way to ensure you don't master the art of mixing, but the only way to mix without them is to practice using the most essential tools the mix engineer has, the engineer's ears. Engineer's ears are a bit different than the ones you have now; they are attuned to sound differently than other people's ears and have a few distinct qualities essential for anyone who hopes to learn how to craft excellent sound. Many of you have probably begun the journey to acquiring them, or were born with a lesser amount of certain qualities that define them, but you aren't born with them and they aren't bequeathed on you just for becoming an engineer; in fact there are plenty of engineers that don't have them. To attain a fully functional pair takes intentional practice, and it isn't the kind of thing that happens overnight.

The Engineer's Ears

If you want to further your mixing abilities and be able to take your mixes into the realm of art, you need to move beyond ears that only hear. No chart, recipe, or memorized rule of thumb will help you identify both the sound of the notes associated with each frequency, and the effects on the mix made by cutting or boosting each one. No list of tips telling you how to avoid feedback will be able to prepare you to locate the frequency feeding back in the five seconds it takes for everyone in the club to be done with it and ready to head for the nearest exit. That takes practice and time. If you try and learn it solely while mixing, it will take a lot of practice and time, and then only if you are intentionally concentrating on it; if you aren't paying attention you can mix for years and not learn enough to become a great sound engineer. However, there is a way to make your mix sessions take you much farther on your journey to getting your own pair of engineer's ears, as well as to speed up the process in general: and that way is *ear training*.

Enhancing Your Assets: Ear Training

For those who know the incredibly involved process of ear training for musicians, we are talking about a similar process, but focused on training the ear to be attuned to things, like frequency, that are most useful for engineers. You won't need perfect pitch, and engineers' ears are nothing like the mythical "Golden Ears," so anyone can train to get them; there are no innate abilities or special talents you need. Almost anyone who wants to put in the effort can succeed; all that's required is the desire to be the best live sound engineer you can be, and average or above average hearing.

Ear training for musicians can be an arduous process depending on the methods they use, but for engineers the methods should focus on steady progress not a pre-set timeline. Cramming is counterproductive so there's no need to set aside your technical training and toil over your ears. Learning to hear sound as a sound engineer is a skill that should be developed over time and in conjunction with all your other skills. The activities are easy, and the time you spend ear training should be stress-free and enjoyable. There is no need to hurry because the process can't be rushed anyway. It does take a while, however, so why not get started sooner instead of later? Getting started on it now instead of waiting around will benefit you in three important ways:

- Training your ears as you learn about the P.A. and how to mix will reinforce learning globally. As each branch of study exposes you to information about frequency, you will be able to connect what you are learning in one area to what you are learning in another. Each time you expose yourself to frequencies in ear-training exercises your mind will naturally and without effort review what you know about frequencies from class as you hear those frequencies. In turn when frequencies are discussed during your other studies, instead of having no idea what those frequencies sound like, you mind will naturally begin to recall the sounds in that range, with the range recalled becoming narrower and more accurate over time. In this fashion, both learning processes benefit from learning that occurs in the other, making it easier for you to understand and remember concepts and get the most out of lessons all around.
- Developing your skills enables you to include more than just your training in audio theory and X hours mixing on a school mixing board on your resumé. Being able to claim a year or two of self-guided ear training indicates dedication and discipline and also adds a valuable and verifiable skill, the ability to name that frequency with more or less margin of error. This ability is invaluable for a mix engineer and most take many years to acquire it. Any achieving it early in their career will stand out among the many other live sound newbies who studied their P.A. guide too, but have not yet engaged in ear training. It is not absolutely required to land a position as an audio tech or even as small-venue FOH sound, but having skills that set you apart can make finding a position happen sooner and be more likely to involve a paycheck when it does. Nor is it only useful for landing your first job, but will continue to be a big plus on a resumé for several years, until other accomplishments and job experience surpass it; even then it won't be something you'll need to consider dropping from your resumé until you need room.
- Developing your ear won't magically give you the ability to mix quality sound any more than any other amateur without much experience can. It will mean, however, that as you do slowly advance your skills, your learning curve will be far easier. Ear training is also the key to moving beyond the paint by numbers style of mixing without having to subject audiences to as much of the "Goldilocks" approach to mixing, which is the required intermediate step between the formulaic mixes of the live audio student and the more advanced mixing indicating a live sound professional who knows their craft. Ear training incorporates the experimentation of the "Goldilocks" approach to mixing (i.e. "How about tweaking this frequency?... Nope, that's too soft. OK, how about this one? Nope, losing too much low end ... reset. OK, well maybe if I cut this a little? Oh that's alright, good... OK now, what if I tweak," etc.) into its activities, so less of this needs to occur at the mixer where unexpected results and uneven sound could prove distracting to the performers and audience. Ear training and experimentation at the mixer are the only ways to move beyond dependence on mixing formulas to produce your mixes. Making use of ear training makes the transition occur much faster than learning at the mixer alone, and is also a far more pleasant and stress-free way to learn.
- The ability to know what it is you are hearing in a mix that sounds "off," whether it can be fixed at the board or not, and if so how to get right to the issue, is as much a matter of ear training as knowing your system. Without ears ready and able to tell you exactly where in the mix a problem lies, you are left having to resort to tweaking bit by bit around the entire board. Knowing your frequencies will make ringing out the monitors more efficient and accurate, helping avoid feedback as well.

Because we keenly remember our own time as students, and as educators have spent a bit of time working with them since, we are pretty sure that some readers hit the sentence about many live audio students not engaging in ear training in the bullet points above and mentally applied the tag "elective" to it, knocking it down closer to the bottom of a very crowded list of priorities. Although ear training is not a requirement at many programs this is only because it's not the kind of process that fits well in a short-form training program, as it isn't something you can or should finish in a four-week module or even a three-month class. Many short-form training courses concentrate on technical knowledge and skills that can be completed and tested in under a year, and more often less than six months. Many of these programs don't teach more than the most basic mixing concepts either; for the same reason, it's a skill they can only start you on, but it's obviously not an elective skill for a live sound engineer. Where ear training is incorporated into the curriculum it is at two- and four-year audio programs, as these are the only ones equipped to spread activities out over several quarters.

Ear training is increasingly recommended and its methods even being taught at programs that don't offer it as a class, at least enough to model the methods. Though historically structured ear training was not seen as too important and was only briefly mentioned, or omitted, the same can be said about the topic of hearing protection, which until relatively recently only received a few sentences of coverage in many live sound texts, and just as often was not mentioned at all. We understand the need to juggle priorities when there is so much that you must learn that is absolutely required. So to students who remain on the fence, we only ask that you look over the chapter and see for yourself how easy the process can be, and consider it again when time permits, especially if you observe how much sooner other students are able to mix with confidence.

Before we talk about some basic ear-training resources and activities you can use to begin on your own now, we want to remind you that developing your ear isn't something you cram for. Being consistent and sticking with it is what's important. The "ear" many professionals have developed took them years, or entire careers, to develop. Intentional ear training will speed up that process for you, but it's just a head start, not a finish line, and it won't happen overnight

Perfect Frequency: Engineer Ear Training

In the link directory for the chapter will be several links to websites where you can download free test tones. They will be in wav. format. Please don't convert them because wav. is a much higher-quality audio signal than MP3, which uses a lossy compression format. These test times tones will range from 20 kHz down to 20 Hz in even intervals. Put them on your iPod or burn them on to a CD, noting the track number for each, and just listen to them whenever you have some free time. Make some mixes and quiz yourself. This is only one of the frequency activities included in the chapter, but even though the others seem more elaborate, make sure to follow up on this one. It is very handy to be able to slip in 10 or 15 minutes of ear training on days when you don't have the time to do the more advanced exercises.

Critical Feedback

There are many who advise sitting at your board or a small board at home and creating controlled feedback along the frequency band, noting the frequency and tone as you go along. For those of you with boards at work, and have convenient access to the board, this is a great method and links leading to instructions for the procedure can be found both in the directory supplement online or with a Google search. Most of you will probably not have access to a board in your off time, or enough access to practice regularly. Until you can "name that tone" or identify what register an instrument is playing, print out any visual aids and keep them handy as you mix. In the meantime you can practice on your computer with an open-source program designed for sound engineers called *The Simple Feedback Trainer*. Again, it has an excellent reputation, works on PCs and Macs, and is free!

Critical EQ

- Choose two or three CDs from your favorite artist you love and are very familiar with.
- Next choose two high-quality recordings which you are less familiar with (preferably from a different genre).
- Finally, choose a high-quality classical recording, another recording, which could be a classical opera or a very complex, well-recorded world beat album. The reason for selecting different CDs is because the exercise will be different each time it is done with a different recording. Each CD will have different instruments and a different style. Being able to compare how frequency adjustments affect different recording will provide us that much more insight into frequency. Also, being able to change the disc periodically will help prevent mental fatigue, which is really just boredom, and keep our ear fresh because excess repetition of the same sound causes ear fatigue and loss of sensitivity to detail.

EQ is short for equalization. When you cut or boost the various frequency band controls on a sound systems equalizer, you can adjust the frequency response of the system to correct the frequency balance of the sound coming from the speakers, ideally to make it sound more like the source sound. Learning how frequency boosts and cuts sound in various recordings can get you prepared to make choices of what EQ adjustments to make when you begin mixing live. If you don't have access to a mixer or already have a program with a good selection of

EQ bands, there are links to reputable open-source programs for both PCs and Macs in the directory. Now set up your graphic equalizer with all the frequency bands turned or slid all the way down to maximum negative gain. Play one of the C.D.'s you selected through that equalizer. Turn it up to a clear enough intensity, but not too loud—you don't need to upset the neighbors or hurt your ears. Definitely be careful to watch your volume levels when wearing headphones (more on that later).

As you play the music start with the left frequency band and slowly start to raise or turn slider up. Listen closely to how the character of the sound changes as you raise the gain on that band, paying close attention to each instrument or voice in turn. Note the frequency value for that band, so that you can begin to associate the changes in the sounds you are hearing to adjustments of that frequency value. Reduce the gain for that frequency and repeat the process with each frequency band in turn, taking the time to listen and observe carefully on each band. When you run out of frequencies to test, go back and start all over again with new songs. Don't forget to use your notebook to jot down your observations as you listen.

To get the most out of this exercise follow these guidelines:

- The same time spread out 10 or 15 minutes at a time over a week or so is better than an hour at once, especially if half of that hour was wasted by forcing it after your brain (and ears) have checked out.
- Do the exercises as long as you want, up to 30 minutes or so, but when you lose interest, set it aside till next time. Regularity and consistency are key and that won't happen if it turns into a chore.
- Do the exercises a couple times a week, and at least once every seven to ten days. In this case you will get more out of 10 minutes a day than you will from spending 30 minutes twice a week, but over time you will progress either way.
- Change CDs every now and then to swap the music styles. This change provides a greater sampling of how the various frequencies sound with a variety of music. To learn a frequency, you need to hear it expressed in lots of ways. (Knowing the frequencies in a band only when you hear them played on piano notes is not much better than not knowing them at all, unless you are the dedicated mixer for a band of pianists.) Over time you will get to learn how different frequencies in different songs respond to different adjustments.
- As you get better at recognizing frequencies start changing up the order of the frequency bands you listen to.
- When ready, have a friend test you by randomly boosting the frequency bands while you call out the frequency value you believe it to be based on what you hear.
- Continue until either you can learn the rest using a real mixer, or get good enough at identifying the frequency of sounds you hear to only need a refresher now and then.

Practice Critical Listening

Practice your critical listening skills. Critical listening is simply listening with intent. Instead of being passive, you are deconstructing the mix and examining every aspect. Start by choosing a song you enjoy and want to examine more closely and allowing yourself at least 30 minutes to devote to it. The process may take around this time for you, or it may take more or less. After the first few attempts you will know how much time you need to allow for it. When you have read the instructions that follow and are ready to listen, pay attention to mix elements that matter to you as a live engineer in addition to the ones mentioned here. Use your imagination as you listen to the song and imagine how you might mix this group live. As you do this think about the elements that "catch your ear" in terms of:

- *Frequencies*: Is the mix bass heavy or does it have too much high end? Are all frequencies represented somewhat equally? If you notice any problems like the sound being too bass heavy, or the singer too nasal, imagine what you would do to account for that in a live situation. Are there frequency ranges that seem empty or too crowded?
- *Balance*: How many instruments are there in the group? Do all the instruments have appropriate weight in the mix? That is, do they each have appropriate volume to be heard clearly without drowning out any other important elements. Are some more prominent than the others, and if so are they emphasized for the entire song or just pulled forward for specific parts? Do you think others are being lost in the mix, or can you pick out each individual instrument? Are some instruments so subtle that you only notice them when you really pay attention? Is there any element you thought needed more or less emphasis? Does it need more cowbell?
- *Dynamics*: Even though the effects of compression on recorded material has created many recordings

with limited variation in volume levels, not all recordings are over-compressed and many do still leave room for variation in relative sound levels. Pay attention to the sense of dynamics (variation in volume from loud to soft). Do you notice any variation in dynamics that seem due to frequency or emphasis instead of amplitude? If so, what element is being used to create a sense of dynamic?

- *Dimension*: Does the mix take full advantage of the stereo field (meaning recorded in two channels or more so that on systems with more than one speaker elements come from different directions, creating sense of directionality and depth) or does it sound mono? How close or far away from you are the instruments? What's the panorama like? Can you notice various points in the stereo field catching your attention? Does the distance at any time create a sense of movement? Close your eyes and see what impression you get about the spatial relationships. Try and imagine where you would put them on stage.
- *Timbre/Tone*: Are there any unique qualities to the styles of the instrumentalists? Is there anything you might have to take account of in your mix like a very powerful drummer matched with a more subtle vocalist? Use your descriptors for timbre. As the song progresses through stages decide which word fits best.
- *Dedicate a notebook to use for critical listening*: While the associations are fresh in your mind, jot a few notes about the timbre, mix, and any other insights you may have about the music. Did you notice anything you never had before? Don't write more than you feel inspired to or force it; it might take some practice just listening before you find something to say about it, or you might have a ton of ideas the first time, so just have it there in case you need it.
- *Try critical listening in other situations where the sound calls for it*: Mother Nature is a sure bet for awe-inspiring mixes, and no one does it better than her if you sneak away from the noise of traffic and chatter to hear the forests or crashing waves. If you are also geographically located to catch her more terrible moods, stop and listen to her howl and crash with new ears; we promise it'll be worth your while. Small blues and jazz venues and other alternatives to the over-amplified prime time fare are also great options. Links are included to sites specializing in sound scenes, stories built from nothing but sounds we take for granted; these can be used as primary sources and as a wellspring of ideas of places to visit and spare a few moments to sit and listen.

Talking about Timbre

When we express a sound's timbre, we can only do so in subjective terms, such as grating, harsh, shrill, and/or mellow, rich, and bright. Since the words are intended to describe the tonal quality of musical sounds, they are usually associated with a specific frequency band, so for each one, you will need to learn the kinds of sounds they are associated with or they are effectively useless to you.

Audio engineers need to understand how to use timbre descriptors to allow them to meaningfully communicate with musicians about tonal quality. With no uniform standard to refer to, this is sometimes tricky, because terms so far removed from the thing they are used to describe will have variable meanings depending on who is using them. Some terms, like muddy or bright or tinny, have proven to be used consistently, and to convey roughly the same thing, no matter where we encounter them used. Others seem to be more variable, but when using them with new colleagues, all you need to do is double-check the meaning if a term is used and you aren't sure everyone is on the same page about what it means. In this way you can recalibrate your definitions to make them useful even if the lingo isn't standard. Remember, in addition to the charts in this chapter, there are some terrific timbre and frequency resources linked to that are worth taking a look at. Check these out before starting activities, as they can make your tasks easier.

- Another exercise you can do here is to see where you stand in relationship to how the timbre qualities in Figure 7.3 are generally used. You should be able to split songs and sounds you hear roughly along three to five bands from low to high frequency, in your mind, and that's enough for this exercise. Think back on the song you just listened to and try to estimate where the instruments tended to fall along the frequency range. Are they dynamic, using frequencies all up and down the range, or do they stand firm in the mids and just dip their toe in the highs and lows?
- Map the song out in a bit more detail, as much of it as you remember, in relation to the general frequency bands (for example, the drums may have stayed solid in the lows and mids if there was a lot of kick and toms but not many cymbals, and the singer could be mid-high if she stayed mid with regular swoops up to the highs in each chorus). Don't worry if you can't map the whole song in your in head; map the parts you remember and like, whether that's one part of the song, or one instrument through the whole song and

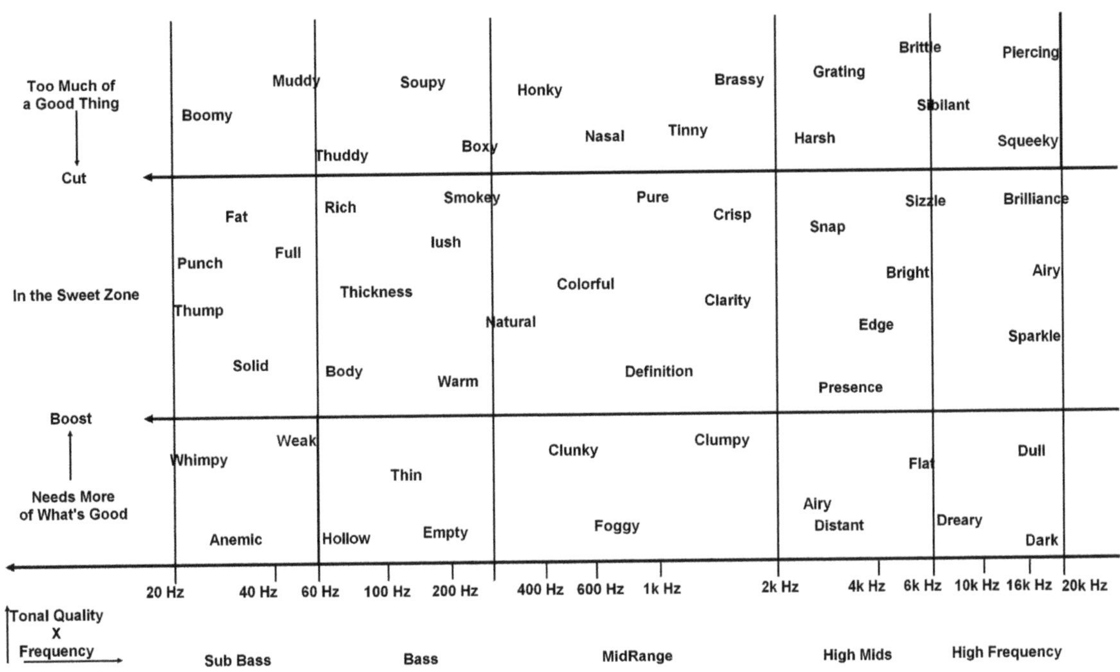

Figure 7.3. Developing a vocabulary to talk about sound can make communicating with bands and other sound engineers more effective.

just the vocal chorus. It's normal to just remember a song in terms of your favorite parts.

- Now with that in your mind, look at the figure of descriptive terms above (copy it to your notes or photocopy them if that's better). Which terms would you use to describe the different parts of the song and elements within the song? Does the frequency they are supposed to describe match the general frequency of the element you are using them to describe or do they seem way off? Are any of the terms you use to describe sound qualities missing from the sheet? Which terms there do you see yourself having an easy time using as is to describe sounds in the same range they appear to describe. Where would you put the words you couldn't stop using to describe music if you tried (already down on your mental map) but are missing from the sheet. If there are any, fill them in the general area they should go to reflect how you use them. Which words on the sheet are not words that fit in your mouth to describe any sound? You might as well set them aside. Cross those off and deal with them if you run into someone who uses them. Which words are there that you might use, but not to describe the same sounds the sheet indicates they do? Move those to the right spot to line up with how you think of them.
- It might take a few listens to different songs to figure out the vocabulary you are comfortable with, so take your time if you can only map out a few words that resonate with you the first time. If you get stuck, visit the directory and the links leading to more descriptions of the way people talk about texture or tone. You'll see there is no "dictionary" of timbre, but that some words seem to be generally agreed on and others used differently. Keep working on your own basic glossary until you get enough of it done to begin using the same words to describe music so you begin to use them more consistently, and with this you will start standardizing your lingo for describing and analyzing sounds.

The point is to work on assigning the words you use to permanent associations and their matching frequency ranges—pretty much in line with the way others use them when you can, but most importantly, with whichever way will aid you in transferring them to long-term memory where you can make use of them. The above is also a system to assist you as you start to learn and remember sounds all along the frequency band. At the same time you will develop a useful vocabulary to discuss the more difficult to quantify qualities of sound, both with others, but also in notes to yourself. Learning the frequency band can be easier for some if they use tools like word associations. If you remember better in terms of numbers alone and words get in the way, then adjust the exercises to your style.

Become a Genre Genius

Start with the music you know well and enjoy, but once you get a hang of the process, branch out and practice on other genres, and practice comparing between them as well. Aside from contributing to your ear for sound, getting to know a number of genres will make you a better mixer in your own genre, and prepare you to branch out and mix new genres which can increase your marketability and ability to find work. If you limit your listening to only one genre of music, your ears will be exposed to a very to a limited set of sound styles. That's like learning to paint, but in only one color. An engineer needs to be familiar with the widest array of possible sounds. The more we expose ourselves to the wide variety of music, the more educated and sophisticated our approach will be when it comes time to analyze a recording, regardless of the recording's genre or style. The more cross-genre music we train our ear to, the better we will be able to hear and understand the sound when listening to our preferred genres as well.

Don't just add a few jazz mixes from a Dixie Chicks CD from your collection and figure you are set. Most music lovers have their pet albums from outside their main genre preferences, so any cross-genre CDs belonging to you will only score you a point more than your standards. Again make use of the links to visit sites where fans of each genre have selected the must-hear sounds and let the ones who know best get you started. Take advantage of web radio sites like Pandora and others linked to so you can listen to a variety of sources without having to buy them first (though if you are turned on to any songs or artists you like, buy a few tunes too).

Also, be sure to make the most of your time by *paying attention* and *taking notes*.

If you listen to ten jazz artists and still can only say saxophone six times fast if asked about the five or six

Figures 7.4a and 74b. Jazz.

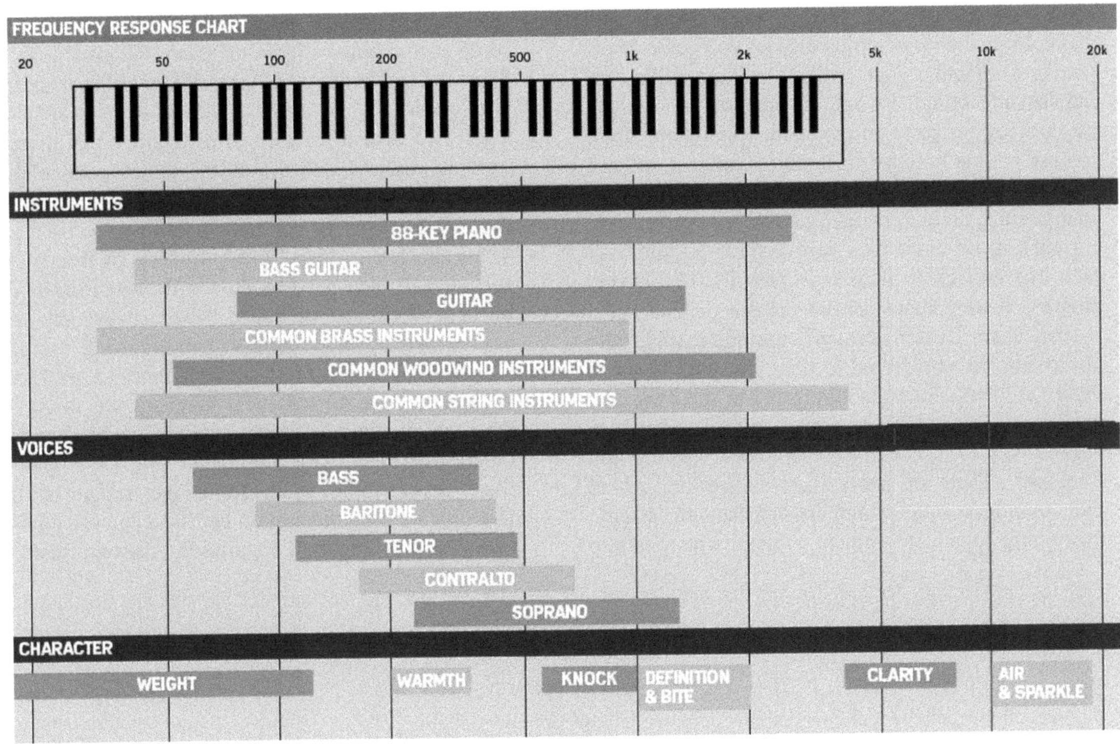

Figure 7.5. Frequency range instruments.

most common brass instruments used by jazz ensembles, you aren't getting everything from the adventure that you could be. Same goes for listening to as many blues albums and not knowing which examples were Chicago and Memphis blues and which ones from the Delta. The resources on genre at your fingertips make it easy to double check online and see what major *subgenres* you should keep an eye (or ear) out for and if any of the artists you are sampling are typical examples of them. Just as important, don't just learn the name of the genres/artists you sample; to be able to talk about genre in general you'll want to learn the *conventions of the genre* for each area you check out too. These conventions include the obvious features that are easy to pick up on like the specific mix of instruments or the band makeup, but include the more subtle factors like how they play their instruments and construct their songs, and even the shared history honored by so many of them too. By the time you have given a good listen to any genre, you should know at least the two to four most relevant subgenres, a little of what distinguishes them from the main genre, and be able to discuss the most essential features that are hallmarks of that particular genre and are the qualities that we use to identify which songs and artists are grouped under one genre instead of another.

Use All Your Resources

Practice at the board is the best way to practice ear training for someone with enough of it already to know how to keep listening, but the prep still starts before. If developing one's ears happened that easily without instruction, there wouldn't be as much bad sound out there. Anyone who can't identify a sound by frequency range, musical note, and source, or who can but can't separate them out from the background when hearing them in a mix, can benefit from ear training:

- Take advantage of the many online aids available, from blogs and tutorials to online and open source tone generators. You can find them in the directory. A very notable one provides interactive blind audio tests covering every frequency bandwidth, peak, and notch and as well as the diatonic scale, perfect pitch, dynamic range, and listening for polarity. Another site has chord and rhythm recognition exercises for music theory and several EQ, spatial, and channel/mix interactive (that also let you get to know genres) to practice your critical listening skills. Try and visit a few times a week and practice, more if you are noticing progress and excited, less if you sense you are getting bored with it. But keep at it, at least for training your ear to identify frequency. Mix up your practice times between sites to maintain interest, but stick to no more than two or three so you visit each one enough to note your progress. Observing progress is what keeps us motivated.

- If you learn better through communicating and discussion, enlist a friend to help you with some of these activities. Someone interested in the topic and wanting to learn too is best. Online forums are also a good way to discuss ideas about ear training and frequency. There are many audio sites online where live sound pros and hobbyists gather to discuss everything sound. In a pinch, a friend who wants to spend time with you and is cool with letting you use some of it to bounce ideas off them will also work. If they play an instrument, even better—they can test your pitch.
- There are also a number of different types of ear-training resources in the online directory available to you; if you find activities you like, add them. If they work better for you, use them instead.
- Many musicians do not think in terms of frequency, so you will also need to do a little listening to the musical scale to recognize where frequency and octave overlap to be able to communicate usefully with those who talk about sound in those terms. Other musicians have no technical language for music or the tonal quality of sounds, but you can be sure that most of them are well acquainted with the ranges of their instruments, so learning the range of the instruments used most often by those you are mixing is also useful. Figure 7.6 shows the diatonic scale of

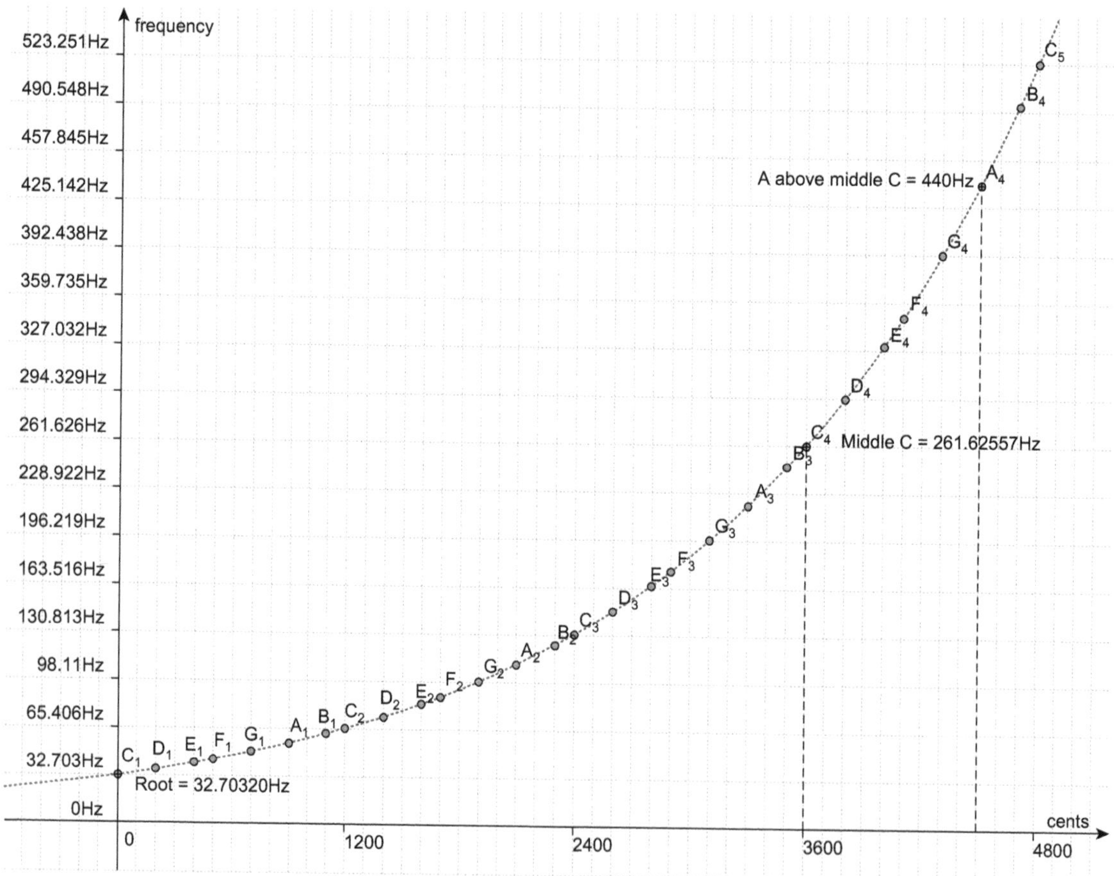

Figure 7.6. The octave scale of musical notes and corresponding frequencies.

musical notes and corresponding frequencies; learning them will make you a more effective engineer in the long run.

If you don't get to spend much time at a live mixing board you should take care to do these activities regularly, so that you are giving your ear some work to do every day. It's OK to push yourself, but ear training takes a while, so don't burn out on it! Have fun with the process, and use it as a basis to make self-guided learning a part of your normal routine. Whatever you decide to learn next, you'll already have some good habits to fall back on. Between the activities here and the links to more information and exercises online, we hope you keep at it and make steady progress long after we part ways with you on the last page of this book!

Part 3

Tools of the Trade:

The Public Address System

These chapters focus on the most obvious tools of the live audio engineer—the audio technology of the P.A. system. While the earliest audio technology was developed with sound production/reproduction in mind, it didn't take long for performers to bring their mics onstage with them whether the show was being broadcast or not, giving birth to the live sound audio field. Live sound, and the audio industry in general, has always been defined by the audio technology of that era, and changes, large and small, in the live audio field have usually been preceded by changes in the capability and cost of the audio engineering technology it relies on. Even industry changes that seem, at first glance, to have been fueled entirely by cultural shifts in the landscape of live music and performance, almost always involved new technologies emerging somewhere in the chain of events if you look closely enough, either as precursor to or natural result of changes in the musical and sound engineering practices of the time.

To succeed in live audio, to thrive, or simply remain happy and sane through the many changes ahead, rookies would be wise to take what lessons they can glean from the experiences of the industry elders who have gone before them. The lesson that can be learned from the engineers who stayed on top of every development in the field for over half a century is the proper perspective an audio pro should maintain concerning the technology that is so essential to all a live engineer does. As important as the P.A. is to the field of live sound, despite how we may learn to appreciate its features anew as we come to know each P.A. we work with regularly, to the point where we can identify the unique quirks and personality traits of each one, we should never see it as a partner, pet, or anything more than a tool we use to accomplish a task. We should consider any tech component or P.A. as easily replaced by any other tool of the same type or able to perform the same function. When the collective choice of the pros who work in the field of live audio has deemed the next technological replacement as being enough of an improvement over the current tech that the consensus is clear in the numbers choosing to adopt it, we should look to the future rather than insist on sticking to older tech just because it is easy and familiar.

While it's fine to use the best gear we can when the situation allows, no gear should be seen as so essential and irreplaceable that we should waste time complaining that we can't mix without the right type of this gear for each channel or any less than eight of that gear type, when those situations arise where we must do without. And gear should never be used to function in ways not intended, such as to make up for a lack of confidence, act as a symbol of career success or lack of it, or proof of one's ability as an engineer. The best outlook for audio engineers is one that simply accepts as a given the high probability of the eventual *obsolescence* of much of the current technology they use; given a long enough time line any gear will become obsolete as better technology is developed. Learn to love the abstract P.A. in any form, but don't become dependent on one type of P.A. or grow too attached to any specific tool.

Chapter 8

The P.A. System

More than the Sum of Its Parts

Block Diagramming the Signal Chain

A *block diagram* of a sound system is a simplified drawing of the system components and their connections that functions as a "road map" of your sound system. It's important to note that a map may be no more than a couple lines, squares, and an X, scribbled on paper or it can be a very detailed rendering of every aspect of the environment, showing all access points, traffic flows, and landmarks. Much like roadmaps, block diagrams can come in versions that are more or less detailed and complete, depending on the purpose.

Whether it simply documents each link in the *signal chain* from one component to the next, or follows the entire path of the *signal flow* through the system, block diagrams can be read from top to bottom and left to right, or by following arrows in the map. As symbolic representations of the P.A., each sound system component in the block diagram is represented by a symbol or picture. In the most basic diagrams, symbols are most common, such as a mic, triangles for amps and speakers, and rectangles for controllers and processors. Cables are represented as lines between the audio components. Some commonly used symbols are shown in Figure 8.1.

Figure 8.2 shows a block diagram of the simplest P.A. system. Following the diagram left to right, we begin at the first link in the signal chain, a pair of microphones, where the sound source is captured and converted to a signal, or a pair of signals to be precise, one from each mic. The mic signals then go to a small mixer, where the two signals are *summed* (combined) together, before leaving the mixer as a single combined signal and moving to the end of the chain. Finally, that signal feeds into the loudspeaker (one with the amplifier and all other necessary components packed inside the speaker box) that represents the final link in the signal chain. Here the signal is converted back into sound as it is emitted from the speaker.

Most P.A. systems are more complicated than this, but all will include these basic components (only the simplest personal P.A.s forgo a mixer and plug the mic right into the speaker). As you can see in the next image, it only takes adding a few more components; now our diagram starts to portray a more typical band P.A. system.

Figure 8.3 shows three stage mics on stands. The stage mics are plugged into a stage-box full of connectors, and a multi-core cable snake coming from those connectors goes to the mixer. Also connected to the mixer in this image are three signal processors (any component that affects the sound in a manner other than a simple gain boost or cut) that lie along the chain after the mixer and before the speakers. For the first time, arrows are included to show the signal flow, showing the route of the signal as it is sent through the signal processors before continuing on the way to the pair of speakers at the end of the chain.

Things get more complicated (and more realistic) in Figure 8.4. In this system, the components are represented by clip art whenever possible, there are many more links in the signal chain, and arrows allow you to visually follow the path of the signal between the components, allowing you to see where it is going from beginning to end, including where it loops between the processors and mixer, and where it splits off to go to the monitor mixer. This diagram has a version of all the components that are represented in the following chapters, and each component is labeled by chapter as well as name, so don't let it bother you if you see any you don't recognize, just continue reading this section to see how the components work. This is not the most detailed type of block diagram, but it is typical of most basic signal flow diagrams, providing an overview of the P.A. system and the signal flowing throughout. However, you will see in the section following this one that, like street view on Google maps allows one to zoom in on details not otherwise visible, block diagrams can contain diagrams that zoom in on specific parts of the system and show even more detail; for example we will be looking more closely at the signal flow inside of components like

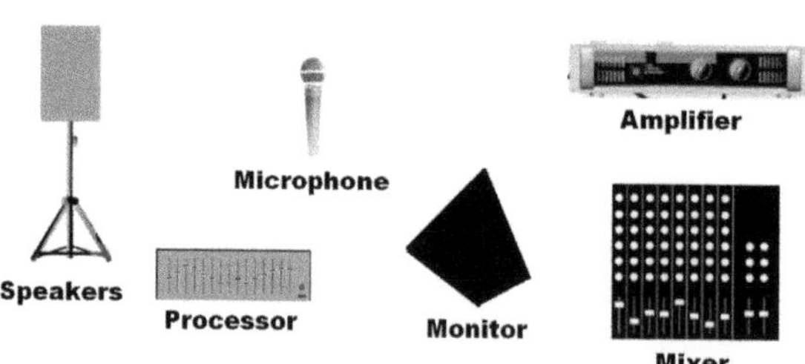

Figure 8.1. Diagram symbols.

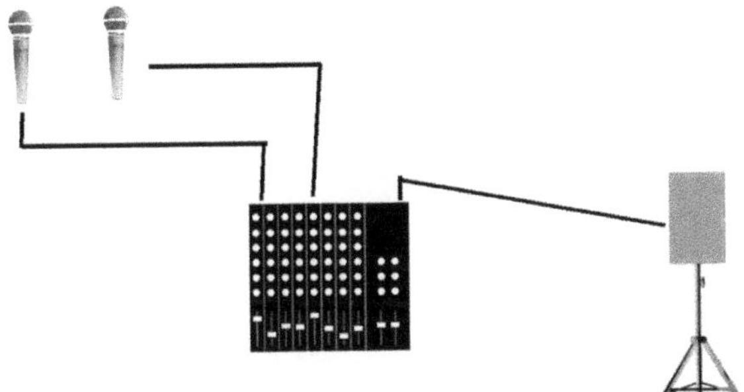

Figure 8.2. Mixers allow you to mix signals together, and send the resulting signal where you want it to go.

the mixer/processor, and between components, such as at the amp/crossover/speaker portions of the signal chain.

For now it is enough to provide an overview of how all the components we will be examining fit together to make up a P.A. system. As we introduce you to each of them, you can refer to the block diagrams to see where each fits, or you can wait until after discussing all the components, as eventually we will revisit the signal chain again to fill out the details and complete the portrait we have started here. What is most important now is that readers keep in mind that none of the individual P.A. components exists in isolation. Nor do any of them need to be the perfect type, or even present at

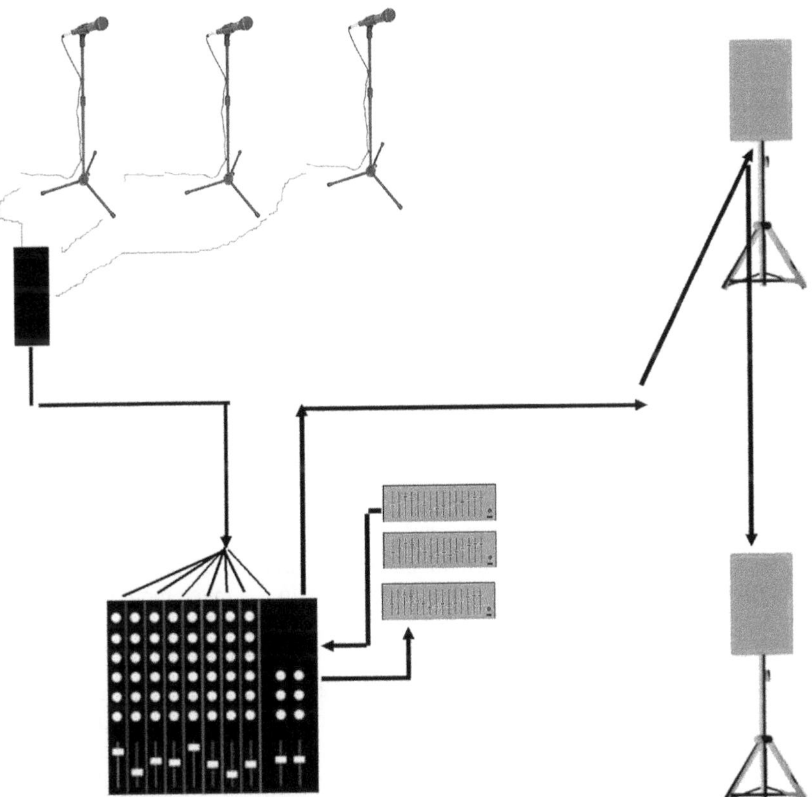

Figure 8.3. More signals, and a few more components has our diagram looking more like a small band P.A.

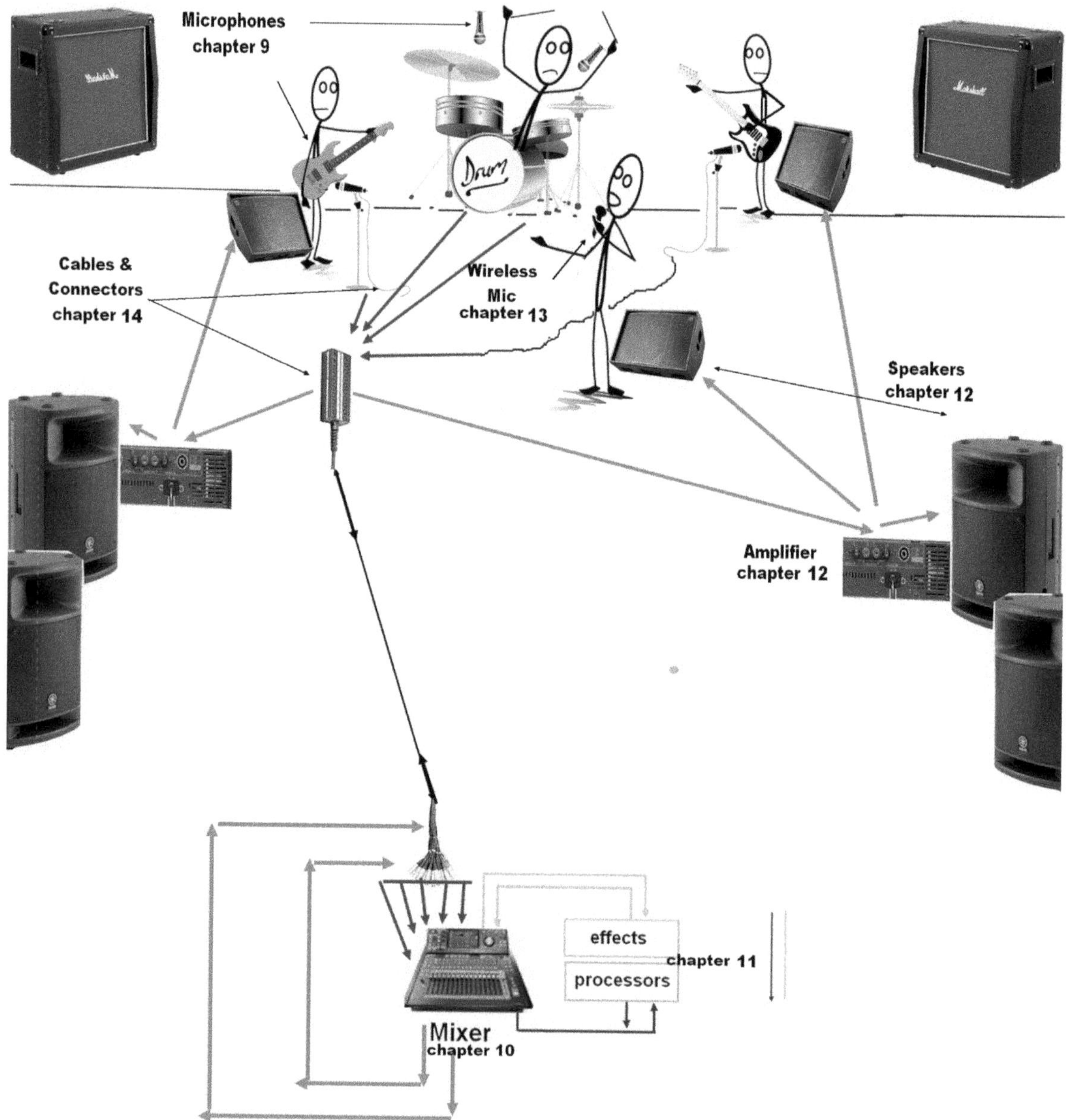

Figure 8.4. Clip art has replaced symbols in this diagram, and arrows show the direction of the signal through the wires, and the entry and exit points of the mixer are shown. This block diagram is a much more detailed rendering of a P.A.'s signal flow.

all beyond the three basics shown in the first diagram, for an operator to be able to manage the system and produce better sound. Each P.A. will have a unique combination of components, both in number and design, and with them its own sound and quirks. Nevertheless, the overall functioning of the P.A., as complex as it is, operates on a stable set of principles and best practices that can be understood and followed to ensure an acceptable experience for audience ears, whether the separate parts of the system are brand new with all the bells and whistles, or a more "vintage" system that's a bit rough at the edges. No one component, so long as it is performing its basic function, will make or break things unless you forget to hook it up to the chain, and then discover its usefulness by having to do without it.

Feel free to draw up your own signal-chain diagram as we examine P.A. gear over the following chapters, as it will help you keep track of what you learn. Overall, knowing the signal chain well enough to sketch one up at will is going to make your training and early time as a sound tech so much easier. If you are designing a sound system from scratch, a block diagram is the starting point. You can use it to generate a list of necessary components and to keep track of your research, and then your purchases. As you set up a P.A., a block diagram helps you easily see where to wire in new pieces of equipment. Perhaps most important, a block diagram is a tremendous aid in troubleshooting the system when it all goes wrong.

What We Use the P.A. to Achieve

What the P.A. allows us to do that none of the parts alone can provide us is control aspects of the sound moving through the system well enough, so when we shape it and send it out the far end it has the best chance of surviving the changes the environment will inevitably create in it, at least well enough to be *adequately loud* and *adequately understandable* for the task (which is usually to extend the reach of sound to include a much larger audience than could be managed otherwise). This minimum is almost always doable. Sometimes it is all that can be done, but that only means the sound without your efforts would be that much worse, so try not to get too discouraged. With experience each operator learns how to coax *good sound* from the P.A. more and more often, and some even will be able to wrangle *excellent sound* anywhere from occasionally up to a good deal of the time, though by no means always. But before good sound or excellent sound can even be a consideration, we need to make sure the P.A. and sound it produces, first meet some minimum requirements.

Meeting Live Sound Minimum Requirements

In order to think beyond providing the minimum adequate sound we need to be able to amplify and/or distribute the sound without creating signal distortion, like **clipping**, *noise*, or *feedback*. To that end we need to set up our P.A. to ensure there is enough **headroom** to allow for the *peak sounds* in our program without exceeding the *maximum amplitude* the P.A can output and retain *fidelity* (ability to reproduce sound accurately and without distortion). To make this possible we need to see that we coax enough *gain before feedback* from our system to allow us to raise the volume without creating feedback, and we need to keep the *signal to noise ratio* under control. Only then can we hope to have the basics covered well enough to take the time to shape the more subtle aspects of the sound with our mixes.

Minimize Distortion

The most commonly understood definition of **distortion** is the simplest one. This view holds that distortion is essentially the unwanted alteration of the original shape of a sound waveform. In some cases, distortion (such as electric guitar) is desirable, but in most cases it is unwanted, ranging from an annoyance, to actively interfering with intelligibility. Under this definition, the addition of noise or other extraneous signals like hum or interference is not considered to be distortion, because they add to the signal but do not change the shape of the original waveform.

A less common view of distortion starts with a similar basic definition, but is more inclusive in that the sine wave is not the only attribute in which a change is labeled distortion. This definition counts *any unwanted change in the composite sound that occurs during transmission* as distortion. While this broader definition includes things such as noise under the broader umbrella definition of distortion, like the first it also recognizes a range of distortion that might be seen as neutral or enjoyable depending on the circumstances:

- *Overdriven sound* situations are when, for example, air-pressure distortion and *transients* at the microphone and/or clipping at the mixer or amplifier distorts the sound.
- *Harmonic distortion* (HD) comes in two forms, symmetrical (odd order) and asymmetrical (even order). Harmonic distortion occurs when related frequencies (multiples of the fundamental frequency combine with the original signal). Can sound like soft clipping.
- *Intermodulation distortion* (IM) is the interaction of two or more frequencies creating new frequencies that are non-related, highly dissonant frequencies.
- Some phase shift and delay artifacts count as distortion, including effects like high-frequency attenuation

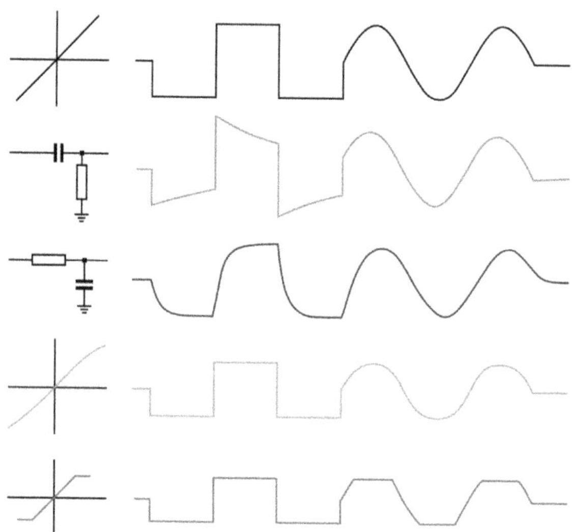

Figure 8.5. Compare the first undistorted signal with the distortion seen in the following three.

(which is time lag that varies by frequency) and comb filtering.

- Analog magnetic tape and LPs are subject to their own set of distortions not applicable to digital. When the playback speed varies from the recording speed it causes a change of pitch and tempo. *Wow* is a slowly wavering deviation from the proper speed, and *flutter* is a fast wavering. Wow and flutter are issues more common to DJs, and with a practiced DJ are usually well controlled and mostly inaudible.
- Digital distortion types include aliasing and quantization noise.
- Non-flat frequency response and compression are both sound distortions.

Sound issues only called distortion under the second definition, and counted as different issues by the first, include such items as:

- resonance, which can be considered temporal distortion or time dilation because it stretches the signal out for a longer duration than the sound could sustain on its own power—stored energy is released as unwanted sound, lengthening the decay, or ringing;
- feedback, another form of distortion according to the second definition;
- noise distortion, the addition of any unrelated and often broad spectrum frequencies, e.g., tape hiss, vinyl surface noise, and electrical noise, including 60 or 50 cycle ground hum.

Distortion is sometimes added intentionally because many listeners have a strong preference for sound that is warmed by small amounts of low-order harmonic distortion which adds a pleasant fullness and depth that is often missing in digitally reproduced sound. The attractiveness of this effect is reflected in the continued interest in vinyl records which are known to add moderate amounts of the right kind of distortion; adding some warmth "soul" to otherwise cold and clean digital sound.

Others don't want to smooth out the hard edges of reproduced sounds so much as break them up, like a serrated knife compared to a razor's edge. These folks enjoy the kind of distortion easily recognized from punk and alternative concerts and recordings, and are created with a distorted guitar sound and vocal distortion produced by shouting/singing loudly into the mic at close range or even with lips touching the microphone. These are the same overdriven sounds described as distortion earlier, but when used strategically and in moderate amounts, they are desirable hallmarks of the sound of some genres.

Finally non-flat frequency response and compression are forms of distortion often sought after, although whether any mix achieves a pleasant, just right, amount of either is usually up for debate by those with differing tastes. Note that they only are considered debatable when they are intentionally done and not the result of poorly set compressors, which can be heard chugging if set too fast or too slow, or improperly set filters, such as crossovers feeding the wrong frequency ranges into speakers. Neither of these effects would be likely to be perceived as pleasant by listeners, no matter what their preference.

Note that the most common understanding of distortion is the first one, so harmonic distortion, clipping, and other unwanted sound that occurs due to a difference occurring in the sound wave is counted as distortion, but feedback, noise, and other unwanted sounds that do not are considered separate issues. The only difference is what each labels as distortion; however, in both cases all the sound problems that can happen are understood by the same names and definitions, the only difference being that the second point of view counts all of them as forms of distortion, and the first is more specific about what problems fall under distortion or do not. Regardless which view you take, to minimize all the negative artifacts above (except for those we intend to include), we need to maximize all the conditions that keep the distortion monster away.

Maximize Headroom

Headroom refers to the margin in dB that you leave between your *average working signal level* and the *maximum signal level* your system can handle (the point beyond which distortion and 'overload', usually heard as clipping, will begin). Leaving this leeway is crucial in order to allow for the inevitable peaks which momentarily raise the signal's level much higher than the average value. If there isn't enough, then transients and peaks in the music will cause distortion known as clipping. A *transient* is a quick spike in signal level, typically percussive noises, or the T noise in spoken applications. When these spikes go above the volume levels your speakers and amps can handle, the result is the wave form getting clipped off at the outer edges, like something went at them with scissors. Not only does such *clipping* result in a loud ugly noise that the audience won't appreciate (or the band for that matter), it can also blow out your speakers.

The maximum level before clipping in a particular piece of gear varies according to manufacture, but most pro systems allow for an operator to maintain headroom of between 16 and 22 dB. (When your average signal level is +4 dBu and your peak signal level is +26 dBu, then 22 dB is your headroom.) You cannot do anything about the system's dynamic range (that's a function of how clean and noise-free your tech can manage to be at the bottom end and the limit of what human ears can handle at the top). What you do need to worry about is maximizing headroom. A commonly used point of reference to figure out how much you need starts with measuring the **crest factor** of the most dynamic sound your system is likely to encounter (which is music, no surprise there). Crest factor measures the ratio of the **peak** (crest) **value** to the root mean square (average) value of a waveform, and music's crest factor of 4–10 translates into 12–20 dB. This means that generally the *peaks* of music are 12–20 dB higher than the "average" value. So you need 12–20 dB of headroom to avoid clipping.

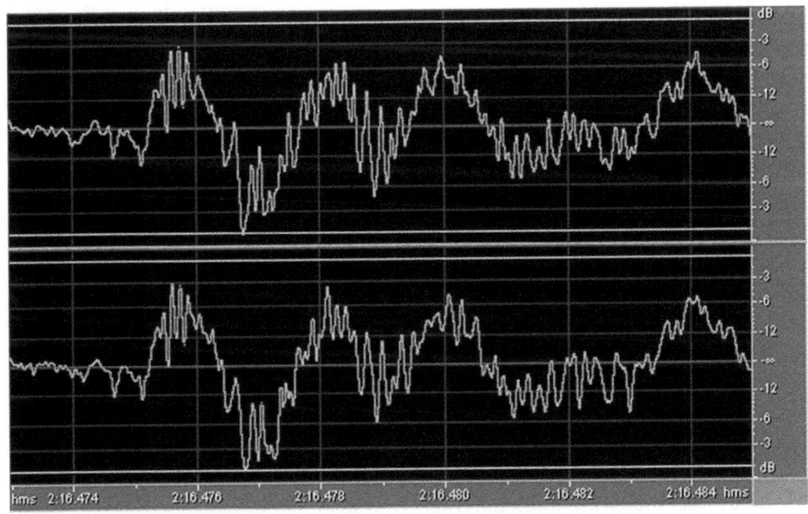

Figure 8.6. Notice the spikes in the signal created by transient drum sounds.

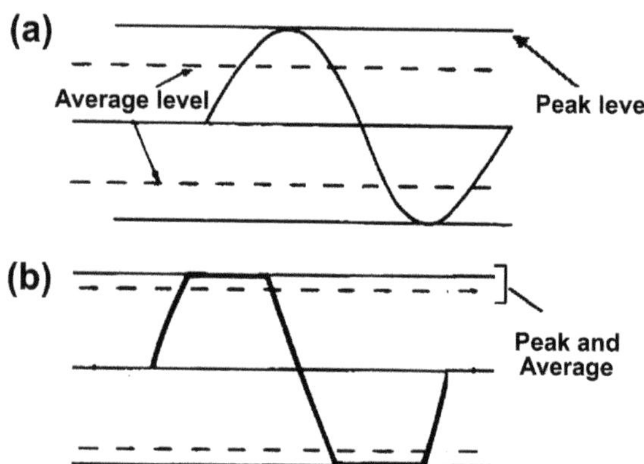

Figures 8.7a and 8.7b. Compare the first undistorted signal with the distortion seen in the following "clipped" one.

Sound Source	Average	Peak
Percussion	90-105	125-146
Amplified guitar (on stage using in-ear monitors)	100-106*	118
Amplified guitar (on stage with wedge monitors)	105-112*	124
Drummer	105	144
Bass guitarist at indoor music festival	101	133

Figure 8.8. The peaks of music are around 20 dB higher than the "average" value.

Taking into account an interesting feature of live music, out of that range it's smarter to lean toward the figure that gives the most headroom. This is because of something that always happens when a band plays a bunch of songs at a gig: *they get louder as they go along*. (They also get faster as they go along too, but we don't need to pencil it into our plans like we do the increasing loudness.) To anticipate and account for this increasing volume in your initial P.A. settings, just add 3 dB to the minimum headroom you should leave, making it 15–23 dB.

In order to ensure you leave enough headroom throughout the signal chain, it is crucial to adjust the level controls of each component and at each channel in your mixer to allow for enough overall headroom.

Maximize Gain/Gain before Feedback

Gain refers to how much (additional) amplification is applied to a signal within a piece of gear, or can be measured in the signal at a particular point along a signal's path, and is expressed as a voltage value in decibels. When there is no amplification added to the signal within a component, and no change in level, we refer to this as a *0 dB gain* or *"unity" gain*. The amount you can increase the gain level of a signal before your system starts to feedback is called your *gain before feedback*, and the amount of it you can manage to arrange for is directly reflected in how much headroom you can achieve.

Just so no one makes the mistake of underestimating the care it takes to keep this parameter calibrated, a few facts bear mentioning. It is important for any one operating a P.A. to understand that most P.A. components include a way to adjust gain, either at input or output, if not both. Therefore, the value expressing the amount of gain applied to a signal at any specific point in the signal chain doesn't directly match the volume of the sound that will be emitted by the speakers at the end of the chain. Added gain at any point in a signal chain will result in an increase in sound level, but exactly how much "volume" results from that particular gain adjustment is variable, because sound level is the overall result of all the gain adjustments made throughout your system. If the initial signal level is extremely low, if any of the gain applied at that point is accompanied by a loss in signal elsewhere in the chain, or if there is a lack of gain set further along the chain, then the resulting volume increase created by a large amount of additional gain will be much smaller than would otherwise occur. So setting gain at any point is ideally done with both the component and the entire system in mind.

How you adjust gain in each component is one part of the task referred to as setting the *gain structure* of the system (the way gain is distributed throughout your system). Gain structure is also determined by many choices made throughout the entire process, from selecting a P.A. with enough power for the given application, to proper placement of the system, to setting gain to maximize headroom during tuning and sound check, and finally to the choices made by the engineer as he/she brings together the mix.

Maximize Signal to Noise Ratio

As already described in the discussion describing digital sound, the signal to noise ratio is the amount of desired sound noise compared to the amount of undesirable system noise contained in the signal which directly effects the final sound coming from the speaker if it is too low a ratio (contains too much noise). Even though noise at the bottom is often masked by ambient sound in live sound situations, it's still a factor the live engineer needs to pay attention to. While not enough here is less likely to have disastrous effect on your sound than being short on the qualities described above, it can get in the way of having the clean awesome sound we at least should always try for, even if the realities of acoustics and other variables out of our control mean good sound is the best we can manage in some situations.

Most of what is done to account for these factors will occur with proper setup. Once the system is going, all you can do at the mixer to help out is to make adjustments sparingly and thoughtfully, as opposed to just boosting frequencies on a whim.

We will revisit these topics in detail again during the chapters on setup and discuss how we avoid or achieve each one. In the meantime, when the terms come up now and then in the following chapters, you will know what they are referring to. Now that we've taken time for a quick overview, we can zoom in on each link in the chain and look at the P.A.'s components in more detail.

Chapter 9

Microphones

Sound to Signal

Microphones

Any device that converts one form of energy to another is called a **transducer**. Microphones are transducers which convert acoustic energy (sound pressure waves) into an electrical signal which gets processed and reinforced as it travels through the P.A., and finally gets converted back into sound waves by the system's loudspeakers (which are also transducers). Microphones are used as the input for all sound sources on the stage that don't generate their own electric signal, as well as for micing the amplifiers of powered instruments that will not be plugged into the P.A. directly or with a DI box. Microphones should be connected to the mixer XLR channel inputs with balanced cable. You should *never* plug a mic into the mixer's line inputs.

Microphones are mainly classified *according to their method of converting sound energy* into electric current, *by the size of their* **diaphragm** (the thin vibrating membrane that captures the vibration of incoming sound), and *by their* **directional pattern** (the angle and distance from which they pick up or won't pick up sound). These classes all overlap, so any type of mic conversion method can come in all different types of directional pattern, and with any size diaphragm (note any useful combination will be available, but not all may be useful and so won't be as commonly manufactured, if at all). The distinction shared by all three of the features used to classify microphones is that these design characteristics have the most influence on the two features that matter most in microphones: the frequency response/quality of their sound and their ability to pick up and/or block the right ratio of sound vs. noise (how much sound they will pick up, and from where). These are the heart of a mic's function; they define what a mic is meant to do. There is no standard for what is the "best" sound for a mic, or what sound all mics "should" pick up, only what combination works best according to the application they will be used for and the personal preference of the people who will be using them.

A fourth classification also makes a distinction along lines of design, classing mics *according to the overall physical shape and size of the microphone body*. While this feature is not as tied to the core function of a mic, it does have a lot of bearing on what application the mic is best suited for. There are too many applications and mic varieties to name here, but not too many that we can't mention a few of the ones most commonly used in live sound. The main types of mic you will run into are *hand held*, those meant to be held in a person's hand, or held in place by a mic stand. Other types, such as *lapel mics* or *headphone mics*, are also commonly used, though mostly for speech reinforcement or mixed applications like theatrical productions. Headphones have also become more popular in music productions, at least those with a performance focused on dancing as well as on music. *Contact mics* are occasionally used for acoustic sound reinforcement, and are small mics that are designed to be placed against the resonating body of a musical instrument (like on the resonating wood face of an acoustic guitar). Last but not least are the wireless mics, which are included here mainly for convenience as wired/unwired is an entirely new distinction making up a topic of its own (so much so that any more to say will have to wait for Chapter 13 on wireless technology). It should also be noted, that *this classification, like all the others, also overlaps with all other classes.*

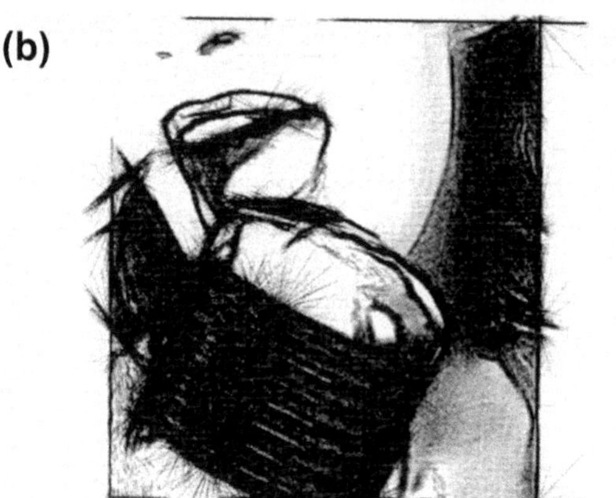

Figures 9.1a and 9.1b. It all starts with the microphone.
Source: Photo 9.1a by Jan Mehlich.

With several ways of converting energy, three diaphragm size distinctions, and five common patterns for picking up sound, not to mention the more practical consideration of body shape, there are already countless choices to select from. When you consider that the manufacturer and model also influence how a mic functions, and in addition that there are almost a dozen other minor features a mic may have or not, in any combination, it doesn't take much imagination to see how one could argue that of all audio equipment microphones are the component with the most dizzying array of options to choose from. This is why in person and in forums, probably the single most discussed audio topic, the burning question shared by almost everyone in audio or music, at one time or another, and regardless of industry niche or genre of preference, is the simple inquiry "Which mic is best for X?"

Transduction Method

The two main types of mics in this category are *dynamic mics* and *condenser mics*. While the differences in their design are interesting and contribute to their unique qualities, if you are interested in your mic's innards refer to the chapter directory for links to everything you ever wanted to know about microphones and you can find detailed information on mic brand names, mic selection, which mic will best survive a 15 foot drop off the stage and crazed fan pile on, and yes tons of details about microphone design and build as well. Our concern here is to give you an overview of the main categories of gear and the most notable features of each type, with the understanding that even with these imposed limits, there is more variation than can be fully covered in a quick and dirty run down.

Dynamic Microphones

Dynamic microphones are the most commonly used mic for live reinforcement applications. They are simpler in design and traditionally more durable than other types. While they are often less effective than condenser microphones at accurately capturing very high frequencies, this difference is not crucial for most live applications, though where the difference matters, condensers can be used instead.

1. *Power*: No power requirements.
2. *Durability*: Traditionally considered the more durable of the two designs. Advances in the design of condenser mics make this less of a factor for newer mics. For ability to withstand extremely rough handling, dynamic mics are still the champions.
3. *Sensitivity*: Dynamic mics are not as sensitive as condenser mics but that is fine, and even preferable for most live sound applications. A highly sensitive mic will be more likely to capture unwanted noise, also known as *spill* or *bleed*, or cause feedback due to lack of mic isolation on stage. *Mic isolation* is the condition where a mic performs its function without picking up unintended, unwanted sound even with other loud sound sources in the immediate area. When the accuracy of a condenser mic is important enough to the sound, it's still worth considering using one; condenser mics designed to function live will have the ability to reduce sensitivity if it is required, often in the form of a manual switch that allows sensitivity levels to be selected. This is a good example of how a mic's value is in the ability to do a given task best, not in a given spec rating.
4. *Frequency response*: A microphone that has a flat frequency response produces equal output at all frequencies. Dynamic mics have a reduced high-frequency response compared to condenser mics. As the choice with less of a flat frequency response, dynamics are less desirable when absolutely accurate high frequency sound reproduction is the goal, however, less response at higher frequencies is why the sound of dynamic mics is sometimes considered "warmer" than the sound of condenser mics. It's notable that this quality of "warmth" found in many dynamic mic models isn't only a consequence of less sensitivity at high frequencies. With a number of dynamic microphones already able to provide maximum possible accuracy within the limitations of the design, dynamic mic designers often opt to work with the strengths of the design and maximize on its ability to add a unique quality of "character" to sound reproduction. This is part of why dynamic mics tend to have more models that are tailored for specific purposes/sound and have more models that are known for a "signature" sound based on the quality of frequency output unique to those models.
5. *Address*: Refers to what position on the microphones sound is captured from. If you have ever witnessed a bewildered looking singer singing into the top of a microphone and wondered along with them why they sounded so thin and far away, then you saw a performer accustomed to a front address mic having their first experience with a side address mic. It's

a rare "singers in the their natural environment" kind of moment, only found in bar venues of the smaller, hole in the wall variety, because hole in the wall bars are naturally where the performers and audio engineers alike are inexperienced enough to make such a mistake possible to catch. Barring the few who regularly perform while inebriated, there is only one chance per performer that you might luck out and witness this special moment. In terms of address mics come in two design flavors: *side address*, usually with a helpful mark to show which side, and *front address*, also called *end address*, which should eliminate any question about where to find the business end.

6. *Condensation effects*: You can breathe into a dynamic mic all night without damage to the sound or equipment, though in most cases, this isn't a "deal breaker" spec for deciding to choose a dynamic over a condenser. There are a few exceptions. Most venues are air conditioned but unless there's a surplus, any vents or fans will be aimed toward the patrons, stage lights are hot-air conditioning or not, and any performer who's worked on a raised stage or riser can verify that heat rises, so under the right (or wrong) conditions breathing may not be more than a condenser can handle, but the amount of sweat pouring off the band may pose a real risk. Should you work at a venue where a significant enough percentage of the talent is known to get more sweaty or frothy than common, and they will be using the house mic, then this might be a very important specfication to consider. (Of course you will want to make it a point to disinfect your mic baskets and accessories routinely, especially when they are shared.)

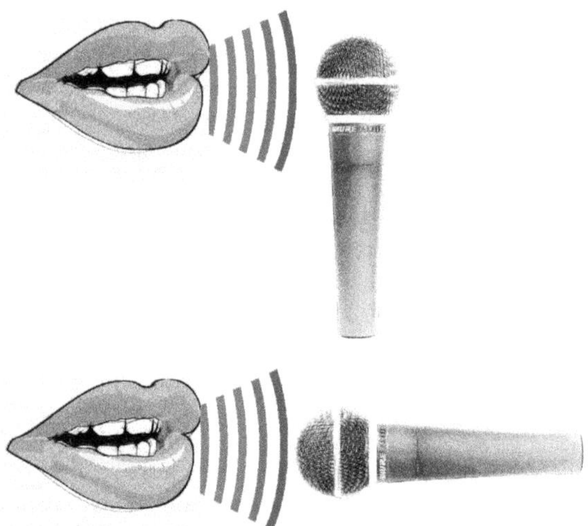

Figure 9.2a. Side address and front address.

Applications

Dynamic microphones are most commonly used for sounds that are close to the mic, reasonably loud, and with frequencies predominantly in the mid-range or below. A few examples include:

1. vocalists who hold the mic close and belt it out;
2. instruments that don't need much high-frequency response, such as bass amps and drums (except for cymbals or bells);
3. instruments that produce high-intensity sound and where the mic can be kept close to the sound, such as guitar and bass amps, kick drums, and brass instruments (as long as the players don't swing their horns too far from the mic);
4. situations where the mic could be dropped or badly beaten (i.e., when used by clumsy vocalists and gonzo drummers).

Figure 9.2b. The Shure SM58 microphone is one of the most popular microphones for live performance. *Source: Image derived from original by Iain Fergusson.*

Condenser Microphones

Condenser mics, also called *capacitive mics*, have traditionally been less common for live applications due to lack of durability. On the one hand, design improvements have made durability less of an issue, and this combined with the fact that their greater sensitivity and high-frequency response are desirable for certain applications, means that they are becoming more common. On the other hand, even though the price difference is not as pronounced as it used to be, and some condenser models can match comparable dynamic mics for affordability, condensers are still more expensive on average than dynamic mics. In addition, it's true that durability issues are less of a concern, but that's not the same thing as "not a concern": mics used in live applications do suffer from rougher handling and can break more often, and a broken condenser is usually more expensive to fix or replace. Those issues combine to ensure the dynamic mic will still be the reigning champ in the live sound arena; they just are no longer the only contender around worth putting your money down on. Even bands or sound engineers on a modest budget can afford one or two condensers nowadays; at least they can if they are the sort of folks who can refer to what they do with their money each month as "a budget." Musicians or live engineers who in the past would have hesitated to use a condenser for most live applications nowadays are more likely to include one or two condensers in their stage kit; the added expense and care involved in doing so is no longer excessive compared to the payoff of being able to reproduce higher frequencies.

1. *Power*: Condensers always require a DC voltage to work, usually in the form of phantom power from the main mixer. Most mixers can supply one or more channels of phantom power, but some older mixer models lack this feature. Some condenser microphone brands use a battery instead, but this is less common. Additionally some tube element mics require a dedicated power supply, which is supplied by a "brick" that is plugged into AC power. Such mics will have a non-standard cable from the tube mic to the brick, and then the brick has the normal expected XLR output that is then run to the mixer.
2. *Durability*: Condenser mics are more fragile, though less so than in the past. Design improvements have been dramatic enough that depending on the brand and the application they are designed for, some condensers in today's market are pretty rugged (for an affordable condenser known for this, check out the *Shure Beta 87A*). The most rugged condenser mic still can't stand up to the toughest dynamics in a mic versus mic contest for durability, and the dynamic with the best word of mouth rep for toughness is referred to as "damn near indestructible" (the dynamic heavyweight referred to here is the *Shure SM57*). Whether the difference between rugged and indestructible is one that matters depends on the individual. It would go completely unnoticed for a lot of people, since in the environment their equipment is used, gear is generally not subjected to conditions beyond those that rugged can easily withstand. There is still a significant minority that would have no difficulty breaking a "rugged" mic in less than a week on the road, however, and for them anything less than practically "indestructible" just won't cut it.
3. *Sensitivity*: Condenser mics are usually more sensitive than dynamics, and so suitable for quieter sound sources. They are also especially valued for their clean, high-frequency reproduction. However, condensers all contain preamplifiers which may distort at high dB SPL (such as with drums). Make sure to check that any condenser mic purchased for loud stage applications have a gain reduction switch (known as a pad), as these are able to provide the same sensitivity for low-level sounds but can also be used at reduced sensitivity with very loud sources or sources with great dynamic range. Also be aware that a condenser mic must be designed to be handheld to do well in live sound situations, as their greater sensitivity makes them prone to excess *handling noise*, the unwanted noise generated when the cable or mic are moved. Those designed for live applications will usually have an internal *shock mount*, a stretchy nest of elastic bands that hold the delicate body of the mic suspended in its center leaving room for it to vibrate when disturbed—the elastic absorbing all the force that might otherwise cause the clicks and rustles typical of handling noise. The Shure condenser mic named above is one made for being handled, and another is the *Rode NT1-A*. Both have a reputation for being sturdy and able to meet the challenges of being on stage, and both are under $250, a very reasonable price for the product.
4. *Frequency response*: Condenser mics usually have better high-frequency response, and often have a smoother frequency overall as well. They are known for having a crisper and more transparent sound than dynamic microphones and prized for their ability to capture the breathier high-end

Figure 9.3. External shock mount fixes issues relating to handling noise. Note that for live applications the same principle is applied internally.

harmonics that dynamics often can't reproduce as accurately.

5. *Address*: May be either front (or top) address or side address. Most mics used for live sound applications use front address.
6. *Condensation effects*: Large amounts of moisture can cause issues that affect the sound condenser mics produce. With enough time (a day or two) and a dry environment a condenser will dry out and return to normal. This is not much consolation when the lead vocals are starting to sound muffled and the mic begins to produce an annoying click halfway through the second set, though if someone is smart enough to keep a spare around it's a minor inconvenience. While humidity can be an issue, it's more likely to occur in a cold studio than a warm stage, and either way in most climates a good pop filter and proper vocal technique will see most condensers to the end of the show with no problem. In addition a type of condenser mics called RF condensers are not susceptible to these issues and can be an option in "deal breaker" cases where normal condenser mics can't hold up.

Applications

Condenser microphones are most commonly used for quiet sounds, and sounds with frequencies that reach into the upper register. A few examples include:

1. situations where quieter instruments mean more sensitivity is required (i.e., acoustic guitar, percussion, piano, some vocalists);
2. situations where more high-frequency detail is desired (i.e., acoustic guitar, percussion, piano, cymbals).

It bears reminding that when a class of mic is described according to limitations and strengths these are generalizations based on trends; there are usually exceptions to most of the tendencies listed. The large group at the center of any bell curve is not the majority, only the largest homogenous group that is considered average. For each bell curve one should keep in mind the numbers falling outside the very center of a bell curve are greater when added together than the group described as the average. So the rule is usually that most things don't fit the rule exactly, and the same goes for audio equipment. There are rugged condensers and dynamics with smooth frequency response; the tendencies are only the "most likely," not the "only available," characteristics for each class. As with any audio component, the real proof of its value is in how it works for a particular situation. Try as many examples of each mic type as you can whenever you get the chance; only then will you have the reservoir of microphone knowledge required to know exactly which mic is perfect for any situation.

Diaphragm Size

No matter what method a mic may use to convert sound to signal, a feature they all share is the diaphragm, a flexible membrane of metal, paper, or plastic so thin that as the sound waves in the air come in contact with it, it responds by moving and forming a shape that captures all the details and textures of that sound, details retained in the signal once it's been converted and enters the P.A. system. However after the diaphragm captures the shape of sound, the method used to move it into the system is dependent on that class of mic.

- *Large diaphragm*: Microphones with a diaphragm larger than $^3/_4$ inches are called "large diaphragm" microphones. In general, Large diaphragm microphones tend to have a "big" sound desired for vocals. While not as likely to be as accurate at high frequencies as a well-designed small diaphragm, greater surface area means they are usually more sensitive than small diaphragm or medium diaphragm mics. Many claim that large diaphragm mics are the best for capturing the fullness of low frequencies. Large diaphragms enhance tonal characteristics in sound; this may create more low-end bass under the right circumstances and account for some of the richness in bass response.

- *Medium diaphragm*: Until recently, there have been only large and small diaphragm mics. More recently the medium size became a separate class, though the precise size limits are argued over. Not everyone agrees on the proper definition of a medium diaphragm but even if they can agree on the sound most associated with this class, most agree that mics with a diaphragm near $^5/_8$ to $^3/_4$ inches can be characterized as a medium diaphragm. Many feel that medium diaphragm microphones are accurate with high-frequency content and catching transients (similar to a small diaphragm) without sacrificing the warmer sound of a large diaphragm.
- *Small diaphragm*: Any diaphragm smaller than $^5/_8$ inches can be called a small diaphragm. Small diaphragm microphones are good at capturing high-frequency content and transients. They are also considered to produce a more airy sound with less coloration than medium or large diaphragm microphones.

Directional Pattern (Pickup)

Microphones can be distinguished by their directional properties, i.e., how well they pick up (or block) sound from different directions. Most microphones fall into two primary groups: omnidirectional and directional.

Omnidirectional

Omnidirectional microphones are the simplest in design compared to others, and pick up sound equally from all directions. Size is a factor influencing sensitivity, and larger omni mics tend to become slightly more directional to higher frequencies sounds because the larger body blocks the shorter wavelengths of high-frequency sounds arriving from the rear. The smaller the microphone body, the closer to truly omnidirectional the microphone is.

Omnidirectional mics are not often used in live applications, but when they are used, it is usually for close-up use on loud sound sources, because no proximity effect (see below) means there is no increase in bass response.

Directional

Designed to respond to sound from the front (and rear for bidirectional), *directional microphones* tend to reject sounds coming from other directions. The more common directional microphones display a heart-shaped polar pattern, and, as a result, are called

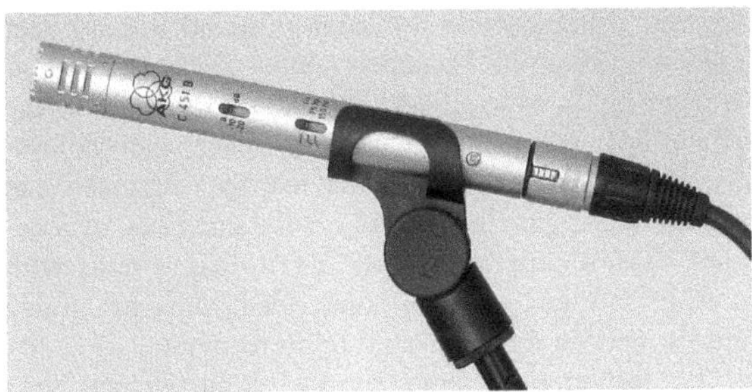

Figure 9.4. Mics come in all shape and sizes. This one is a C451B small-diaphragm condenser microphone by AKG Acoustics. *Source: Image by Harumphy at en.wikipedia.*

"cardioid" microphones. Common directional mics include *cardioid*, *super cardioid*, *hyper cardioid*, and *bidirectional*.

Directionality means that in a noisy environment the cardioid picks up less environmental noise from around

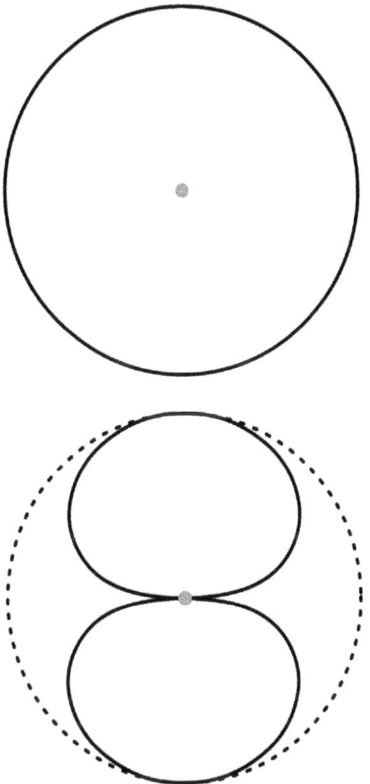

Figures 9.5 and 9.6. Polar patterns showing directional sensitivity of omnidirectional and bidirectional microphones. *Source for both: Image by Omegatron under a Creative Commons BY-SA 3.0 license.*

the mic, which enables it to capture sound with a better ratio of wanted to unwanted sound. Because of their greater ability to suppress unwanted noise, reduce the effects of reverberation, and increase gain-before-feedback (another way of saying the engineer can raise the P.A. gain higher without the same risk of feedback than is possible for performers using mics with a different type of pickup), directional mics are almost always chosen for live sound reinforcement.

It's important to remember that polar *patterns* tests are run in an anechoic chamber simulating an ideal environment (i.e., with no walls, ceiling or floor to reflect sound). In the real world sound readily reflects, so off-axis sound can bounce off nearby surfaces and into the mic. Because of this normal use rarely provides as perfect a directional response. So cardioid microphones do help reduce unwanted sound, but rarely do away with it entirely. Nevertheless, cardioid microphones generally reduce off-axis noise by about two-thirds.

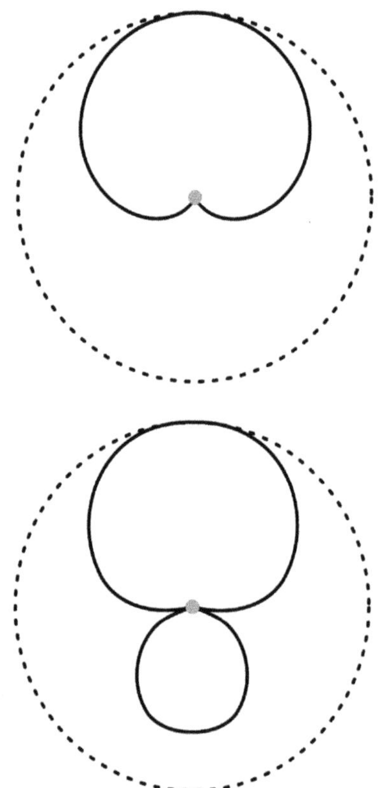

Figures 9.7 and 9.8. Polar patterns showing directional sensitivity of cardioid and hypercardioid microphones. *Source for both: Image by Omegatron under a Creative Commons BY-SA 3.0 license.*

Directional pickup also varies with frequency, with some cancellation of lower frequencies unless the microphone is very close to the source. Because they are more sensitive to low frequencies when used closer to the sound source, they are prone to the proximity effect (see below) to some degree. Generally the better the microphone, the more uniform the rejection over a wide range of frequencies.

Most microphones designed specifically for vocals are cardioid or hyper cardioid, and have frequency response tailored to the vocal range and reduced bass sensitivity to counter the proximity effect. It is important to remember that improper mic handling can greatly reduce the performance of a mic, and in some cases cause major sound issues. In most directional mics, vents in the rear allow the microphone to respond to the difference in pressure between the front and back of the diaphragm. Covering the vents reduces or cancels the microphone's directional properties essentially making it omnidirectional; and *cupping a cardioid microphone basket in your hands should be avoided as it instantly causes feedback.*

Recently, manufacturers have taken much of the dilemma out of mic purchases by offering products with switchable polar patterns like the condenser mic in Figure 9.9.

1. *Cardioid*: Picks up best in front of the mic. Partly rejects sounds approaching the sides or rear of the mic but rejects sound best toward the rear.
2. *Super cardioid*: Picks up best in front of the mic and partially rejects sounds approaching the sides or rear of the mic. Has a narrower pickup pattern than regular cardioid microphones.
3. *Hyper cardioid*: Picks up best in front of the mic and partially rejects sounds approaching the sides or rear of the mic. Has a narrower pickup pattern than super cardioid microphones.
4. *Omnidirectional* or *omni:* Picks up equally well in all directions, with no proximity effect.

Proximity Effect

As you get close to directional mics (about 2 inches) an increase in bass response becomes obvious. This is known as *proximity effect*, a characteristic not found in omni microphones. Proximity effect can be unwanted or desirable, according to how it is used. A singer can make creative use by lowering their volume as they move in close to achieve a deeper, richer sound. This can provide a more dynamic overall vocal sound, but takes some practice. Some

Figure 9.9. Neuman U89i—large diaphragm condenser microphone with five switchable patterns.

vocalists prefer to sing very close to a cardioid throughout the performance to "beef up" their vocal style with more bass. On the other hand, vocalists moving the microphone in and out without adjusting the power of their voice create changes in the overall mic level and issues with tonal balance.

Proximity effect can be used effectively to cut feedback in live reinforcement applications. When you boost the low frequencies of a vocal mic (for example) with EQ, it becomes more sensitive to "bleed" from other low frequency sound (such as a nearby kick drum) and more susceptible to low-frequency feedback. If a performer works very close to the mic without needing the additional tonal "beef," an equalizer can be used to reduce bass response on that channel, making the microphone less sensitive to low-frequency signals arriving from more than a foot away (such as are produced by a kick drum). This technique will help reduce feedback at low frequencies, as well as reduce the effect of any handling noise.

Microphone specs

1. *Self-noise*: The electrical noise or hiss a mic produces is known as self-noise. Measured through a filter to correlate more closely to its annoyance factor as well as to simulate the frequency response of the ear, 14 dB SPL or less is excellent, while 30 dB is OK; Because a dynamic mic is passive, it doesn't produce much self-noise. This is why spec sheets for dynamic mics do not specify self-noise.
2. *Signal to noise ratio*: The difference in decibels between 94 dB and the mic's self-noise. The higher the S/N ratio, the less noisy the signal.
3. *Sensitivity*: How much output voltage a mic produces at a given SPL. A high-sensitivity mic puts out higher voltage (a stronger signal) when exposed to loud sound, so needs less gain at the mixer, meaning less potential noise. When you record quiet music at a distance (classical guitar, string quartet), use a mic with high sensitivity to override mixer noise. When used for loud sources or close to the source, sensitivity matters much less, except for perhaps being too high to handle spill.
4. *Frequency response*: Range of frequencies that it will reproduce at an equal level (and within a tolerance, such as ±3 dB). This is how accurately the mic picks up the complete frequency range. A good mic will have a full graph of the frequency spectrum on the X axis and the slope of the lines show how well the mic will "hear" at each frequency. If the response of a mic is a flat line across the middle of the graph (only test microphones are even close to this) that would mean that the mic picks up low-end frequencies with the same accuracy as it picks up high-frequency sounds. Some microphones are more sensitive to a specific range as part of their design. For example a "presence peak" around 5 to 10 kHz sounds crisper for some vocalists and sound systems (but can be too harsh where there is already enough high-frequency response) this type of response is called tailored or contoured. Be aware of specs that only state the variance of the mic response (±6 db) without providing the reference points that tell you what frequency range the response occurs under (±6 db @b 20 Hz–20 Khz). *Some microphones have switches that alter their frequency response.* These microphones should include information about the frequency response at each setting.
5. *Impedance*: This spec is the mic's effective output resistance at 1 kHz. Mic impedance between 150 and 600 ohms is low. Live sound applications

always use low-impedance mics so long mic cables can be utilized with less risk of picking up hum or losing high frequencies.

6. *Transient response*: Microphones' ability to capture a very sudden or percussive sound, with fast attack and release. Smaller diaphragms have a better transient response than larger diaphragms and condenser microphones have a better transient response than dynamic microphones (no matter the diaphragm size). A decent transient response is essential for capturing percussive sounds like drums, and piano. The "attack" of these instruments is a larger part of their timbre.
7. *Sensitivity* (at 1,000 Hz Open Circuit Voltage): Microphone sensitivity is measured by playing a 1 KHz tone at a set dB SPL level originating from a source that is 3 feet away. If you compare the dBV levels produced by each microphone, the mic with a higher dBV output (from the same sound source) is the more sensitive one. Note that the standard test used to determine microphone sensitivity has changed over time. It used to be that the 1 KHz tone was set to be 74 dB SPL at one meter, and then tests began to use a 1K tone of 94 dB SPL. When comparing the sensitivities of an old classic microphone and a more recently released one, make sure to check which standard the reported results used.
8. *Max SPL:* If the maximum SPL spec is 125 dB SPL, the mic starts to distort when the instrument being miced is putting out 125 dB SPL at the mic. A maximum SPL spec of 120 dB is good, 135 dB is very good, and 150 dB is excellent. Dynamic mics are less likely to distort, even with very loud sounds. Some condensers are nearly as good with the switch used to lower sensitivity and reduce chances for distortion. Because a mic pad reduces signal-to-noise ratio (S/R), avoid using it unless the mic distorts or picks up unwanted sound.
9. *Polarity***:** The polarity of the electrical output signal to the acoustic input signal. The standard is "pin 2 hot." That is, the mic produces a positive voltage at pin 2 with respect to pin 3 when the sound pressure pushes the diaphragm in (positive pressure). Be sure

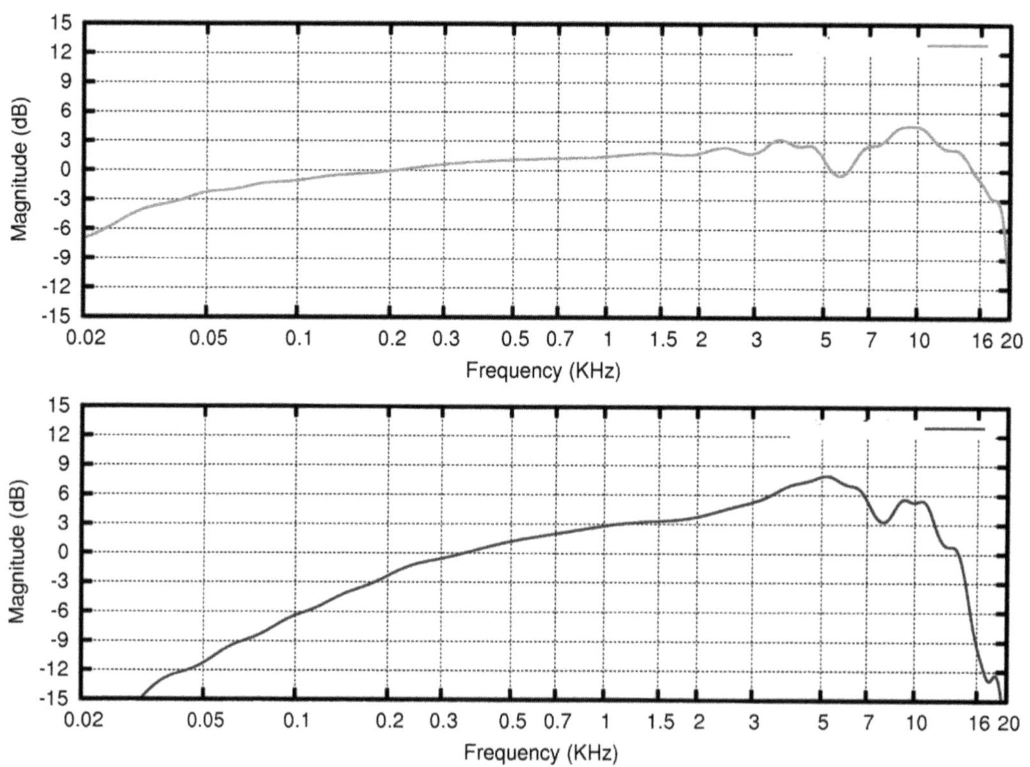

Figure 9.10. Not all microphones have the same frequency response. Choose the mic with the best frequency response for the application.

that your mic cables do not reverse polarity. If some mic cables are correct polarity and some are reversed, and you mix their mics to mono, the bass may cancel.

10. *Distortion*: The THD (Distortion and Total Harmonic Distortion) rating given reflects the percentage of distortion for a certain dB SPL value. For example: THD = 0.001% @ 120 dB SPL.

About Choosing a Mic

While general characteristics of mics and specification sheets are a good way to narrow your choices down, and reviews are a great way to learn about products you wouldn't have known to consider without being turned on to them, the only sure way to get to know mics is to try out many different ones. As you get used to using them and grow to understand the different qualities of each kind, don't be afraid to try them in new ways not already mentioned in a guide; that's the only way cool tips are discovered, and when you find one, you can share what you've found. In the meantime, the mic is where your sound starts, but the next link in the chain is where you will make many of the choices that shape the sound it captures.

Chapter 10

The Mixer

Shaping the Signal with the P.A. UI

While every part of the P.A. is equally important, the mixer is where much of the magic happens. Of course there will always be factors that will impact the sound and are out of the control of the live engineer. One example includes the acoustic qualities of the room that can't be accounted for with careful setup and tuning of the P.A. system. Another is the quality of the band's equipment and talent. Aside from such factors which can't be helped, and being sure to make the proper setup choices to maximize the system's function, once the P.A. is connected and tuned properly, it's what happens at the mixer that most determines the kind of sound that the P.A. will produce. Before we can set it up and mix with it, however, we need to map out the lay of the land and learn what each control is supposed to do.

Mixers: The Brains of the Operation

At the simplest level, an *audio mixer* is a device which *sums* or combines two or more separate signals then *routes* the resulting signal, providing instructions that direct the signal where you want it to go, though the mixer does so much more than simply directing traffic. The signals typically all originate from the mics and instruments of the performers on stage (via instrument pickups/DI boxes, and microphones), and are routed to the mixer from these sources using a stage box and snake.

While mixers include some functionality that goes beyond the simplest tasks of summing and routing signals, including the ability to add or subtract from sound intensity, it is not any particular feature but the weight of everything that these basic tasks can achieve when combined that makes the mixer essentially function as the "brain" or control center of every P.A. system. This is where gain structure is established and crucial processing choices occur, with channel controls allowing the operator to adjust the tonal quality of each incoming signal with the channel EQ controls, use the pan controls to adjust for a stereo or mono mix, and adjust the levels of incoming signals as well as the levels of outgoing signals. The operator controls the signal's routing both within the mixer and upon output allowing for more complex signal manipulation as well. For example, the signal of any channel can be routed to outboard processors and then returned to the mixer so the processed signal can be adjusted further. Signals may additionally be duplicated and the duplicates assigned to their own channels; this allows for a duplicate signal to be processed for one mix, and the original signal kept clean to be used in a separate mix, or for a copy of any signal to be processed before being recombined with the original.

Finally, after all signals are adjusted as desired, they can be summed into various mixes and appropriately routed to continue on their journey to the end of the chain. In short, it is where the mix comes together. While we describe it as the brain, it is equally correct to view the mixer as it is more commonly described, as the beating heart of the P.A. system.

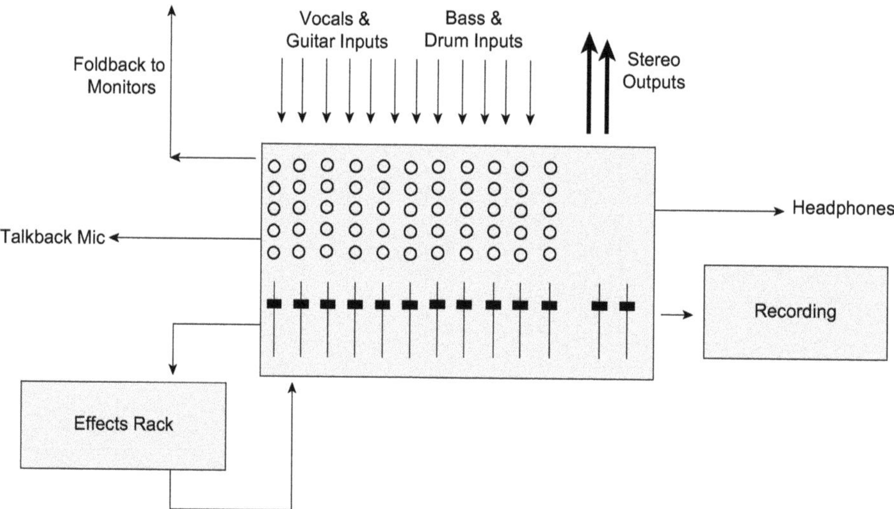

Figure 10.1. Mixers allow you to input many signals, adjust them in key ways, mix them together, and finally send them where you want them to go.

Mixer Types

In line with the varieties that are available, mixers are also referred to by the terms console, board, desk, or deck, though there is little agreement about whether any of these terms apply to specific types of mixers and they are often used interchangeably. However, very generally, the largest mixers are usually referred to as desks and consoles, while mid-sized and smaller mixers are more often called mixers and boards, and DJ mixers are called decks. By any name mixers come in a variety of flavors and vary greatly in their size and mix of features.

Mixers' features are largely dictated by the applications they are intended for. In addition to live audio mixers, there are mixers built for other audio applications. While most of this chapter is dedicated to live audio mixers, it is worth it to point out some of the features that distinguish other audio mixers from live audio mixers. Since all audio boards share many features, the best way to tell them apart is by the small differences. The ability to do this can be useful, as often when researching topics online, for example "feedback" or "headroom," the articles that address such issues don't always say if they are referring to live sound or studio sound applications. Being able to tell if an article is about live or studio mixing by looking at the mixer photos in the article or looking for identifying keywords within the article will save you time when researching and allow you to better evaluate audio information you find online. (This doesn't mean a live engineer can't find valid information in articles on studio engineering, as many facts will apply to either mixing application, but before you decide if you can count on information, knowing what application is being discussed is an important first step!)

Studio and Broadcast Consoles

Mixers built for music production and reproduction are known as *studio consoles*. Studio and live consoles share many of the same features, though each also has features geared to how it is used. No matter which design, studio consoles are tailored for recording. They will have an additional insert on top of line and mic for each channel, called tape returns or direct out, so each channel can be recorded to tape. They will also allow the operator to monitor the signals being sent to their recording device, and listen to the recording play back. The monitor section will have pots or faders to adjust the level of the recorder's outputs, and some desks will have basic controls on the monitor inputs, like pan and EQ. Studio boards will also provide controls for the studio monitors used by the musicians.

Studio consoles will typically have many more inserts and channels than live mixers to allow for *multitrack recording* or *tracking* for short. Essentially, tracking is

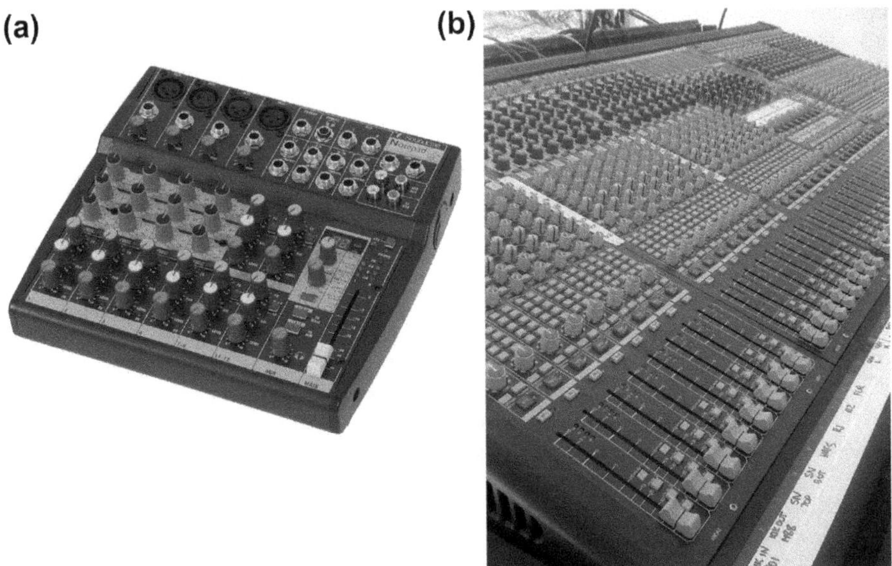

Figures 10.2a and 10.2b. Mixers come in all sizes! You could fit the Soundcraft Notepad124FX on the left in your backpack. The mixer on the right would typically be referred to as console or desk because of its large size.

a way of recording that assigns one track to each channel and allows for the simultaneous recording of many sound sources at once, while keeping each source on its own track, separate from the others. It is also a convenient way to record multiple versions or tracks of selected sound sources, either simultaneously with multiple mics capturing a single source, or over multiple separate *takes*, which is often how vocals are recorded.

Studio consoles can be designed as split consoles or inline consoles. Split consoles spread the mixing functions out across a pair of boards that look a lot like a live board and monitor board placed side by side, while inline consoles conveniently group controls together. Inline mixers have additional features designed to make recording even easier and are very flexible in how you can assign controls. They allow one to use both channel and monitor paths as separate inputs with separate faders/pans. Inline consoles also have the channel section and the monitor inputs on a single channel strip. Each channel strip has two faders, one for "to tape" levels and the other for monitor levels.

Broadcast mixers for pure audio are specifically radio broadcast mixers. Because they need far simpler controls, live boards are often used for radio broadcast, especially since many broadcasts transmit live shows any way, making live audio boards the best for taking care of all mix tasks at once. Often on a broadcast board there are fewer auxiliary (aux) controls, it's assumed that there won't be a large number of mixes needed (for instance no need for monitor mixes). The EQ section may not have as many options, and the metering is also usually simplified since the broadcast level leaves so much headroom there is little danger of overload. There often aren't channel meters or clip lights for the same reason.

Some features unique to radio broadcast consoles include:

- the presence of a tone generator so an agreed upon "zero" level can be established for tape machines and other sources receiving the mix;
- a *detent* mode with the fader all the way down, pulling back on it a little more sends the input of that channel to a cue speaker so the operator can hear what is coming in on that channel before they bring the levels up;
- faders often inverted (upside down);
- a feature called a *mix/minus* which is basically a "program" mix, minus the incoming signal of whoever is going to have the mix returned to them through the phone. A caller's input signal is flipped 180 degrees out-of-phase before it is added to the total mix going down a phone line to the caller (such as a listener or a field reporter calling in). Note that this is different from the polarity reverse switch at the top of a channel strip as it only reverses the signal to a specific mix bus output (instead of to the whole mix). It is used to send a mix back along a phone, minus the caller's signal, so they won't hear their own voice delayed on the return mix.

Two features of live boards that won't be found on other audio mixers are VCAs and "matrix" controls. These features are both targeted at making it easy for the live audio engineer to create and manage multiple groups (bundled channels that can be manipulated as a group). Grouping is an important feature in live mixing because the task requires the engineer to manage a lot of information in real time, especially when handling FOH sound and mixing monitors at one time. By grouping channels the operator reduces the number of faders they must deal with; this allows them to respond to the sound in real time with much more efficiency than would be possible otherwise. Both features will be discussed in more detail later in the chapter and following chapters, but for now all you really need to know is that if these grouping features are referred to or seen on an audio mixer, you can be sure the reference is about a live audio application.

Rack Mounted Live Mixers

Live audio mixers include more than the FOH mixers that typically come to mind, and can range from very small boxes with just a few knobs and switches, all the way up to the massive and complicated units utilized for the largest multi-act festivals. *Public address mixers* have a mic, phone handset, or other voice input and the ability to select channels that route the announcement to specific rooms or areas of the building or to the entire building. Another type of live sound mixer includes simple *installation mixers* and *auto mixers*. Installation mixers and/or auto mixers are small rack-mounted mixers for installed systems such as can be found in conference rooms or retail establishments. Built with auto features and/or pre-sets that will allow anyone to use the mixer for limited applications, such as simple speech applications or prerecorded music. Features like auto delay for public speaking, auto ducking to lower sound levels when a specific mic is used and turn the mic off when it is idle, and auto feedback suppression to

Figure 10.3. Yamaha IMX644 rack mount digital installation mixer. *Source: Photo courtesy of Yamaha.*

ensure the system can be used for basic applications even without an engineer on site.

FOH Mixer Anatomy

All analog live mixers will follow a similar pattern, even as each will include a different mix of features and unique layout within that common pattern. One reason why analog mixer layouts can be counted on to follow this similar pattern is because each control on the surface of the mixer is directly coupled to the circuit inside the unit that is performing the desired function.

Hands on Control of your Sound

Each of the knobs and faders on a mixer are potentiometers (called pots for short) that directly control the signal's content and direction in the mixer circuitry position directly below. Pots looks like plain knobs, but the business ends of them are actually circuit components—sliding resistors that act as adjustable voltage dividers which either add or subtract from the circuit voltage and change the content of the signal moving through *that* circuit. Pads act much the same way; when we engage or disengage them we are opening and closing switches that lie directly along the signal's path. As we block off one path, we open another, thereby routing the signal along the path we've chosen. Each time you make an adjustment on an analog mixer, picture the signal pulsing and changing its content or direction only an inch or two below your hand on the other side of the mixer's outer panel, because that's essentially what's happening. That's why analog mixing is as close to molding sound with bare hands as you can get!

Regardless of specific controls, all mixers will follow this familiar pattern:

- a variety of plugs and jacks for connecting the mixer to other P.A. components with cables, which serve as entry and exit points for the analog signal to be routed through—will mainly lie along the top of the mixer's face panel and along the back panel;
- a number of channel strips for individual control of each incoming signal—these will run vertically from top to bottom, and be placed next to each other in a row running from the left side of the mixer to the right;
- a main control panel on the right side of the mixer split into two parts, the group fader controls and main fader on the bottom right, and the main controls and metering on the top right.

A common feature shared by all mixers, and a term you will encounter in most discussions relating to mixers is the bus. One of the central tasks of the mixer is to take signals from different sources and combine them into one or more signals before sending those signals to the proper output or outputs. The place where they come together is the bus. At the same time, another task taking place in the mixer is the duplication of individual signals to be sent alone to one processor or another. The pathway along which a signal's duplicate is created and splits off to be processed is also a bus. The important thing to recognize is that every bus can have several inputs but only one output, though that output may go to more than one destination (including another bus).

The number of buses is one measure of a mixer's complexity. In the simplest monophonic mixer, all of the input signals are sent to a single bus, to which is connected the output. In a stereo mixer, incoming signals are combined to a pair of buses, corresponding to the mixer's left and right outputs. Beyond these most basic situations, as the number and variety of routing possibilities increase, so do the number of buses, one for each additional pathway/destination.

Inputs and Outputs

Until we hook it up to the P.A., the mixer is no more than fine sculpture for gear heads (or a paperweight depending on the mixer's size). When describing a mixing board according to its components and functionality, the first and last words need to be about the mixer's ins and outs, where signals are routed in and out of the mixer. Mixers have two places where most of these are clustered. A mixer may have the majority of their ins and outs along the back of the board, with a small number along the top of the board, or they can

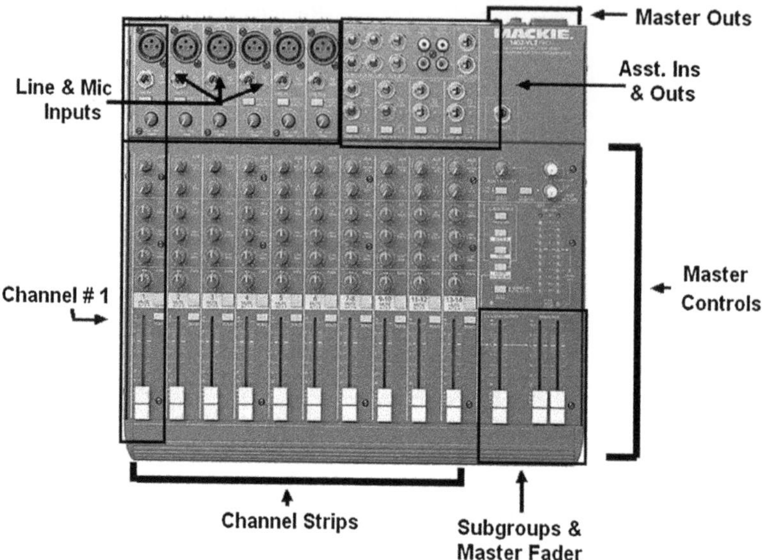

Figure 10.4. Basic analog mixer layout (Mackie-1402-VLZ-PRO).

have the inverse arrangement, with the majority of these along the top strip, and a few along the rear.

Inputs and outputs may be balanced or unbalanced; those who are interested in learning more about balanced audio can refer to Chapter 14. In the meantime there are three ways of identifying whether an input is wired for balanced audio or not:

- The use of balanced connectors (see Chapter 14) usually means the insert is balanced, while the use of unbalanced connectors means it is unbalanced. This method is not foolproof, as some connectors can be wired to be either balanced or unbalanced.
- The insert is labeled as balanced (BAL) or as unbalanced (UNBAL).
- If you are in doubt and the insert is not clearly labeled, refer to the product documentation to find out if the insert is balanced or not. This method is a sure thing as the manual or specification sheet (or both) will always include this information.

Inputs are also differentiated by their *signal level* (see Chapter 14 for more info on signal levels) which refers to the voltage level of the signal. Signals come in three general levels:

- mic-level (low voltage): microphones produce a mic-level signal (−50 dBu standard microphone level, −60 dBu common lowest microphone output);
- line-level (a little higher): almost anything which runs on AC or DC power produces standard line-level signals of −10 dBv (.03162Vrms), while pro audio gear line level is higher +4 dBu (1.228Vrms);
- speaker-level (very high): speaker-level signals are produced by amplifiers and are only meant to plug into a speaker. *Never plug a speaker-level signal into anything other than a speaker.*

Figure 10.5a. Rear panel of a mixer with rear panel inputs/outputs (Yamaha-MGP16x). *Source: Images courtesy of Yamaha.*

Ins and Outs and Mixer Models

Mixers are generally described with respect to the number of inputs, sub master outputs, and *stereo* or *mono* outputs they come with. So a 16 x 4 x 2 mixer will have 16 inputs, four sub master outputs, and two outputs (a stereo pair). Many mixer models have this information included in their product names, with the initial number indicating the number of inputs and those that follow referring to the number of sub buses

Figure 10.5b. Face panel of Yamaha MG206C-USB—notice the input configuration is reversed, with most inputs on front panel. *Source: Images courtesy of Yamaha.*

or combined outputs. Examples include the Mackie Onyx 1640 which has 16 ins and four subs, as well as the Yamaha MG206C with 20 inputs, and six outs (four subs plus two stereo outs).

Common Ins and Outs

- *Microphone inputs*: Mixers must be able to admit both mic-level and line-level signals for each channel, but will only use one (mic-level signals must be sent through the mixer preamplifier). Most mixers have a switch near the input or on the channel control strip which will allow you to select between inputs and/or signal levels. In some mixers, inserting a jack plug disconnects the XLR socket. Microphone inputs have female XLR (balanced) inputs, and are where microphone signals.
- *Line input*: This is a channel input specifically designed for line-level signals, and where you would plug in the line-level output of a signal source.
- *Main inserts*: Inserts are ¼-inch jacks which are bi-directional, i.e. the jack functions as an input and output simultaneously allowing the signal to flow in both directions. They require the use of a "Y" cable: one side has a single ¼ TRS plug where the other side of the Y cable has two ¼ TS plugs. The single side of the Y cable plugs into the insert jack while the other side of the Y goes into the input and output of the processor. Main inserts are used when processing 100 percent of the audio signal, as processing is applied to the mixed signal just before it leaves the mixer out the main output.
- *Channel insert*: Similar to main insert, but processing applied to single channel signal (routes signal to an outboard processor, then routes processed signal back). Allows outboard equipment like compressors and noise gates to be introduced into the signal path immediately after the channel preamplifier. Note that the insert actually interrupts the input signal, so if

Stereophonic and Monophonic Sound

Most mixers nowadays are stereo mixers. Stereo sound is dual channel sound, where the entire sound output is divided between two channels. Because each channel contains only half the signal content, both channels must be played back together (typically with each channel assigned to one of two loudspeakers, left and right), in order for listeners to hear the entire sound content. Mono sound is single channel sound; 100 percent of the sound content is contained in one signal, requiring only one channel to transmit and reproduce it. It is easy to turn stereo sound into mono sound by simply summing the single channel of sound referred to as a *summed mono* signal. Splitting a mono signal to produce a stereo signal is not actually possible, though there are ways to duplicate a mono signal and process the results to create an imitation of true stereo, mono sound with a *stereo effect*.

a cable is in the jack and not plugged into a processor there will be no signal on that channel.

- *Group insert*: Works as the inserts above, except here processing can be applied to a grouped signal prior to metering.
- *Main output:* The main output on most mixers is a stereo output, using two output sockets (one for each stereo channel). The connectors are normally three-pin XLRs but can also be ¼ jacks or RCA plugs. This is where the mixer connects to the P.A. power amps and house speakers, and where the final house mix is routed to the output.
- *Sub/group outs*: Sub/group outs are group aux sends that carry the sub-mixed signals that are routed into those sub-mix channels. This enables the operator to process a grouped signal prior to routing it to the main output. For example, rather than processing each backing vocal signal individually, all backing vocals can be assigned to a group then sent to be processed via this output. The connectors can be three-pin XLRs but more commonly are ¼ jacks.
- *Auxiliary sends/returns*: Situated in the master section, the auxiliary inputs (aux return) are always line level and in stereo. Whereas auxiliary outputs (aux sends) are line level but are most commonly mono and are found in each channel as well as in group and master controls, auxiliary inputs and outputs (commonly named as the aux path) are essential tools for routing and processing that can be used in a variety of ways. One common use is to create a looping path between mixer and signal processors to send signal on for processing (commonly used with time-based effects). Auxiliary outputs are also commonly used to send custom monitor mixes to the stage musicians.
- *Effects sends and returns*: These jacks are designed as dedicated jacks for sending and receiving signals to external effect units. Very similar to the aux sends and receives jacks, except lacks the added functionality of aux sends/returns (such as the ability to send custom mixes to chosen locations).
- *Amp in or P.A. in*: Powered mixers often have a direct input to the built-in power amp. This is a special ¼-inch jack socket with a switch that cuts the signal from the mixer to the power amp as soon as you plug into it. Now only the signals entering through the amp in or P.A. in jack will be amplified.
- *Tape inputs/sends*: Tape inputs are usually RCA ("phono") type and are not balanced. For connecting to recording devices like two-track tape recorders, they usually can be found among the main connectors. The tape outputs are generally RCA (phono). A tape send or tape out control could be among the masters to modulate the level, though it is possible that the main or sub stereo masters control them; if in doubt, check your mixer's documentation. If a tape in level control is included (to adjust sound so recording can be monitored), it will be located in the master controls.
- *Talkback*: The talkback input is normally a standard female XLR mic connector with a level control and a line-level output. It may sometimes be routed to a choice of outputs, or include an on/off button. The talkback mic allows the engineer to talk to performers through the stage monitors.
- *Headphones*: The headphone output may be the same as the monitor feed, or you could be allowed to listen to selectable sources.

Channel Strip

On most sound mixers, input channels take up most of the space. All those identical rows of knobs starting on the left side of the mixer are channel controls. Exactly what controls each channel allows depends on the mixer but most mixers share common features.

The channel controls a typical mixer can include any of the following controls (not necessarily all the same, or in identical order):

- *Input gain/trim*: This knob determines how much amplification the channel pre-amp will apply to the incoming signal. Because of the range of voltage levels of the signals entering the mixer, preamplifiers at the beginning of each channel enable the operator to boost low-level signals like microphone signals and commercial audio line-level signals to a level the mixer can use, before sending the signal on to the rest of the channel. Input gain is normally set once when the source is connected to the mixer and left at the same level. It is not a volume control—volume adjustments for each signal are made through the channel fader instead of the gain control. It also ensures all signals start and remain at the same input strength; any difference in signal strength will be evident as it will be reflected in the position of the faders.
- *Channel meters*: Even very basic mixers often have a single LED *clip light* to indicate channel overload. Others have a row of signal lights: one for signal present (usually green), one for intermediate levels (yellow or orange), one for overload (will flash red). Higher-end mixers will sometimes even include full channel metering in the form of multi-segment LED meters alongside each channel fader, or a meter bridge: comprehensive metering from a separate meter section above the channel strips.
- *Phantom power*: To allow for the use of gear (most DI boxes and condenser mics) that requires a small amount of phantom power to function most mixers can supply 48 V phantom power a system whereby the mixer supplies a DC voltage to the mic's electronics through the same wires that carry audio. This will either be switchable on a per channel basis, or in the form of a single switch that turns phantom power on globally.
- *Attenuation PAD*: A pad button which reduces the input level (gain) of a signal that is too "hot" by 20–30 dB. This can be useful for plugging a line-level source into the mic input when emergencies require that mismatched gear be used.
- *Low-cut filter*: Many boards will have a button allowing you to engage a filter blocking all frequencies below between 80–100 Hz. These filters are often only included on the first few channels and are intended to be used with vocal signals.
- *Equalization*: Most mixers have at least *two fixed bandwidth* EQ controls (high and low frequencies) for adjusting the tonal quality of the signals. Good mixers will also have a couple of sweep able and/or

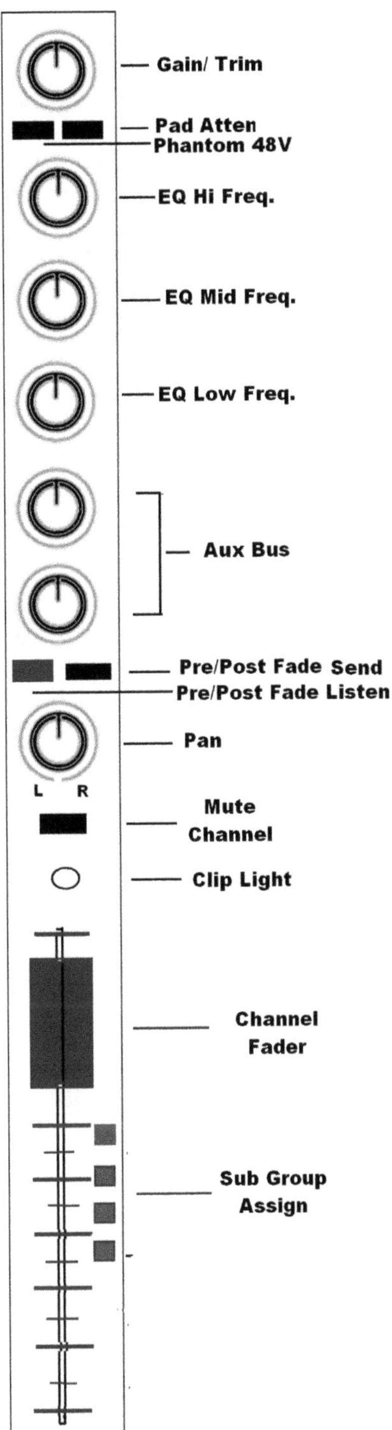

Figure 10.6. Typical channel strip controls.

parametric EQ controls (see Chapter 11 for more on equalization). These enable finer controls because, unlike fixed frequency EQ controls, these allow the user to either set any bandwidth for EQ adjustment, or select one from a range. EQ is there primarily for adjusting each channel's frequency balance (the monitor amp/speaker system has different frequency response than the PA, so usually you will want to send signals to the monitors before EQ adjustment).

- *Auxiliary send*: This enables routing of the channel signal to various *buses* on the main controls (reasons include multiple monitor feeds, private communication, incorporating effects, recording different mixes). Each channel includes anywhere from two to eight selectable sends, as well as switchable controls for each send, or group of sends, that determines if the routed signal will be affected by the channel's fader position. *Pre-fade sends* route the signal before the channel fader. *Post-fade sends* are routed after the channel fader. A number of mixers also include the choice to route some sends before or after the channel's tone controls (*pre-EQ send* or *post-EQ send*).
- *Monitor send/effects send*: For those less expensive mixers which do not have aux sends or include only two, dedicated controls on each channel (usually pre-fade for monitor sends and post-fade for effects) sometimes perform these functions instead.
- *Pan control*: Pan positions each signal along a left to right axis in the stereo soundstage. Leaving all pan pots in the center position allows a stereo mixer to output a mono signal. Pan pots may also be used to control the signal's routing for group assignments (for example, assigning signals to one of two paired (1–2, 3–4, etc.) groups).
- *Subgroup assigns*: These determinine where the channel signal is sent (either to the main mix, or to any of a number of subgroups). Assigned signals are routed to a *bus* and each group of summed signals can be managed as a single signal using the group controls and faders located in the master control section. Each group is then routed back to the main mix.
- *VCAs* (Voltage Controlled Amplifiers): These do the same thing as subgroup assigns, but VCA signal routing allows for a cleaner signal. On boards that have these, they are in addition to groups rather than instead of them.
- *Mute/channel on*: Switch that either mutes or unmutes the channel.
- *PFL/AFL* (Pre-Fade Listen and/or After Fade Listen): This determines the routing of the signal to the headphones and also to one or more of the master meters. PFL is the signal after the EQ and before the channel fader. AFL is the signal after the channel fader, and may be called solo instead.

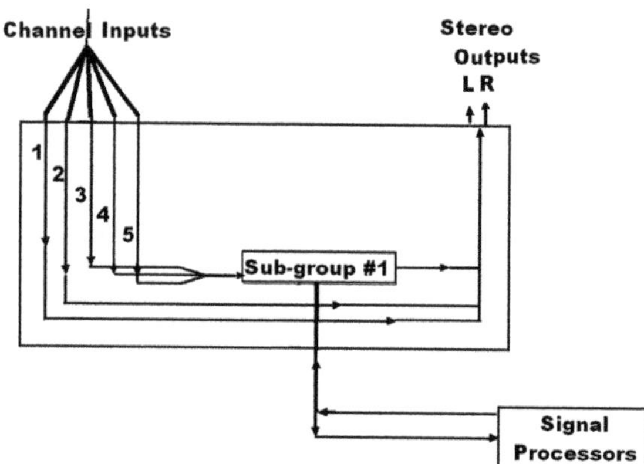

Figure 10.7. Sub-groups help you mix efficiently.

- *Channel fader*: On cheaper mixers this may be a knob. This sets the final level of the channel signal in relation to the level of other channels. May go straight to the main mix or if channel is assigned to subgroup(s), the signal will be routed there.

Master Controls

The master control section is usually to the right of the channel controls and is the control section allowing for control of the main mic, as well as the various sub-mixes. The master section may include:

- *Talkback*: The talkback mic can usually be routed from here to a choice of outputs (for example the monitors of individual performers).
- *Headphone/monitor selection*: This allows the engineer to listen to different parts of the mix, with routing determined either by controls here or by PFL/AFL buttons (may also have a separate output for the engineer's monitor). Always use headphones with their own volume controls as sometimes these outlets are quite loud.
- *Aux send masters*: This sets the overall level of each auxiliary send. There may be a PFL/AFL button for each fader or knob.
- *Mono sum/sub out*: This is an extra fader on stereo mixers for sending a *summed mono* output allows for extra control of the sub-bass level.
- *Aux returns:* These extra inputs for effects returns may be routed to the main mix by default, or

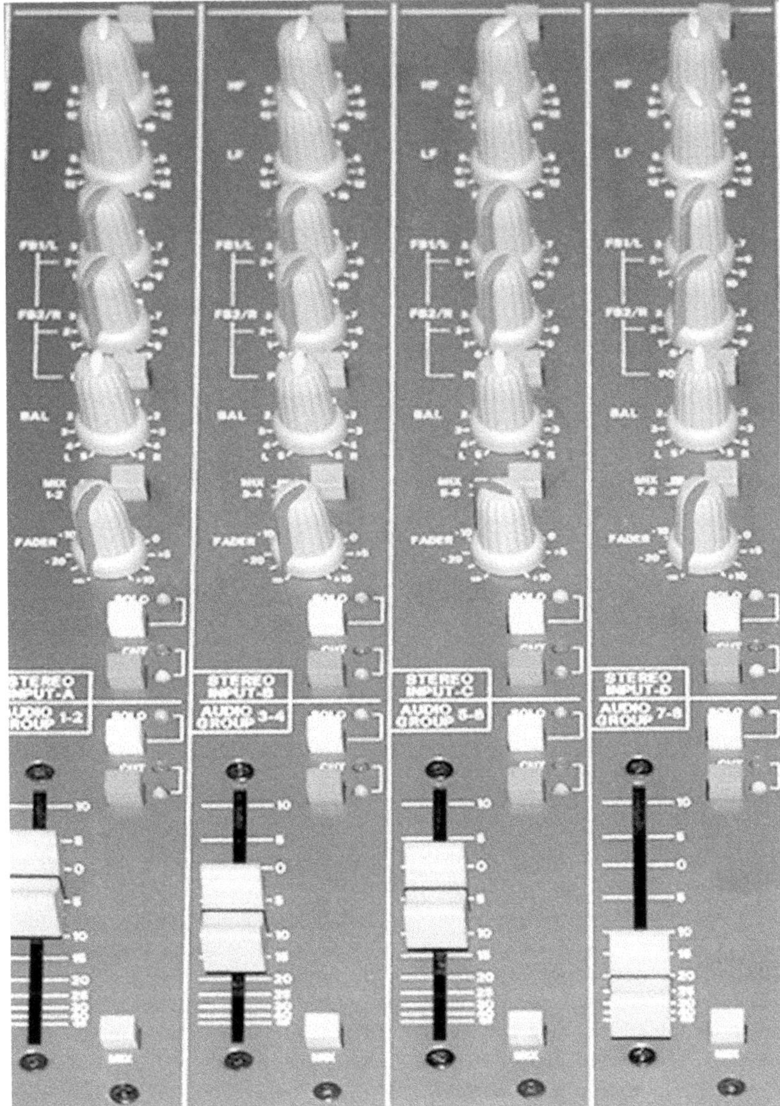

Figure 10.8. Main control panel on Yamaha's MG24/14FX analog live sound console.

may allow you to select routing options for each return.

- *VCA faders*: Fader controls for the level of any channel or group routed there.
- *Group masters*: Each group will have its own fader and channel controls for controlling output from the mixer, and will also have a switch allowing its output to be routed to the main mix.
- *Mix matrix*: Some mixers have a matrix section allowing for even more sub mixing options.
- *Main faders*: Volume control for the main mix (left and right stereo, or single fader mono).
- *Mute*: On/off: switches mute on or off for the groups and main faders.

Meters

Meters may be placed or attached to the top of the board on a *meter bridge*, or integrated in the board. Most

Figure 10.9. Mix Matrix controls.

mixers provide only limited metering functions, with meter bridges available as an option. The two most common types of meters used in live applications are *VU meters* and *peak program meters* (PPMs). VU meters measure average sound levels based on RMS measurements. Designed to represent the way human ears perceive volume, their main drawback has historically been an inability to accurately represent peaks in sound, as well as being too slow to register the fastest transients. VU meters can have analog or led displays, but true VU meters are analog whether the display is or not. RMS meters exist in software form but they tend to read higher as the relationship between RMS and VUs isn't exact. PPMs are designed to measure and display the peak levels of audio signals instead of the averages. This makes PPM meters very good for reading fast, transient sounds, and makes it the meter of choice for most situations—the more useful when pops and distortion are a problem.

While VU meters follow a scale that is fairly standardized, PPM use a number of accepted scales, depending on type. Some meters also try to take a "best of both worlds" approach, displaying both types of measure in a single bar display. In these cases, either the entire bar represents averaged levels with a single flashing led above to register peak levels, or the green and yellow portion of the bar represents averaged levels and the red portion represents PPM measures. Finally, the last type of metering that must be mentioned is one as common as the rest, but which has features quite distinct. This form of meter is the *digital meter*, a method of metering which measures using a decibel scale referred to as dBFS (decibels full scale). This scale counts 0 as the top of the scale, representing the maximum level digital can function without distortion, so all decibels will be negative.

Considering the variety of metering approaches that manufacturers can take, to be sure you understand the metering system on any mixer you will be using, always consult the mixer's user manual to be safe. We will examine how to read meters and introduce the tasks required to safely make use of metering (such as calibrating our meters, and converting between commonly used meters) in future chapters. In the meantime, learning to recognize the three styles of metering in Figures 10.10a, 10.10b, and 10.10c will enable you to recognize 99 percent of the meters that will come standard with live audio equipment.

Monitor Mixers

Most monitor mixing occurs using FOH boards (which work fine as long as they have sufficient channels and

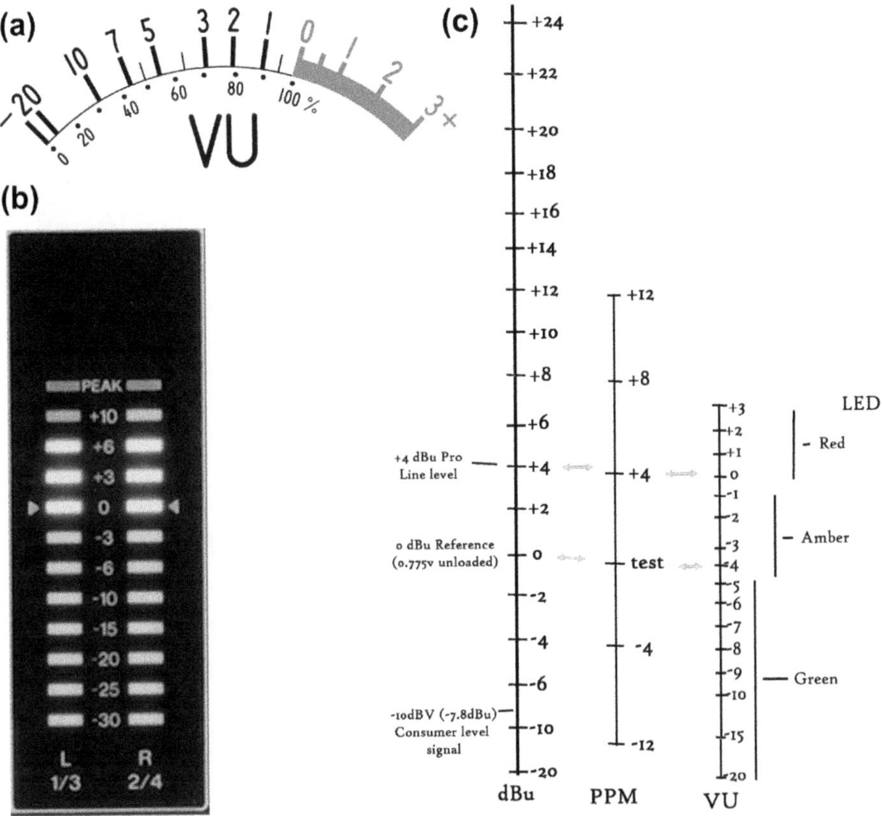

Figures 10.10a, 10.10b and 10.10c. VU meters with analog display, a LED PPM (peaks registering in red at the top of the meter), and a comparison of the most common analog scales with a digital meter using a dBFS scale.

sends), and many manufacturers don't even make boards dedicated solely to monitor mixing. Monitor mixing desks differ from regular studio or FOH mixing boards in the number and type of features included in the design, though the controls shared in common will work much the same way they do on FOH mixers For example, a monitor mixing board will typically have many auxiliary buses and more extensive aux controls, to allow the operator to easily assign, fine tune, and send different mixes to different monitor speakers because the most essential functionality a monitor mixing board needs is lots of room for individual mixes. Similarly, to streamline operation, many of the controls not relevant to the task of monitor mixing will usually not be included in the design of a monitor mixing board. For example, there will often be minimal global controls and no main faders (there is no need for extensive global control and routing for a master mix when the user will not need to send any signals out of the main bus).

Digital Mixers

Earlier digital mixers tended to appear quite different from analogs, with the screen menu as the focus; however, newer products have adopted a more analog layout, accompanied by an on-screen menu allowing the operator to assign the channel's strips and controls. Once only found in studios or the largest stadium venues, recent reductions in the price of digital mixers is making them more common for small and medium venues. Many sound providers also carry them, so you are bound to run into one eventually, even if they are still not as common as analog mixers for live application. Additionally there are now also a number of hybrid boards, like the Yamaha MGP series, which function mostly like analog boards but include select on board digital effects and/or other features usually associated exclusively with digital boards.

Even with digital boards following a more traditional interface design for their visible controls the most

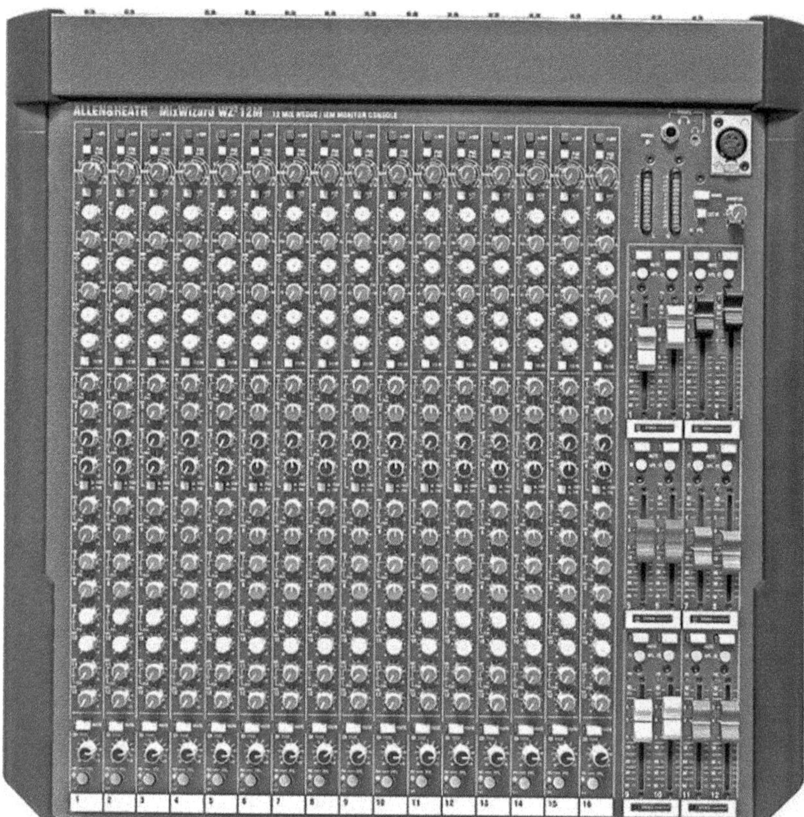

Figure 10.11. Allen and Heath WZ3 12M monitor mixing console.

obvious difference between digital analog systems is still in the layout. Whereas analog mixers have a fixed layout with controls coupled to the circuitry that performs the functions of analog mixing, digital boards allow the operator to assign any input to any set of controls, and to save all board settings as well. For example, the operator could choose to assign channels in the order that is easiest for him/her to access, putting higher priority channels where they can be more easily found. This means much greater freedom for the operator, though it can also add to the confusion of any operator who can't remember the order of their layout.

The assignable and layered control interfaces means digital boards can provide the same control functions in a smaller space than can be achieved with analog boards. This is achieved by using a touch-screen menu where "layers" of channels can be programmed to the fixed controls via this screen. By accessing the menu to choose different layers, many more channels can be operated using a single set of controls. For example, a digital board may have only 12 "channel strips" but, by accessing the menu, each group of inputs can be pre-programmed to attach to those 12 channel control strips in layers. So selecting "layer" one in the menu gives you control of channels 1 to 12, and selecting "layer" two means the controls are now attached to channels 13 to 24. Additional layers can be programmed with aux and group master levels, as well as an infinite number of alternate layouts.

Digital mixers use A/D conversion (often at the stage box and connected with an ethernet cable and not a snake) which converts the analog system to the digital world of zeros and ones. Digital console manufacturers typically incorporate gated channels and frequency controls and compressors on every channel. They also allow for the use of "plug ins" (software expansion programs that can be purchased to provide added functionality). This ensures digital mixers also usually allow for on board access to all the same processors and effects that can be used by analog systems,

Figure 10.12. One of Yamaha's MGP analog boards which incorporates a number of digital features. This blending of analog and digital is very likely one of the audio trends only likely to become more common in the future.

without the added cost or inconvenience of having to buy additional hardware (or the racks used to store/transport it).

There are downsides to digital as well, however. With analog consoles, going into the red on meters during a mix isn't necessarily going to immediately ruin your sound and can even be pleasant. Digital systems offer no headroom; as soon as they are out of bits and bytes, the feedback is loud and unforgiving. You need to avoid digital distortion at all costs by giving yourself a little more headroom than you otherwise would. In addition, the conversion process to and from analog is still not achievable in real time, so digital systems do involve a small delay. Even the best systems include a delay of at least 1.5 ms.

Analog vs. Digital: The Debate

There are also disagreements in the audio realm as to sound quality. Some operators feel that analog boards produce a warmer, richer sound. Conversion is another point of contention. Any digital device will have to turn the analog signal to a digital conversion. Some audiophiles argue that even with the best conversion rate, the result is still not the highest quality signal and feel they can detect this. This is usually not noticed by normal listeners, however, and in live sound there are so many other factors messing with the sound quality that conversion often goes unnoticed even by audiophiles. On the other side, there are artists who don't like digital boards, claiming that with a digital board you're limiting

Figure 10.13. Yamaha LS9 16 Digital Live Sound Mixer.

the talent to never be analog in their sound. So if one of the performers is using a Mini MOOG or other bass analog device you're taking away the advantages of an analog device which is more noticeable in the bass frequencies. After all, who doesn't want more bass?

Digital Advantages

What can a digital console do that an analog can't do?

- Operators have more control, for example the ability to change which fader controls which control function, or to program one fader to control multiple channels without duplications.
- Digital signals are often cleaner and less likely to pick up noise.
- You can transfer much more information digitally than you can with analog, and the cable is much shorter. The ether sound maximum distance is 100 meters. The analog cable snake alone is over 100 m and would weigh more than anyone would want to carry!
- Noise gates on every channel are a given, so inputs that often just create noise are kept closed.
- Plug-ins allow for much more functionality at lower cost, after initial increased cost of the board.
- Smaller footprint makes digital mixers easier to transport and set up, after initial programming.

Digital Disadvantages

What problems are associated with digital consoles that are not present in analog boards?

- It is easy to forget to change things from one saved group of settings to another saved group settings.

- The operator needs to remember his settings throughout a mix; digital boards offer no holistic view of the board—with any one view some information will be missing because the user won't be able to see all their settings at once.
- It is not as forgiving as analog—when the user runs into brick wall and overloads, the feedback is very ugly. This also means that digital meters have different scales to account for the different headroom needs of digital audio.
- A loss of power to the board will lose all digital configuration data, with boot-up time and other configuration data a loss of power would be catastrophic at a running show. (To combat this many digital boards have backup power sources that kick it when the main power line drops out this way you can keep your comfit data.)
- There are too many options. It's great having all these amazing options for changing the routing, patching effects, etc., but there is something to be said about having too many options. Ever heard the expression "shoot oneself in the foot"? This is all too possible with a digital board.

Regardless of personal preference, today's audio engineer must learn to function using both types of audio technology, the future will find both types to be well represented. In the long run, systems which blend the best of both technologies are the ones most likely to win out, meaning the engineers of the future will need to be comfortable with both. In the meantime, the many processing functions that go hand in hand with live mixing will continue to be most often found in separate units mounted to rack as will be examined in the following chapter. We will revisit the mixer again when we look at how to use the features introduced here; however, first we need to look at the processing functions that are as essential to mixing as a mixing board.

Chapter **11**

Processors, Effects, and More

Rack-mounted Equipment

It could be argued that everything an audio engineer does while running the P.A. is "audio signal processing" (in fact different kinds of signal processors can exist throughout the P.A. setup, not only at the mixing stage); it is at this stage that the majority of sound adjustments take place, and where the largest array of audio signal processing technology resides. With analog mixers these processors are mostly *off board* or separate from the mixer, usually in the form of *19" rack mountable units* which are built to fit into a dedicated audio rack (a portable one for live audio applications). All but the simplest analog P.A.s include at least one rack containing a wide variety of processors and effects that allow the live engineer to adjust for sound issues and enrich the overall quality of the sound. Digital mixers usually include a number of effects *on board* (included in the mixer's functionality) that are typically rack mounted in analog systems, but systems with digital mixers will often still include a rack as well, since digital mixers rarely have the exact number and variety of features that a complete off-board selection can supply. In both types of system one of these racks will also typically include at least one unit for recording, and often more than one.

While there is a wide variety of processors and effects that are used in audio signal processing, the majority of them will fall under three broad categories: spectral processing, dynamics processing, and temporal processing. *Spectral processing* includes all processors and effects which are used to adjust the frequency content of a signal, including some of the basic ones that define much of the essential work a live engineer performs at the mixing board, such as *equalization*, but also including tuning effects such as *auto tuning* and *pitch shifting*, as well as *modulation* and *synthesizer effects*. *Dynamics processing* refers to any processes which alter the dynamic range of sounds, which includes all *compression* and *expansion* processes. Finally, there is also *temporal processing*, which encompasses the wide array of *time-based effects*, including *delay and phase effects*.

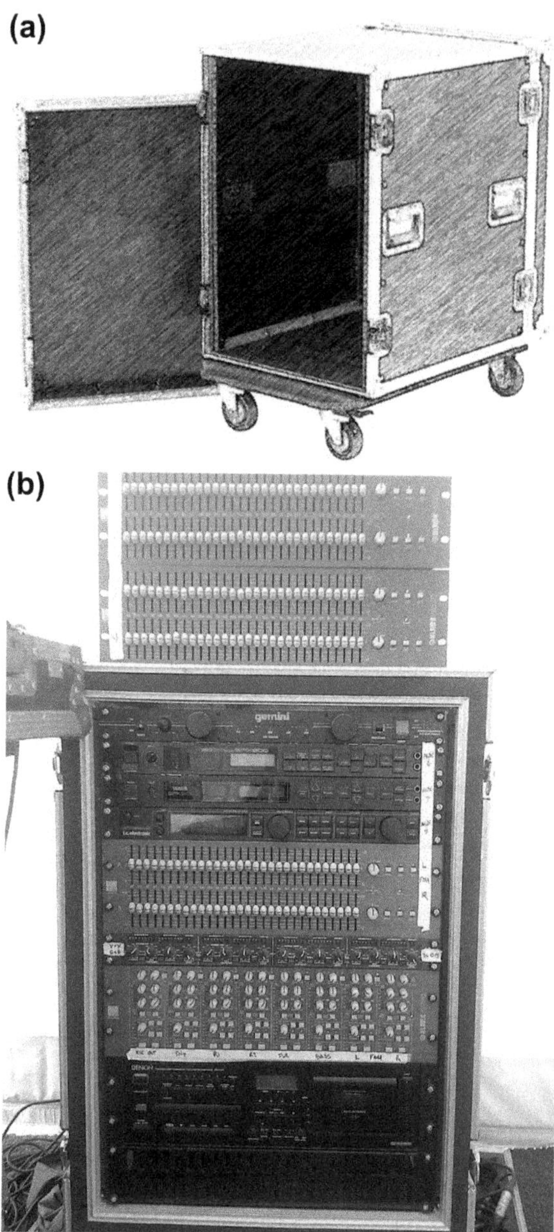

Figures 11.1a and 11.1b. An empty rack mount with room for six or more units, A full rack with multiple EQ and sundry signal processing units. *Source: Photo by Duncan Underwood.*

Audio Filters

In order to understand the processing tools audio engineers use to alter the frequency content of sounds for both corrective and creative purposes, as well as most other processes and effects one needs to know a little about *audio filters*. *Filtering* is an essential process that the live engineer uses to alter the power and range of frequencies that can be heard in audio signals, also referred to as the *spectral content* of the sound. Without this ability, the job of providing good live sound with the P.A. would be impossible. Audio filters are a very important part of the P.A. system, and are not only part of the P.A. components discussed here, but are also essential to important mixer functions as well as for ensuring that the amplifiers and speakers function properly. Yet because filters accomplish their task indirectly, the way they function isn't as straightforward as many assume.

In terms of results audio filters work much like any filter, and do their job by separating out and removing the unwanted portion of whatever is directed through them, while leaving what is desired to flow through unhindered. Filters are essentially analog electronic circuits which contain frequency sensitive elements, such as capacitors or inductors which combine with a resistor to attenuate certain frequencies contained in the signal while allowing the remaining frequencies to pass through the circuit. For *digital audio filters*, algorithms perform the same function instead of physical circuits, but the principle is the same: as some of the frequencies in a signal's spectral content are "filtered" out of the overall sound, the result is to alter the frequency balance of the remaining sound. Since the end result of using audio filters is to change the frequency content of the signals passing through them, it is easy to assume that they do so by allowing the user to directly alter the frequency of a sound; however that assumption would be incorrect.

Filter Function

If you recall from Chapter 4, sounds are the sum of a number of different sine waves which can be described objectively according to frequency (f), amplitude (A), and phase (φ). As we mentioned, these sine waves vary from each other in frequency and amplitude, with the fundamental frequency determining the pitch and the harmonics combining to determine the timbre of the overall sound. However, these component sine waves are always *phase locked* in relationship to each other, which is how they result in a single coherent sound instead of just sounding like a bunch of separate tones played at once.

As the audio signals representing these sounds move through a filter's circuitry, the component sine waves of the signal are either conducted easily through the filter and pass with little resistance, or they encounter resistance to their passage through the filter because of their frequency. As a result the power contained in the component sine waves moving through greater resistance is dissipated to a greater degree than the rest of the signal, reducing the amplitudes of those waves and attenuating them in relationship to the sine waves in the signal that didn't pass through resistance. In addition, the process slightly alters the spatial relationship between the original signal and the filtered one, creating some degree shift in the phase of the sound passing out of the filter compared to the signal entering the filter. Some filters rely on this shift exclusively for their effect—the filter's circuitry shifts the phase of the signal by a specified amount so that when it is combined with the original signal the phase difference between the two results in an output where sounds partially cancel at some frequencies, thus creating an altered frequency response. In still other more complex designs multiple filters are used so these effects can be combined.

Filter Response

One should note that most analog audio filters never actually act on frequency directly, however; all analog filters can do to the signal passing through is to alter the amplitude of its component sine waves to some degree and create some amount of phase shift in the overall output. Even though the filter has altered the phase of the entire signal, and reduced the amplitude of a portion it, the shape of the component sign waves (and of the filtered signal) remains the same throughout, meaning their exact frequency has technically not been changed. The only change is that some frequencies have had their amplitude reduced enough that they can't be perceived (it is possible to create a sine wave with 0 amplitude). An audio filter circuit's capacitors or inductors will respond to frequency, but can only alter other parameters. The timbre of the remaining signal is only changed because a portion of the original frequencies have been rendered too low in amplitude to remain a coherent part of the output signal's frequency response. Thus, what is typically referred to as a filter's *frequency response* is actually an expression of the combined results of the *magnitude response* and the *phase response* of the filter. (This also should make it clear that filters are useful for dynamic and time-based signal processes as well.)

The reason it's important to keep this in mind is because it explains why filter outputs take the shapes they do, which we will look at soon. More importantly we hope that this understanding will always serve as a reminder to use filters sparingly. We noted at the start of the chapter that signal processing with audio filters is an essential part of mixing audio, live or studio. Without it, shaping the sound to suit the situation and producing the best possible mix in the face of the many difficulties of reproducing and reinforcing sound would not be possible. However, an important caveat, which is left unstated nine times out of ten, is that the above is only true if it is exactly the amount of processing required for good sound and no more. As important as just the right amount of processing is to produce the best mix, it is equally true that if you are going to miss the target and provide the wrong amount of processing to produce the best mix, the sound will usually be much worse off if you over-shoot the target than if you don't process it enough.

When students observe that the sounds resulting from their processing choices have a different, improved, sound quality, it makes sense for them to assume, if no other explanation is provided, that the process has allowed them to create a more perfect sound wave, with a natural shape to suit the new sound. In that case, what could be the problem with making as many adjustments as you want, as long as the balance between frequencies is maintained and you don't run out of headroom? How much you adjust shouldn't be an issue with the result being a fresh new wave form.

However, to create a brand new sound with a different frequency mix, filters would have to somehow break the phase lock of the incoming signal, neatly pluck out the portion of component sine waves with the offending frequencies, then reapply phase lock, or the filter would need to somehow grab hold of the signal and squeeze a bit here, stretch a bit here, and sculpt the sound wave into the right shape with the right frequency content. This is beyond the ability of typical analog filters, so instead filters allow us to do the next best thing—alter just enough to end up with a signal that keeps its original form, but has a different sound quality. Used sparingly, the overall result is a lovely illusion to compensate for acoustical issues. Overdo it, however, and the end result is a rather unpleasant, freaky sound effect that bears little resemblance to natural sound. Much like a woman can improve on nature with a little makeup, or end up looking more like a drag queen if she overdoes it (no offence meant to drag queens or women who aspire to look like them, but the comparison assumes the intended effect in this case is a natural-seeming one).

Filter Examples and Parameters

One of the most basic filter types are the pass/stop filters: *low pass*, *high pass*, *band pass*, and *band stop* (also called *band reject*). The first two split a signal into frequencies above or below a specified *cutoff frequency*, while the following two split a signal according to frequencies that fall within or outside of a bandwidth defined by a specified pair of cutoff frequencies. The cutoff frequencies that determine the dividing lines between frequencies act as each filter's *threshold*, the point beyond which frequencies no longer pass freely and are attenuated. The attenuated range of a filter is called its stopband, and the unattenuated range is the filter's passband.

These filters include a few notable variants such as the *all pass filter* (which alters the phase of all frequencies without attenuating any) and the *notch filter* (a band stop with a very narrow bandwidth). Though each is used alone for specific situations, what makes audio filters so important to live sound is the many ways they can be combined, either simultaneously or in a series, to produce so many different processes and effects. Because the basic filter response of simple filters is very imprecise, for any type of base filter type or filter combination, various designs exist to create improved filter responses aligned with the particular type of task that the filter is meant to be used for. In general each filter design's frequency response has a distinct enough shape that for most designs, a filter's design name can be identified by its bode plot and vice versa.

As mentioned, the basic filters described above are not perfect, and in fact even the alternative designs that manage some improvement over the basic design don't reach perfect precision. If filters worked precisely, they would have complete magnitude response at the assigned threshold frequency. In the case of the basic filters described above, the spectral output would look like the filter response depicted in Figure 11.2b because low pass filters would allow *all* frequencies below the cutoff frequency to *fully pass* and would *fully block all* those above. Conversely, high-pass filters would fully pass all frequencies above the cutoff frequency and fully block anything below. In a similar vein, perfect band pass filters would pass everything within some frequency interval between a specified low cutoff frequency and specified upper cutoff frequency, while band reject filters would do the inverse, fully block everything between these intervals and let frequencies above and below to pass freely. This perfect response would look like a wall holding back any and all unwanted frequency content.

In reality filters only approximate their cutoff points, and are not nearly as precise as the ideal. Look at the more realistic portrayal of filter output below and you will notice that these filters do not suddenly cut off all sound at a specific frequency. Rather, they show a *slope* where the frequencies *"roll off"* gradually.

Because of this slope, when we talk about a filter's *cutoff frequency*, we are not referring to an unattenuated frequency beyond which the rest of the

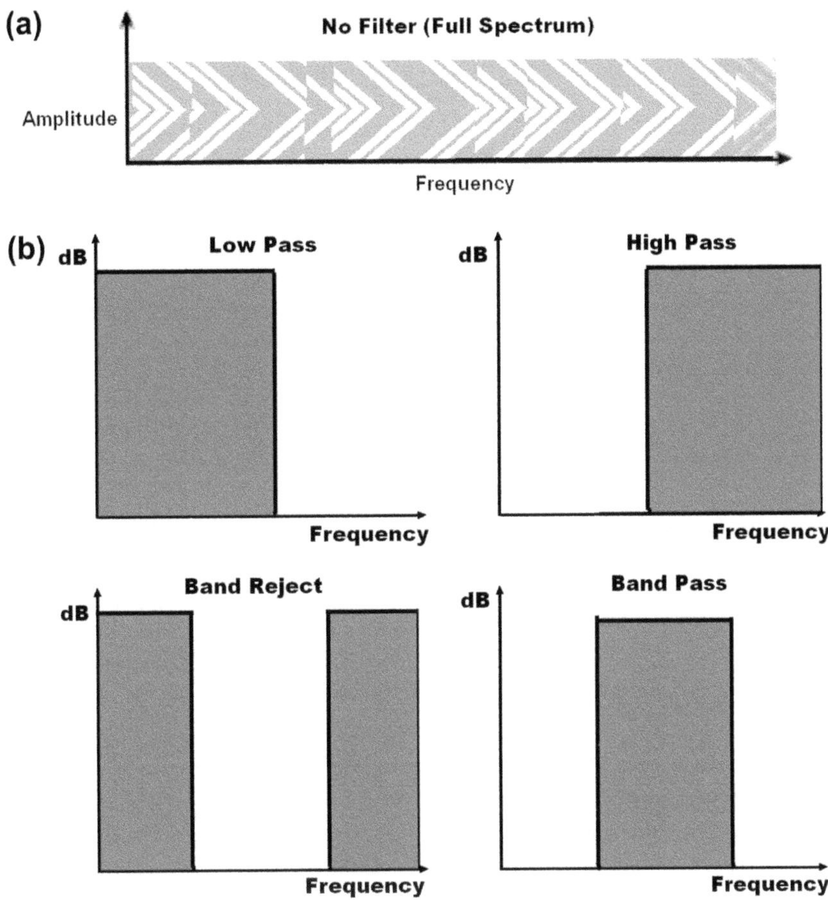

Figures 11.2a and 11.2b. Full frequency (all pass) output, and idealized (atypical) filter output for the four main spectral filters.

frequencies fall off, but instead are referring to the point along the slope where the output level is half the power (−3 dB) of the maximum dB level within the passband. This is important to take into account for any applications that allow the operator to choose the cutoff frequency, since for any given cutoff frequency, there will be a region within the pass-band where the amplitude of the frequencies immediately adjacent to the cutoff will be partially attenuated (some that ideally should pass at full power, others that ideally would be attenuated more). So if a frequency that needs to be fully passed or attenuated lies too close to the cutoff value, care will need to be taken to adjust accordingly and set the cutoff sooner or further along than would be necessary if not for this *transition region*.

The slope of each filter depends on the number of frequency responsive elements used in the circuit (analog), and determines the amount of gain reduction per interval beyond the cutoff frequency. These intervals are usually measured in *octaves* (a ratio of ½) or in *decades* (a ratio 1/10). As a general rule, slope increases by a minimum −6.02 dB per octave for single element, or *first order* filters, with an increase in attenuation and therefore a greater degree of slope for each additional filter order. Order also determines the amount of phase shift created by a filter.

Another filter parameter is "Q," which refers to the quality of *resonance*, and is a ratio of a filter's bandwidth. In the case of band pass and band stop filters; Q is often used to express the sharpness (narrowness) or broadness of the filter band. Q is a parameter that you will need to set when tuning the P.A. and as you mix, and you will want to alternately set it narrow or wide for different adjustments. Luckily, the formula

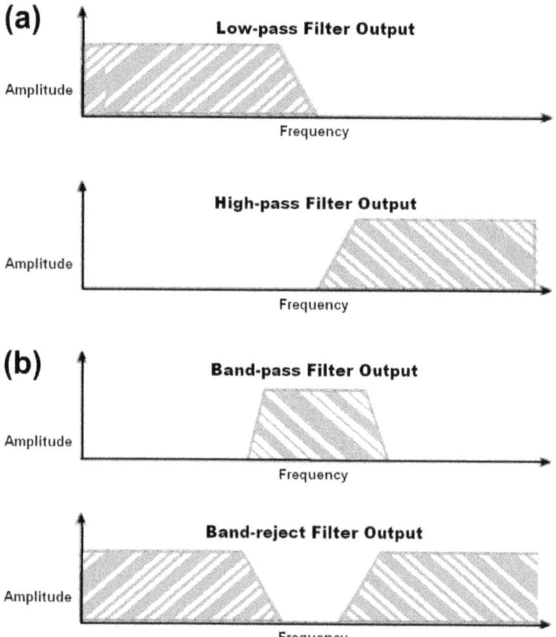

Figures 11.3a and 11.3b. Typical filter outputs show frequencies attenuated in a gradual slope, not a sudden cutoff.

for determining Q is simple enough even for the mathematically averse: filter Q = center frequency/bandwidth.

The filters described above are the simplest form of filter, and make use of unpowered circuitry to filter sound. Because they are *passive circuits*, they are limited to processing sound using subtractive synthesis only. In other words, they can only attenuate frequencies, but cannot boost them. To achieve an EQ boost, you have to include active circuitry in the design if you want the power to amplify frequencies as well.

Spectral Processing: Equalizers

As with most audio designs, filters are not restricted to passive design models; active versions are used as well, including powered circuit components which can amplify selected frequencies. The original powered filter design was the bass and treble equalizer, which was capable of cutting and boosting both low and high frequencies using two independent controls. This basic active circuit still forms the basis of many mixing consoles' high and low equalizer sections and is still found in the form of high pass channel presets on many mixers. Equalizers have naturally developed to allow for a great deal more control than the original active filter designs allowed for, which for all their single button convenience are too limited in scope to really compare to the most commonly used equalizers today, which form the two most fundamental processing units in an audio engineer's arsenal and are the only processing units found in the typical analog P.A. rack that can be called indispensable (this only fails to apply to digital systems because analog to digital, digital to analog processing units often reside with the rack as well, but aside from this the same can be said of the primacy of equalization in digital audio processing as well).

Equalizers include small amplifiers in their circuits to perform their boosts. If these worked like the passive reactive elements used to attenuate frequencies, and boosted them by 6 decibels or more per octave, the typical engineer would find themselves running out of headroom almost immediately. Luckily, *"shelving"* equalizer designs with minimal slope are typically used instead; after a small slope, the curve flattens out—or shelves—and the gain of the desired high or low section of the audio spectrum can then be adjusted by the same amount.

Graphic Equalizers

Generally, graphic EQs are used to account for acoustic deficiencies in the room (as much as can be), to manage feedback, and/or for general tone shaping. While a graphic EQ can limit the damage of resonances arising from the P.A., it cannot eliminate them, or remove room resonances. Graphic EQs cannot completely correct the worst acoustic issues; however, they can mitigate many of them, making the difference between an enjoyable show or a constant struggle against outbreaks of feedback.

A graphic EQ comes as a box with vertical faders, with each fader moderating a specified frequency range. Each fader controls the level of a single band pass filter circuit (used in a cascade formation), and each controlling a particular frequency range. Moving each fader boosts or cuts a signal in one or more parts of its frequency range. The aggregated result of the filters is to modify the overall balance of frequencies in the room. The position of a graphic EQ's faders gives a graph-like view of the frequency balance, which is how this kind of equalizer came to be called a graphic EQ.

Graphic EQs will usually have two channels, each with ten (1 octave), 15 ($^2/_3$ octave), or 31 ($^1/_3$ octave) *frequency bands*. For live applications 31 band EQs are preferred, as EQs smaller than $^1/_3$ octave (three

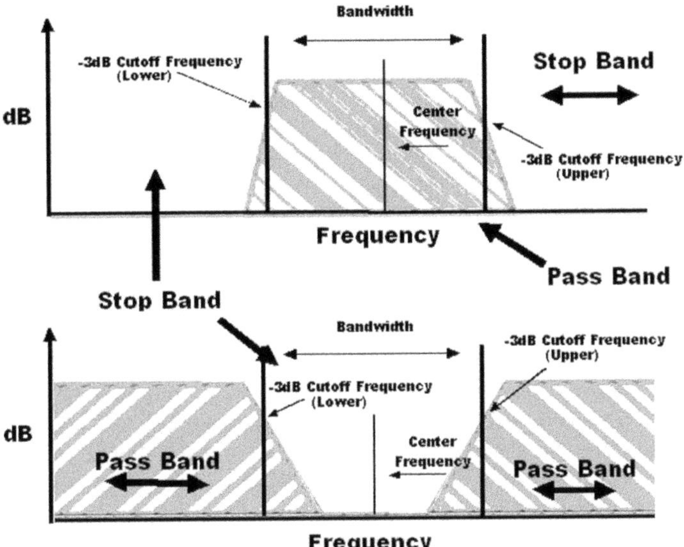

Figure 11.4. Typical filter outputs, frequencies below and above the cutoff frequency are also affected.

(a)

Filter Order	Attenuation: Db/Octave	Attenuation: Db/Decade	Degree of Phase Offset
1st	6	20	90
2nd	12	40	180
3rd	18	60	270
4th	24	80	360
5th	30	100	360+90
6th	36	120	360+180

(b)

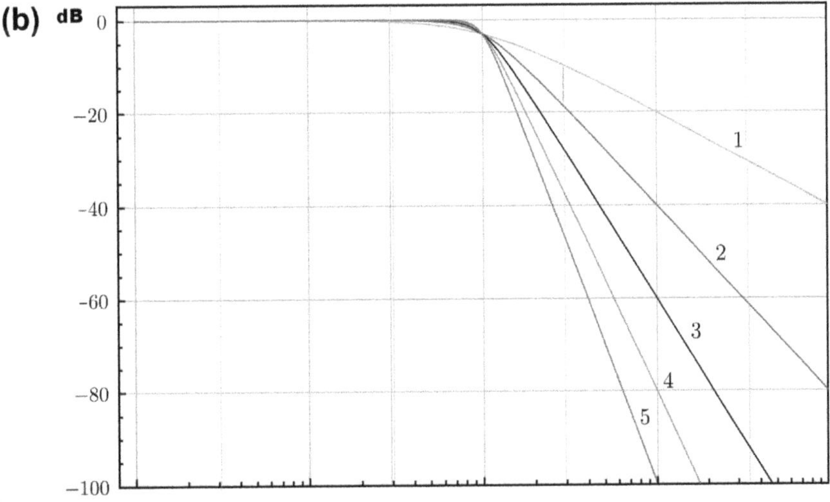

Figures 11.5a and 11.5b. Attenuation and phase offset by order and six orders of Butterworth, from low to high degree of slope.

steps for each doubling of frequency) are not as useful for fine tuning the P.A., as each frequency band will be too broad. The center frequency of each band will often be set to an *International Standards Organization* standard frequency; these frequencies are: 20 Hz, 25 Hz, 31.5 Hz, 40 Hz, 50 Hz, 63 Hz, 80 Hz, 100 Hz, 125 Hz 160 Hz, 200 Hz, 250 Hz, 315 Hz, 400 Hz, 500 Hz, 630 Hz, 800 Hz, 1k Hz, 1.25 kHz, 1.6 kHz, 2 kHz, 2.5 kHz, 3.15 kHz, 4 kHz, 5 kHz, 6.3 kHz, 8 kHz, 10 kHz, 12.5 kHz, 16 kHz, 20 kHz.

A graphic EQ can be connected to a P.A. system either in-line, along the path of the signal flow between the mixer and the monitors or amps, or using the main (left and right or group) mixer inserts. When connected via the main inserts, any changes to the graphic settings will be seen on the channel meters, and heard on headphones at the mixer (or on the monitoring wedge if you use one) whereas when graphic EQ is connected in-line, changes will only be heard through the main or monitor speaker system, i.e. the mixer's meters may not accurately represent the signal strength at the amplifier inputs.

Most graphic EQs also have a *master control panel*, with controls that might typically include controls for input and output levels, high- and low-pass controls, and bypass controls, including:

- *Input gain*: Some graphic EQs have meters or overload lights, and allow some adjustment of the input signal to bring it within the EQ's nominal operating range.
- *Output level*: Over time, adjustments to EQ frequency bands can accumulate and create a noticeable difference in the overall volume. The output level control allows you to reduce the output gain at the equalizer so that the graphic EQ affects only the tonal balance of the sound, not its overall volume level.
- *High-pass and low-pass controls*: Many graphic EQs include *shelving EQ*, which are fixed high-pass and/or low-pass filters like the treble and bass filters described above with a simple switch control for cut or boost. Some of these filters may have variable frequency and variable gain, allowing users to assign the values used by the preset buttons. Use these (rather than the faders) when you want to cut or boost the highest or lowest frequency ranges.
- *EQ in/out*: Most graphic EQs allow you to bypass the EQ (on some, high-pass and low-pass controls may also be selected independently). This is useful for instant comparison: the ability to compare the original sound with the output sound allows you to make sure you are making the best use of equalization. Some units also include the ability to choose level matching for the output; because volume can influence the impression of the sound that is not in line with the real sound quality. Therefore, comparing and evaluating two mix sounds is easier when the output level is adjusted so that switching the EQ in or out has no apparent effect on the overall volume.

Parametric EQ

Named for the fact it allows the user to set parameters of frequency, bandwidth or Q, and gain, this type of equalizer allows for very precisely tailored adjustments to be applied to the sound. Many mixers have one or more parametric EQ sections on each channel strip. However, single- and multi-channel parametric EQs (with variable bands) are also available as rack-mountable units. Parametric EQ works like a normal band-pass EQ (gain) control, but the frequency and bandwidth are also adjustable, allowing the operator to make very fine tuned, precisely targeted adjustments to the frequency levels. The shape of the band-pass

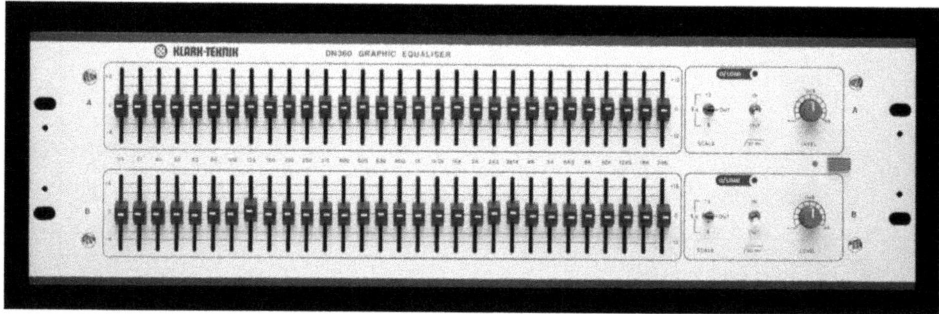

Figure 11.6. The iconic control face of the dual channel, 31 band graphic EQ.

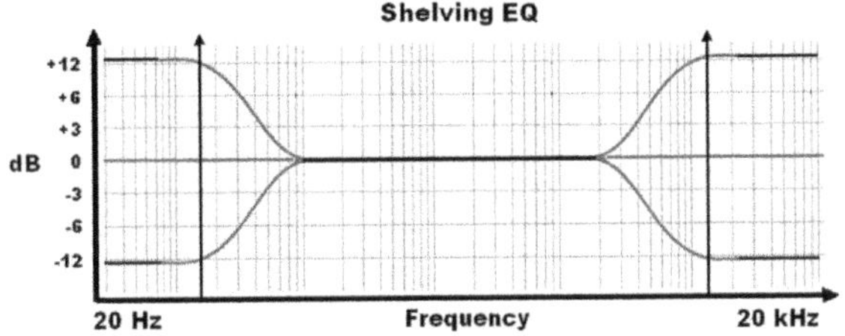

Figure 11.7. High- and low-pass shelving EQ.

curve is bell-like, and is often referred to as a *peaking EQ* because the maximum cut or boost occurs at the center of the bell and gets progressively smaller along each side. The center frequency of the filter divided by its bandwidth (defined by the −3 dB points on each side) gives the Q of the filter; the higher the Q value, the narrower the filter response

Sweepable mid-range EQ, which has only frequency and gain controls, is sometimes incorrectly described as "parametric" and found included in units advertised as parametric. However, the term "semi-parametric" is more often used for this type of EQ. Like many other EQ types, the name refers to an aspect of the appearance of the filter display; because the bandwidths are preset, moving between them creates the appearance of a sweeping motion.

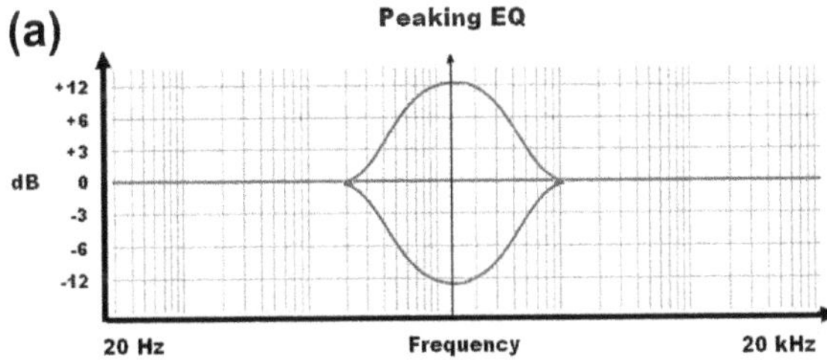

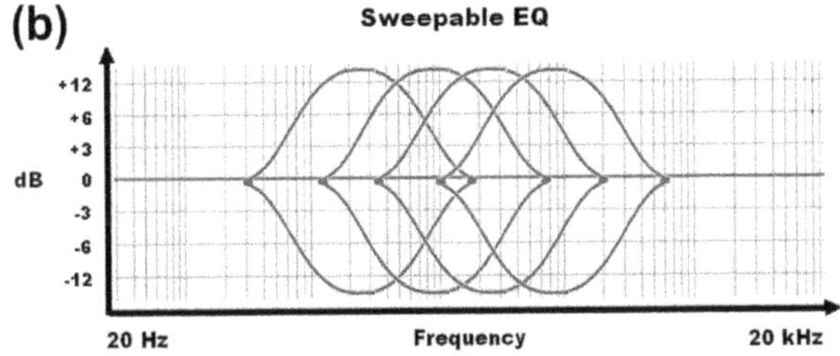

Figures 11.8a and 11.8b. Peaking parametric and Sweepable semi parametric EQ.

Dynamics Processing

The Compressor

Because P.A. systems rarely can achieve a dynamic range of more than around 60 dB, unless you apply some compression to yours, your system won't cope with sources that have too great a dynamic range. A compressor places a cap on the maximum level that can pass: it essentially turns the signal down when it's too loud and then raises the system back up to original levels when the loud sound source ceases. In small systems, compressors are used to reduce the distance between average and peak levels, allowing the operator to raise the signal level without increasing peak output and risking clipping, which can make the system seem louder. In larger systems compressors can also prevent the P.A. from distorting. Used alone on vocals, a compressor can reduce the level of backing instruments during vocal sections, bringing forward a quiet singer.

Despite allowing the operator to play a set louder than would be possible if they had to maintain headroom for music peaks, compressors don't boost levels and bring the music up to the peak levels; they lower levels by squashing the peaks in volume—this reduces the dynamic range (difference between the quietest and loudest sounds) of the signal.

Compression is the main type of dynamic processing; other types include expansion and limiting (however, limiting is really just a high-ratio compression, and expansion is a compressor with a ratio value of 0–1). If used correctly, the compressor is a valuable tool for controlling the dynamic range of a signal, which has the effect of subjectively stabilizing that signal, increasing system headroom, and reducing system noise. Like all processes that audio filters allow us to apply to our sound output, compression works best when used sparingly and only when needed. Care should also be taken to set the unit properly for the type of sound it will be compressing; otherwise it can be heard as a faint breathing or chugging sound each time it engages. If it can be heard when it engages, or because an extreme and unnaturally reduced dynamic range can be noticed in the sound quality, then you are using it incorrectly. Misusing compression can also have undesirable effects on P.A. system performance, including greater sensitivity to feedback and noise.

The main compression parameters that must be set anytime compression is used are:

- *Ratio*: This setting determines how strong the compressor will be, or how much the compressor will compress a peak that goes above the threshold parameter. It's given in the form of a ratio in relation to one; for instance a 5 to 1 compression will be shown as 5:1. Ratio level also determines what kind of dynamic process it is:

- a ratio of 1:1 is no compression;
- a ratio of up to 10:1 is normal compression;
- at ratios *higher* then 10:1 (sometimes stated as ∞:1) the compressor becomes a *limiter*.

- *Threshold*: This setting determines the peak level a signal must reach before the compressor will affect that signal and is given in terms of dB. Any peaks above this level will engage the compressor. When the peak again drops below the threshold, the compressor will begin to release.
- *Attack*: This setting is given in ms (.001), and determines how fast the compressor will compress peaks once they pass above the threshold setting. If this is set too fast, you'll get a pumping sound and transients will be lost. Set too slow and it will seem as if no compression is happening.
- *Release*: This setting is the opposite of attack and is also given in milliseconds. This determines how fast the compressor will "let go" of the signal once the peak drops below the threshold.
- *Output/makeup gain*: Once a signal has been compressed, we need to raise the whole compressed signal back up to match the signal level immediately prior to compression (or we would also lose clarity due to lack in volume/amplitude). Some devices perform this function internally and don't let the user choose how much, some don't even have makeup gain, and others allow for the operator to choose this setting. If allowed to choose this setting, set the makeup gain to a level close to the level a signal is right before reaching the compression threshold.
- *Knee*: Some compressors have this setting, which only affects signals that are near the threshold level. A "hard knee" only compresses signals as they exceed the threshold, when the gain suddenly starts being reduced by the exact ratio dictated. A "soft knee" works a little differently. As signal level nears the threshold it begins to be reduced slightly, and this reduction slowly increases as the signal gets nearer and crosses the threshold. Only after the signal is slightly over the threshold does the compressor use the full ratio value of compression.

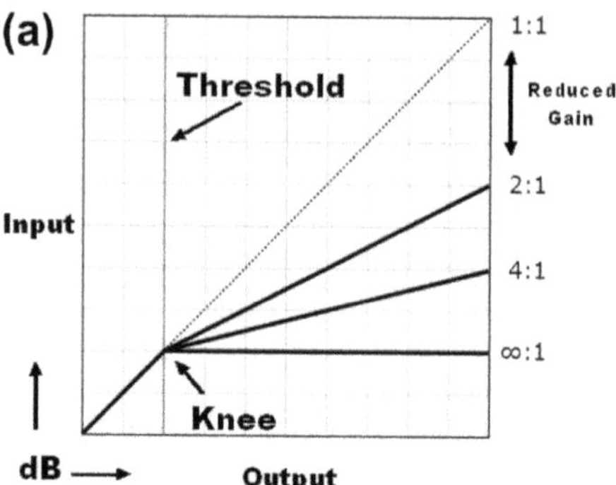

Figure 11.9a. Ratio determines the rate of compression.

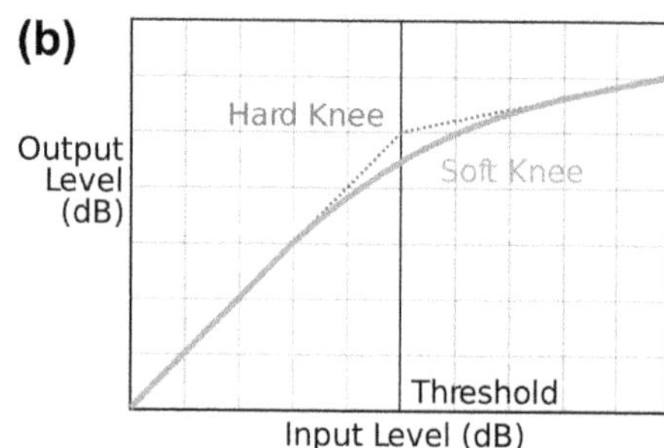

Figure 11.9b. Like other filters, compression has a slope too, but it runs uphill and is much less gradual!

Common Uses of Compression

- To control the level of vocals. Because it takes vocalists years of training to achieve the degree of breath control necessary to maintain a consistent level, compression can assist singers in achieving a consistent and even tone and to help produce vocals that sit correctly in the mix.
- Applying compression to the signal reduces dynamic range to produce a more controlled and usable sound, and allow for greater system gain. (By lowering the peak signal level the system gain can be increased without compromising headroom).

The Brick-wall Limiter

A limiter is basically a compressor with a very high ratio (10:1 or higher) that prevents a signal from exceeding a threshold, and some compressors include a separate limit function. When a signal is too big for an amplifier, the result is "clipping" of the signal. Clipping in the main amplifiers can kill your speakers, and always sounds bad. Limiters are used to prevent the signal from exceeding the capability of amplifiers, and to ensure careless users don't destroy loudspeakers. The limiter is hooked into the system in line as the last device before the power amps that drive the main P.A. Some speaker crossover and amplifiers have their own limiters and if you mix in clubs, most installed systems have limiters on them automatically.

The brick-wall limiter is the device in an installed sound system most likely to be locked away to preserve its settings. This is because for venue owners who want to preserve their costly speakers, the limiter is the first and final defense against a mixer operator, or DJ, turning levels up too loud and overloading the system, which can end with them blowing speakers and burning out amps. Most club owners or installers who know what they are doing will have already set the limiter at a safe level to ensure no one can set dangerous and potentially damaging amplitude levels and then locked the unit away so no yahoos can mess with it.

If you are looking to protect your own P.A. system from some yahoo possibly turning up your system to levels loud enough to blow out your speakers, or simply want to prevent the ability to turn it up to levels that are far too loud then you'll need to get a limiter or device that functions as a limiter (such as a DBX Drive-rack).

Sidechaining

Sidechaining is the process of triggering an effect on one channel by using the signal (or key input) from another. Sidechaining is useful for many different tasks; what you can do with it depends on what your key input controls (compressor, gate, etc.) are. Both the examples below, gating and ducking, are different types of side-chaining, and useful anytime you want to ensure one sound is heard over another. Other examples of side-chaining for a live applications are discussed in Part 5, where we detail how to use the processes described here.

Ducking

Recall the last sporting event or concert you attended or watched on TV. Imagine an announcer telling the TV audience what is happening in the game or a performance. While this happened you likely could hear the crowd at the event cheering in the background and reacting to what was happening at the event. Yet when the crowd surged in volume at times, as some event on field or words from the stage that the crowd especially liked set them to making a lot of noise, did they manage to completely drown out the performer or announcer? Of course not! Have you ever noticed that when the announcer is not announcing, the crowd gets even louder?

This is accomplished by a process called ducking. For signal ducking, a compressor is put on a channel, but is not triggered (turned on and off) by the channel that it is on, but is triggered by another channel's signal. In our example above the compressor is put on the crowd's channel, and it is triggered by the announcer's channel. Therefore, when the announcer is talking, the compressor pushes down the signal of the crowd, making sure that the announcer is not drowned out. When the announcer stops talking, the compressor lets go of the crowd signal again and the crowd's signal then gets louder. Similarly, on every radio channel you listen to, there is a ducking compressor on the music channel that is triggered when the radio announcer speaks.

Noise Gates

Sometimes referred to as just a gate, a noise gate is a dynamic process whose purpose is to turn off any input source that isn't being "used" at the time. In short a noise gate applied to a track will keep that track silent, until there is enough amplitude to "open" the gate. While the gate is open, all sound from that track passes through the gate, until the channel's amplitude drops below the point where the gate opens. The gate then closes again, and the channel it's on is blocked at the gate—no sound comes through—until the gate is triggered to open again.

One example of gating is when we use a compressor as an on/off switch for system microphones. Think of a gate that sits somewhere on the amplitude scale. When a particular microphone is not in use, then the sound entering it will remain below the gate (below the threshold), and the gate stays closed and no sound is let through (such as the ambient noise an unused mic would pick up and relay if it were left on, even if not as loudly as sound captured within its polar response range). However, when the sound's amplitude goes above the threshold, such as when someone speaks into the mic within its operating range, the gate then is pushed open and the sound is let through. Gating is useful for speaking applications when the laypeople using the mic can't be relied on to turn them off during pauses.

A similar use is to keep the low-level noise coming from stagnant instrument's effects and "open" mics from interfering in the sound of the mix. We can't always expect the talent to turn their distortion effect off when they aren't playing, so a simple solution is to set a noise gate to be closed while the guitar isn't played, and when the guitar is played (sending a higher amplitude level) the gate opens and the distorted guitar is heard.

Gating is also incredibly useful if you have a channel that has a lot of noise on it, but the channel's signal is strong enough to "cover" the noise, as long as it is present (i.e., snare drum). Use a gate to keep the channel silent until the snare drum is hit, and when the snare is hit, the signal gets loud enough to open the gate and allow the snare signal through. When the drum is not present, the gate closes, the channel is muted, and with it the troublesome noise—what noise?

Since the noise gate is a dynamic process, the parameters that need to be set are the same as any other dynamic processes: ratio, threshold, attack, and release.

- *Ratio*: This determines how far the gate reduces the amplitude when the gate is closed, usually this would be set to INF:1 for complete silence when the gate is closed (a gate doesn't have to be 100 percent silencing when it's closed).
- *Threshold* (set as a VU or dB level): This determines at what amplitude the gate opens and closes again.
- *Attack*: At the moment the amplitude level is high enough to open the gate, the gate takes the attack time to fully open.
- *Release*: Once the sound has dropped below the amp level of the threshold, the release time determines the time till the gate closes again.
- *Hold* (optional): This is the amount of time the gate will hold open once the attack time has fully been accounted for.

Temporal Processing: Time-based Effects

While processors are used to correct or control sound, as well as creatively shape it, effects are exclusively used to creatively enhance or add to a sound. Additionally, the processors already discussed can generally be used in line to treat the whole sound, though we may choose to apply them to a single channel. With time-based processing, processing all of the signal is no longer an option. Instead, we split the signal and send a portion to the effects unit to be processed. Then the effected portion of the signal (the wet) is returned to the mixer where it is combined with the unaffected portion of the signal (the dry) by the mixer. Since the signals are often split into two parts as part of the effect, time-based processes are sometimes referred to as parallel processing. As with the other processes examined, many time-based effects are made possible because of audio filters, only here they are combined to produce their greatest effects in the time domain. Some commonly used effects are as follows.

Echo

An **echo** unit copies the original signal and replays the copy one or more times after a selectable interval or range. As the original signal and the delayed copy combine, the result stimulates the way a sound can be reflected by a single acoustically reflective surface (a cliff, for example). Although it is often also called delay, in practice the effect is invariably that of an echo: the delayed sound is mixed with the original sound, producing one or more discrete repeats. However, while delaying the whole signal isn't used as an effect, a delay function is quite commonly included in crossovers and system controllers to time-align the sound from separate loudspeakers or loudspeaker drivers.

Most echo effects will include controls allowing the operator to set:

- *Interval*: the length of time between the original sound and when the signal copy is mixed back into the original sound producing the first and subsequent echoes.
- *Count*: the number of discrete signal repeats or echoes.
- *Level*: how loud the echo is compared to the original sound.
- *Feedback*: whether the echoes are themselves echoed, and in what proportion.

Reverb

Reverb is an effect created by hundreds of sound reflex ions bouncing around in an enclosed or semi-enclosed space. Reverb is always present in sound made inside buildings; it is a feature of indoor sound. Each time you hear someone speak, you are hearing a different kind of reverb depending on where you are sitting in the room, for each place in the room will receive different reflections from different surfaces, and each location will receive different reflex ions. Reverb is not a way to cover a sound, or to make a sound more robust. It's actually meant to make the sound *less* in your face by making its dynamic space *bigger*.

If you have played with reverb before today, you know that too much reverb will make the signal sound wishy-washy and lose definition. This is because if you use too much reverb in a signal, it will mask and distort the signal and you lose the clarity of the signal you were processing in the first place. This loss of clarity due to reverb is usually referred to reverb wash.

We avoid this problem with reverb, by using parallel processing. Since we are keeping some of the unaffected part of the signal, that unaffected part will keep the definition of the original signal, while the wet (effected part) will give the signal its "space." Only by using your combinations of dry and wet signal can you keep definition of the signal and get the "space" you want into the signal in question.

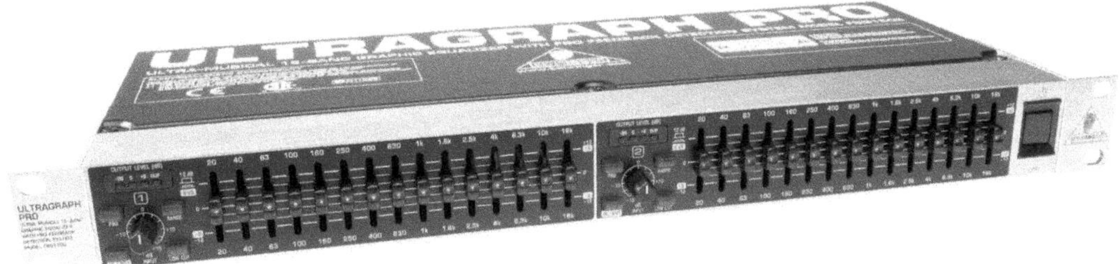

Figure 11.10. Many effects, but only one rack space!

Multi-effects

A combination effects or *multi-effects* unit usually incorporates a variety of delay-based effects, including pitch-shifting algorithms. Multi-effects units may also include some sound-processing functions such as compression. For live applications, you want a unit that gives you useful live effects and is not too complex. All effects should always be used sparingly. A multi-effects unit typically will include:

- *Flange or pitch-shift*: Shifts can be used to create artificial harmonies.
- *Enhancer or exciter*: Adds high-frequency harmonics to the original sound, and mixes some back in with the original sound. It can make sound from a cheap mic sound cleaner and clearer even though it is actually a form of distortion.
- *Phaser*: Takes the input signal and splits it into two channels, adding a slight delay, or phase shift, to one channel. This creates a moving comb filter effect when the two signals are recombined. The phase shift is modulated by a low-frequency filter to enhance the characteristic whooshing effect.
- *ADT* (artificial double tracking): Simulates an effect pioneered by Les Paul. It adds a small varying delay and pitch to a sound and then mixes it back in with the original sound. ADT is used to "thicken" a vocal part by making it sound like the vocalist is singing the same part twice.
- *Chorus*: Basically a multi-layered version of ADT. Can add texture when used in moderation.

And More...

Of course, the outboard rack is where any other useful equipment goes that isn't attached to anther larger part of the system, notably signal conversion processors and digital recording units, although automatic feedback reduction units are becoming more common to supplement traditional methods of controlling feedback. It's never a bad idea to save a rack space for the next must-have item; if you don't you may just end up with a nearly empty rack begging to be filled.

When we look at how to set up our system in upcoming chapters, we will revisit these processes and begin learning how to set up and use them for some basic live applications. We still have a few more essential P.A. components to discuss first, however—learning to use any particular component is easier after acquiring an understanding of the P.A. system as a whole. With that in mind, let's turn to the next chapter and look at the final group of components in the signal chain. This is where we apply power to the signal—the amplification stage that puts the loud into live sound reinforcement.

Chapter 12

Amplifiers, Crossovers, and Speakers

From a Little Signal to a Big Sound

The three components discussed in this chapter represent the final links in the signal chain. The right combination of components here can take the sound engineer's carefully crafted signal and reproduce it faithfully, emitting clear, well-balanced sound evenly along a specified path, or can distort that signal, emitting a range of less desirable sounds with more or less similarity to the sound contained in the input signal. This makes the equipment found in the final stage of sound reinforcement just as important as any other part of the system.

Power Amplifiers

A speaker amplifier takes a line-level signal and reproduces it at the much higher power needed to drive a loudspeaker. To distinguish them from the other various types of amplifiers found along the signal chain, such as backline amplifiers and pre-amplifiers, P.A. main speaker amplifiers are referred to as *power amplifiers*. In many compact systems you may not even be able to see the power amplifiers, as they can be placed within other system components. For example, *powered mixers* come with a power amplifier built right into the mixer, which is intended to be plugged right into the passive speakers needing amplification. Another component that may come with a built-in power amplifier are the speakers themselves. As you will see later in the chapter, some types of speakers come with the amplifier and the speakers in one self-contained package. Naturally these are referred to as *powered* or *active speakers*.

When included as separate components in a P.A. system, power amplifiers come as 19 inch rack-mountain units and are intended to either plug into rack-mountable crossover units which then direct the signal to the speakers, or to be plugged directly into speakers that have internal crossovers. The system amplifiers and crossovers in this case will be mounted together into a portable rack that remains near the stage with the main P.A. speakers. Power amplifiers usually are stereo, meaning they split their power between two channels meant to power the paired left and right speakers of a stereo P.A. system. They all have a volume knob for each stereo channel (or only one for mono amplifiers) on the front panel and most also have a variable number of switches for controlling other functions on the rear panel. There are three modes that all power amps can run within, and the operator has the option to switch between: normal, parallel, and bridged mono.

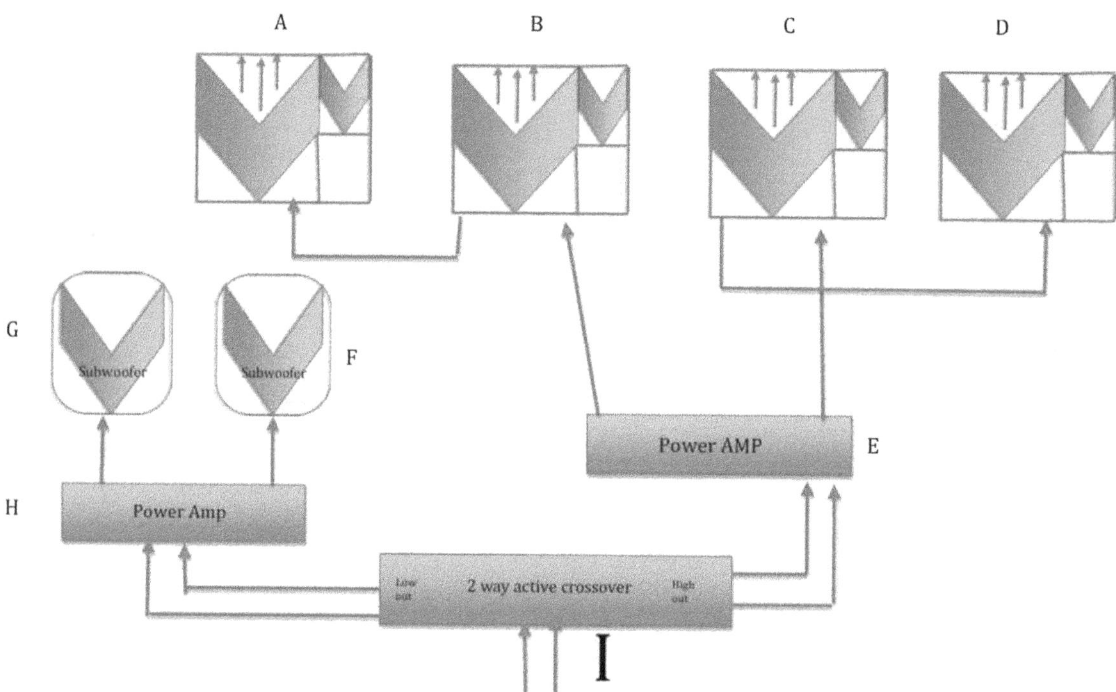

Figure 12.1. The last stage of live sound reinforcement is no less important than the ones earlier in the chain.

- *Normal mode*: the normal stereo signal in, stereo signal out, with both channels operating at the amplifier's stated power levels.
- *Parallel inputs*: routes the signal on a single channel to both channels. Useful if you need an identical signal to go to more than one speaker or speaker cluster.
- *Bridged mono*: this mode takes a stereo power amp and turns it into a mono power amp. It does this by adding the power available on both channels and sending it to one mono-loaded output for powering a single mono speaker.

In addition to the three operating modes available, there are other optional controls and features that some amps may have, including:

- *Line-level matching*: the level of input that will yield an output equal to the amplifier's stated average power. Usually described in volts or decibels (dBu), some amplifiers have switches enabling the amplifier to be matched with either a professional or consumer line level source (0 dBu (−10d BV) or +4 dBu).
- *High-pass filter:* switchable high-pass filter (the frequency may or may not be adjustable/selectable). This protects bass speakers from subsonic frequencies, and midrange speakers from low bass frequencies.
- *Limiter on/off (aka protection circuit)*: for when amplifier has a limiter built in the threshold is set by the manufacturer to prevent the amplifier being driven into clipping. While most power amplifiers specify their undistorted output power, *they are actually capable of producing many times that amount of power when driven to distortion.* Speakers will fail quickly if driven by more power than they are designed to handle; the limiters are there to protect your equipment, and only operate when maximum safe levels have been reached. Therefore, except for rare exceptions, there is usually no good reason to turn these off.
- *Ground lift*: This is a switch which will disconnect the amp's connection to the electrical ground (analogous to putting an adapter on a three-prong wall plug to enable it to plug into a two-prong wall socket). A type of noise that can infect P.A. systems is a constant background noise at 60 Hz, called a ground loop. Ground loops are created when a circuit has multiple paths to different ground references, and one common fix for this problem is to lift some devices in the circuit from the disparate ground references. This control allows user option of disconnecting the amp's ground reference.
- *Volume controls (gain)*: In amplifiers, these are attenuators, so they are not making the signal bigger, just determining how much of the signal is allowed to pass. Turned to maximum they allow the full input signal to pass to the output amplifier stage which has a fixed amount of gain (full voltage gain will be applied to the input signal). This allows you to use the mixer's maximum undistorted output if your amplifiers are linked directly to the mixer's outputs. Set to any other position, they reduce amplifier output. The scale on amplifier volume controls will usually be stated in decibel values from maximum attenuation to none (0 dB). Intermediate points indicate negative gain (in decibels).

Amplifier Specifications

- *Output power*: the power (in watts) that the amplifier can produce into a stated load (impedance) (i.e. 800 W at 8 ohms). The two most reliable ratings of this are those given by the Federal Trade Commission (FTC) and Consumer Electronics Association (CEA) because both tests state the total harmonic distortion (THD) created during testing (since an overdriven amp can output twice as much power, stating the THD verifies the power rating as below distortion levels). Where FTC or CEA ratings are given, capability can be compared with others that use the same standard.

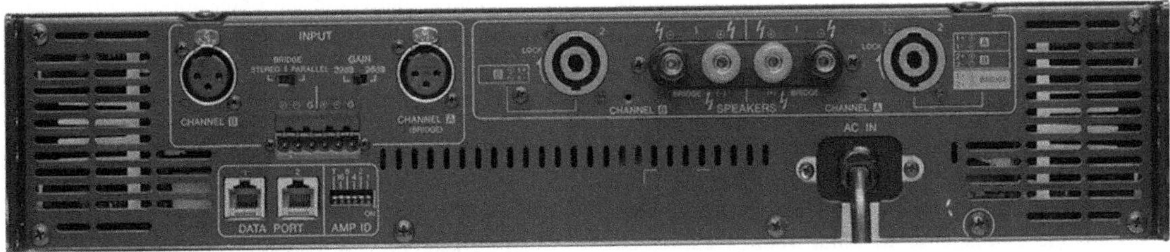

Figure 12.2. Power amp rear panel.

- *Headroom*: the difference between the rated power and audible clipping. Headroom is a condition where some unused power output is available for those signal peaks that demand a little more than is usually required and leaving headroom is a way to provide a margin of safety for both the health of your gear and the quality of your sound. A good pro loudspeaker can handle spikes that go above their rated capabilities if the signal from the amplifier is clean. Making sure your amps have a little extra power above what is absolutely required to drive your speakers will ensure that, even at the highest peaks, your amps will be able to send a clean undistorted signal to the system loudspeakers. This rating will rarely be above 3dB.
- *Damping factor*: determines the ability of an amplifier to control a speaker cone's movement, and is usually given as a number representing the ratio between the load impedance and the amplifier's output impedance. The dampening figure given only represents its nominal load, but higher is still better than lower, especially for bass speakers, because a lower damping factor can cause unwanted resonance (booming or ringing). Speaker impedances vary, however, meaning damping factor will vary by speaker and frequency. The amplifier's actual damping ability is also affected by speaker cable resistance: keep cables short as possible and use larger gauge when in doubt.
- *Gain*: the ratio between input and output voltage; may be selectable on some amplifiers. Note that output power is determined by the amplifier's power supply, not by gain. An amplifier with low power and high voltage gain can be driven to clipping with only modest input signals. Amplifiers with very high power output and low gain require a larger input signal to produce full power. However this is set, what matters most is that systems using more than one amp have each amp set with the same gain settings, and that gain, sensitivity, and crossover setting be set in conjunction.
- *Sensitivity*: the level of input, usually stated in decibels (dBu), or in volts/millivolts, that is required in order to yield the rated output; may also be selectable but usually 0 dBu or +4 dBu. Lower sensitivity values increase the level of output for a given input.

Less important values are:

- *Frequency response, noise*: These specifications are similar from power amp to power amp and realistically don't affect the overall quality of the signal.
- *Slew rate*: Describes the rate at which output voltage can change in response to input voltage expressed in volts per microsecond (V/μs)—a high slew rate indicates sudden transients could be reproduced faster than any limiters could deal with, not a desirable capability. However, slew rate is an overload condition, and should not happen—even in low-budget amps, no slew should ever occur with real signals in real-world applications. Tests measure for it by inputting artificial signals that are too fast for the amp to deal with, signals which don't occur naturally. The amplifier-buying guide on one of the webpages of the online retailor Sweetwater sums up slew ratings best by noting that slew "should not happen at all for an audio amplifier. Therefore, being proud of a slew rate is very strange indeed."

Amplifier Class

Class refers to the power supply and output-stage configuration. Generally, the class of amplification refers to the different methods used to amplify the signal. Each class type has a different ratio of quality of sound reproduction and power efficiency. There are four main class types, A, AB, C, and D, but new ones are always being developed by speaker manufacturers. Included below are some of the classes most commonly found in P.A. power amps:

- *Class A*: uses a configuration now more commonly found in hi-fi amplifiers than power amps. If you find one in a P.A. it is safe to wager it a "vintage" component. In class A amplification, the output signal is driven by a single DC power supply. The output-stage transistors are half-on even when no input signal is present, so the amplifier is drawing current (usually the most current) and creating heat when no signal is present. The least efficient amplifier class, these amplifiers, they usually waste at least four–five watts as dissipated heat for every watt of power output.
- *Class B*: The output signal is driven by a split DC power supply (ground, plus, and minus). Paired transistors are used, with one half of the pair driving the positive half of the output signal, and the other driving the negative half. Draws very little current, because when no signal is present transistors are off. There is a risk of distortion, called "crossover distortion," as the signal is passed from the transistors handling the positive half of the signal to those handling the negative half. These amplifiers are not typical in audio circuits, but help explain class AB amps, an amplifier class commonly used in power amps.
- *Class AB*: cross between class A and class B amplifiers. Designed with a small overlap in the middle, the paired transistors pass a small amount of current at

idle (though much less than class A amps). This is done to eliminate the crossover distortion associated with class B amplifiers, while retaining the greater power efficiency advantage over class A designs. Most amplifiers in P.A. systems are class AB or AB design variations (classes G and H).

- *Class G and H*: both class G and class H amplifiers are similar in design to class AB, but use more stages in the power supply. These amplifiers use higher voltage power supply rails to deal with bigger signals. (Low-level signals are dealt with in the same way as in class AB.)
- *Class D*: the most recent advance in amp technology, unlike the amp classes above, which are "linear," class D is a switch-mode class (transistors are paired and trade off functions so one or the other always acts as a switch). While the amp is operating there is either no current (when no signal is present and switched "off") or no voltage (when current is being drawn to apply to signal and switch is "on") across the transistors. The upshot is that whether the amp is sitting idle or during amplification almost no power is wasted as heat, allowing these amps a high degree of power efficiency (80–90 percent). In light of the explosion of digital tech, it's not hard to understand why people commonly mistake the "D" in class D as representing "digital." However, class D amps are based on analog technology and there is no digital coding of the signal.

In general don't buy amplifiers of a class other than AB (G/H), or D unless you have a clearly defined reason why.

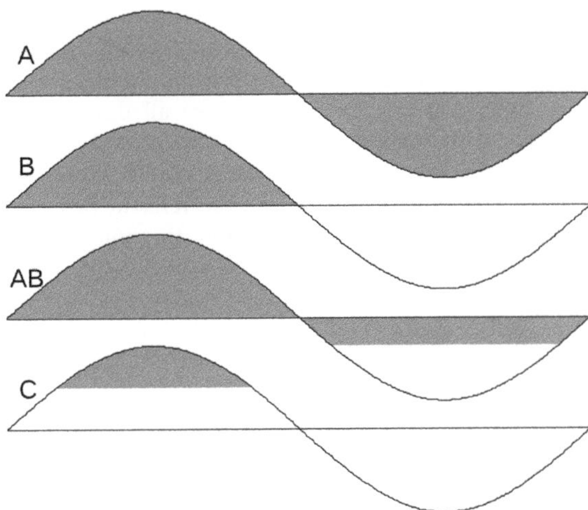

Figure 12.3. View of the transistor area of conduction (in grey) for different classes of amplifier operation. *Source: Graphic by Yves-Laurent Allaert under a Creative Commons BY-SA 3.0 license.*

Circuit Safety Features

Amplifiers all need some kind of circuit overload protection because they are capable of pulling enough excess current to produce outputs well above their continuous ratings, and also have the potential to create so much heat even when used at their rated levels. Just as important is doing so without peaks causing "false-alarm" shut downs, or allowing safety features to limit performance at normal loads. Because these features are essential to amplifier safety, try to use amps that incorporate at least some of them in their design; they can spell the difference between having to pause and reset (after taking care of the problem) or having to douse your flaming amp with fire retardant foam, and being left with a steaming pile of charred debris.

- *Switchable mute*: Rather than being forced to use the less convenient options of lowering gain settings or shutting down, this feature allows for muting during actions (plugging or unplugging components while the P.A. is live) that could create outputs with the potential to damage amp and speaker.
- *DC protection*: A relay that disconnects the output terminals if DC power is detected on them.
- *Over-current protection*: Limits functionality or shuts down if the amplifier tries to draw too much current.
- *Short-circuit protection*: Short-circuit protection prevents the amplifier from trying to drive a short circuit which can create enough heat to burn out important circuitry.
- *Thermal overload protection*: To prevent component failure or fire, amp will shut down when temperatures exceed a safe threshold.

Setting Amplifier Gain

There are many live audio pros who feel it is easiest to set amplifier volume controls to maximum (0 dB), and just control output levels from the mixer. As long as speakers and amps are properly matched (the output is below the peak rating of the speakers), or are otherwise limited so there's no worry about the system output being too loud, it is safe to set the power amp's front gain controls to maximum. While safe and easy are a fine way to go when those are the priorities for that day, some applications require you use your system to its maximum performance capabilities, which may require you set amplifier gain more precisely. Ideally amplifier volume controls should be coordinated with sensitivity and crossover settings, so all devices can achieve their highest input and output levels at the same time (this maximizes system output and headroom). For more on Amplifier gain, refer to Part 5 of this book, when we talk about how to optimize gain throughout your system.

Amplifier Current Draw

Power amplifiers are the part of the audio chain that use the most current, so it is important to be sure that any load is less than the supply's rated current capacity. When more than one amplifier is used their total current consumption should not exceed 13 A when they are being supplied through a single outlet, or draw more than 80 percent of the total current rating of the circuit (often 20 A in small venues) if they are supplied through multiple sockets on the same circuit. Fortunately, P.A.s never run at full power when driving a musical signal so power draw is not as bad as you'd expect, though because some power consumption goes to waste (dissipated as heat), draw will always exceed output. Many specs will provide current draw information for some or all of the following circumstances:

- *Idle (no load)*: sometimes referred to as quiescent, this rates current consumption when no signal is present.
- *$^1/_8$ or $^1/_6$ power*: represents current draw under normal use conditions. Basically, the draw required for the average power of a musical signal (including occasional peaks and clipping).
- *$^1/_3$ power:* represents the power used by a system that is being driven very hard (including periodic hard clipping caused by being overdriven).
- *Full power*: current required to drive a continuous sine wave at maximum level. Thermal failure would occur quickly in such a use scenario.

When planning for an event, you should at least plan for requiring the $^1/_8$ power rating, though to be absolutely sure there's enough power, the $^1/_3$ power rating is the safest bet.

Amplifier Power

As noted, power amps are rated in watts, which tells you how much power the amp unit will deliver to a specific load (rated in ohms) Power ratings are numbers you need to pay close attention to if the longevity of your last stage components is a priority; knowing what the maximum power levels of your amplifiers are and understanding how they relate to the power-handling capabilities of your speakers will help you keep your sound system running properly. Ideally you should try to use amps that are matched to the peak capacity of your loudspeakers, and limit the average signal voltage to the speakers' continuous average capacity with a limiter. This ensures continuous power levels do not exceed the speaker's continuous average power-handling capability, and that instantaneous peaks are handled cleanly. However, when the ideal is not possible, as a rule of thumb an amplifier with an average output of between 1.5 and 2 times the speaker's continuous capacity is usually adequate. Before we can talk about matching amps to speakers more specifically, it would be helpful to learn a little more about the speakers that power amps are built for.

P.A. Loudspeakers

Loudspeakers are the most visible part of the P.A. system other than the microphone at the opposite end of the chain. Like the mic, speakers are a kind of transducer, a device that converts one type of signal into another. However, while the mic captures sound and copies it into an analog voltage signal, speakers do the opposite, converting a voltage signal drawn from the amplifier into variations in sound pressure.

Though most people would be able to instantly recognize the outer shell we put speakers in, called the *speaker enclosure*, fewer folks would be as familiar with the speakers inside, also referred to as *drivers*. While there are a dozen basic driver designs, the most common loudspeaker drivers used in live audio reproduction are *direct radiator dynamic drivers* and *horn loaded compression drivers*. Both operate on a similar principle, using a magnetized coil which moves in response to the audio signal feeding it, and in turn driving the diaphragm's movement. The movement of the diaphragm then pushes the air in its path, reproducing the air pressure fluctuations of the input sound's waveform. If you have ever seen the cone of dynamic speakers moving, you will recognize the description.

Used for sub bass, bass, and mid-range drivers, *dynamic drivers* are attached to the magnetic *voice coil* at the base of their *cone*, and the diaphragm is driven back and forth with the coil in response to the

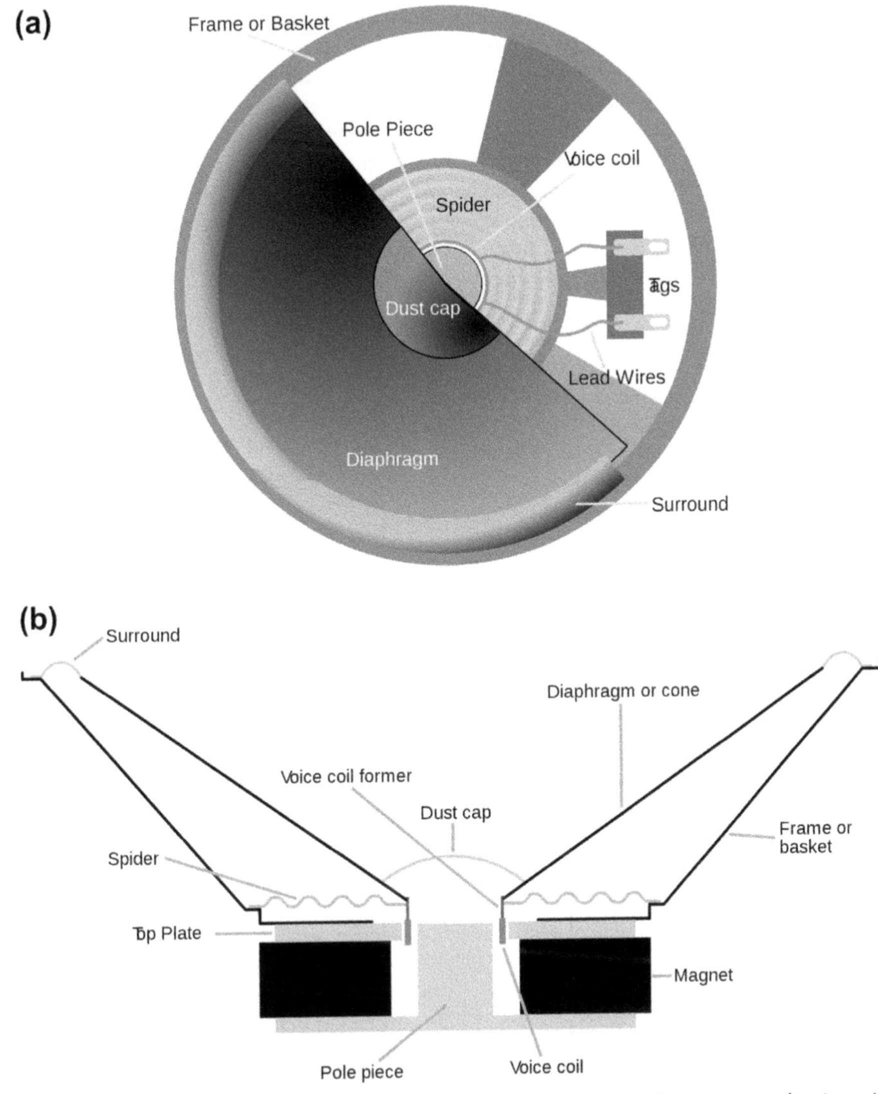

Figures 12.4a and 12.4b. Side view and cross-section allowing a view of loudspeaker anatomy that is typically hidden by the enclosure.

signal voltage feeding the coil. To reproduce high amplitude, low-frequency pressure waves, the large diameter diaphragm must be driven with significant thrust and range of motion. The signal being sent to the coil needs to contain a great deal of energy, and the cone diaphragms of these drivers needs to be rigid, yet light, and damped enough to move vigorously without rattling after the signal ceases. Since there is not yet a material that can fulfill all three requirements, designs using standard materials and methods base their choices on the best balance in these features or trade off, and to compensate for what the materials can't provide, these cones are *front loaded* in their enclosures with a rigid *basket* held suspended by a flexible web called a spider which allows for the large amount of movement in the driver without transferring vibrations to the wooden enclosure.

The high-frequency version of this basic design also has a magnetized element that drives a diaphragm creating the smaller wavelength air pressure fluctuations that reproduce the high-frequency and mid-range sound waves of the P.A. signal. This *compression driver* design encloses its coil and small diaphragm in a rigid shell, leaving only a small front portal which is positioned at the *"throat"* of a long enclosed *horn* within a loudspeaker enclosure. By directing the compression energy from its imperceptibly vibrating diaphragm directly into the horn, this small amount of movement is amplified and reproduces the driver's output sound waves at high amplitudes with much more efficiency than the direct radiator can. Because of the horn's length, a horn-loaded driver is located at the back of the enclosure instead of the front.

Because of the inability for one driver to reproduce the full spectrum, P.A. speakers need to include a range of speaker driver sizes. Of course, these each must be enclosed too because speaker anatomy is no more complete without an enclosure than human anatomy is without skin. So, with the understanding we need a range of driver sizes to get a good sounding system, and that those drivers are not complete without being placed in an enclosure, it becomes a matter of how to organize those drivers.

Speaker Driver Size

With multiple driver sizes needed to cover the full frequency range, most cabinets will contain more than one size of driver, though there are a few exceptions to this rule. To avoid confusion and allow for people to more easily make reference to drivers by size/enclosure, each driver can be named according to its relative size. Though these drivers all come in various sizes, some even overlapping with each other (for example both large diameter driver types can come in 12-inch diameter versions while the two smaller drivers can both come in 5-inch sizes), they each are a specific size in relation to the others, the order determined by the range of frequencies that each is intended to reproduce.

- *Woofers* are the kinds of loudspeakers used for the reproduction of low frequencies. They have a relatively large membrane (10 to 15 inches in diameter), allowing them to reproduce the lower frequencies well. In smaller systems woofers handle all the low-frequency reproduction, though for larger, high-power situations a special type of very low-frequency driver type is also used.
- *Sub-woofers* are a subcategory of woofers built for handling very low-bass frequencies (20 Hz–150 Hz). Because of the frequency response of the human ear, sounds at these frequencies are often felt before they can be heard, and must be reproduced at much higher dB SPL than higher frequencies before we can hear them. To reproduce the lowest frequencies these drivers have the largest diaphragm sizes (10 to 18 inches). They are not used for every type of music genre, but anyone who likes beats knows what subs penetrating your senses can bring to a dance floor or music performance, and why subs are now an indispensable aspect of sound reproduction and reinforcement for many popular genres of music. They allow us to feel the sound as vibrations in the air and through the floor under our feet, and for anyone who's been moved by them, no sound system can be called complete without them.
- *Mid-ranges* are smaller than woofers (around 5–8 inches in diameter) and are used for the reproduction of middle-range frequencies.
- *Tweeters* are small diameter speakers (1–4 inches), used for the highest frequencies, but can also reproduce some mid-range sound as well, and in systems with only two drivers, they reproduce most of the frequencies above 1,000–3,000 Hz. Tweeters are the only one of these drivers that will typically be a compression horn-loaded driver, as this design is too expensive and bulky to use with large diameter sizes.

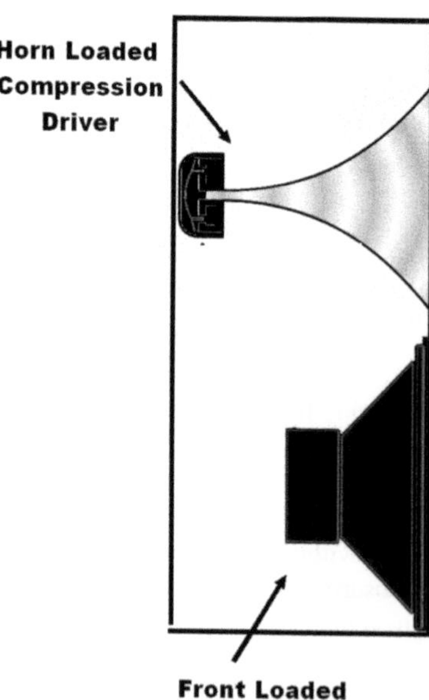

Figure 12.5. Direct radiator and horn.

So now that we know the design, the relative size and frequency response of the most common driver types, and can even identify them by name, getting them organized and assigned to enclosures should be simple, but we aren't ready yet. We still need to make sure that each driver is only fed frequencies that are outside a driver's optimal frequency range are not sent to that driver. Unless each size is sent a signal in its optimal frequency range, all the efforts to organize our speakers is for nothing. Fortunately we have a tool for this job, called a crossover.

Crossovers

A *crossover* is a device that ensures that the system's speakers each get the right division of frequencies for their optimal operation. Its main function is to divide the entire frequency range into subsets of the full range and then send each frequency subset to the speaker or speaker groups that can best produce that range of frequencies. Crossovers do this by splitting the audio signal into two or more frequency bands using bandwidth *filters* like those we learned about in the previous chapter. Because they work in partnership with the P.A. amps and speakers, crossovers are also called *crossover networks*, and each is named for the number of pass bands available to feed to (two-way, three-way, and four-way).

Crossovers are used with all driver and enclosure types commonly used in live sound, and come in two types: active which will be located before the power amp in the signal chain, or passive (unpowered), which means it will be located after the power amp in the signal chain. *Passive crossovers* are usually installed inside the P.A. speaker enclosures and use filter circuit elements to passively split the incoming frequency band into two or three frequency pass bands and route a portion to each of the driver types in the cabinet. Since they are often incorporated into a speaker's cabinet and not somewhere easily accessible, their settings are preset by the manufacturer, and are not intended to be changeable.

Active crossovers are external devices that reside outside the speaker enclosures in the form of 19-inch rack-mountable units located in the rack case with the power amps. Active crossovers come before the amps in the signal chain, and contain powered electronics instead of using passive filters, so the unit's settings can be altered to facilitate optimum frequency reproduction for any type of speaker combination and setup. Depending on design, settings may be set without limits, or may be restricted to a selection of setting options. The parameters that may be adjusted can include:

- the cut-off point for splitting the spectrum into its separate bands (the crossover frequency);
- the degrees of dB roll off (the slope/filter order);
- type of filter design topology (Bessel/Butterworth/Linkwitz-Riley, etc.) used by the crossover.

Because active crossovers allow for a greater degree of operator control of their settings they are preferred by most sound companies for their added flexibility and control over a speaker's sound.

Crossovers improve system efficiency and headroom by ensuring each speaker driver only receives its optimum operating frequency range. Most importantly, many crossovers have programmable limiters on each band to protect the drivers from abuse. In addition to the parameters stated above, many active crossovers also allow the engineer to modify other aspects of the signal passing through the device. If they have enough functionality, such units are often called *controllers*, as they are designed to make using an active crossover

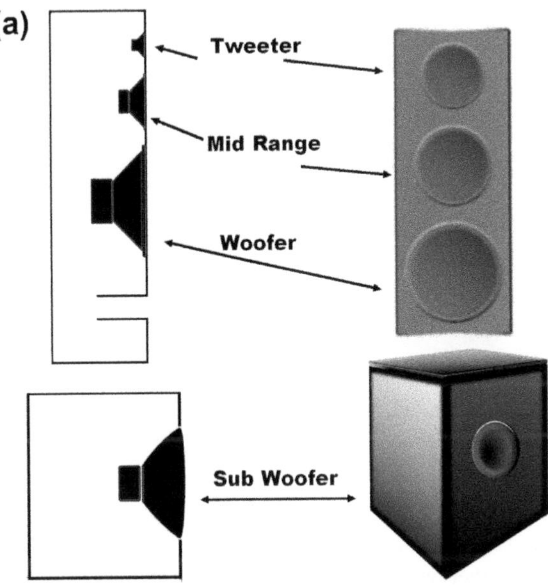

(b)

Speaker	Diameter **(in)**	Frequencies (Hz)
Woofer	12	50-2,000
Mid-Range	**6**	2,000-5,000
Tweeter	**2**	5,000-20,000

Figures 12.6a and 12.6b. Subs, woofers, mids, and tweeters have the frequency range covered.

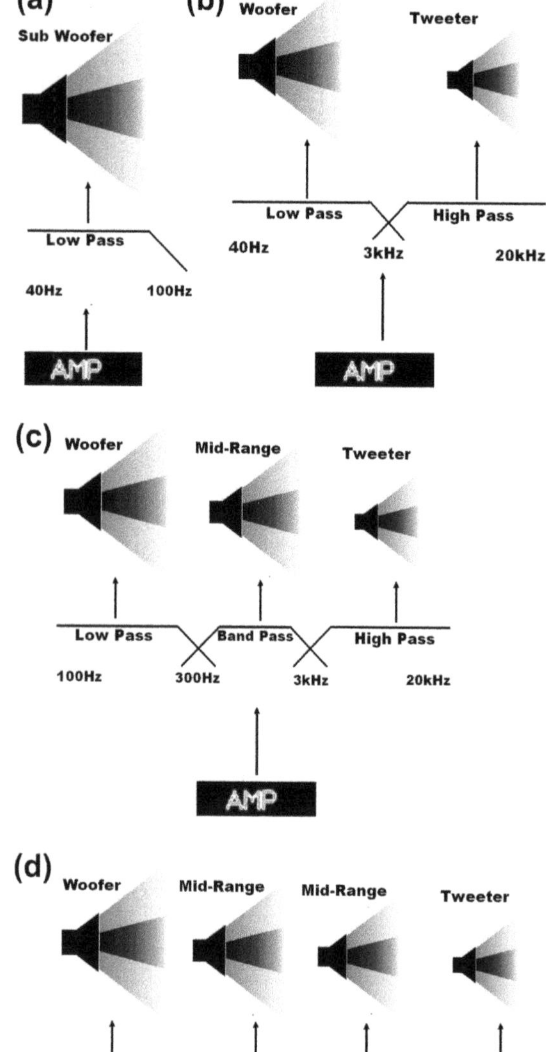

Figures 12.7a, 12.7b, 12.7c and 12.7d. Several sizes of crossover networks with filter crossover frequencies shown.

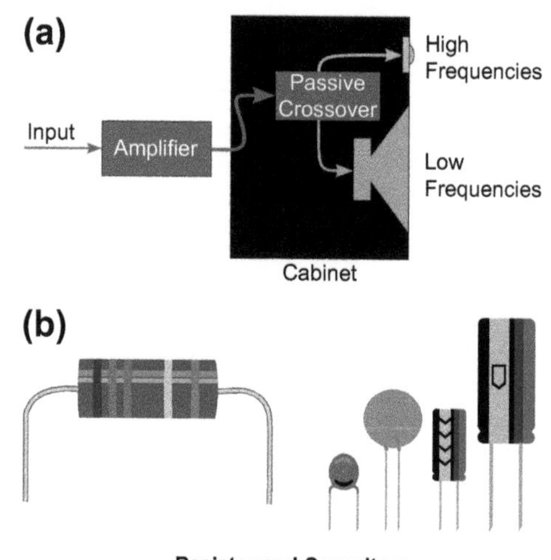

Figures 12.8a and 12.8b. Passive crossover and passive filter elements *Source: Crossover graphic by Ian Ferguson under a Creative Commons BY-SA 3.0 license.*

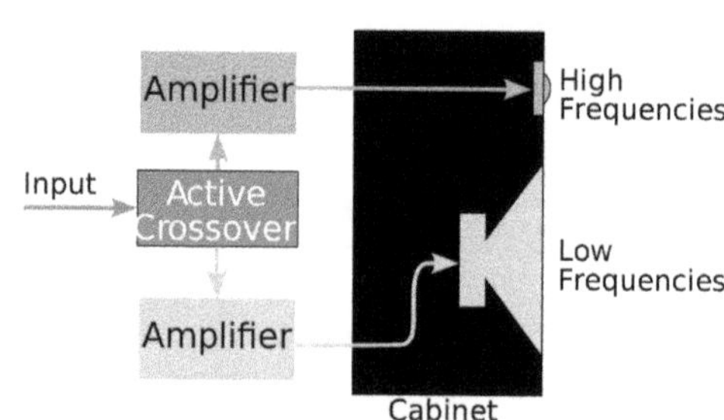

Figure 12.9. Active crossover *Source: Crossover graphic by Ian Ferguson under a Creative Commons BY-SA 3.0 license.*

network easy even for beginners. Other functions and settings such units may include are:

- *Input level/gain*: line-level matching (−10 dBV and +4 dBu).
- *Delay*: where some speakers in an array are mounted further back or further forward phase differences can cause undesirable interference. Using this delay parameter in the crossover can be used to fix any time arrival anomalies due to differences in speaker distances from the audience.
- *Limiter/compressor:* to protect overloading the power amp at the input stage (over driving the power amp will produce horrible distortion).
- *Gain/gain adjustment*: may be adjustable for each output band (adjustment range may or may not be limited).
- *Summed mono*: allows you to add the sub-bass frequencies from left and right channels together into a single output.

If using an installed P.A. system with the crossover/controller already configured, don't touch their settings if you don't know what you are doing. Setting crossover/controller limiters requires precise understanding of power amplifier and driver capacity. You must read your loudspeaker manufacturer's literature to establish recommended crossover points, appropriate filter types and slopes, recommended gain adjustments, etc. At best, hooking up the wrong speakers to the wrong pass band output from a crossover will sound horrible; at worst, it could do damage to costly equipment.

One requirement for using active crossovers is that each *output band* uses its own amplifier. One way of doing this is to use one crossover to split each output band to a single enclosure containing multiple drivers of the same size. However, that requires you be very careful to ensure you use the right number of drivers for each amplifier's impedance, and also uses many more amplifiers and units than the way we will examine here, making it best for higher powered applications. For now will look at the generally agreed upon "best practice" for using active crossovers to connect your network up, and the one most commonly used for smaller scale sound reinforcement, a configuration called *Bi-amping.*

Bi-amping is a technique which uses one amplifier for the low frequencies, and another for mid and high frequencies. If you look at Figure 12.9, you'll see the simplest type, which uses a two-way crossover to split the signal into low and high bands within a single enclosure, with each driver powered by its own amplifier. A slightly more complex version shown in Figure 12.10 uses the two-way crossover network to feed high-, mid- and low-frequency drivers in separate enclosures, but in addition uses a passive crossover to further split the high band into high and low.

The choice of crossover frequency can vary, and which one is chosen is not too critical, provided that the amplifier powers are properly balanced, and the drivers are operating within their frequency and power limits. As a solution to just about any amplifier–speaker

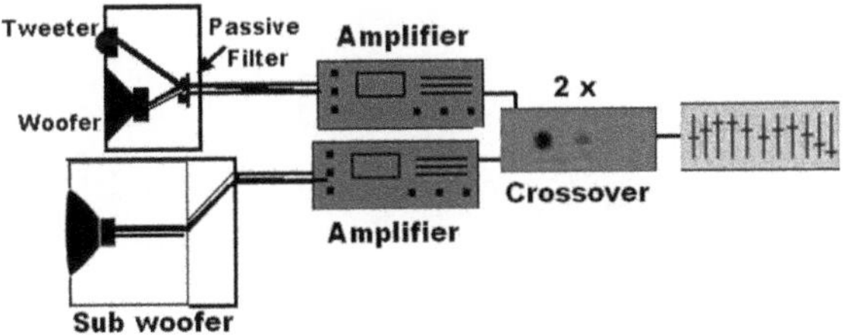

Figure 12.10. Another bi-amping configuration.

combination, bi-amping is at the top of the list—easy to set up and easy to use, providing the best frequency response and plenty of headroom. No matter what kind of P.A. you have, low-end commercial or high-end pro, for low- to mid-power applications, a bi-amped system will sound better and function better than a setup using conventional passive crossovers with a single amp powering all frequencies. Tri-amping is another version of this setup; just split the signal into three enclosures instead of two.

Assigning Speakers to Enclosures

Now you can assign speakers to enclosures in any configuration you want, though as in real life, once in an enclosure, drivers are meant to remain there. For most people, assigning drivers to enclosures is a choice made when they purchase the speakers. Then they live with the choice and can't do much until a major P.A. overhaul. This is why the best bet is choosing a configuration allowing the most versatility.

For loudspeaker and system designers, the main principle is to match the internal space requirements of the enclosure with the physical space that each driver takes up according to parameters for producing the best sound. Acousticians have a set of rules that govern the size requirements inside the enclosure that ensure optimal sound generation and efficiency based upon the frequencies that a particular cabinet will be reproducing. This determines the minimum size of any cabinet, while the maximum size is determined by practical considerations (the largest speakers are still kept below a set size to allow for portability). Beyond these conventions, a few simple rules govern how speakers are assigned to enclosures:

- Enclosures that function by using multiple driver sizes to produce full-frequency output, as well as those that only put out one or two bands of output are called *multi-way enclosures* and are labeled according to the number of different driver sizes used in the enclosure (not the total number of driver units).
- Enclosures that contain two speaker sizes are called *two-way enclosures*; in a similar fashion enclosures that contain three speaker sizes are *three-way enclosures*, and those containing four sizes are *four-way enclosures.*
- The largest size speakers reproduce the lowest bass frequencies. These should be put into their own enclosures rather than mixed with drivers of different sizes, due to the many differences between very low frequency sounds (20 Hz–150 Hz) and the

rest of the full range spectrum (i.e., more power required, interaction with the environment, placement within the venue, etc.). They always come enclosed in their own boxes and are called one-*way enclosures.*

- So long as the enclosure does not exceed the maximum size for portability, there is no restriction on how many individual driver units can be put together in one enclosure.
- The power needed to drive that cabinet changes depending upon the number and size and type of the drivers contained within (the amount of power will be the sum of the load of each driver within), and the industry standard is to wire cabinets so all drivers can be powered by a single external connection jack.

The exact range of frequencies used by a particular driver will vary according to what other drivers it is paired with. Each type will split up the frequency band according to its capabilities— two-way speaker systems have at least one woofer and one tweeter, three-way speaker systems have at least one woofer, mid-range and tweeter, and four-way speaker systems may divide the mid-range into two segments for coverage by different drivers. Figure 12.11 shows some of the common enclosures available from Yamaha. We now have the vocabulary to pick out some common types pictured. Notice the stage wedges have both a mid-range (direct radiator) and a tweeter (horn) driver within the enclosure, and so can be called two-way speakers. In the same picture there is a speaker on a pole above a single large speaker box. The bottom enclosure is a one-way subwoofer, while the cabinet on top of the pole is a two-way P.A. speaker.

Self-Contained Speaker Systems

There are two kinds of speakers that are exceptions to the driver and enclosure conventions we've examined here: full-range speakers and powered speakers. *Self-contained* speakers require no thinking about how their components are organized, as each speaker box simply covers the full spectrum. They offer the advantage of simplicity and portability but at the cost of reduced flexibility, power, and sound quality. For live sound, where these are found in use, they will usually only be used for the smallest event rentals or venues. *Full-range drivers* are a special kind of driver intended to cover the entire frequency range (with an accepted trade off of fidelity for the low cost and convenience of the approach). These go in their own box and are called *full-range enclosures*. The other kind of self-contained speakers are powered speakers.

Powered Speakers

The fundamental difference between *powered (active) speakers* and *passive speakers* is that the power amp is

Figure 12.11. A small set of some common cabinet types: one way sub-woofer cabinets, two-way stage wedges, two-way mid-speakers on pole and main P.A. two-way mains. *Source: Photo courtesy of Yamaha.*

contained within the cabinet of a powered speaker, whereas a passive speaker must have a power amp driving its input (amp right before the speaker in the signal chain). Identifying the difference is easy, active or powered speakers require an AC power connection directly into the speaker itself. So if a cabinet connects to a power source through cable and plug it's a powered speaker.

While powered speaker systems can come with drivers assigned to multiple enclosures and in any arrangement found in passive speaker systems, nine times out of ten they will be self-contained, and at most include a separate sub for the lowest frequencies. This makes the most sense as the main advantage they offer over passive systems is ease of use and portability, which are both undermined by a powered system that requires the user to deal with multiple enclosures. Whatever their arrangement all the enclosures in a powered speaker system will contain their own amp and a passive crossover to limit the frequency spectrum sent to the drivers in each.

Active speaker advantages include the assurance that the power amp driving the speakers is as closely matched as possible, as well as less bulk since there is no need to carry the amps separately to and from the gig. They require less knowledge of power, voltage, and impedance matching to use, and there are fewer ways to make mistakes in system setup and use. Disadvantages of active cabinets are decreased flexibility in sonic coverage and frequency dispersion control, system expansion is more complex and costly, not to mention when an active speaker blows out its amp, it is more difficult and costly to repair.

Speakers are the element of the P.A. that are most beloved by the public as well. Both in pro audio and commercial audio, it seems at least half of the discussion surrounding audio is all about speakers. Since the gear heads have made evaluating every aspect of speaker performance a field of study of its own, no discussion of the topic would be complete without mentioning speaker specifications. Since books have been written on the topic, and even more material is available online, look in the directory. If you crave more speaker information than we cover here, there is more than enough to satisfy any audiophile or aspiring audio pro.

Speaker Specifications

- *Frequency response*: manufacturers using graphs provide the most useful information about this. This value represents how the speaker responds in its range, and its range under narrower definitions. This is seldom constant at all volumes, so any true value must include a margin of error −/+.
- *Response pattern/dispersion/directivity*: the angles along the horizontal and vertical axis at which the speakers produce a response. Some specifications derive a value using a −6dB reference, and some a −10dB, so be sure the spec includes this value. Also, outside of high-frequency horns which have a consistent response, lower frequencies and front-loaded drivers of any frequency are not as consistent. This means that when looking at full-range enclosures, this spec may only reliably refer to the high-frequency horn, and not be as accurate for the rest of the speakers in the enclosure.
- *Sensitivity*: value in dB SPL of the relationship between power and acoustic output. Of the tests to measure this, AES, IEC, or EIA specifications are the ones allowing for the most useful comparison, if the spec gives no details which standard was used, look deeper—the value is probably not a true indicator of the speaker's performance. Also the tests are normally taken from an on axis position, so values

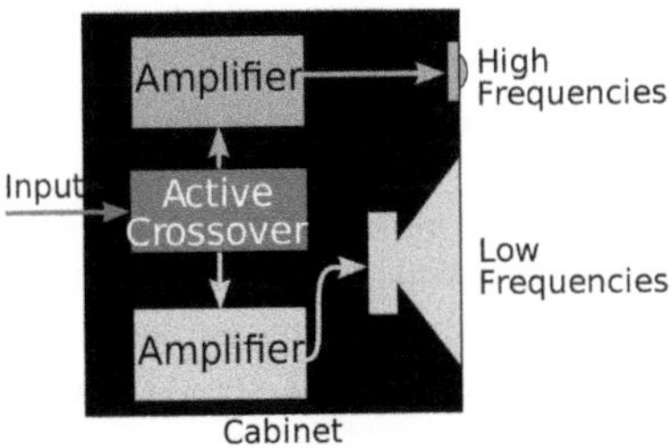

Figure 12.12. Powered speakers; just plug in your mixer and plug these into main power. *Source: Graphic by Ian Ferguson under a Creative Commons BY-SA 3.0 license.*

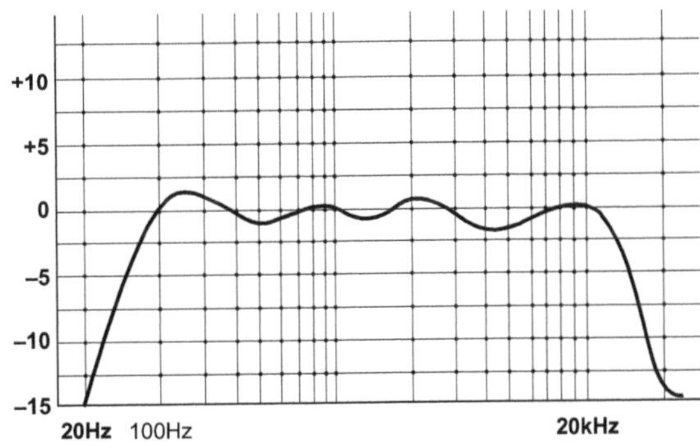

Figure 12.13. Bode plot showing the frequency response of a speaker.

favor directional speakers, and those with broader, more even converge don't rate as high. Depending on your needs, keep this in mind: if broad coverage is your focus, look for speakers with AES ratings, they are taken at a distance and give a truer value for this quality.

- *Power handling*: how much juice your speaker can handle before things melt or crack. This value is not an indication of sound quality, just survivability. Again the ratings awarded by AES, EIA, or IEC are the most reliable. Generic ratings in terms of watts RMS, program power, peak power, or PMPO aren't as reliable (though the amount of skepticism to use with these can range from "take with a grain of salt" to "use the entire shaker"). Assuming you have no intention of overloading your system, this rating doesn't mean much in this format (*but you can use it to calculate some important information typically left unspecified, like efficiency*), so just refer to the links in the chapter directory. Don't clip your amps too badly, don't overload your speakers, don't buy the worst of the bottom-end speakers, and you won't need to give too much thought to this spec!
- *Efficiency*: value of how easily the speaker turns volts to sound. Efficiency is mainly determined by motor efficiency, the freedom of the coil and cone to move. Most speakers are very inefficient, managing no more than 5 or 10 percent efficiency (i.e., for every 10 units of power used to create sound, 90 more are wasted, usually dissipated as heat). There are some new technologies allowing for more efficiency, and live sound reinforcement gear is pro audio, so it's more efficient, from 10–20 percent, which is twice as much as consumer speakers. Many "tips" on audio claim that the higher the dB value provided for sensitivity, the more efficient the system will be even if you don't know the exact percentage. This is dubious, and a better measure is to compare the ratio of output capability to power handling. Hopefully efficiency will be something that gets more attention, as it can make a huge difference all other factors being equal (consider that a system that can output 3 dB more than another using the same wattage is twice as efficient).
- *Output capability*: measure of how loud the speaker is at max power. Not an indicator of how that loud sounds. Good value for "My system is meaner than yours" contests. Useful for narrowing down a list if you need a minimum dB output capability to handle the venues you work. This value becomes less useful for evaluating final purchases, however, because it says nothing about sound quality. When deciding between two systems that both meet minimum requirement, we want to know which one reproduces sound loud with the least distortion. If using this rating to choose between a system with a 120 dB output and one that can only push 115 dB, beware that if the 115 dB system doesn't begin to distort until at the maximum, and the meaner 120 dB speakers start sounding like a chainsaw after 105 dB, you could end up regretting being so dazzled by decibels if you choose the first system merely for being 5 dB louder.
- *Impedance*: the AC equivalent of resistance. Stated nominal values not fixed because it is frequency dependent. Beware of buying speakers with an impedance value too low for your amplifier to safely handle. Generally, a single amplifier channel will comfortably drive one 4 ohm speaker, two 8 ohm speakers, or four 16 ohm speakers. We will cover this topic in more depth when we demonstrate setting up a P.A. system, but if in doubt, ask around online or seek input otherwise (there are resources in the web directory as well). Better to take an extra week making a choice on a purchase than buying on impulse and regretting it.

Considerations not Specifications

When deciding which speakers are right for the application, there's no better yardstick to use than starting with the application (what do you need to do with them?). To make a choice, you need to know exactly what you need. No matter what the application, a sound system needs to be loud enough for the task, cover the sound field evenly, and sound good at all volumes you will be using them at. Specifications can be a place to start, but often can't supply important details. Speakers sufficient for public speaking may not be able to handle a kick drum, even if they are powerful enough for the room (if the low end sounds bad played loud, and you need it to sound good, then those loud-enough speakers are useless to you if they distort). Once you know what you need, and narrow down what looks good by looking at specs, you are only halfway there. Before deciding on speakers you always need to *go listen* to some of your options before you can know which will work best.

It does defeat half the joy of free shipping, but your ears are the only real guide, so whether buying online or at the store, you need to listen to speakers to be sure they are what you need. As you listen, make sure to walk around in front of the speakers. Does the sound quality drastically change the moment you move off axis? Does the sound distort at certain frequencies? Is it loud enough, but also is it clear at the level you need? If you can't get a hold of the exact speakers you are looking for, then look for speakers from the same manufacturer that are rated as closely as possible and listen to those. Also refer to the following chapter for suggestions on how to get hold of speakers to demo.

Chapter **13**

A Word to the Wireless

Wireless microphones use radio waves instead of wires to transmit audio signals to the P.A. The transmitter is either built into the body of the microphone or into a body pack that plugs into the microphone XLR socket and is worn by the performers, while the receiver takes up a line of site position off stage. Wireless mics come in most of the classes found in standard microphones, and can be handheld, headset, or *lapel lavaliere* (clip-on style microphones that are often used in theater, spoken reinforcement, and broadcast applications). Transmitter/receiver combinations can also be used with other signal-generating equipment, which for live sound applications is most often for *wireless personal monitoring systems*, also referred to as *in-ear monitors* (IEMs). Wireless personal monitoring systems operate on the same principle as wireless mics, only with the signal traveling the opposite direction: The monitor mixers' mixes are transmitted from the mixer to receiver packs worn on the musicians.

Anyone who becomes a live audio pro is bound to work with wireless technologies sooner or later. While many people have not yet made the switch over to wireless, the steady growth in popularity and current technological trends make it nearly guaranteed that anyone working beyond the next few years will have the experience using some sort of wireless technology as part of their P.A. system. If anything, the only issue getting in the way of wireless's continued expansion has been the problems caused by its meteoric rate of expansion. For a technology that was rarely in use only a decade ago, except for those who could pay for systems many times more expensive than the wired standard, wireless's growth in the past decade has been extremely rapid. When you combine the traditional wireless users of broadcast television and radio with the live entertainment and music production industry sectors that have adopted the technology, as well as the many churches and schools who have adopted it, it seems like the only thing capable of slowing its further expansion would be the dwindling number of free radio waves created by its recent popularity (well. that and the loss of bandwidth to the recent hordes of smartphone users filling the airwaves with their own signals).

Like sound, we describe *radio waves* according to their frequency and amplitude. Radio frequencies are a little further along the frequency spectrum than audio, ranging from a few hertz to the gigahertz range, and most pro wireless RF systems operate in the range of *megahertz*. With so broad a spectrum one might initially assume there should be plenty of room on the airwaves, but radio waves, already highly regulated, are now a hot commodity and even more rigidly allocated than before. The short "wild west" period of wireless is at an end, so even with a lot of bandwidth along radio frequencies, users can't roam freely without paying mind to the assigned frequencies they are allowed to use, or they risk trespassing into airwaves they are not allowed to be transmitting on. The most notable confusion over bandwidth allocation has been the government's designation of all 700 MHz frequencies as off-limits since June 2010, effectively limiting users to the *white spaces* (those free spaces between broadcast transmissions) in the range between 470 and 678 MHz (the spectrum shared with TV broadcast stations).

This ruling was controversial with users because many systems owned in the US only operated within that suddenly off-limits bandwidth and were rendered immediately illegal to use or sell. Considering the cost of such systems, many who were not eligible or did not meet the deadlines to get partial reimbursement or trade-ins with manufacturers offering reimbursement were understandably cranky about the turn of events. These users found that with a decree of the government, their high-cost systems had been effectively turned into paperweights, or if they chose to continue using them then they had been turned into criminals. Many users feigned ignorance and chose the latter, becoming frequency rustlers for a time. With the number of churches and schools represented among these desperados of the airwaves, it seems unlikely that the motive was rebellion, but more likely for lack of any options,

Figure 13.1. Wireless systems share the radio waves.

since these groups are among those least likely to have the money on hand to immediately buy replacement wireless systems. At this late date, there are few risking this practice anymore; the first heavy fines were slapped on users caught doing this around a year after the "no trespassing" signs went up, so ignorance, not even ignorance combined with non-profit status, will no longer get users a pass. Continuing to use these frequencies is no longer possible anyway; they were cleared to open up space for 4G networks for cell-phone service providers, a popular service with transmissions so powerful when interference occurs there's no chance of an individual's tiny wireless systems winning out.

Unfortunately, with nowhere else to roam, and with wireless systems gaining popularity daily, not only with performers, but in churches, schools, and businesses, finding free space between the TV signals can be a difficult task, especially in urban centers. The situation is the same in the UK where live production event industries recently fought to prevent the sale of a broad portion of their own bandwidth for fear it would have the exact same effect of crowding users into a smaller bandwidth and rendering systems who use frequencies outside that bandwidth useless. Organizations like Save Our Sound UK argued that the situation is much like somebody being kicked out of their residence without compensation or a chance to find a new place to live.

While situations like these have caused confusion for wireless users and made some users hesitant to spend money on a technology with an uncertain future, the only uncertainty is what the solution will look like, not the future use of wireless itself. For current live sound students this means they should expect wireless to be part of the future landscape of live sound, although in the meantime they should be cautious to double check that the frequency ranges on any systems they are looking to buy are usable in their location before putting down money. TV channels from 40 to 51 representing the space between 626 and 698 MHz are in the spectrum with the least congestion at the moment. Even that bandwidth may be auctioned off down the line, though you should be safe investing in systems using that range for now and in the near future. Links included in the online directory for this chapter lead to a detailed discussion of this topic, as well as tools to assist in locating free frequencies provided by Shure and Sennheiser, two major wireless manufacturers. In the long term, alternative solutions will likely be technological or in the form of a shift to digital wireless technology. Until then, an understanding of the overview of the two main ways we employ wireless systems is recommended for anyone entering the audio field.

Personal/IEMMonitors

While many bands have gotten more accustomed to using wireless mics, IEMs have been adopted more slowly and have met with more resistance by bands. Recently, more performers have decided to get on board, and there are good reasons to, though there are also some legitimate issues pointed out by performers who haven't. For those who are undecided where they stand, there's a lot to weigh. Wireless monitor systems have several advantages for both the band and engineer, as well as a few disadvantages. Since there are distinct benefits for the overall sound, many engineers are keen for bands to adopt them, but don't feel they should wear them too, which we disagree with. We feel it is as important for mix engineers to wear IEMs as their bands: IEMs are not hearing protection per se, and with careless use or abuse can harm hearing, however together with some good ear plugs, they are the best way to control stage noise, and if used correctly can protect the hearing of both band members and sound engineers.

Should any readers find themselves in a position to be required to make the choice to switch to wireless or not, or conversely be required to talk musicians they work with into seeing the value of wireless systems, the first thing they will need to succeed at either task is to understand the pros and cons of wireless systems, but before we can do that we must look at some problems related to traditional monitor uses. Monitor systems are used because they enable each musician to hear themselves apart from the overall sound of the music and also to hear those other parts of the music that will help them do their best. Being able to hear themselves and each other is what allows musicians to stay in time with the music, but also connect and "gel" with each other as musicians. Being able to clearly hear themselves and their band also increases their ability to perform on a psychological level, helping them to sound their best by providing them the assurance and confidence created when they can hear for themselves that they sound good.

Traditional Monitoring

Monitor systems are made up of three parts. First, is the backline amplification of the electronic instruments, the amps and speakers for each of the guitarists, bassists, and keyboardists. Next are the floor wedges, so named because they have a wedge shape, that sit at the feet of each musician and play the mix tailored to their needs produced by the monitor mixer. Finally, there are the side fill speakers on each side of the stage, in the wings, and out of sight of the audience that fill in the stage sound on the largest stages. The problem with such traditional monitoring systems tends to be that sometimes, in the process of seeking to help musicians hear themselves and perform their best because of it, monitoring systems can have the opposite effect, resulting in a wall of overlapping mixes on the stage in which the musicians are subjected to maximum decibel levels, and a net reduction in their ability to hear themselves.

This has also created problems for off-stage sound since the whole problem on-stage is that sound bleeds into areas it isn't necessarily wanted, and this escalating sound on-stage naturally bleeds into the audience where the FOH engineer must turn up the power on his system to compensate as well. Some of you may have experienced an unpleasant side effect that is not uncommonly experienced as a result of this phenomenon if you've ever left a concert with your ears ringing and feeling uncomfortably full.

Worse is the friction created between band members all trying to hear their mix above the others, each contributing in turn to the escalating problem by turning up the volume of their amps and asking for more volume from their monitors. The notorious friction known to occur sometimes between performers and sound engineers is also often a direct result of struggles over this phenomenon. Even personal relationships can be strained to the breaking point. More than one band has broken up due to musical differences that had nothing to do with songwriting or the music itself, and everything to do with building resentment created by some of these on-stage "loudness wars." Even bands where the disagreement is not severe enough to cause actual harm to their relationship have experienced days where the low-level irritation at one another affects performance.

Advantages of Wireless IEMs

For bands experiencing something like the friction depicted above as a result of excessive stage sound, IEMS are an easy sell. Monitors and wireless systems have few problems of their own, but they are insignificant compared to the ones above. On the other hand, traditional monitoring is not perceived as problematic and does not produce problems as large as the ones above in all cases. Many bands even revel in producing enough decibels to come down like a sonic hammer on their audience, from the stage or the front of the house, if they can talk the engineer into it (and unfortunately there are engineers who hold the same ethos, though they should know better). For these bands the solution to excessive stage noise produced by monitors and wireless monitor systems is less enticing, and the cons related to IEMs bear more weight. Should you work with musicians who may be more open to the idea of IEMs than the above example, there is plenty of information that might help the discussion along. Some of the benefits of taking monitor mixes wireless and ditching those wedges and fills are:

More freedom of movement on stage:

1. no more wedges and wires for set up crew and musicians to trip over;
2. musicians aren't chained to one spot just to be able to hear their monitor mix;
3. allows musicians more freedom to move about on stage, with maybe even room for some showmanship;
4. also gives musicians more freedom to move up to edge of stage and interact with audience at will.

More harmonious feelings among band and engineers:

1. no more escalating battles of "turn it upsmanship" between band members trying to hear their monitor mix over the others;
2. eliminates a big area of contention not uncommon between artists and audio pros (FOH wanting to manage stage volume vs. musicians turning it up to max volume trying to hear their monitor mixes)—this relieves the monitor mixer stress from being at center of this tug of war and reduces the overall likelihood of friction between musicians and their live sound professionals;
3. eliminates friction or resentment that can occur between band mates who disagree over stage noise levels, a situation that is also not uncommon.

Better performance from those at the mixers and on stage:

1. reduced friction between parties raises the comfort level for all, which correlates with better performance for all;
2. band members each control volume of their own mix, can hear themselves at all times, and therefore will perform better overall;
3. the sound engineer who doesn't get signaled every few minutes by one or another band member

Figure 13.2. Wireless mics only set you free if you aren't chained to a monitor. Wireless monitors and wireless mics are complementary technologies.

requesting more volume can better focus on content, and is likely to produce better mixes.

Eliminate the negative effects of stage noise on the sound:

1. less chance for reflections from stage noise to contribute to room resonance;
2. less risk of bleed or spill leading to feedback;
3. house sound no longer needs to adjust for excessive stage noise, more gain before feedback.

Eliminate the negative effects of stage noise on the people subjected to it:

1. less risk of damage to sensitive ears of all present and, by extension, less likelihood of long-term damage to the valued careers of musicians and engineer alike.

Disadvantages of Wireless IEMs

There are some cons to take into account as well, both for the sound engineers working at the mixers and the band members. Wireless systems in general can cause more nuisances for the sound engineer due to:

1. fear of radio interference weakening the signal and causing loss in quality;
2. worries over battery life;
3. worries over the transmitter and receiver communicating properly.

Most engineers will willingly deal with such nuisances, however, if wireless monitoring produces an overall benefit to the sound by reducing excess stage sound. The cons most mentioned by band members are a bit harder to overcome, although with careful planning and smart choices most of these complaints can be mitigated as well. The disadvantages of IEMs most cited by musicians are:

Isolation and being closed off:

1. from the world around them;
2. from the audience;
3. from each other.

Initial discomfort and a sense of lost hearing sensitivity:

1. *Occlusion-* Occlusion is the low frequency sound accompanying our own voice due to the bone conduction that occurs from having plugged ears. Normally, this sound exits our ear canal, but with plugs in our voice reflects back toward the eardrum instead.
2. *Reduced bass response-* This is the biggest complaint leveled by drummers, bassists, and FOH engineers about wearing IEMs. Improvement in IEM drivers recently have made it possible to get models that are tailored to give a better bottom end sound to those wearing IEMs. Another solution that works for some drummers is to attach a butt kicker or shaker to their stool, as the vibration can supplement the bit of bottom end missing from their hearing.

Solutions and Workarounds

There are a few universal tricks to try, and taken all together there are ways to reduce all of these but not eliminate any of them completely. Usually, reducing the effect as much as possible and getting bands used to wearing their monitors slowly is enough to convert hesitant musicians over time. There are some that are adamant that any of these annoyances is enough to drive them nuts, and IEMs are not an option for them as long as any of these cons remain. *The irony of course is that each of these complaints is about a side effect that mimics what they would experience going deaf*, which they will have a good chance of doing if they take no action to lower their noise exposure early—and unlike ear buds, you can't remove deafness once you've bought it!

To reduce the negative effects felt during a transition to personal monitors:

1. Take a few hours to select the IEMs that sound best to you This means don't grab the first one that's good and in your price range, but take an afternoon

to compare the most recommended ones two at a time and carefully narrow down the options until you've found the best fit. If making the purchase for the entire band, as many of them as is possible should join the shopping trip to help choose.

2. Better-quality systems sound better and better fitting plugs produce the least discomfort. If possible, anyone looking at transitioning to IEMs should get their ear buds custom molded. If that's too expensive, they should get the best pre-made ear sets in their price range and put the same care into shopping for proper fit as I advise they put into shopping for drivers.
3. For example, the only cure for occlusion is fit; you want the least space possible between the IEM and the eardrum and custom molded plugs will provide the most comfortable fit. If you must get off-the-shelf plugs, consider them temporary until you can get better ones. Proper fit makes all the difference in both your ability to adjust to and the overall sound quality you enjoy while wearing IEMs.
4. We advise not making a practice of using only one IEM as many people do, although if you must, then switch back and forth between ears. Going deaf in one ear is probably worse than losing a smaller amount of acuity in both.
5. It is also important to take the time to acclimatize to wearing them. Wearing IEMs in rehearsals several weeks before playing live shows is the best way. Until acclimatized, always be careful to increase volume slowly; it is very important in order to avoid turning the volume on your IEMs up too high. Getting used to wearing them a little each day at practice before you wear them at a show can keep you from listening to them with a setting that is too loud for safety and defeating half the purpose of choosing to use IEMs.
6 One thing that all receivers have in common and that is essential for any in IEM system is brick wall limiting. This will prevent spikes in the audio signal from doing damage to your ears because this is a significant risk when monitoring audio with IEMs and can occur within a split second. Unless you have immediate control over the volume, or you know there is a brick-wall limiter installed in the signal path of the P.A., never plug your IEMs directly into the headphone output on a mixer, instrument, or DI box.

Wireless Microphones

When a sound engineer and/or musicians they work with do choose to adopt in-ear wireless systems, the kinds of features and controls that are included in their system will depend on the variety of its make and manufacturer. Hooking up the system will be relatively easy as much of the information that follows about wireless mics applies to IEMs as well.

Engineer IEMs

Engineers should seriously consider using ear monitors as well. Just because they are not on stage and subject to the worst sound levels created by escalating stage sound, doesn't mean their hearing is not just as at risk as the musicians. Engineers are often exposed to well beyond the healthy dose of daily decibels and those who argue that they should not wear adequate protection simply because musicians suffer more are ignoring the fact that deafness or some degree of hearing loss is equally likely for them if they fail to take proper safety precautions. The online chapter directory has links to articles and blog entries where sound engineers talk about their experience of making the switch.

Features to Consider when Choosing a Wireless System

1. *Band*: Wireless systems come in either *VHF* (Very High Frequency) or *UHF* (Ultra High Frequency). UHF systems usually have more possible transmission bands, greater range, and less danger of interference and dropout. Dropping prices mean recently most systems are UHF. VHF is traditionally cheaper and when range and line of site isn't an issue, and VHF channels aren't as crowded as they tend to be in urban centers, VHF systems can be adequate.
2. *Diversity*: Diversity systems have more than one receiver channel. Because they use the strongest signal automatically, they are less prone to drop out when the microphone is moved about (which is why we use wireless to start with). Always use diversity systems unless the range is short, and the geographic area where the system is being used has minimal competition for air waves.
3. *Compansion*: Compansion determines the quality of the compression and expansion of the signal required to fit it into the smaller bandwidth that is

available for a radio signal. A higher compansion rating is preferable, and usually increases in proportion to the cost of the system.

4. *Frequency*: Frequency can be *fixed* or *agile.* Fixed frequency systems have a pre-set operating frequency which cannot be changed by the user. Agile systems allow selection from a range of frequencies to provide the user a greater degree of control over interference. Within agile systems there are greater or lesser degrees of frequency agility (more or less frequencies to choose from) depending on make and manufacture. Frequency agility in systems is almost always preferable, and more is better than less.

Controls on a Radio Mic and Receiver

1. *Output level*: If the receiver has a level control labeled "Output," or "AF Level," this sets the level at the receiver output. Set this so that the level of the signal at the mixer input is at the best range for the channel gain control.
2. *Sensitivity*: Some transmitters have the equivalent of a gain control, labeled "Sensitivity Control." There may also be a sensitivity control on the receiver. Too much sensitivity at the microphone can cause compansion issues and increase the risk of feedback, so use the lowest setting you can that gives an adequate signal.
3. *Frequency*: This is for systems that allow control of the radio frequency. The transmitter frequency needs to match receiver frequency; if you change the frequency on the microphone, remember to change the receiver frequency to match.
4. *Squelch*: This is basically a noise gate acting on the radio signal, muting the receiver when there is no signal. Setting squelch too high can cause signal dropout. To set squelch threshold, set the control to the minimum setting. Then, with the transmitter off, gradually increase the squelch setting until no hiss can be heard. Finally, switch the transmitter on and check signal strength (no squelch is better than no signal).

Using Wireless Systems to Maximize Range and Avoid Interference

1. Mount receiver antennas as high as possible (preferably 8 to 10 feet) above the floor or stage.
2. Keep antennas away from metal objects, such cables, pipes, cabinets, scaffolds, and acoustic ceiling tile supports.

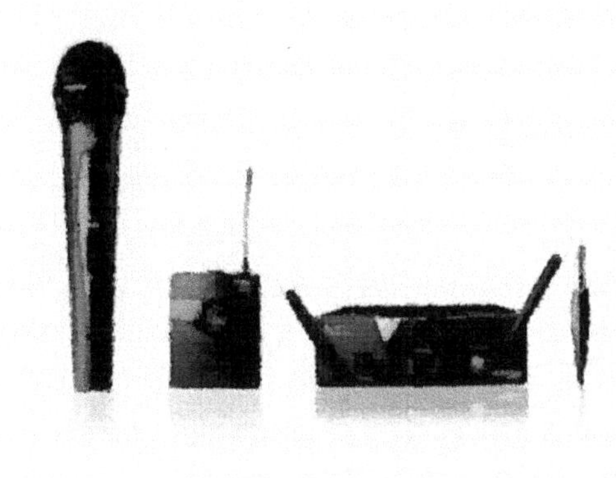

Figure 13.3. Wireless system: mic, transmitter, and receiver.

3. Make certain that there are no obstructions between the transmitters and the receiver antennas; there should always be a line-of-sight path between them.
4. Keep all transmitters' and receivers' antennas at least 10 to 15 feet apart.
5. Keep diversity antennas apart and off axis. Running diversity antenna parallel increases the chance they will receive identical signals, which defeats their purpose.
6. Keep digital devices or computers at least 3 feet away from the wireless receiver and its antennas. If possible, turn off any you don't need.
7. For UHF body-pack transmitters, keep the microphone cable away from the transmitter antenna. For VHF body-pack transmitters do not wrap the microphone cable around the transmitter body or bundle it near the audio connector. Doing this will increase the range of your systems.
8. Make sure your wireless frequencies are not set to a local AM radio or TV channel.
9. Keep squelch settings as low as possible for maximum range. If interference problems require a high squelch, see if changing frequency allows for a lower setting.
10. Use only high-quality alkaline batteries for transmitters. Other batteries have inadequate capacity or too low a voltage for the transmitter to achieve full power output.
11. If using remote antenna, use only coaxial cables designed specifically for RF signals.

Other Things to Consider when Using Wireless Systems

1. Make certain that all necessary supplies and accessories are on hand. Always keep extra batteries, spare cables, and spare remote antennas. In addition, always have a spare mic for Murphy's Law type situations.
2. If traveling with a system, check ahead to make sure you are prepared with a list of the available frequencies. It is common for a frequency to work well in one location but not in another. Links to a number of frequency finders, as well as to other wireless tools provided by Shure and Sennheiser, are included in the online web directory.
3. Remember, proper maintenance of gear is much easier than troubleshooting it down the road. Keep silica gel desiccant packs in the box you store the transmitter packs in; often when they come back to you, they will still be damp from a performer's sweat, and if you are in a hectic teardown with no safe place to let them air out, the silica packs will make sure they dry out even when closed in their case. When you put them away make sure the units get stored with the cord wrapped neatly (not wrapped around the unit).
4. The higher the number of wireless systems you intend to use, the higher the likelihood of encountering interference. If you intend to use many systems for the same production, use higher-performance systems—the better the system, the better it will be at rejecting interference.

Important Things to Remember when Using More than One Wireless System

1. Make certain that no two wireless frequencies are too close together (1 MHz is the recommended minimum bandwidth that should separate any two frequencies).
2. Keep antennas from different receivers as far apart as possible. Do not allow receiver antennas to touch each other.
3. Be certain that both transmitters are not turned on at the same time when two wireless systems are on the same frequency. The same holds when using combination systems (handheld + body-pack) with two transmitters on the same frequency.

Troubleshooting Interference

Doing all the above is the surest way to avoid interference, but not all audio system noise in systems using wireless is caused by interference, or by interference due to wireless equipment. Because a known problem with wireless is interference, it users are often quick blame audio noise on the wireless immediately, without checking anything else. Sometimes, however, the real problem lies elsewhere. Recognizing the true source of the problem can help save time trying to correct a nonexistent wireless problem:

1. The first step is to determine where the interference is entering from. Try turning off all wireless receivers and disconnect the audio cables. If the problem is still present, the trouble is not the wireless but elsewhere in the audio system.
2. If you have eliminated the possibility that noise is getting into the audio system elsewhere, test for intermodulation interference by turning transmitters in the area one by one. If turning off a particular transmitter will completely eliminate the problem, the problem is likely due to intermodulation. Most intermodulation involves two transmit frequencies and one receiving frequency. Changing any one of them will fix the problem; try changing the frequencies on each wireless system one by one.
3. If interference persists, try turning off all wireless transmitters at once. If the "signal" indicators on the receiver stay lit, there may be low-level interference in the system.
4. If low-level interference only occurs when the transmitters are off, and otherwise is no problem, simply mute the receiver. If not, make sure you have taken all the set-up precautions above and also turn off any unnecessary electronics. If possible move the necessary electronics that can't be turned off even further from wireless receivers.
5. Make sure all batteries are charged. Double check that you have properly adjusted the squelch levels.
6. If the signal indicator does not go out after taking all the precautions above and with the squelch control set to maximum, it is likely that serious interference is present and the problem may not be due to problems with the system.

If Interference Persists, Especially If It Is Still Audible When the Receiver Is Squelched, It Could Be:

1. *AM radio interference* (distorted speech or music in the audio): rare with stations farther than 2 miles away. To track down the particular station creating the interference, tune an AM radio across the local stations and listen for audible changes.
2. *TV station interference* (buzzing sound of varying intensity): can occur at distances of 2 to 3 miles, or

as far as 4 or 5 miles in the case of UHF. To track down TV interference, watch local TV channels listening for audible changes coinciding with scene changes on a particular channel.

3. *Radar systems* including airport radar systems and weather radars (high-pitched buzzes or noise bursts that occur every few seconds): radars that turn a few times per minute only cause interference when they are pointed your direction. As with AM radio and TV interference, radar interference is likely only when the radar is fairly close, usually when no more than one to three miles away.
4. *Electrical interference* from heavy machinery and other electrical equipment (buzzes or hum at the power line frequency, random static, and noise bursts are the types most encountered). Electrical interference: travels through power lines and is more likely near industrial areas where arc welders and similar equipment is being used.
5. *Random interference* (brief bursts of noise of various types): This is the result of random high energy electrical events, such as lightning or wind causing arcing power lines. Random interference occurs over a very broad range of radio frequencies. Try increasing the signal available to the wireless receiver. If the problem is temporary, wait it out. If it is a repeating issue and could be entering through the power lines, make sure you are using good surge protectors with RFI filters. These types of interference are hard to deal with, short of changing to a higher-quality wireless system. The only possible fixes to try are:
   - If it is less expensive than upgrading your wireless system, try getting a power conditioner or UPS unit.
   - Improving equipment grounding.
   - Shielding or filtering audio cables at equipment inputs (especially mixers and power amplifiers), coiling.

Chapter 14

Parts to P.A.

Tie It Together with Cables

Professional Audio Cables

These are used to carry electrical signals from one place to another within the P.A. All cables are rated according to what they can be safely used for, and the main factor that determines each class of cables used for audio production is the amount of **power** (**current**) and **audio signal** (**voltage**) they are designed to handle. In analog wires, *signal cables* are designed to carry audio signals and used with a wide variety of sources. Because they are used with both low impedance sources like keyboards, mixers, and signal processors (50 to 600 ohms), and high impedance sources such as electric guitar or bass pickups (20,000 ohms and above), the electrical current flow in signal cables is limited to a few thousandths of an ampere (milliamps). In contrast, *power cables* are designed to supply power to the different parts of the P.A., as well as to the audio signal, and are able handle a large amount of current (hundreds of watts).

Balanced and Unbalanced Audio Cables

Signal cables can vary by the *signal level* (how much voltage accompanies the signal) they typically handle, and by whether they are designed to carry *analog* or *digital signals*. No matter what signal level they carry, analog cables can be identified as being either *balanced* or *unbalanced.*

Unbalanced audio cables (commonly referred to as *instrument cables*) are low voltage audio cables used for a wide range of sources. The voltages these cables typically handle range from a few millivolts (electric guitar), to over 10 volts (line-level sources such as mixers) which represent power levels of less than a thousandth of a watt. While they are not noisy when used to carry a signal on very short runs, their own weak signal means they are prone to signal degradation over cable runs longer than six meters. This is compounded by the fact that beyond that limited distance, unbalanced audio cables easily pick up noise from just about any contact with a source of *electromagnetic interference (EMI)* For example, wherever the audio cable crosses a power cable at 90 degrees. Electromagnetic interference emanates from fluorescent lighting, lighting dimmers, and similar devices sharing the same circuit as the P.A. and can result in hum in any power lines sharing the circuit. These EMI signals can penetrate into an unbalanced audio cable and become noise in the signal itself, not just in the wire, thereby interfering with the sound quality emitted by the speakers.

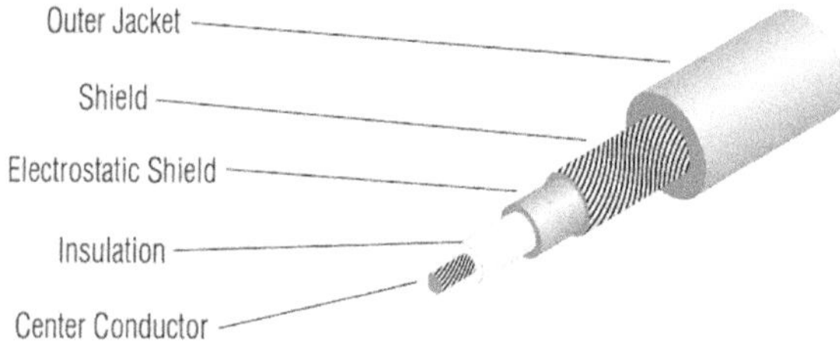

Figure 14.1. Unbalanced instrument cable anatomy. *Source: Image Courtesy of ProCo Cables.*

Unbalanced lines are built using a *coaxial configuration* consisting of one *"hot" center conductor*, which carries the signal current from the source and is separated by *insulation* from a surrounding shield, which acts as the *ground reference* (the current return conductor necessary to complete the circuit). In addition to these three components unbalanced lines are augmented by an *electrostatic shield* to reduce handling noise and an *outer jacket* for protection. Unbalanced lines are standard for home audio gear and as *instrument cables* for connecting instruments like guitars, bass, and keyboards up to backline amplifiers. (See section on connectors for a discussion of connector types and classifications.)

There is more than one way to balance lines but the *balanced audio cables* described here employ a method of canceling out noise picked up from the outside environment, and keeping it from being passed on with the signal into the next component in the P.A. This method, known as *common mode rejection*, also takes care of the problems caused by signal loss over runs longer than five to ten meters. Balanced audio cables can use this method because an extra *"cold"* conductor is added to the *"hot"* signal carrying wire and the shield ground. This construction is referred to as a shielded *twisted pair* because the conductors are twisted together.

Phase vs. Polarity

Before we can talk more about a procedure used to balance an audio cable, we need to discuss a feature of sound we've already mentioned that can be mimicked in wires by an electric analog of sound. Recall from Chapter 4 that when identical waves become superimposed in space, their relative time in relationship with each other will either be fully in time, or *in-phase* (which will reinforce the sound and make it louder), or they will be as much as 180 degrees offset in time, or fully *out-of-phase* (in which case they will interfere destructively and cancel each other out). As an analog signal is simply a copy of a sound wave, and has all

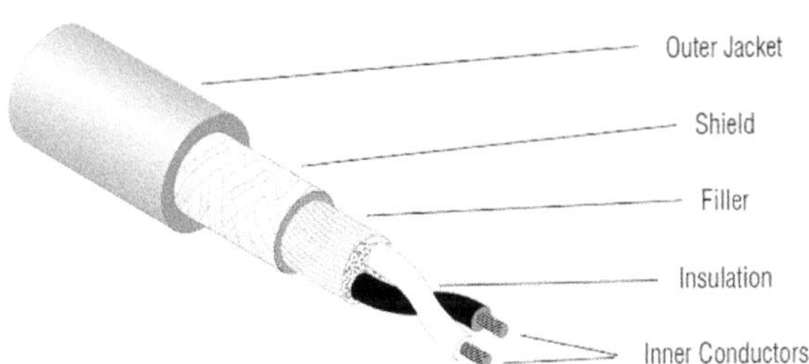

Figure 14.2. Balanced cable anatomy. *Source: Image Courtesy of ProCo Cables.*

the physical qualities required to have a similar interaction with a paired signal. This phenomenon is often referred to as *due to phase*. However, when two signals reinforce or cancel within our wires, it is not actually because of a *phase interaction* based on *time difference*, but is due to a *polarity interaction* based on *voltage difference* between the two signals. In other words, the signals in wire that seem to be completely out-of-phase are actually *opposite polarities* (i.e., one signal is effectively flipped upside down rather than being delayed 180 degrees). Some of this confusion could stem from the fact that in a drawn sine wave (see Figures 14.3a and 14.3b) the result of flipping the polarity of signals in wire looks the same as two sine waves with 180 degrees phase shift.

When balanced audio cables use an extra conductor, the audio signal is transmitted on both the hot and cold lines, but the voltage in the cold line is inverted (i.e., the polarity is changed so the cold signal has a negative polarity when the hot signal has positive polarity). Then, as the signal travels along the line, the usual noise from external sources, such as electromagnetic interference (EMI) from power cables and *radio frequency interference (RFI)*, hitches a ride. As it arrives on board the cables after the polarity flip, this noise will be identical on both hot and cold lines, and is known as a *common mode signal*—a signal which appears equally on both conductors of a two wire line. Now the balanced cable carries a wanted signal with opposite voltage on each line, and unwanted noise which is the same on both lines. If these were left as is at the end of the cable, the effect would be the opposite of what is wanted, as the wanted sound voltages would cancel each other out and the unwanted voltages would reinforce each other. Luckily, this is where the magic of balanced audio kicks in. At the input stage of the component receiving the cable, the polarity is flipped once more: The unwanted noise is now inverted and cancelled out, leaving only the identical original signal on each line which combines and reinforces the wanted sound (and also eliminates the effects of any loss of signal power).

In this technology-driven world, where our entire environment is filled with a plethora of different interference sources, the best plan will always be to keep your wiring tidy and consistent. The best way to see to this is to use balanced cables as much as possible, and minimize the length and number of any unbalanced cable runs.

Signal Cable Types

- *Microphone cable*: The signal levels carried by microphone cables are so faint (−50 dBt to −60 dB) that they must be boosted by preamps at the mixer's inputs to be useable. These signals are too weak to transmit over an unbalanced cable of any

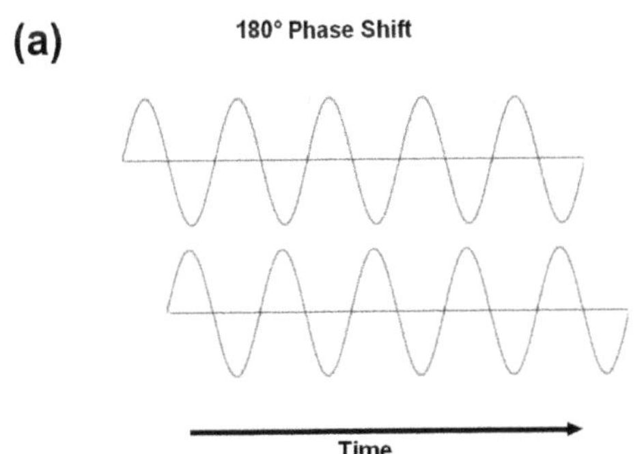

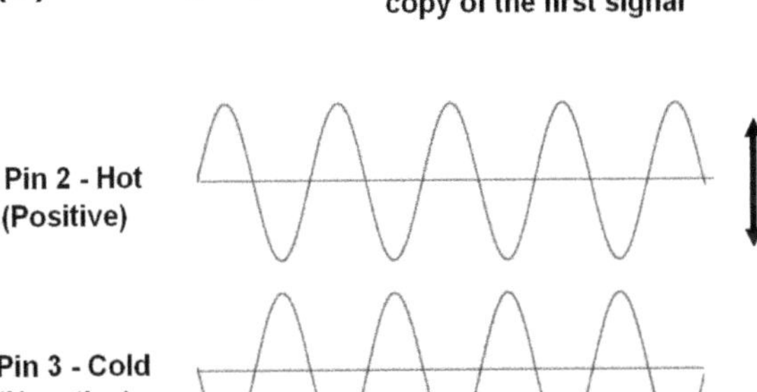

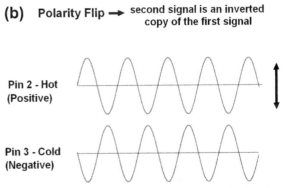

Figures 14.3a and 14.3b. These two signals are often referred to as being 180 degrees out-of-phase with each other but this is technically incorrect—the signals are not actually out of phase, they are opposite polarities.

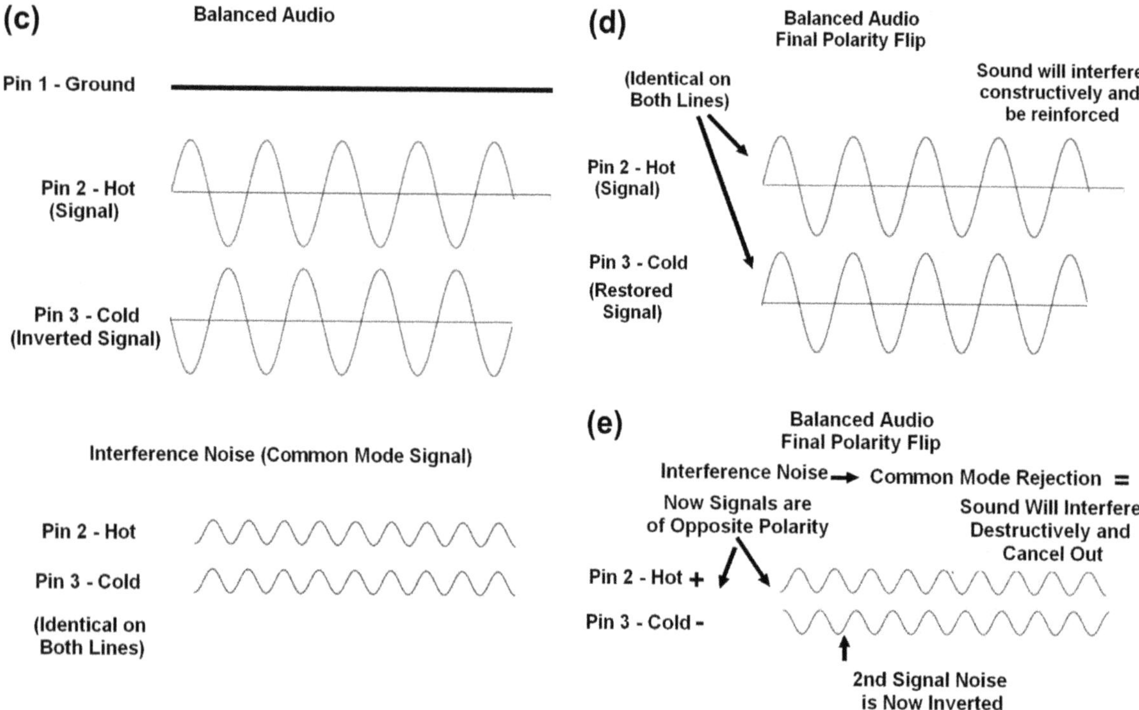

Figures 14.3c, 14.3d and 14.3e. With a final flip, the user is left with a stronger desired signal and no noise.

reasonable length without degrading completely due to power loss and noise. This is why the length of the average microphone cable run is kept short and why microphone cables always use balanced wires regardless of length. Mic cords used for live sound applications must endure more handling than most cords, so they need to be both rugged and flexible.

- *Instrument cable*: Unbalanced cable that connects a guitar, bass, keyboard, or other electronic instrument to its amplifier. It has a positive wire and shielding that serves as a ground. It is designed to carry low voltage instrument signals and most often uses ¼ inch phone plug connectors.
- *Patch cable and line level*: Line-level pro analog audio cable has a signal strength rated at +4 dBu (1.228Vrms). Line-level cable is used to link various components in a P.A. or recording setup, or to link effects pedals to each other and to the instrument-to-amp circuit. Though typically kept balanced for consistency, line-level cable can also be unbalanced (unbalanced line-level cable is essentially instrument cable). Patch cables have many of the same features, but are very short and are distinguished by their purpose which is to connect components to patch bays. Both types of cable can have various connector types (XLR, ¼ inch phone, TRS, RCA).
- *Insert/Y cable*: In pro live audio is often used to route signal away from and back into the insert jack in a mixer allowing it to function as a direct out (which depending on mixer type can be in short supply). This is possible because the Y-cable is bi-directional and can be used in either the transmit (Tx) or Receive (Rx) direction, at either end of a cable span, depending on the function it is serving. More generally the Y-cable can also be used in any application where it is desirable to split a source signal.
- *Multicore cable*: Many wires bundled together tightly surrounded by a rubber jacket. One example, called a *stage snake*, connects the *stage box*, where all the onstage cables are plugged into, to the mixer at the FOH position and is much more convenient than bundling many separate XLR cables together. Multicore cables such as snakes ensure a neater stage and facilitate a faster setup and teardown. Another multi-core type of cable is that used to transmit digital signals, using digital cables/connectors but otherwise similar to analog versions in overall design and function.

Audio Connectors

Connectors are the elements found on each end of your cables that match up with the various connector inputs found on your audio components. They enable temporary connections to be made and broken quickly and safely. There is a variety of different audio connectors available, and the type used for any application depends on whether the attached cable is balanced or not (connectors are usually only used with one type or the other) and the signal level of the cable (which depends on the components to be linked up). Some connectors are found on several types of cable; others are only found paired with one or two specific types of cable. In form, connectors usually follow a familiar design: *male connectors* are outies with exposed contacts; *female connectors* are innies with insulated internal contacts hidden inside a shell with holes which the contacts of male connectors plug into. The direction of signal flow is dependent on the mix of cable/connector used for an application and isn't as standard. The most commonly used connectors for live audio applications include:

- *3-pin XLR*: One typically sees XLR cables connecting microphones to mixers, and connecting various outputs to powered speakers. 3-pin XLR connectors are mainly used for balanced audio signals. When dealing with XLR connectors, female connectors are always inputs, male connectors are always outputs, with signal flowing the same direction, from female to male. In general (though not for older adaptors, so check when in doubt): pin 1 is the earth (or shield); pin 2 is the hot; pin 3 is the cold.
- *¼ inch jack (6.5mm jack)*: The two types of ¼ inch jacks used in pro audio applications are referred to as TS and TRS connectors.
- *TS (tip, sleeve)*: A specific type of ¼ inch or ⅛ inch connector that is set up for two-conductor, *unbalanced* cable operation. One insulator ring separates the tip and sleeve. The tip is generally considered the "hot," or the carrier of the signal. These can be visually identified by the single ring found after the tip.
- *TRS (tip, ring, and sleeve)*: Looks like a standard ¼ inch or ⅛ inch plug, but with an extra "ring" on its shaft. TRS cables have two conductors plus a ground (shield). Used to connect balanced equipment, or for running both "left" and "right" mono signals to stereo headphones. On the stem of *Y cables*, these are used for mixer insert jacks where the signal is sent out through one wire, and comes back in through the other.

(a)

(b)

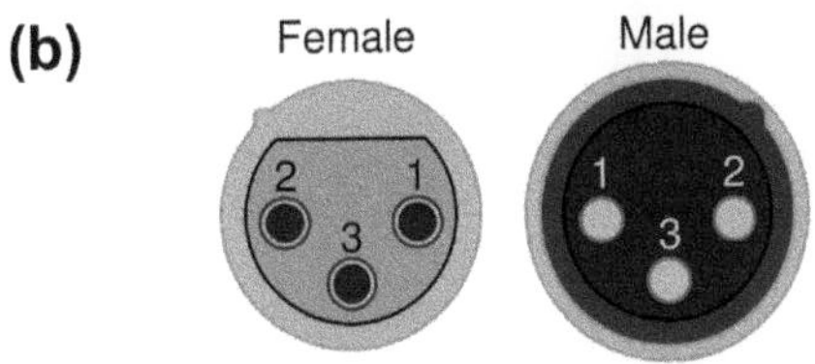

Figures 14.4a and 14.4b. XLR plugs and pin outs.

When a balanced jack is used for a stereo signal, the tip is the left signal, the ring is the right signal, and the sleeve is the shield/ground. *Unbalanced* => tip is the signal, sleeve is the ground.

Balanced => tip is the hot, the ring is the cold, and sleeve is the ground.

- *⅛ inch jack (3.5mm jack)*: Sometimes called mini jack, a smaller version of TS /TRS connectors, common in both commercial and pro audio.
- *Banana plug*: An electrical connector designed to join audio wires such as speaker wires to the back of power amplifiers.

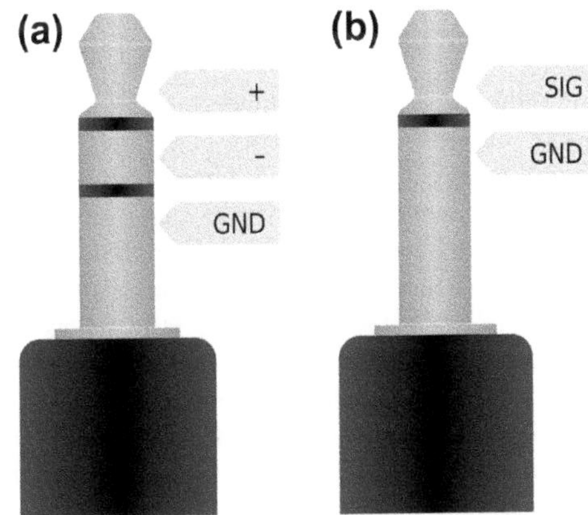

Figure 14.5a and 14.5b. TRS/TS connectors.

Figure 14.6. Mini jack connectors alongside a larger ¼ inch Jack, for comparison.

- *RCA*: RCAs are used a lot for home stereos, TVs, DVDs, etc. The RCA can carry either audio or video, and is wired similarly to other mono jacks (center pin is the signal, outer ring is the ground reference).
- *Speakon*: Speakon connectors are used to connect power amplifiers to speakers and stage monitors. The ability to lock into place makes them preferred over ¼ inch TS connections. Its unique look also serves as a safety feature; with a TS connection on your speaker wires, the visual similarity between your speaker wires and line cables makes it all too easy to make a costly wiring mistake.

Digital Audio Cables

Digital audio is becoming common enough that digital encoding and playback systems can now be found outside of our computers and laptops and are also becoming commonly used in our home stereos and boom boxes, and even in our tiniest handheld personal audio systems. In pro audio more and more digital equipment is being used, as prices fall and as those who prefer the specific benefits and drawbacks of digital over those of analog are able to make the switch, depending on budget. Now that every audio function/process can occur in digital format, the trend in broadcast and studio production is one of steadily increasing use of all digital systems. Even mostly analog systems may include digital components, although these must be limited or kept together in the signal chain to avoid the need for repeated conversion of data from analog to digital and back again. While adoption of digital systems has not been as quick or

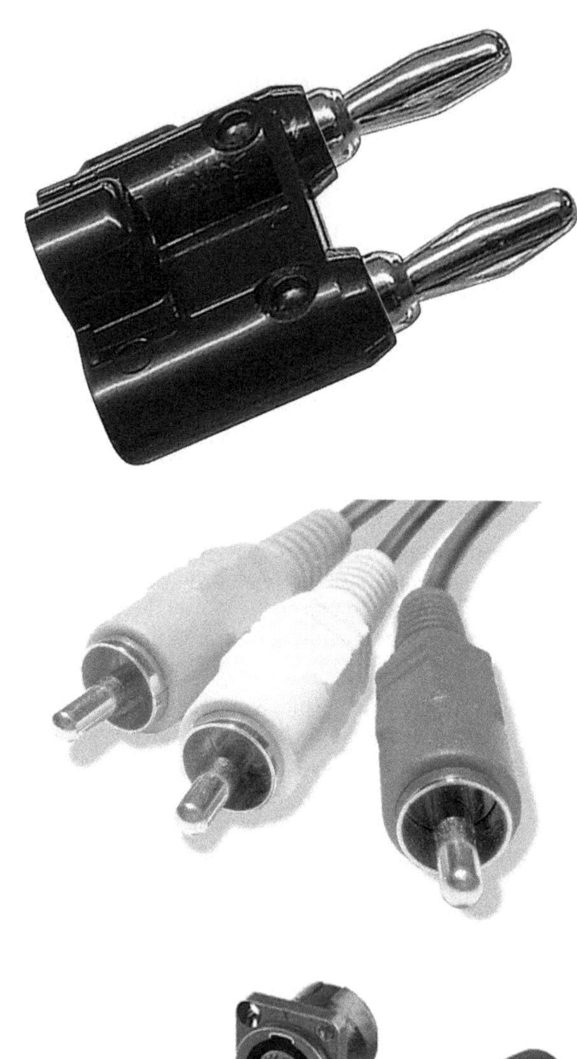

Figures 14.7, 14.8 and 14.9. Banana, RCA, and one style of Speakon connector.

AES/EBU Digital Standard Data Rates

Sampling Rates (kHz)	Bandwidth (Mbps)	Application
32 kHz	4.096	Pro. broadcast transmission (recorded voice, reporting)
38 kHz	4.864	Music, FM station quality
44.1 kHz	5.645	CD Sampling rate for music
48 kHz	6.144	Cable broadcast (pro video recording audio tracks)
96 kHz	12.228	Top of the line pro audio devices
192 kHz	24.576	Double the quality of 96kHZ

Figure 14.10. AES/EBU digital cable data rate standards.

AWG Gauge	Diameter (inches)	Diameter (mm)	Resistance (Ohms / 1000 ft.)	Max Current (Amperes)
8ga.	0.1285	3.2639	2.060496	24
10ga.	0.1019	2.58826	3.276392	15
12ga.	0.0808	2.05232	5.20864	9.3
14ga.	0.0641	1.62814	8.282	5.9
16ga.	0.0508	1.29032	13.17248	3.7

Figure 14.11. Common gauge sizes.

widespread in the field of live audio, that is changing. The typical freelance audio engineer can no longer avoid learning to mix on digital systems, even if analog is still the most common system in use.

AES/EBU Digital Audio Standards

To help in the orderly adoption of and transition to digital audio technology, the *Audio Engineering Society (AES)* along with the *European Broadcast Union (EBU)* has developed the *AES/EBU digital audio standards* to avoid the kind of confusion that a lack of one set standard has caused in analog audio, and these standards include parameters for the cables used for digital signal transmission. To efficiently transmit digital square wave signals, digital cable must have specific impedance and capacitance characteristics that set it apart from analog cables, even if many digital cables are almost identical to analog cables in terms of appearance. The AES/EBU digital audio standards that pertain to digital cables include their appropriate *impedance* as being 110 Ω +/−20 percent (88−132 Ω) depending on application. Also indicated are the proper *data rates* according to application, which include both sampling rate and related bandwidth, and are listed in the chart in Figure 14.10.

The AES/EBU standard is one of the oldest, along with MIDI, and is currently the most popular format. Most consumer (CD players) and professional (DAT decks) digital audio components featuring digital audio ins and outs will support AES/EBU cables. Nevertheless, rapid advances in technology such as optic data transmission means digital audio is continuing to evolve. As is natural in a time marked by large-scale shifting from one technology format to the next, and a more general culture of rapidly emergent new technologies, the final format that will become the digital standard in the long run is far from being set in stone.

Digital Interface/Cable Protocols

In the meantime a number of different *proprietary* and alternative open formats all continue to duke it out right alongside the AES/EBU standard, including MIDI which along with AES/EBU is one of the oldest digital formats, in continual use since it was introduced in 1984. Proprietary technologies are privately owned technologies with the patent belonging to an entity, usually the designer, who can exact a fee from all other manufacturers that wish to use the format in their own designs. Rather than try to list all digital cable types available, I will instead limit our list to the four most important generic choices (open and/or non-proprietary):

- *AES/EBU*: An interface format for digital signals developed by the Audio Engineering Society (AES) and European Broadcasting Union (EBU) in the early 1980s. It uses AES Type 1 cabling—a three-conductor, 110-ohm cable with XLR connections. It transfers two channels through one connection and is the transfer protocol which S/PDIF is based on. Because of differences in resistance, XLR microphone and line-level cable will not work as an AES/EBU cable, despite visual similarities.
- *Coaxial digital audio cables*: One cable type used for S/PDIF (Sony Philips Digital Interface Format). Similar to analog RCA type cables (and using the same RCA connectors), the signal in unbalanced S/PDIF coaxial cables is carried along a conductive wire, though it has a greater bandwidth than analog audio and is better insulated against outside noise. As most digital systems don't have signal quality greater than coaxial capacity, and coaxial the more affordable and durable choice of the two S/PDIF cable types, this will be the digital cable most appropriate for the bulk of commercial and pro audio applications

using S/PDIF. Despite their visual similarity with analog RCA cables, they are not interchangeable.
- *FireWire*: A data transfer protocol first developed for video, FireWire cable is now also widely used for audio applications. There are three kinds of FireWire connectors: 4-pin, 6-pin, and 9-pin. The 4- and 6-pin connectors are also known as FW400, while the 9-pin is known as FW800. The 6-pin and 4-pin have the same data-transfer speed, but the 6-pin can also supply power. The 9-pin is twice as fast as the others and also can supply power.
- *Optical digital cables*: Categorically different than analog modes of transmission, instead of conductive wire, optical cable uses pulses of light to transmit the audio signal along a fiber optic line. The signal on an optical cable will not degrade and is not subject to interference. However, the disadvantage is that without need of shielding optical cables tend to be less rugged than coaxial. In addition they cost much more than other cable designs. Fiber optic technology is used by more than one kind of format. One example format is called ADAT light pipe, which transfers multiple channels of digital audio using a format specific cable and connectors. Optical cables are also the other type of cable used for S/PDIF and optical cables designed for use with this format use TOSLINK connectors, which were developed by Toshiba. Many S/PDIF compatible digital devices have dual ports, one for coaxial, and the other for optical cables.

Digital cable is very stable and reliable if used in purely digital systems, and when used to transmit digital signals in hybrid analog/digital systems, using digital cable reduces (signal adjustment) equipment costs significantly. The advantage of digital formats is the fact that the signal retains the quality of the original source throughout the signal transmission process. With the use of well-constructed digital cables for digital signal transmission, the degradation of signals is basically eliminated, and noise resistance is greatly improved. However, these advantages hinge upon using the proper cable for the format.

That said, there are distinct disadvantages to making use of digital data transmission protocols. The first is simply the danger of accidentally switching data and analog cables that in an unfortunate series of design choices has ended up with the most commonly used digital and analog cables looking exactly alike. If running a system that uses both digital and analog cables, or having the option to run either analog or digital must store both types of cable, it is imperative to organize and label them so that they do not get mixed up and plugged into the wrong inputs.

The other is a quirk of digital signal transmission akin to the phenomenon that occurs to the signal and sound output when the 0 dB FS limit at the console is exceeded by the FOH engineer. This quirk has to do with the difference in the way the signal degrades, which is very different for digital connections than what we are used to dealing with in analog cables. With analog signals any noise that makes it onto the line will begin to degrade the original signal, slightly at first, and getting progressively worse the longer the exposure occurs across a single transmission or a sequence of transmissions along a noisy cable. Digital signals react differently when exposed to interference; the cable will not degrade from analog interference that gets onto the cable, which is a highly valued feature of digital signal transmission.

However, it is important to remember that even if the digital signal on a noisy cable stays clean, the interference is still getting onto the cable. It is important to use well-shielded digital cables and keep cable runs short with digital as well as audio because, much like the world of difference in the digital sound that occurs when the perfect digital sound of a decibel below the 0 dB FS threshold and the terrible distorted feedback that begins the moment that threshold is exceeded, digital signals are impervious to noise only up to a certain threshold but not beyond. A digital cable will transfer the signal perfectly up to a point, however once a certain threshold of degradation is reached along a digital cable transfer, the signal goes from nearly pristine to utter and incoherent static. So once again care must be taken to observe the absolute limits of digital signals because either a digital signal is kept close to perfect, or it is not kept from exceeding limits where it will become complete noise.

Two more features of note and that bear mention are that cable length is also important even when noise is kept off the wire or is not enough of an issue to reach a threshold right away, because unlike an analog signal transmitted along a cable that is too long, which can still be heard but with lesser strength and signal quality, digital audio will sound great until the last inch, after which it will not transmit and won't be heard at all. Live audio learners with a studio background and any familiarity with studio digital applications should be reminded that even though digital connectors are designed to have error correction abilities, error correction does not apply to live audio, because at this point error correction is not possible with real-time transfers over a digital connection.

Power Cables

Unlike the analog signal cables described so far, current carrying cables like *speaker cables* and *power cables* are intended to carry more power than just the small amount of voltage required to drive the analog signal to the next component, and have different requirements than signal cables. To deliver the right amount of current to the critical components that need it, power cables must have low impedance to allow enough current to flow, and must be able to handle the heat that builds up from the large amount of current flowing through the resistance of the cables.

Cables, Current, and Heat

All wires have resistance, and as power flows through a wire, the friction created as current moves through that resistance causes heat. This is why the amount of power a wire can safely carry is determined by how hot the cable can safely get without melting. The more power running through a wire, the hotter it gets; unless you make sure to select cable that can handle the amount of power running through it, the heat can reach a critical point resulting in undesirable consequences. If a power cable gets too hot for too long, the insulation starts to break down and can melt away, leaving the wire exposed to other wires, grounded metal, or even people—not a good situation. The heat can even ignite nearby materials in some cases, such as the rugs which are often used to cover cable runs which must run across walkways, both to protect the cables from the edges of hard soled shoes, which can be very sharp, as well as to protect people from tripping on exposed cables.

Ensuring your cables can carry the power you need them to without overheating starts with buying quality cables that are intended to be used for the application you will use them for. This doesn't mean you need to buy the most expensive cables on offer, there are plenty of quality cables that are affordable enough to be appealing even to budget conscious buyers. However, it does mean that if you are buying cables for a live sound situation where they will inevitably be stepped on or potentially even have flight cases and racks rolled over them, you can't skimp on the construction or the amount of insulation. Beyond buying quality cables, there are two things you need to account for before you choose cables for any purpose: how long they need to be and what gauge they should be.

American Wire Gauge

Any power cord, extension or otherwise, contains an inner metal wire, called a conductor, which carries electrical current from one end to the other. The thickness of any cable's conductor is indicated by a number which is referred to as its **gauge** or **gage**. Wire gauge is an index which shows (indirectly and logarithmically) the area of a round wire. A wire's thickness directly affects the amount of impedance of the wire (thicker wire = lower impedance), and by extension the amount of current (or wattage) it can safely carry, which also defines that wire's associated power capacity. Cable gauge is an important factor in determining if your power cables will get too hot to handle, or if they are able to carry all the current you need them to and still keep cool in the process. This is more cut and dry when measuring solid wire, less so when measuring *stranded* wire which has air pockets that increase its diameter. It's important to remember that gauge is only a measure of the wire at the core of a cable, the overall width of two cables of the same gauge may be slightly different depending on the relative amounts of insulation they both have. Using well insulated cables with a gauge size that insures the wires have enough current carrying capacity is essential if you don't want your cables overheating.

The American Wire Gauge (AWG) system is based on 44 standardized wire sizes: 0–40, as well as 00, 000, and 0000 gauges (the thickest). It may seem a little counterintuitive, but with gauge, the lower the number, the thicker the wire is. This is because each gauge is named after the number of sizing dies the wire needs to be drawn through to reach the correct diameter (for example, a 24 gauge wire is drawn through 24 different sizing dies. Even though 44 different wire diameters are recognized within the AWG standard, they're not all widely used, and most people are likely to encounter only a small range of them. One feature of AWG worth noting is that *with each reduction of three in AWG the area of the conductor doubles* (i.e., a 13 AWG conductor has twice the copper of a 16 AWG conductor, a 15 AWG twice the copper of a 18 AWG, and so on). Knowing this will make it easier to match your speaker wire to your amps and speakers when the time comes. The range of gauges most common for pro audio applications is 16 through 8. The approximate size of common audio gauges is shown in Figure 14.11.

Power capacity is not just determined by the wire gauge, but how far you need it to go as well: *the longer the wire, the greater the impedance of the wire.*

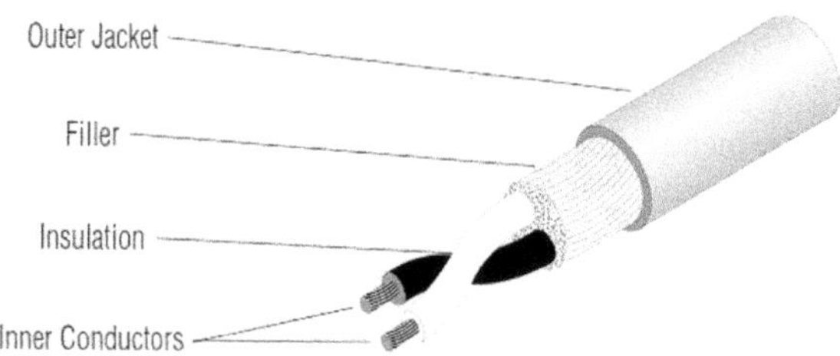

Figure 14.12. Speaker cable cross-view. *Source: Image Courtesy of ProCo Cables.*

This means that for longer cable runs you need to use a larger gauge wire than may be indicated by current carrying capacity alone, to compensate for the added impedance resulting from the length of the cable run.

Extension Cables

The power cables you will use in live audio are the same extension cords you may be familiar with from other uses. The number of watts an *extension cord* can safely transmit (given its length and gauge) is its *wattage rating*. Before plugging into an extension cord, be sure that the power demand (or pull) of any device doesn't exceed the cord's wattage rating. On the same note, if you'll be powering multiple devices from one extension cord, calculate their combined energy requirements and make sure that the total isn't higher than the wattage rating for that cord. Never use a cord to supply more wattage than it's rated for or overheating may occur.

Determining Power Requirements

It isn't difficult to figure out how much electricity is required to run different devices. For the most part the information will be included in the manufacturer's instructions. Another good place to look is on the tags attached to a device's power cord. And if all else fails, a targeted web search or a quick call to the product's manufacturer should yield answers to questions. In some cases, you may find that wattage requirements are listed in amps and volts instead. For these situations, there's a simple formula that can help you calculate electricity requirements: *multiply the number of amps by the number of volts, and the resulting number equals that appliance's wattage*. For example: a device that uses 5 amps at 110 volts uses 550 watts (5 x 110 = 550).

Extension cords usually come in 16 ga., 14 ga., and 12 ga. widths, and in various lengths between 5 feet to 100 feet long. If determining exact power requirements is not convenient, the safest bet is to follow these rules of thumb; for components like the mixers and processors 14 or 16 ga. is fine (depending on length, 16 for shorter, 14 for longer), as these components usually draw less power. To be sure your cords have ample current-carrying capacity to provide power for amplifiers (including powered mixers) always use 12 ga. regardless of cable length.

Speaker Cables

Speaker cables connect amplifiers to speakers and carry the amplified signal that results from the original signal plus the amount of power applied to the signal by the power amp; the amount of the actual current in a speaker wire varies according to the power amp's settings. A speaker cable has two cores and is usually the same thickness as mains power cables. Unlike a power cable, a speaker cable doesn't need to be double insulated (unless specified for high power) because even though it can conduct the same amount of current as a power cable, it carries current at a lower voltage. Nevertheless speaker cables carry a much higher voltage signal than line level, instrument, or mic cables and have bigger gauge wires, so it is very important that other audio cords are *never* used in the place of speaker cables. In an emergency even zip cord (or lamp cord) can be used as a safer replacement for speaker cables than any other kind of audio cable. Speaker cables can have ¼ inch phone, banana clip (also called MDP connectors), or speakon connectors.

With a speaker cable, barring odd or shoddy construction practices, far and away the most important aspect of the cable is its wire gauge. Even though speaker cables are usually thick enough to handle heat from power loss without melting, as they don't get quite as hot as power cables, speaker cable that has excessive impedance due to being the wrong length/gauge for the application and amp rating will have negative effects on the performance of your system.

For example, a speaker cable that's too thin will lose amplifier power as it is converted into heat and result in a loss of decibel power in the sound output of the system. Figures 14.3a, 14.3b, and 14.3c show a sampling of cable gauge and length combos and their typical power loss depending on amplifier ohms. For more extensive power loss charts, refer to the chapter directory, which links to charts online.

(a) Power Loss @ 8 Ohm Speaker Load

AWG Gauge	Cable Length (Ft.)	Power Loss (% of Watts)
12ga.	25	1.67
14ga.	25	2.75
16ga.	25	4.35
12ga.	50	3.29
14ga.	50	5.39
16ga.	50	8.41
12ga.	100	6.41
14ga.	100	10.34
16ga.	100	15.76

(b) Power Loss @ 4 Ohm Speaker Load

AWG Gauge	Cable Length (Ft.)	Power Loss (% of Watts)
12ga.	25	3.9
14ga.	25	5.39
16ga.	25	8.41
12ga.	50	6.41
14ga.	50	10.34
16ga.	50	15.76
12ga.	100	12.21
14ga.	100	19.10
16ga.	100	27.98

(c) Power Loss @ 2 Ohm Speaker Load

AWG Gauge	Cable Length (Ft.)	Power Loss (% of Watts)
12ga.	25	6.41
14ga.	25	10.34
16ga.	25	15.76
12ga.	50	12.21
14ga.	50	19.10
16ga.	50	27.98
12ga.	100	22.24
14ga.	100	33.08
16ga.	100	45.43

Figures 14.13a, 14.13b and 14.13c. Speaker cable power loss.

Another example of system performance reduction is one even more detrimental to the quality of the sound output; using cable that is too thin (too high a gauge number) also results in the loss of low-frequency performance due to a loss of *damping factor*, the rating that measures an amplifier's ability to stop (or damp) the movement of the speaker's cone after a signal terminates. A poor damping factor allows a speaker's woofer cone to continue vibrating after the signal ends and this rattle will interfere with the following low end signals. In the worst case this can even create a state of continual resonant rattling that turns your low end sound into a *"one note base"* sound, or the equivalent of bass mush. Have you ever pulled up next to a car with the bass turned up so much louder than the system's capacity that you couldn't even try to identify the bass line because the notes had lost all definition and just sounded like pulses of the same rattling low end note? Then you have an idea of how terrible "one note bass" sounds.

Higher damping factor numbers are better than lower and a simplified way to determine a system's damping factor is to add the damping factor of the amplifier (listed in its specifications) and the damping factor of the cables (see the chart below). Specifically, a damping factor of 200 is ideal (higher numbers are possible but

(a) Damping Factor (4 Ohms)

Cable Length → AW Gauge ↓	10ft.	25ft.	50ft.	100 ft.
16ga.	45	19	10	5
14ga.	69	30	16	8
12ga.	102	46	24	12
8ga.	--	--	56	30

(b) Damping Factor (8 Ohms)

Cable Length → AW Gauge ↓	10ft.	25ft.	50ft.	100 ft.
16ga.	90	38	21	10
14ga.	138	60	31	16
12ga.	201	91	48	25

Figures 14.14a and 14.14b. Speaker cable damping factor.

don't result in much audible improvement), while a number of 20 or below is the range where bad becomes awful, such as the one note bass effect described here.

In addition to differences according to gauge and cable length, the exact amount of impedance in speaker wires also varies according to the resistance created in the components they are connected to. To determine the best gauge for your speaker wire you also need to account for your amplifier/speaker impedance in ohms, as well as for the length of the wire needed for your particular setup, which is why we will wait until we talk about setting up and wiring your speaker system to look more closely at how to select the correct gauge for your needs.

The Long and Short of Audio Cables

While we've covered most of the cable categories you are likely to encounter, there are a few things we should mention about the materials and construction typically used in audio cable. Nowadays there are many different "esoteric" cable designs and materials, many of them created to satisfy the audiophile market, and many probably offering some improvement in performance over ordinary twisted-pair type cables. However, not only are these cables extraordinarily expensive, most of them are also very fragile and not made for portable situations. Fortunately, the improvement any of these might afford is not enough for most people to be able to hear, so there's no reason to regret not using them in your P.A. However, as far as standard construction methods and materials go, some are better choices for live sound than others. Live applications require cables that are strong and flexible.

Cable Shielding

Except for speaker wires and digital optical cables, all audio cables are shielded to block interference and noise from entering the wires and contaminating the P.A. signal. There are several types of shielding, but each kind surrounds the insulator for the center conductor(s) and serves to isolate the signal from all types of noise interference. Radio chatter, static, and hum that makes it into the signal often means that the shielding around your cables is inadequate or compromised. Good shielding can save you a lot of headache in the quest for maximizing system performance and also may serve as a ground. The three main types of shielding are:

- *Braided shield*: The most common is the braided shield. Small wire strands are braided to form a sheath around the insulation of the current-conducting wire. This type of shielding is flexible and durable. Onstage mic and instrument cables are constantly being bent, pulled, and stepped on, and braided shielding holds up best under these conditions. This is also the most expensive kind of shielding.
- *Spiral-wrapped shield*: The second most useful type of shielding is the spiral-wrapped shield. This sheath is formed by wrapping thin strands of wire around the center wires in a spiral. Though not as strong as a braided shield, it is more flexible because it stretches when the cable is bent. One downside is that as the cable is bent, gaps in the shielding can be created which allow interference in. Spiral shielded cords are also less resistant to radio frequency (RF) interference, because the shield actually acts as a coil and has inductance. This kind of shield is usually less expensive than braided shield.
- *Foil shield*: The foil shield is a mylar-backed aluminum tube with a copper drain wire connected at each end. Generally too fragile for most live sound applications. This is the kind of shielding used in thin patch cables such as the RCA-ended connecting wires that link stereo system components. Foil shielding is cheap and can be an effective shield, but is fragile and can easily be damaged by flexing. It is best used in non-critical applications where the cables stay put, such as patch bay cables that aren't moved once they are connected.

Deciding which cable shield is best will have to occur in the context of real-world factors, like how much a vocalist handles their mic cord or how much interference is likely. Both braided and spiral shields have situations where they shine and situations where they should be avoided. While braided is the strongest, tightly braided shields can be literally shredded by being kinked and pulled, while a spiral-wrapped shield will stretch easily without breaking down. Of course, that opens up gaps in the shield which can allow interference to enter, which can be a problem in high-interference areas since the inductance of the spiral shield already makes it vulnerable to RFI.

More Preferred Methods and Materials

Solid conductors are made of a single wire, and are the cheapest and easiest to produce but when bent or flexed are quick to break. A *stranded conductor* is

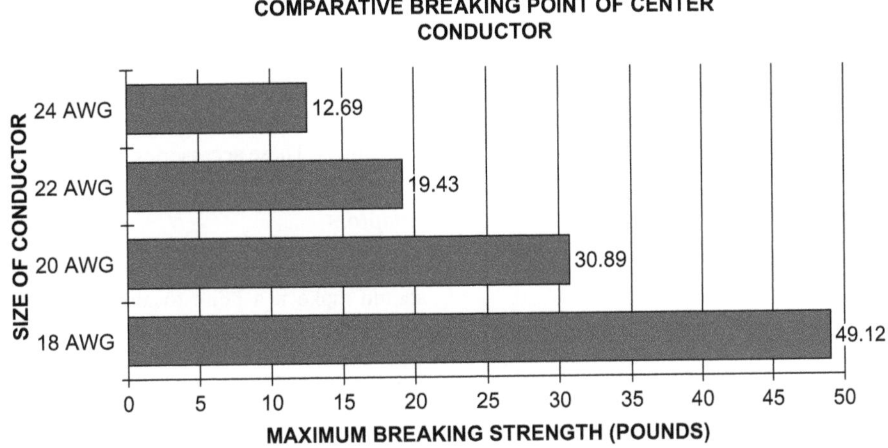

Figure 14.15. Stranding and tensile strength. *Source: Image Courtesy of ProCo Cables.*

composed of a number of twisted strands of copper wire bunched together to form a larger wire increasing the flexibility and the flex life of the cable. The finer the strands, the more there are for any gauge and the stronger the stranded conductor will be, but also the higher the cost. Because the most common cause of failure for instrument cables is broken center conductors using the finest strands possible is often worth the price.

The gauge of instrument cables is also a factor to consider. In order to be compatible with standard ¼ inch phone plugs the diameter of the cable is effectively limited to a maximum diameter of about .265 inches. Because of this, and because of the lower power running through instrument cables, many people assume gauge doesn't really make a difference like it does with power cables. 20 AWG center conductors, with a *breaking point* of approximately 31 lb has become the standard. But for applications where rough handling is likely, it's worth considering using a lower gauge. Reducing conductor size to 22 AWG reduces breaking point to about 19 lb (a reduction of 39 percent); while increasing it to 18 AWG increases the strength to over 49 lb (an increase of 58 percent).

Another way to increase cable performance is to look for a tighter twist in balanced cables. The distance between the twists in twisted pair cable is called *the lay* of the pair. Shortening the lay (increasing the number of twists) improves its common-mode rejection, and also improves its flexibility. The typical pair lay in microphone cables is about ¾–1½ inches. Shorter lay uses more wire and cost more, but can be worth it.

Capacitance and Skin Effect

Capacitance is the ability to store an electrical charge, but is something you want to avoid in your cables. Combined with the source impedance, cable capacitance acts as a low-pass filer between the instrument and amplifier; i.e., it cuts high frequencies. *Skin effect* is a phenomenon that at higher frequencies causes current flow to be concentrated more to the surface of the conductor, almost as if it were a hollow tube. This increases phase shift and cuts high frequencies and is also something you want to avoid. The question is how?

- While there is no point at which high-frequency loss suddenly appears or disappears, increasing the length of the cable increases the high-frequency loss. So even with signal cables shorter is better.
- High-quality insulation materials such as *polyethylene and polypropylene* have low *dielectric constants* making them preferable to cheaper materials, at least when long cable runs are necessary. (Polyethylene insulation tends to have one-third of the capacitance of a cable insulated with the same thickness of rubber.)
- For microphones, *four-conductor-shielded* cables (two twisted pairs twisted together also called *"star-quad" cable*) produce lower inductance and lower-phase shift.
- Avoid tinned copper in mic cables as it increases skin effect.

Handling Noise

Handling noise is usually negligible except in cases involving very low-level signals. In, low impedance

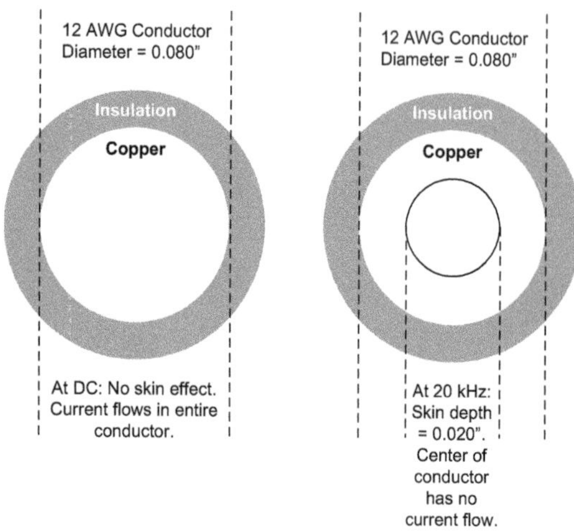

Figure 14.16. Skin effect. *Source: Image Courtesy of ProCo Cables.*

applications, soft, impact-absorbing insulation and jacket materials with ample fillers to insure that the cable retains its shape is the best way to avoid handling noise. Note that a cable with no termination essentially presents infinitely high source impedance, so handling noise should only be evaluated in cables that are plugged in on both ends.

Audio Cable: Essential Accessories

Before moving on to the chapters of the next section, which look at the first part of the production process, we need to mention the two most essential cable accessories, as they are almost more than accessories; without them your cabling will only function with considerable extra expenditure and planning. Therefore they are almost as vital to setting up your P.A. as a good cable selection is. These accessories are adaptors and DI boxes.

Adaptors

There is little to be said about adaptors other than you should make it a point to always have a few of each type. They are connectors that have an input of one type that leads to the connector type needed to connect to a specific audio component. A cable with a connector of the type that matches the adapter input can be plugged into the adaptor and then connected to a component input that it would otherwise not fit with via the adaptor. As an added point of failure along a cable run, the common advice is to only use them in emergencies and not as part of a regular setup, but just as common is the advice to keep many of them on hand, because with the variety of backline changes and stray components added to a system by visiting engineers and/or performers, short of having a few of every possible cable/connector combination, which is unreasonable in terms of cost and storage space, the only way to ensure one is prepared for every cabling situation that may occur is a plentiful selection of adaptors.

Direct Injection (DI) Box

A DI box converts an unbalanced high impedance signal into a balanced low impedance signal (required to allow

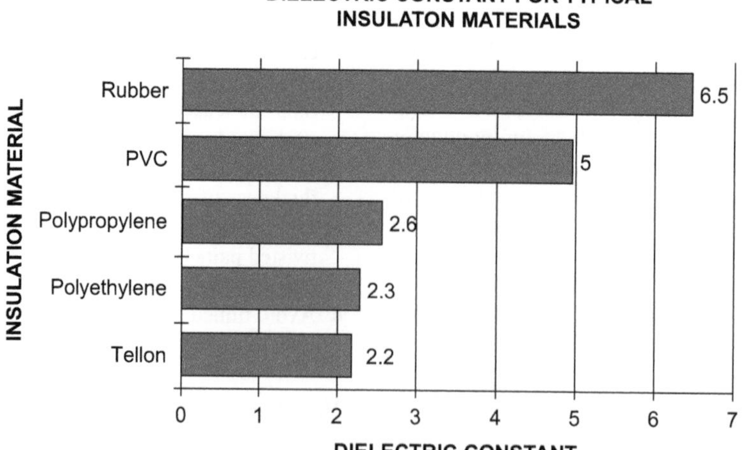

Figure 14.17. Dielectric constant for typical insulation materials. *Source: Image Courtesy of ProCo Cables.*

the instruments signal to make it along a long cable run and to plug instrument cables into most mixers). It also isolates the output signal from the input signal, so in an emergency DI boxes have a ground lift switch which can be used to cure ground loop problems (only a temporary solution). Passive DI boxes are usually cheaper than active DI boxes, and require no power, but will have a very small amount of high-frequency signal loss. This will not usually affect the signal enough to be a problem. Active DI boxes always require power. A few run from separate power supplies, but most use batteries or phantom power (and some allow you to use either source).

Use a DI box when:

- the signal from an instrument is unbalanced and/or high impedance (which is the case with all electrified instruments)—instruments that you will typically use a DI box with include electric guitar, electric bass, and keyboard;
- mixer inputs are farther than 10 feet in cable length from the source of an unbalanced line input.

Connecting an electric pickup-based instrument directly to the mixer can cause a ground loop (some mixers have special circuitry on a couple of inputs that allow for directly connecting instrument pickup). The above describes the most basic function of DI boxes; many now come with a variety of possible additional useful features.

Wrapping Up

We advise you to get a handle on the information in this chapter first and worry about the rest later, but should you find yourself needing to buy audio cables sooner rather than later check the chapter directory for links to several excellent cable guides with additional information on the topics in this chapter, and several other topics not mentioned. The same is true of all the chapter topics and associated chapter links. If any of the information about P.A. components in this part of the text or linked to in the companion chapter directories has only made you want to know still more, there are even more links to all things audio gear-related in the main directory. Before you start wading into the main directory too deeply, however, we strongly advise you to finish the rest of the chapters in this book first as you'll get more out of any in-depth exploration if you finish getting an overview beforehand.

Part **4**

Getting Ready for the Show

The many different areas of knowledge required for a live audio pro would surprise most people, and even many students already committed to entering the field don't expect the exact extent knowledge required in one hoping to step into the shoes of a live sound engineer. The duties of the soundman are commonly understood to be relatively straightforward, both by outsiders who see them as being little different than gear mechanics, and by insiders who see them as mechanics who also hop in and drive the machine (like taxi drivers or like race car drivers depends on the outlook of the observer).

In Part Four we look at the many tasks that may go into preparing to mix a live show, and offer a first glimpse of some of the ways all the separate fields of knowledge and skills start to combine to get the job done. The complex system we work with in the live sound field, includes elements like sound, signal, and gear, but also people and place. As with all systems, the individual components that seem simple when looked at in isolation, become exponentially more complex when interacting with the rest of the system.

Chapter 15

Preproduction

Preparations in Advance

A *production* is the generic term used to refer to any event staged for an audience or guests, whether it's a play, a concert, a panel of speakers on a topic of interest, a party, or a convention (and the last two often will include events where the guests also play the role of audience). The term *preproduction* simply refers to the process of planning and preparing for a show—all the things that happen before a production in order to make it happen. While everyone knows that it takes preparation ahead of time to throw any event, most people who haven't been involved with planning for a production do not know exactly how many people coordinate or how many hours go into pulling off a typical production.

The "business" aspect of the audio field may get the least attention from those just learning the art of live sound engineering, and even in articles and blogs aimed at audio pros usually gets less coverage than topics revolving around gear, sound science, and mixing techniques. Without the business end of the live sound business, the time we spend tweaking the tech and mixing the sound would be spent dabbling at a hobby, not working as professionals. Because planning and preparation in advance for each event is as essential a part of the job as any tasks live audio pros do, in this chapter we will look at the many layers of preparation that the live engineer begins taking care of well before the day of the show arrives.

Advancing the Gig

Before a live engineer becomes busy with the physical tasks of procuring or packing a P.A. to be used on the day of an event, or can complete any other tasks that are their responsibility, they must first make sure they have all the information they need to make the right choices and to focus on the right problems in order to do their tasks for that particular show. Just as important, they will need to provide information they know to others on the *production team* who need it to do their jobs. This process of information gathering and exchange on the part of those involved in any production is the specific set of preproduction tasks that many in the audio and music industry are referring to when they talk about *"advancing the gig"* or show, though some also use the term to refer to all preproduction tasks.

Even the smallest of bar gigs require coordinating the efforts of at least a half a dozen people or more (venue owner, house sound crew, band members for each band playing, band support) to pull off. Larger events may require 25 or more people to coordinate their plans. Now imagine if even only the 12 people involved in a small gig had to each discuss all the necessary details with each of the other involved parties; there'd be so much time wasted tracking each other down and repeating information that no one would have time to get around to throwing a show!

Of course that's what assigning *point people* are for—reducing the 12 people who need to exchange information to three or four people who can then pass on the relevant information to the rest of the people who need it. But if they all need to talk to each other, that's still nine to 12 separate conversations, at least, and then several more as each point person passes on information to those not involved in planning. That represents a lot of time wasted repeating the same information more than once, and also leaves a lot of room for discrepancies between versions of the same information to develop; the kind of *transmission errors* that the children's game "telephone" humorously demonstrates.

Introducing the Production Team

The production team is typically made up of several different individual parties who come together and act as a team for the duration of the planning and throwing of a live event. Because the music industry is rather small, these parties may know each other, or know of each other, in advance and some might even have a well-established history of working together on productions. Each party's particular duties will depend on the scope of the show, the preferences of the person calling the shots for that production, and the generally accepted practices in that regional or national market. In other words, there will be practices that almost everybody follows but because the industry is made up of free agents and contractors, is relatively young, and is not standardized, the exact procedures and practices will vary from production to production and from market to market.

- *Producer*: This is the person that initiates the production whether it is a one-time event or organizing an 80-week tour. The producer will be the one who provides the upfront funds for a show and who will hire and oversee the other managers and contractors involved in a typical concert production. Producers who act solely in this capacity typically are the people throwing the largest limited engagement events, festivals, and tours.
- *Production manager*: The production manager is the person who handles the bottom line and other office tasks, including the hiring of other production personnel such as promoters, tour managers, sound

Figures 15.1. Preproduction for a night of jazz performances at a city theater, for a dive-bar punk show, and for a massive summer outdoor festival gig will share common features, but most of the details will be specific to that particular event's production team, location, size, and type.

engineers, and other technical personnel, caterers, merchandising managers/staff, stage hands, bus and truck drivers, and caterers. Production managers also book the venues, ticket sellers, and sound providers, and handle hotel reservations and travel arrangements, making sure all expenses get paid.

- *Tour manager or road manager*: This is the person who coordinates all those under him or her to ensure that each part of a production occurs on time and is completed satisfactorily. To that end, the tour manager will be present to oversee loading and setup of the P.A. and make sure sound check occurs on time. The tour manager will make sure the performers know the following day's schedule in advance and will be there to oversee teardown at the end of each show, making sure the P.A. makes it safely into the trucks and is headed to the next venue and the tour staff and performers are safely tucked into their buses and on their way as well. The tour manager will also troubleshoot random issues that arise on the road, mediate disputes among other production crew, and report developments or requirements to the production managers to take care of. In short tour managers are much like a combination of whip-cracking chain-gang overseer, automated scheduler, and tour-den mother.
- *Promoters*: Promoters are the people who *market* and *promote* live events using a variety of old and new media sources as well as leveraging their industry contacts. Promoters may arrange for corporate sponsorships, contact record labels for promotional materials, print posters and flyers and arrange for their distribution, send press releases to the news media, and arrange for artists to sit for interviews with local media outlets. Promoters will work throughout to create a word of mouth buzz about the events they are promoting using social media and leveraging networks of fans with boots on the ground to stoke and spread the word. Promoters may be hired by producers to promote a tour, festival, or event for a percentage or they may work on their own behalf while producing their own midsize local or regional events. When acting as producers, promoters will hire the bands and contract with the venues, sound companies, and other vendors required to throw an event.

The exact parties involved in any production will depend on what type of production. While the production-team members described below will be part of a typical production thrown by a producer or promoter, they also can be involved in simpler productions that will not include any of the players described above:

- *Venue owners or venue event management*: These people provide the space for live performance. When contracting with producers or promoters they will typically use a rental contract and charge a fixed fee based on the size of the space and the amenities and services provided. For these events they will provide some or all of the following: security, merchandising space, box office/ticketing, stagehands, sound and light hardware and techs, and refreshment sales, with the value of the services represented in the fee they charge and refreshment sales profits either retained in full or shared according to a contractually agreed upon split. Smaller venues will also throw their own regular events where the contract is directly with performers via a performance contract instead of a rental agreement and will either take a percentage of ticket or door sales, or simply get their profit from the increase in food and drink sales achieved through the increased traffic drawn by the performers. Venue owners will also advertise the show, either assisting the promoters renting their space or providing promotions and publicity for the bands they hire.
- *Venue staff*: The venue may come with a dedicated production staff hired full or part time that may include stage manager and stage hands, sound/lighting techs, FOH engineers, security pros, and kitchen staff.
- *Booking agents*: They secure performance opportunities for bands they agree to represent and negotiate performance fees to be paid to the band in exchange for a percent, typically 15 percent, of the performer's fees or box-office take. They may work with the

bands directly or work with the band's management. Though they will typically represent the same artists over time they are freelance contractors who usually will represent more than one artist at a time and also rely on good relationships with promoters and venue owners as well. While they act for the performers in pay negotiations (as this directly increases their pay too), they also are expected to work with bands who know how to behave professionally, have a good sound and potential to draw concertgoers, or already have a growing fan base. In this way they benefit venues and producers, who often rely on them to provide vetted, high-quality performers. Even if these performers have a little more power in fee negotiation due to the booking agent, venues and promoters who trust a booking agent often see hiring their bands as a better investment than taking a risk on booking unknown quantities.

- *Performers and their representatives*: Whether mega stars or first timers, it's important not to overlook that without them everyone but some of the venue owners and equipment providers would be out of a job. This makes them important team members even if all they do is show up. The performers may use a booking agent or book their own shows, and may have a contract with the producer, promoter, or the venue. In addition to possibly partnering with a booking agent, performers may bring their own stage hands, managers, instrument techs, and/or sound engineers to the production team.
- *Independent service providers*: These are the employees of sound and/or light rental companies, backline rentals, security professionals, catering providers, etc.
- *Audio engineers* (FOH or monitor): Many events will have more than one of these because they may be hired by almost any of the players already listed or one may be provided by all of them, a gaggle of engineers. A production team's sound engineers could be venue or independent provider employee, or contracted by the producer, promoter, venue, or band. To add to the confusion audio engineers often wear more than one hat and may also be the producer, promoter, booking agent, or the soundco owner (as opposed to operative).

Though this list of potential production crew members is not complete and not typically all found working on the same production, the people represented are those most likely to act as fellow point people and production-team members a live audio pro can find themselves working with. In light of this, describing a single preproduction scenario a is being representative of what employment as a live sound pro is like is no more possible here than describing a single position or employer was in the first three chapters. Yet no matter what position a live audio pro holds, they will want to gather their information in written form, and write down what they provide to others as well.

Gathering Intelligence/Getting It in Writing

Since large shows easily can involve the efforts of hundreds of people, and between the promoter and all the vendors and bands involved there can be as much as one or two dozen point people, it's clear that assigning point people is only part of efficient and accurate planning. This is where *contracts, lists, and other documentation* become valuable. These tools allow a few or a few dozen point people to stay on the same page *and* keep everyone else involved in a production updated with far less risk of ending up with conflicting information, and without forcing anyone to repeat themselves. In the case of any point people involved in a production, whether for Saturday night at the local gig bar, or the much larger production team coming together for the local dates for a popular band's first big tour, basic information must be exchanged in order to ensure everyone can prepare and be ready for the show. This information is typically contained in a group of documents that can be exchanged or provided for review by production team members who need information.

These may include, *a production team contact list, a production timetable, a venue site plan, a venue walk-through checklist, set lists, and the relevant contracts and riders that outline the expectations of each party.* For the largest productions where there is so much documentation that sharing copies of it all is only likely to result in a huge pile of loose papers, and even more confusion as some get lost and vital information gets buried, the solution that is becoming increasingly common, once all contracts are signed, riders are agreed on, and travel reservations are made, is having all documents sent to a company that provides *tour books* and having one printed for each team member. Tour books will contain each performing act's contracts and riders, all tour members' contract info, role, and bus assignments, the tour schedule and travel itinerary, maps or any other essential or useful information.

Building the Backline

The first thing that needs to happen for any production is to hammer out contract and rider details between the performer and the producer and/or venue. While the agreement is over which party will supply gear, the producer/venue owner typically will hand the task over to the venue engineer or the sound company engineer. On the other end of things, early on bands will handle this negotiation themselves, but as soon as they can get their own engineer, they too will pass this task on to their engineer. It is good to get this rolling sooner rather than later as many performance contracts include a clause requiring all changes to contacts and riders be finalized no later than two weeks prior to the production date, and you want to have time for some back and forth. What you will need is a copy of the performance contract *technical rider*, including *channel input list* and *stage plot*. Ideally the performers will email or fax most of this information within a day or two of being booked (if booked only a few weeks or one month in advance), or within a couple weeks (if the show is still a few months out).

In reality, often the band will be unaware of the need to get rolling on this, or will just procrastinate and forget until the appropriate amount of time has passed so that they can pass the buck or the venue owner's wife will stack other papers on fax and bury it minutes after arrival, two days after booking, where it will remain safe and sound until six months after the event has passed. So to be safe, if it doesn't arrive within the first week you will need to call or email asking for it.

The contract technical rider is supposed to be in three parts, condensed and to the point, realistic, and without extraneous information (like mixed with hospitality rider requests). The three parts to be included are:

- *Technical specifications*: Gear lists from each band clearly listing every item of backline equipment the band will be bringing, as well as concisely listing the backline equipment they require be provided for them. Other information that may be included are any specific P.A. equipment requirements, specific power requirements, number of monitors and monitor mixes they are requesting, and number of console channels/ins/outs they require.
- *A channel list*: The mixer channels the band will need revealing details that might not be evident otherwise, even in the stage plot, and are essential for planning a smooth transition between acts. Information should include channel number, instrument, microphone or DI box make/model, stand or micing instructions, channel inputs such as compression or effects, SL/C/SR instructions for each stereo channel input, and phantom power needs.
- *A stage plot*: Each band's stage setup and placement. Ideally will include band member names, instruments/amps, power drops, position, placement, such as drum layout, preferred pedal, amps, monitor, and DI box placement, and all other stage details not evident in a straight gear list or channel list, including fills, and stage covered by band members who leave their monitors to ham it up for solos, etc.

Most technical riders will have something missing, or some extraneous info, and be more or less unrealistic in their requests, and more or less successful in their

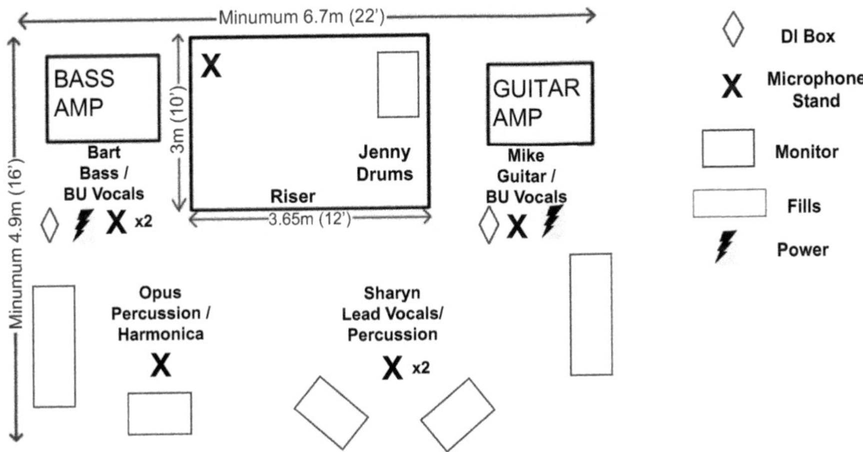

Figure 15.2. An example of an uncluttered stage plot.

layout, but a good technical rider is an art, and most technical riders are works in progress until a performance act has gotten quite a bit of performance experience. In the first set of sample documents provided, note that the technical rider includes a common caveat: for *no-fly gigs*, all the specified gear will be provided by the performers, but for gigs where the band is flown in to perform, the gear will be provided for them. In some cases you may find that all the gear requested is available at one provider and you can negotiate a package immediately; in other cases you may be able to do the same using two, and in others there will be provisions that can't be filled as requested.

Negotiating Alternatives

Make a note of any rider requests that can't be filled, as well as any other vital missing information; for example, the tech rider appears to be missing some instruments, there is no mention of compression or effects on their channel list and the stage plot is missing power drop information. When you contact this band to negotiate alternate rider options you can ask for any information that isn't included. Set them aside for the moment; before you call you may want to make sure there isn't additional information you require.

First, look at the next rider example provided, the one from the opening act. This band's requirements are simple and easily provided, and they leave the selection of brand for the few requested items up to you. Note another frequently used clause regarding P.A. requirements for bands who don't have any specific ones, which is that a P.A. system be provided that is adequate for the size of the venue and the program material. Their channel input list is much better formatted than the opening band's list, and because of their simple setup, the written stage plot is not a problem and looks complete. An example of how a simple stage plot should appear is provided, however, because any bands contracting your services are likely to ask you to create a technical rider for them, and the opening band's example is too cluttered and hard to scan. Because this band included a column for input information on their channel list, the lack of instructions about inputs can be assumed intentional and not an oversight. If you feel compressing or gating any channels is important you can call and ask if they object, but otherwise they cover all essential bases.

Figure 15.3. We know the rider said nothing less than a grand piano would do, but it also said negotiating options was possible; the chances are good that if the Steinway Grand is not possible, the good upright will work.

Before examining other information you may want to ask each of these bands about their own tech riders, and there are a few words worth saying about tech riders in general. First enjoy the humility and simplicity of both of these riders, because many bands, be they ever so humble, will deliver documents that are dizzying in their grandiosity and some that will even concern you over their forcefulness during those portions where it occurs to them that such grand technical documents deserve a strong finish, and settle on a bit of drama by wrapping up with pointed reminders that the rider is part of their contract and dire predictions of how breach of contract might be dealt with. But there's no reason to think a band will be impossible to work with even if their tech rider has a haughty tone, and certainly no reason to worry over any threats. Aiming as high as possible is just part of how the tech rider game is played, and insisting they can't budge or compromise about a particular point is not actually signaling an unwillingness to work on fixing problems together. It's just marking out the strongest position they can for whatever haggling does ensue. Remember a perfect tech rider is an art and you will be getting copies of riders that for many bands are their first attempts, or their engineer's first attempts. So if you are working for a smallish midsize venue or renting your P.A. and get a list of exotic high-end gear and dire warnings about contractual obligations and severability, don't get too annoyed right away.

In fact, go to the web directory for this chapter and look at a few of the many rider examples online that range from excellent to way over the top, and notice how many of the ones that insist on nothing less than a Steinway grand in one paragraph will include a more humble caveat somewhere later, offering comically long lists of alternatives in order of declining cost and prestige and ending with the mid-range standards. (One band with a bit of tongue in cheek humor has a tech rider posted online where the list of acceptable options begins with a Steinway grand and ends with a late model Casio keyboard.) Other bands will be equally insistent that nothing less than the best will do and list no alternatives, but these riders also will almost always include a caveat somewhere indicating they are willing to work with the sound provider to find a solution for any provisions that can't be provided as requested, which says the same thing as a list of alternative options.

So why is it so common to find very precise gear configurations and explanations for how these provide a superior sound, if more often than not what they will be offered is a selection of solid, good quality, mid-range standards as usual?

- First, the tech rider is a convenient platform for showcasing each band's idealized audio "aesthetic" and indicating their knowledge and understanding of gear, so it provides bands with a little calling card that they can use to set themselves apart, or develop a persona that's edgy and unique, while still turning in something that functions as a useful and informative document.
- Second, sure the odds are long that some of these requests will be honored, but the odds of scoring a few hours to gig on their ideal backline and P.A. setup are a whole lot longer for those who don't ask for it. And while shooting for the stars may not land you on the moon, or even make you king of the mountain, it can get you the penthouse suite for bottom floor prices often enough to make it worthwhile risking a no now and then.
- Finally, it's more fun to select and detail a list of ideal gear than to do the same for a list of the same old gear, just like girls would rather play with the doll in a pretty dress than the one in sweatpants, and little boys will pick the muscle car hot wheels toy to play with over the hatchback. So the ideal is listed first and the mundane is the alternative.

No matter how over the top their requests may appear, the audio sound provider isn't asked to arbitrarily guess or assume what is or isn't available or a reasonable option, so your next step should be the same in any case: to check for what is or isn't available and see if what seems unreasonable is in fact all that unreasonable. To claim an item can't be arranged for or is too expensive without checking first is bad faith, especially when you aren't entirely familiar with the item being requested—you may well be wrong. Even if you are pretty sure none of the backline rental companies have a Stradivarius available to rent, take two minutes to check first; that way you won't have to lie, and you can honestly say it's not an option but you tried other local providers to try and get it.

When you can't find a specific item that is important to the band, try and use what you learned of their tastes from what they stressed or highlighted in their initial requests to guide you in selecting agreeable substitutions to include in your counter-offer, even if the ability to offer a tailored alternative solution is limited to reframing what is available to feel more tailored. So if a technical rider mentions the superior sound quality of an esoteric and expensive piece of equipment, don't simply dismiss the request as ridiculous just because it isn't possible to fulfill it at that point. Instead, do something surprising and first put some thought into the meaning behind the choice so you can honor the meaning behind the gear request even as you dismiss the option of filling it as is. Do a quick search and find out what makes that item generally valued and note what choices out of the available alternatives offer the most of that kind of value even if it is only a little more than some other choices you might recommend.

It takes only a couple minutes to reframe negotiating tech rider changes of this type from an interaction with a subtext that devalues, i.e., "that request is ridiculous because that gear is out of your price range, too specific, too old, not pro audio, etc. and therefore it is not available," into one where the subtext confers value: "I am not able to provide that but let's talk about why you value it most and see if we can't find the best alternative by looking at what we can arrange with the gear that is available and what I can do as I mix to help you achieve the sound you are after. I'm sure if we put our heads together we can come up with a solution." If you approach the task of negotiating tech rider changes with this kind of attitude, we are confident you won't run into too much resistance from the bands you work with, who will be glad to make compromises when working with someone on their side, once they know that's who they are working with. It's just up to you to make the first moves at establishing a good working relationship.

Figure 15.4a **PERFORMANCE AGREEMENT**

Standing Room Only (artist), agrees to provide a live musical entertainment performance for:

NAME: ______________________________

ADDRESS: ______________________________

____

PHONE: (________)______________________(buyer). under the following terms and conditions

1. **Date of performance** ______________________________
2. **Time of performance** Reception: From______________to ______________
   Artist shall be entitled to reasorable breaks each bour unless otherwise negotiated
   Ceremony: From______________to ______________
   Cocktail: From______________ to______________
3. **Location:** ______________________________
   ____
4. **Number/Performers**______________________**Attire** ______________
5. **Fee for Services $**______________________Overtime $______________per hour.
6. **Deposit $**______________________Due w/signed agreenent______________
   Made payable to: ______________________________
7. **Final Payment: $**______________________Due:______________
   Made payable to:______________________

Unless otherwise negatiated, please make personal check payments from a local institution payable to name indicated, Payments written on out of state checks must be made one week prior to performance date.

8. **Cancellation:** This agreement serves as the only binding instrument for both parties. Deposit and signed agreement must be returned by date indicated or this agreement is null and void. This agreement may be cancelled by written notice no less than 90 days prior to the date of performance by either pary. Should unforeseen circumstances prevent Standing Room Only from fulfilling this agreement, all deposits and payments shall be returned in full and this agreement shall be considered null and void. Should the buyer be prevented from fulfilling this agreement, all deposits and payments shall be forfeited and this agreement shall be considered null and void.

______________________________ ______________________________

Buyer Date **Band Rep** Date

Figures 15.4a and 15.4b. Performance contracts can be short as the one shown in Figure 15.4a and 15.4b (Continued) 15.4a, or can cover just some of the clauses detailed in 15.4b and can run too many pages.

Figure 15.4b Sample Performance Contract Details

SECTION 1: TERMS OF EMPLOYMENT

Contracting Parties- (Venue/Producer "Purchaser", and Band "Performers") - Including all names and contact info

Governing State Law

Event, Location, Date and Times- Including event title, venue name and address, and all relevant preparation and performance times (load in/out, sound check start/finish, all sets start/finish)

SECTION 2: COMPENSATION

Currency and Payment Type and Amount - one of the following * Percent of ticket sales, * Fixed fee * Percent of ticket sales with guarantee, * Fixed fee and percent of ticket sales, * Other

Additional Compensation- Percentage of bar take or % of merchandising, if applicable

Deposit Amount and Due Date- Including if and when performer has right to cancel if deposit is not paid in full

Balance Paid, How, to Whom, and Due Date- For example Business check or cash, to Band member X or band manager, After performance but before load out ends the same or final day of the performance contract

Paid Expenses- Such as travel, lodging, and or food per diems, if any, (or drink tickets/snacks, at minimum)

SECTION 3: SEVERABILITY

Cancelation Terms- Under what circumstances either party may terminate/cancel without liability, or when each must compensate for terminating/canceling (including compensation details).

SECTION 4: DUTIES OF PERFORMERS

Always - *Render services in a competent manner, *Abide by all reasonable rules and requirements, *Arrive and perform at specified times, *Provide all agreed upon rider information and marketing materials in a timely fashion, *Be committed exclusively to the purchaser for the period of the performance, *Hold and/or obtain all rights to songs performed (If required to perform songs for purchaser, the rights will be obtained for them)

Sometimes- *Pose for photos or attend publicity events as agreed upon in advance, *Mention or thank sponsor's each set, *Provide own liability insurance, *Learn and perform new songs for the event, *Provide backline equipment, *Other

SECTION 5: PERFORMER RIGHTS

May Include- *Merchandising rights (rights to all merchandising proceeds, they provide vending persons and merchandise, venue provides space, table and chair), *Sole authority to make changes to key

Figures 15.4a and 15.4b. (Continued)

personnel (opening acts, engineers, etc.), *Right to be consulted about any changes to key personnel, *Right to be informed of all additional purchaser rules and requirements and rider substitutions a set time in advance or reserve right to sever without liability with full or partial payment, *Other

Usually Will Include- *Right to dictate what reproduction of performance if any will be allowed and retain full rights to performance reproduction and to decisions regarding how they may be used (all permissions to be obtained from the performer in writing),* Some or all rider amenities

SECTION 6: DUTIES OF PURCHASER

Almost Always Will Include Providing ALL Event Logistics - *Sound provider and engineer, *Marketing and promoting for event, *Lighting and effects, *Amenities and vendors, *Ticket Sales and security, *Venue and general personnel, *Liability insurance

May Include Providing: *Backline equipment, *Reservations and lodging, *Reservations and flight, *Some or all other rider amenities,

SECTION 7: PURCHASER RIGHTS

Right to Performer Duties

Right to Set Additional Rules and Expectations for Performers: including limits on language, dress codes (such as for corporate functions), limit what performers may bring to the performance, use during or as part of the performance (for example no illicit drugs, no pyrotechnics, and a one drink maximum per set until completion of performance duties), Set sound level limits in accordance with local laws

SECTION 8: MISCELLANEOUS TERMS

Release of Liability – for example Indemnification and Act of God clauses

Breach of Contract Procedures- for example, Unresolved disputes must first pass through mediation followed by arbitration before other remedies sought, Arbitration final for all disputes under $-------- , etc.

Additional Clauses- Hospitality rider and Equipment rider noted, Contract to supersede all verbal agreements made before or after and to be altered only with the consent of both parties

SECTION 9: SIGNING DETAILS

Signatures, Execution Date

Witnesses- if applicable

Figures 15.4a and 15.4b. (Continued)

Performance Contract

While most of the information you seek is found in the three *performance contract* additions already discussed, there are some topics usually attended to in the contract itself, which is usually not made available to non-relatives for privacy reasons. To know what may be contained in the performance contract that may be relevant to your position as live engineer, you have to look at one. Learning the kind of clauses that are typical in these contracts is a good idea anyway, because live engineers who can double as booking agents or promoters get more work and get their clients more work.

The questions to ask about the contract are listed below; make sure you find out well in advance what is addressed. Ask the artist or venue owner about these—if they don't want to show you the contract so you can extract the info yourself, then give this list and have them pull the information for you:

- What are the times posted in each contract for load in/out, sound check, beginning and end of event and start/end of every set and break time. (You need to make sure these all agree, with no overlap between each of the bands and no deviation with what is stated in your contract to provide services.)

Schedule

Venue:	The Bar Bar
Location:	1015 Oaks St.
Room capacity:	400
Date of show:	27th August 2013
Venue Owner -	Bob Bar – 515-555-1212 (mob)
Bar manager -	Turner Downs – 515 555 1213 (mob)
Sound person -	Ian Eeirs – 515 555 1234 (mob)
Load-in time:	3pm-4pm
Load-out time:	3am-4-am
Entry/exit points:	Side alley off Oaks@ Willow Street, through double doors. Dolly next to stage left.
Power supply:	Three phase, 1 outlet situated on stage and 1 outlet on the exterior wall facing Oaks Street.
Sound Check	4:45 pm-5:15 pm Chucky and the Cheeze
Sound Check	5:30 pm- 6:00pm Genghis Green
Sound Check	6:15pm- 7:30pm The Adenoids (longer if needed)
Doors open:	8.30pm
Drinks supplied?	**YES** ✓ **NO**
If YES, details:	Openers: 1 drink ticket per band member
	Headliners: 2 drink tickets per band member, All: No limit on Soda or Water pitchers

Opening 1: Chucky and the Cheeze	Start 9:00pm	Finish 9:40pm
Opening 2: Genghis Green	Start 10:00pm	Finish 11:00pm
Headliner: The Adenoids	Start 11:30pm	Break 12:20am
	Restart 12:40am	Finish 1:40am

Figure 15.5. Pulling out important info from the performance contracts and creating an easy-to-read production schedule is worth the time it takes to do so.

- Is there a clause indicating any SPL limits according to local law?
- Has each performer agreed to take responsibility to own rights to every song they will perform, or if they intend to perform any song they do not own, then to obtain performance rights in advance? (If not, ask if this clause can be added to the performance contract or a clause added to your service contract

indemnifying you from responsibility for any band that chooses to ignore your advice and perform songs without rights.)

- Are there any clauses concerning when any further rider changes are no longer allowed?
- Are there recording restrictions—clauses limiting the ability to record the show for personal use—and who is in charge of giving permissions for each band?

Evaluating the Site: Venue Walkthrough

The site visit is an essential part of any production, and should happen early in the process by the first person who can make it. Naturally the focus of the visit will vary according to who is handling the task, and each point person has their own primary concerns; however it is easier to have those who can go earliest gather as much information as they can for their fellow production team leaders. The categories will cover areas of interest other than sound, but it is not particularly difficult for an audio pro checking out the venue for his/her own purposes to jot down some notes and gather some info for others on the production team, so you may be asked to fill out what you observe that relates to the duties of others working on that production.

The time to go for a site visit is when it will be the least problem for the venue, which means midweek in the afternoon when few patrons are there and no other outside productions are using or planning to use the venue that day. You will want to bring a flashlight, a clipboard and pen, a tape measure, a stud finder, and an outlet tester. Be sure to introduce yourself before you start measuring and circuit testing; staff can have answers you need, so be sweet to them. The checklist provided along with the chapter's online directory is similar to what you might find in the filing cabinet of any production company, but this one is a shortened version created years ago for a party collective that threw not-for-profit parties. Fill it out as you walk around the venue, but don't be limited to it; use the floor plan to document any features not covered in the checklist but that may be pertinent.

Some things to note that are not on the list include actually checking the outlets with the outlet tester, noting the stage dimensions and other features such as if it is hollow beneath. Using the floor plan to take

Visual Inspections Can Reveal Many Safety Issues

Pro Jobs
are Neat
and Tidy

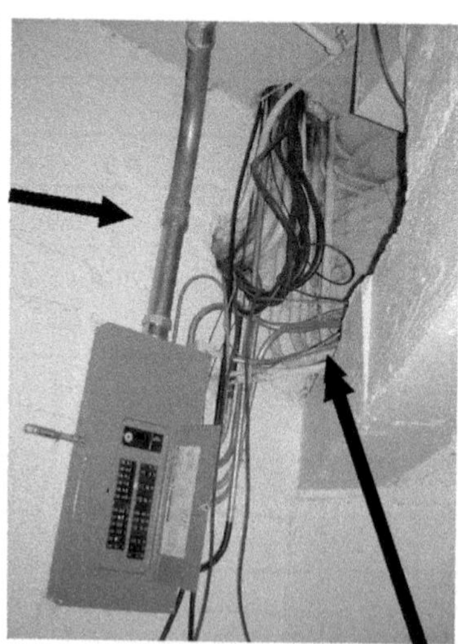
Ask about Cases
that Seem a
Bit Sloppy

Some Situations
Require No
2nd Opinions.
Don't Plug In!!

Figure 15.6. Checking the venue's power set-up and capabilities early enough is important because it allows enough time to actually bring in a contractor to fix any problems you may find.

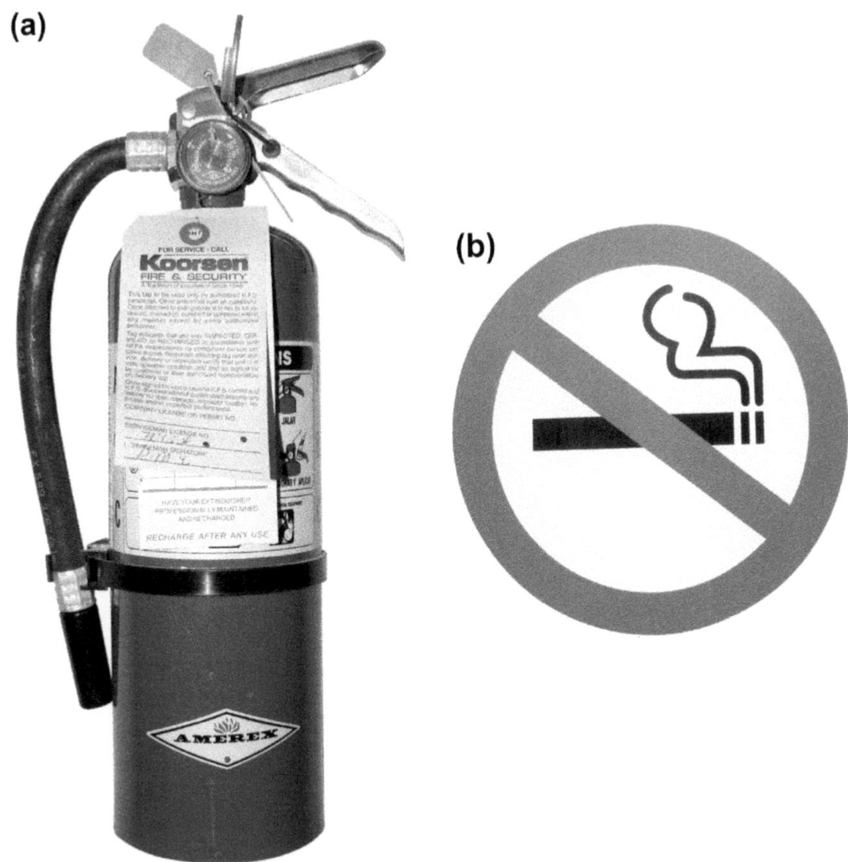

Figure 15.7a and 15.7b. Checking out that important venue features are in place can keep you from scrambling to provide them at the last minute. Features like safety equipment and signage are easy to take for granted and overlook, or to de-prioritize as not being that important; sadly you may only see how important they are if you need them and they aren't there.

notes, indicate where carpeted areas are, estimated ceiling height, room dimension, and materials for room acoustics (clap your hands to check reverberation decay times and warmth). It should take around an hour to check a typical venue, a little longer if you make use of the measuring tape or take time to chat with staff. Make sure to say goodbye before you leave, and to scan all notes into your computer and email a copy to the rest of the production team as soon as you can.

Preparing to Mix the Program

There are only a couple major areas to cover as far as the task of gathering information to plan for a production. Although other preparations like packing will remain, the most important tasks of gathering information are finished. All that remains is gathering the information you need in order to set up and mix the show. Information about the venue has been gathered, the unknowns that remain are the P.A. and the *program*. This early in the P.A. may not have even been selected yet; however the program is set. The term program refers to the performance material that you will be mixing.

Unless you have worked with the production's band before or heard them on more than one occasion, you will need to become more familiar with their material if you want to have a chance to mix them well. To that end, when you call with your last remaining questions about their setup, you should ask each band for sample material to listen to in order to become more familiar with their sound, as well as any input or commentary they have about their vision, their sound, their strengths and weaknesses. To be ready to mix them simply requires listening to their music until it becomes more familiar. Just remember to use some of the methods we mentioned when discussing ear training; you will learn the music that much faster if you listen with intent.

Chapter 16

More Preproduction

Plan It, Do It, Double-check It

General Planning

In addition to gathering and accounting for the details specific to each gig, there will always be the general procedures you will use for every gig. While this stuff soon becomes as automatic as mixing since it applies to every show, you will still need to go over it before gigs to make sure you have it all covered. If you work for the venue, you won't have packing to worry about, but you'll have the venue to prepare, and could have promotional tasks to take care of as well. While those tasks will be different, your order of operations should be the same:

1. think about what needs to happen;
2. write down how you will make it happen;
3. do it—make it happen;
4. double-check that everything is done.

For this chapter we are going to assume the point of view of the live engineer who is acting as/working for the sound company and responsible for providing the P.A. for a private rental event. That means figuring out what to bring, packing it strategically, and making sure we are ready to effectively get it unloaded and set up on the day of the show. By packing smart and anticipating your safety and setup needs ahead of time, you can use your setup time efficiently on the day of the show—by spending it actually setting up, rather than figuring out what to do next.

Use the Venue Parameters to Plan Your P.A.

Of course it'd be hard for even the most scattered person to forget their P.A., but it is not unheard of to bring the wrong P.A., for example to bring one that's not powerful enough or doesn't suit the shape of the venue (one array, two rooms). To avoid this mistake you need to use what you know about the venue to determine your P.A. requirements. Visit the chapter directory and try out the calculators that are designed to help you figure that out. Play with them for a little while too; just make up some measurements and see the results you get for each. We recommend the one provided by Crown Audio, as it comes with a detailed instruction page and is hard to miscalculate if you pay attention to the instructions. It's also important to take more than just power into account, and make sure to plan in some flexibility to your set up.

For example, one amp might easily be enough to power your P.A., but it can't be in two places at once, and by now you should be aware of some of the pitfalls of trying to connect your P.A. using long cable runs. For many small- and mid-sized applications it's better to bring multiple low-power components that can be mixed and matched and placed in a variety of locations/configurations than it is to bring one high-power amp and stack. And once you've planned your P.A., write it out anyway so you can be sure you've remembered to include every cord and stand and adapter.

Everything and the Kitchen Sink—Determine What to Bring

Packing starts with making a packing list. This starts after you have advanced the gig and gathered all the information you need to plan, but don't wait too long. It might only take one afternoon to do as far as minute count goes, but you want to start early and make your lists as things occur to you, because no matter how hard you think about it, you will forget items. If you don't give yourself time for several waves of "Oh yeah, I almost forgot" items to pop into your awareness so you can write them down in time to do something about them, they'll hit you anyway, but only after everything is packed and you are five minutes from the venue.

So start well in advance; keep a copy of your ongoing list somewhere handy (on your phone so you have it as you think of things is ideal). Then as you think of items you need to take with you, write them down. Do not assume you will remember something, no matter how common sensical. If you are bringing a guitar, put strings on the list or you will find you have brought an unstrung guitar. The same goes for any items powered by detachable cords or batteries. If you don't put it on the list you will forget. Make copies of the list to share and pack, and put your copy on a clipboard (otherwise you will forget the list). Conveniently, this provides you a two for the price of one return on your time investment, because when you're finished with your packing list, you will also be equipped with an equally important list—everything you need to bring back when the show is over.

Bring More than You Think You'll Need

You can make your task easier and drastically reduce the odds of needing something by bringing it. Conversely you can seal your own fate and ensure a string of bad luck in advance simply by not being prepared for it. Want to guarantee you'll be mildly flatulent and hypoglycemic? Forget to bring your beano and emergency snack. Conversely, you can feel confident that you won't cut yourself too bad or end up with a headache if you know there's a first-aid kit with band aids and aspirin lined up. Summer show? Forget to bring a sweatshirt and you'll be right under the 50 degree "Arctic blast" air conditioner/snow machine freezing your butt off.

Also, don't pass the buck unless the intended recipient is there with a catching mitt and you clearly hear the thwack of the catch and "I got it" for confirmation. In other words if it is required for the show to run, and you haven't confirmed over the phone or in person, and it isn't specifically indicated on a band's tech specification that they will bring it, assume they will not. The same with whatever it is you think the venue is supposed to provide. If you follow this advice, nine times out of ten the common items bands are supposed to bring or venues commonly provide will be there even if it wasn't on a list. But we can assure you, that's only because you went to the trouble to bring a backup. It's the time you assume you don't need to worry about it that it will turn up missing, even if it was there the last ten times you worked that venue or with that band. If you will need it and don't want to bring it yourself, make sure the person who should bring it has promised it in writing or to you directly. Otherwise, bring your backup to be safe, like it or not.

It Takes More than the P.A. to Provide the P.A.

Once you've listed your P.A. and peripherals, you are a quarter of the way done. While this list is extensive, it's by no means complete: it is good enough to get you started in a pinch. Some other things you'll want to bring include:

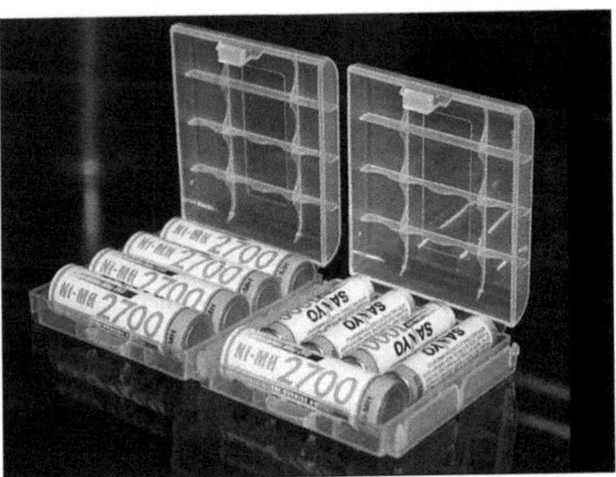

Figure 16.1. Batteries—bring them even if you don't use any battery-powered gear; odds are good someone is short, you can sell them at a premium if you want, but most importantly you can start the first set on time.

- at least one each of the most common cables, tested and known to be good, before the gig: XLR , ¼ inch TS, ¼ inch TRS, Dual RCA, 3.5 mm to TS "Y" cable, XLR cable with the shield/ground lifted on one end (with end clearly marked);
- a good collection of adapters (TS to speakon, TRS to XLR male, TRS to XLR female, XLR female to XLR female, XLR male to XLR male, XLR "Y" cables, and ground lift adapters), spare DI boxes;
- spare workhorse dynamic mics (SM57, SM58, etc.), spare stands, spare mic clips, spare foam mic basket covers, personal talkback mic;
- tools (in a small lockable toolbox): Philips and slot screwdrivers in several sizes, cable/wire cutters/wire, strippers, soldering iron and solder, cable tester, voltage tester or multi-meter, adjustable wrench, Leatherman;
- iPod loaded with genre compatible music (because you're a DJ, too, unless you think going from full bore to dead silent intermissions is good for the vibe), iPod charger, headphones/ear buds, phones extension cable;
- hand-held digital recorder;
- SPL tester and earplugs, spare earplugs for crew-members and friends, CDs for tuning system;

Figure 16.2. Adaptors, because 300 yards of cable won't fit in your pockets.

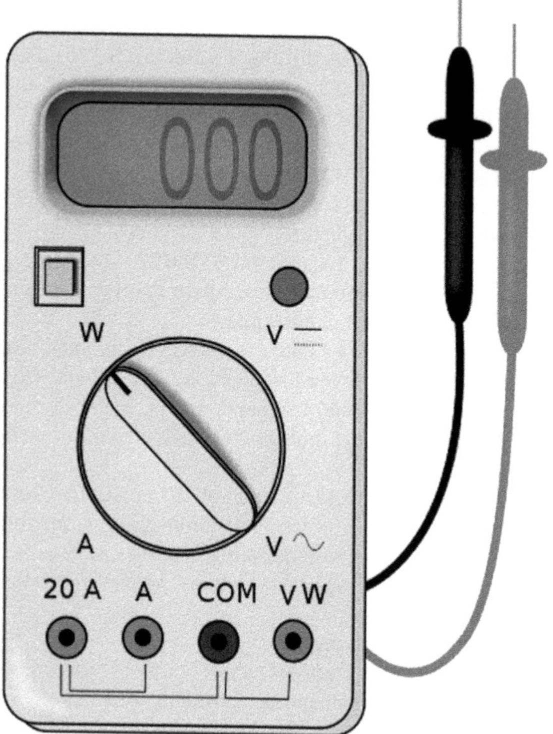

Figure 16.3. Meters and testing devices are tools too!

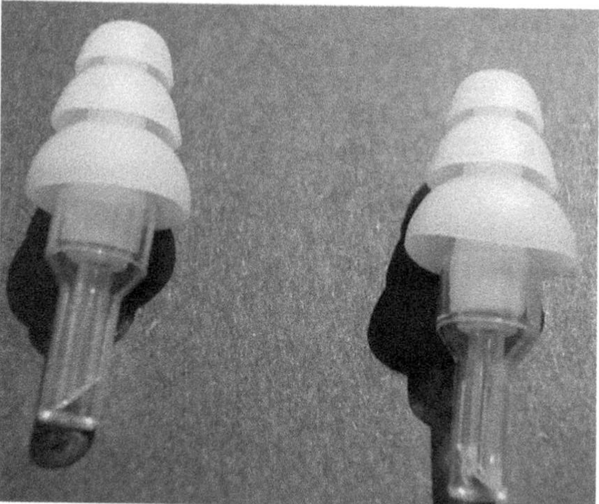

Figures 16.4 and 16.5. Pre-molded earplugs for you (because deaf sound engineers kind of suck), freebies for any crewmembers who forgot their own or for the band when they're not playing (it's worth a shot).

- gaffer tape, board (or masking) tape, duct tape, Sharpies (x 3, for when one gets borrowed and the other lost), clipboard and paper, copies of all handy documents;
- phone charger and spare batteries (in packaging or a battery box, not loose);
- flashlight, sharp pocket knife, spare amp fuses, spare batteries (AA, AAA);
- extension cords in a couple of lengths, power conditioner, power strips with ground fault circuit interrupter;
- acoustics quick fixes (see Chapter 20);
- electronic cleaning spray (fix for crackling connections);
- canvas cover for mixer and rack (if you need to step away, cover it up), caution tape (for legs of speaker stands or any other hazard that might trip someone or get run into if not noticeable);
- weight-lifting belt, casters, leather gloves, steel toes (leather);
- hand sanitizer, toilet paper for outdoor gigs, tissue for all gigs (in a pinch it can do double duty), sunscreen if outdoors (45spf or better), any other personal meds;
- screw-top water bottle (refillable), snacks;
- phone, credit card, keys, cash (including quarters);
- change of clothes, deodorant (unless you want to wear your sweaty load in outfit at the gig), light hoodie, jacket, make-up bag (for the ladies and goths).

Put It in a Logical Order

Once your list is complete, rearrange items to go in the most logical order. If everything packs in your van in a particular order (and it should), make sure your list reflects that. You will want to check items off as they are packed and don't want to slow down the process by

Figures 16.6 and 16.7. A moving blanket makes a great sound curtain, some foam pads for damping. Build an acoustic fix-it kit and you can lower stage sound by 6–9dB—that can make the difference between an OK sound and a really good one.

having to scan your whole list repeatedly to find each item. The list will also help you stick to a logical load order; you won't want to accidentally pack your casters or rolling cart first, as it'll be in the back where it's no use to you. Always pack in reverse order so what you want to access first is at the back of the truck.

Plan to pack for layout not for type. Your XLR cables may look nice, so neat and organized, nestled together with all the cables you own. But that won't make life easier when you are unpacking microphones on stage and realize all the XLRs are tidy and snug and by the mixer only 20 meters away, yet across an orchestra pit deep as the grand canyon.

Knowing What's What: Labels

Label everything. Make sure the labels are well attached and won't come off. A label can convey a lot of information in a short word or phrase. A proper label should tell people:

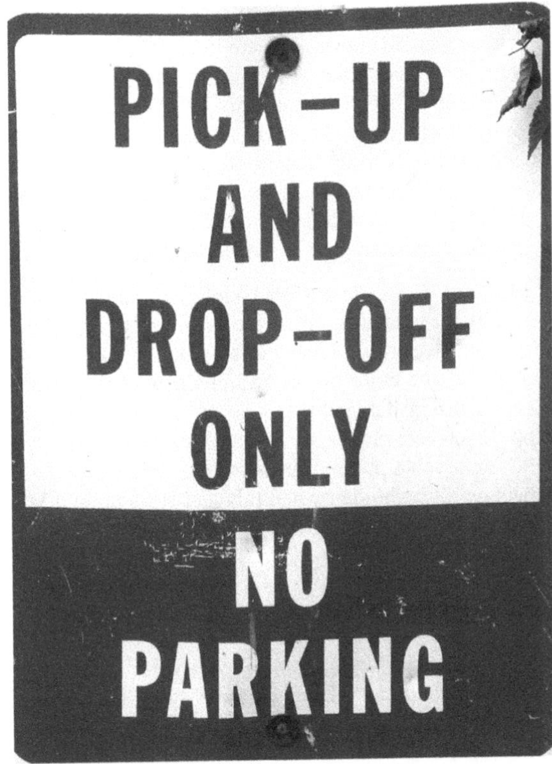

Figure 16.8. Never forget quarters—some parking tickets will cost you more than you get paid for the gig!

- *What's in it.* Unless you buy your storage bins in a variety of rainbow colors, you have storage like most sound/music industry types—a dozen black rack cases of a couple sizes and eight gray storage tubs. While it's a truism that you only find what you are looking for in the last place you look, this will be proven to statistically absurd lengths when you are usually looking for something in eight identical tubs. If the number eight rolled with as much reliability anywhere else, Vegas and statistical science would need to be revamped to account for it.
- *Where to put it.* Venue load in grips are pretty good at simple instructions. They are less able to make simple logical deductions if you aren't there to ask where an item goes. If they have a tub of booms (labeled mic stands) and a box labeled mics in 2-foot letters, they will put one by the mixer and one in the bathrooms instead of the more traditional home of mics, and stands, the stage. If it is labeled "stage" it might even end up pretty close to the target.

- *Which way is up.* Rack cases look a great deal alike from either side unless you are facing the lock side. If you have cases that aren't flight cases, then latch failure is 200 percent more likely when face down, 500 percent more likely when face down and full of approximately 1000 inch-sized adapters.
- *Who it belongs to.* This is handy as it's easy for things to get mixed up with every band's load in items and the venue-owned gear in the same place.

So your labels are always visible and reflect what's currently in your tub, given the different things a tub may hold over time, label the side of the tub not the top, and don't write on the tub or on hard to remove tape. Instead, put your labels on hanging tags or tape plastic envelopes to the side of tubs and label with removable index cards inside. That way your labels won't be hidden when tubs get stacked, and you can edit the labels as the things are stored.

The Night Before: Stage It First

This is not always an option, if you work for a small soundco that doesn't have room to stage in advance, or rent your own P.A. from your garage and your spouse doesn't think the living room is an appropriate staging area, you may only be able to stage it the morning of the gig. However, you can stage most of your small personal items, lay out your work clothes, have your clipboard and lists, maps, truck rental papers, etc. laid out and ready.

It may seem like a small thing to get a half hour or hour head start, but it's going to be a hectic day. Having that time lets you start the day cool and collected, and gives you time to eat and still take a quiet moment to have an unrushed smoke (or sleep in an extra 15, or watch cartoons, or whatever centers you). If you can get it started without needing to skip food and rush out the door shoes in hand, it's already a great day.

New Soundco Gear or Guest Engineer? Either Way Means Preparing to Mix the P.A.

No discussion of gear could hope to encompass every component and configuration available, nor could any one feature set or design model suit the preferences and needs of every user. That said, the vast majority of equipment shares certain commonalities with all the others in their own class, even as they also all vary in their specific functionality and layout. These commonalities are what make it possible to learn the P.A. as a system, and be able to operate any P.A. no matter what components it contains with a minimal lead time and reasonable learning curve. Note that the words minimal and reasonable are not synonymous with no, zero, or none. The fact that any two pieces of gear, even gear made by the same manufacturer, will be slightly different is why one of the first tasks of live sound engineers for any event or at any venue they are hired is to determine the make and model of every piece of equipment they will be using, then refer to the manual to get acquainted with its particulars.

Read the Manual

Before we end our tour of the P.A. system and move on to the chapters covering procedures, we need to remind you to *always consult the manufacturer's included documentation (the manual and any supplementary guides and specification sheets) to become familiar with the unique features of each piece of new gear you will work with*, regardless of how similar it may look to gear you have used before. You can't be positive that you know everything you need to know about a piece of equipment, and therefore should not use it, unless you have:

- worked with that *exact* model very recently and/or very extensively so that you are sure that you are fully acquainted with its functions and controls, and that you fully remember how to access and use its functions and controls;
- recently read the manual to learn how to use the equipment according to the manufacturer's recommendations.

It's important to read the manual when using a new piece of gear because even equipment that may already feel familiar, with features and layouts seemingly identical to other models by the same manufacturer, can hold surprises; that familiar interface could easily be mapped to different commands and/or control a completely different set of functions.

Manufacturers can also differ from each other in the terms they use to label their components/processes with, even though they are identical otherwise. Equipment with the same features will have different layouts and controls. When you have a limited amount of time to set up a P.A. you don't want to waste time figuring out how to use a piece of equipment that is unexpectedly different than other similar equipment from different manufacturers, nor do you want to try mixing live on a board with a completely unfamiliar layout, as many digital boards have even when they are identical in their features.

Equipment may also vary in the type of inputs and outputs they have and the kinds of connectors and cables that they require, often in ways that aren't always visually obvious or clearly labeled on the equipment. The safest way to approach the task of plugging anything into anything else is to *never plug one piece of gear into another until you are satisfied that they are able to be matched.* For example, if the power rating and impedance of your amplifiers and speakers is not indicated on the unit, or if the inputs and outputs on the mixing board aren't labeled to clearly show *line level* and whether they are balanced or unbalanced, then you shouldn't plug anything in and turn power on until you've double-checked the manuals to be sure they are compatible with the way you intend to connect them.

Why the extreme emphasis? Well any audio educator can tell you that most of their students will check the manual faithfully and about a quarter of them will act like they are absolutely allergic to product manuals, or just cheerfully ignore that they exist even if one is placed in their hands. Furthermore, this number stays pretty much the same whether they just tell students in passing that it's good practice to always read the manual before using new gear, or whether they repeat it on a daily basis (or by the third month of class shout it while tearing their hair out). While they don't poll every class on this topic, they can extrapolate this figure from their experience of teaching many different groups of students, because on any given day when setting up or sound checking the P.A. the class is using someone will inevitably ask a question that they wouldn't need to ask if only they'd read the manual, right up to the last week of class.

This phenomenon is due to one particular habit that technology enthusiasts and audio hobbyists sometimes pick up, and in too many cases fail to "kick" when they become live audio students (and many students come from this pool). The habit is simply a refusal to ever check any manuals! Instead these students will acquaint themselves with new gear or software by fiddling with each dial and button until they have stumbled upon and deciphered most of the command functions and settings. This is not something that can be attributed to lack of ability or being bad students; in fact it's often the best and brightest students who are the worst offenders. Some developed this practice on their own, taking a certain pride in having enough technical knowledge to figure out a new piece of equipment or program without really needing to refer to the manual, while others may have picked up this habit under the influence of peer pressure from a group that thinks only "newbs" check the manual, an opinion that's not all that unusual for junior-high-school-aged techies. Others simply aren't big fans of reading unless it's absolutely required; these students may feel there's no reason to read the manual when their experience has proven that they can easily figure out technology without the manual. This attitude is also pretty common, especially among hands on learners, technical and math-oriented people, and anyone who has a hard time sitting still.

When you consider that many of these audio students have used this strategy successfully from early childhood until beginning their formal audio training as adults, it is actually understandable why it is difficult to convince many of these students that they need to read their manuals by simply telling them they should. In the end most students do eventually learn the importance of referring to equipment documentation, but there are always a few who carry on with the habit as they enter into the live audio workforce. Unfortunately, it is a very bad practice if not eliminated before students move on to become interns and professionals. That makes it worth strongly emphasizing the importance of reading the manual, and for those that don't take the admonishment seriously the best way to do this is to explain why it's not acceptable to simply ignore the precept.

Professionalism and Due Diligence

The difference between a hobbyist or student fiddling around to learn how to use their own piece of $300 consumer audio (usually done in isolation and when the component in question is still under warranty anyway) is worlds apart from a paid professional thinking it's a cute trick to do the same with a $3000 piece of pro gear that belongs to someone else and is hooked up to a system worth several times that amount. If the party hiring a live engineer wanted a student messing around with their system, they'd have found one. They hired a professional instead and have a right to expect that a paid professional knows how to use the equipment they were hired to run. This is why highly skilled audio engineers with decades working in the audio field still make it standard practice to consult the manual before using new equipment or equipment they haven't used in a while, even though they could figure it out without it. When they forgo it, they do so only because they have enough experience to know when they can be positive a component will hold no surprises and when they should check the manual just to be certain. Many will check the manual no matter what, even when they know they don't need to, simply as a matter of course.

One of the principles separating a professional from a hack is *due diligence*, or the understanding that hired professionals are expected to act with a certain standard of care and to follow the established professional practices of their field. It's simply understood that if you sell your services as an audio pro, and are hired based on that claim, you are contractually obligated to observe professional standards. Reading the documentation for equipment you are unfamiliar with is standard practice because live engineers are paid to know how to use the equipment from minute one. Barring a limited number of circumstances where it is understood and excused for standard procedures to be temporarily suspended (such as when a live audio engineer made a reasonable effort to locate documentation and found none, or is called in to fill in last minute and didn't have enough lead time), when a hired professional fails to observe professional standards there are only two reasons this can be attributed to: either they are unable to do so (if they don't know how or lack the ability), or they are unwilling to do so. Entering into a contract as a pro when you don't know how to be professional or are unable to fulfill your responsibilities is an act of *fraud*, and failing to follow professional procedures even though you know better and could do so if you chose is an act of *negligence*. Whether one is employed full or part time, or under contract as a freelancer, action taken or not taken for either of these reasons amounts to a *breach of contract*. Due diligence is the reason reading the manual is recommended; except in circumstances like those given above that make it excusable to use unfamiliar equipment without referring to the manual, live engineers are obligated as professionals to do what they can to know about the equipment they will be using before they touch it. Most engineers consult the manual to make sure they do; due diligence is simply a matter of professional responsibility and self-respect.

Keep it Covered with Due Diligence

The next reason to consult the manual is a concept that humorously enough is linked to as a "related term" in Wikipedia's definition of due diligence: CYA (cover your ass). Even people who aren't too worried about professional ethics know how to CYA. Doing your due diligence by reading the manual will help you avoid making mistakes of the sort that can cause you trouble or hold you back. Those risk takers who don't care to CYA, usually point out that the troubles are small, if they occur at all, but that's missing the point entirely. CYA is about watching out for yourself in the long term, and in the long term is where you'll reap the benefits. So while it's true that 98 percent of the time failing to read the manuals will not lead the engineer into making a catastrophic mistake, your chances of stumbling into that 2 percent chance of catastrophic failure go up over time. There also can be undesirable consequences that are far more likely than catastrophic failure, even though they too may require the long term before they will be felt.

Even if you are lucky and the new gear you encounter on the job usually holds no surprises, and even if most of the time your skill and ability to figure equipment out on the fly makes up for your lack of preparation when luck fails you, over the long run, failing to familiarize yourself with equipment in advance will still make you prone to minor mistakes that will inevitably have a negative effect on your sound. Even if the worst trouble you may encounter is stretches of bad sound from trying to keep up in real time while mixing on an unfamiliar board and/or the occasional headache from taking twice as long as is typical to troubleshoot issues because you are unfamiliar with the gear, it will be enough to make you look a bit incompetent when it occurs. Considering that part of the job means enough unavoidable difficulties to ensure that even the greatest engineers have "bad days," adding to your own difficulties with additional "bad days" due only to a lack of preparation will be hard to hide; in the short term they will only hurt your sound now and then, but in the long term they will add up to hurting your professional reputation.

While those "in the know" give live engineers some room to have "bad days" because they are aware of the unpredictability of mixing live sound, or because a venue is known for being difficult, most people don't cut live sound engineers much slack. Either way it doesn't take too many "bad days" above what is average for an audio to develop a reputation as inconsistent even if their good days are great. Bad days when the FOH pro can't seem to keep up with a mix or troubleshoot an issue quickly for no obvious reason will hurt an engineer's reputation even more if those good days are anything less than excellent.

The consequences of many small mistakes are far more likely than catastrophic failure, so though you could mix 1000 shows on unfamiliar P.A.s without ever damaging the P.A., it's much harder to do the

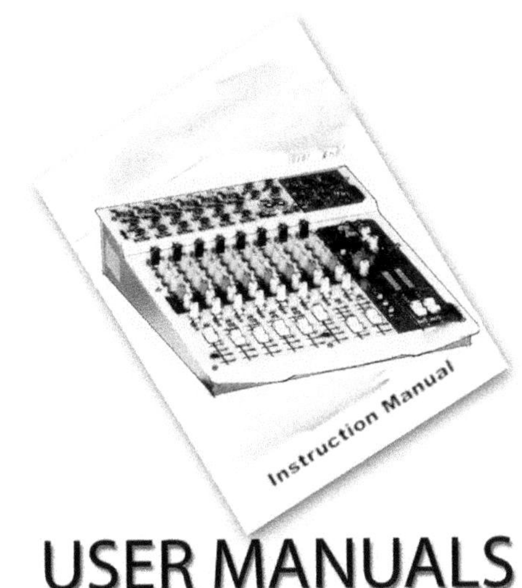

Figure 16.9. RTFM is not only for learning about how the gear works; equally important is reading what the manufacturer specifically says *not to do* when setting it up or using it.

same without a hit to your professional reputation. However, on the upside, if ten or 20 years down the road your consistent lack of preparation finally leads you to overlook something important and make a catastrophic mistake, after a decade or two of running P.A.s blind your reputation won't be good enough to take much of a hit and it also means your income potential will be low enough that, even if you turn the P.A. into a smoking rubble pile, no one would bother to sue you. So read the manual for the long-term benefits too, since neither mediocrity nor catastrophe are good for you career.

All sarcasm aside, we do hope that knowing why reading the manual is important helps persuade those most prone to disregarding the advice when it's given without explanation. The facts are, no matter how long the odds are that ignoring this standard could lead you to damaging the P.A. or messing up a show badly, they are not zero, and no matter how long you can get away with the practice before doing so damages your reputation even if only a little, it won't be your entire career. The reason for due diligence is first because it is the ethical choice, but equally because it protects your employer's interests, and it protects your own interests.

Reading the manual is actually not as big a chore as it might seem from a new audio learner's current perspective. While many manuals for pro gear seem dauntingly large, read the manual actually means consult the manual, but you only need to read specific information and anything that pops out as important while you skim through. It may seem like locating what you need out of those 100 to 300 page tomes that some companies include with their P.A. components has to be difficult, but that's only because at the moment you don't know enough about the P.A. to know what you will be looking for or about equipment manuals to know how easy they are to skim through. By the time you will need to consult the manual regularly, the presumption is that you will know much more than you do now. You will also quickly catch on to the way manuals are written, since most manuals group information in similar ways, and it won't take much practice until you will be able to zero in on what you need much faster; usually you won't need to actually read more than a few pages to get the information you are looking for.

This is only a quick rundown of what you will be looking for, and how much you can skim quickly or should read carefully will depend on your amount of experience using that kind of gear, and the complexity of the gear. You will be able to extract what you need to know from most speaker or amplifier manuals in a couple minutes, but making sure you understand an unfamiliar mixer will always take longer, especially early on when you haven't used many. You may have mixed a hundred shows on an analog board, but if the mixer you'll be using next will be only the first or second different digital mixer you've used since the one you learned on, you'll probably want to read the instructions more closely than you would otherwise. That said, for any piece of gear, you will want to skim for equipment features and for instructions relating to normal setup and usage to make sure it's all similar to other gear in the same category, and to make note of any differences you find so you don't waste minutes figuring out how to set up or use them on site. While you skim, some specific details you will want to find and make note of include:

- Make note of any preferred settings that might be useful, if different than what you would use (such as compressor/limiters/crossover settings) and be sure to note any hard limits or requirements (and to distinguish these from suggested/preferred settings).
- Check what is said about ins and outs and note cabling requirements.
- Note essential parameters for each component (speaker and amp power and impedance ratings, mic polar response, etc.) and any important features (for example if the mic has switchable polar patterns, or

if mixer has on board effects or features such as VCAs).

- Note the scale used for the mixing board's meters.
- Make note of any critical warnings or safety information that is different from those that are standard for that type of equipment. These are usually hard to miss, and will be set apart from surrounding paragraphs, and/or emphasized with one or more methods: italics, bold font, by being written in all capitals, or being preceded by the words "warning" or "attention." They may start with the words "Do" or "Do not", but "always" or "never" are also used. Here is where you will find any surprises that could be critical to not frying the equipment. Usually there won't be anything you don't know, but occasionally gear has quirks specific to that make and model.
- Create a cheat sheet of the mixer's layout to refer to while you learn where everything important is at. (For digital mixers you may need to take a few notes about how to access menus on the touch screen if the mixer uses a method you aren't used to or that seems counterintuitive.)

That's essentially it, though if you need anything more, you'll know what, why, and where to find it by the time you get to the point where you are looking for it on your own.

Locating Audio Technology Documentation

Sometimes you will have easy access to the manuals for the gear you need to learn about. If you are hired at a sound company, they should have manuals for all of their equipment either on file or kept with the equipment and you will want to learn about all the equipment even though you may not be using it immediately. Just remember that as a new worker you may not be aware of all of your company's procedures, so even if you know where the manuals are kept, don't just grab one and start reading. Ask your department lead if you can first; they may have a specific procedure for new hires to check out manuals to learn the equipment. If you are hired to run the sound or work on a sound crew at a venue then ask the venue manager or lead engineer for access to the manual for each of the venue's P.A. components. Even if someone was careless and lost the manual, or you work freelance and can't get the manual from the P.A. provider, you can usually find access to the information you need online.

There will be many times you need to go online and search for documentation of audio equipment. For example, if you are hired as a freelancer for an event, even if the venue or sound company providing the P.A. has the manuals, they may be located too far away to make it convenient to pick up the manuals, or may not want to hand them over to an outsider for fear they will get lost. In that case, simply find out the make and model number of the P.A. components and find the info online. The best place to start is with the company that made the item. Almost all manufacturers have manuals available for download on their websites, and many even keep this information long after an item has been discontinued. If you don't see what you are after posted, don't assume they don't have it. Call or email them to ask how to obtain a copy; often they will have a copy available for those who request one and pay a small fee. Find out the process in case you need to use it, but to save money and possibly time, check all your online options for the manual first. Since requests may need to be mailed with a check and the manual returned via mail as well, also ask how long they take to fulfill requests to make sure you leave enough time for this method; you don't want to wait a week to check for alternative sources only to find out there are none and then also realize you shouldn't have taken as long to send off your request. On the off chance the manufacturer does not have a manual for discontinued equipment, has gone out of business or has been sold to/absorbed by a larger company, or charges more for a manual than you want to pay for it, don't sweat it yet, there are other online options for you to check.

Many online retailers have download pages for the manuals on any of the gear they sell, and for older models that are no longer for sale in stores, there are databases dedicated to archiving manufacturer product info. There are links to many of these in the chapter web directory, but before searching each one individually you might save yourself some time by doing a general web search of the make and model number of the equipment to see if you can get a list of all the web pages that have what you need. If the web search engine that you usually use doesn't get a hit, or only gets hits to archives that charge for their downloads, you might want to try one of the web crawlers linked to in the search tools section of the web directory. Many documents might only be searchable through a specific engine, especially if the equipment is relatively old. Web crawlers are more likely to find listings like these because they allow you to search as many as 100 different search engines at once.

If none of the above works, or only nets you sources that charge you a fee, your last resort for on demand downloads is to check a few of the more popular document upload sites (also linked to in the search tools section of the directory). Some of their documents show up on a regular web search, but many don't, so you will want to check them yourself before giving up and moving on to your next option, or choosing to just absorb the cost of downloading at a pay site. Websites like DocStoc and Scribd reward users for uploading documents, so they are a great place to find all sorts of random documents, including manufacturer documents. They all allow you to search for and save documents, have readers for scrolling through documents and magnifying font, and none of them charge users to browse documents on their site. If you must have a hard copy, most of them do charge you to download or print documents; however, a few will let you download a document free for each document or two that you upload. If you find a manual you need at one of these sites, just find a useful document or two to upload in exchange.

If a component is so elusive that you can't find a manual anywhere, try consulting the audio boards and online communities that are specifically focused on audio gear. Throw a clear description of what you seek in your title and if possible include a good picture of the component in the body of your request for info. If you hit a handful of web boards, nine times out of ten you will get a lead from one of them. There is an entire niche of gear heads who pride themselves in knowing all about any gear good enough to survive in circulation, yet old or rare enough to be missing documentation. This method is definitely not going to be useful if you wait until the last minute to try it, so be sure to allow for enough time for people to respond, which could be as long as 48 hours, though usually 24 is enough. If you get any responses, be sure to thank everyone in the thread for trying whether they are able to lead you to the information you need or not, and especially thank anyone who helps you get hold of what you need. You may need their help again in the future and it will come more quickly and more freely for those who are polite than those who drop requests and never reply to those answering.

The quirky thing about any production is that there is only one way for it to go right, while at the same time, there are so many things that can go drastically wrong and even more ways for the event to merely be ho hum, less enjoyable than everyone expected, even if it manages to skirt total disaster. Murphy's Law states that anything that can go wrong will go wrong—and at the most inconvenient time and place. We don't know if that's true, but we do know that when the job is throwing a great live gig, you need to be prepared for just about any eventuality and packing everything you need is an essential part of that.

Chapter 17

Safe Gear Humping 101

Get It to the Gig without Breaking Your Back

Safety First!

One of the first safety procedures every live audio pro should know about is safe materials handling or, in less delicate terms, how to hump gear without breaking it, or your own back. Though this is a good place to start, it's hardly the end, and with each new set of tasks there are procedures the live audio pro needs to follow to ensure their own safety and the safety of others. Sadly, safety procedures may only be briefly mentioned or not covered at all in some cases, and even where they are known they are often applied inconsistently unless everyone is actively encouraged and periodically reminded to follow proper procedures.

Even for experienced pros who know better than to ignore safety rules, they are all too easy to overlook when people are tired, in a hurry, or distracted by other thoughts, and even more so for inexperienced new workers who haven't been performing a task long enough for safety procedures to become an ingrained habit. Safe practices are the least likely to be followed carefully enough to become a habit by those who don't fully understand exactly why always observing safety procedures is so important, what the risks of a given task are, and how each suggested practice reduces risk.

In this chapter we will look first at several distinct categories of safety procedures of vital interest to live audio pros that will be covered in the chapters of this part and the one that follows. Because the time spent on safety instruction of live audio pros and the depth and breadth of topics covered as part of any course of instruction for live audio are highly variable and the topics covered, as part of internships or the initial on the job training at different types of entry level positions, are even more difficult to ascertain, we feel it's important to provide more safety instruction than just the standard list of dos and don'ts that describe the practices to follow or to avoid while undertaking any task.

To that end, we will begin our examination of the safety of loading equipment in and out of a venue by looking at the associated risks and consequences of the activity. Then, on the way to discussing the specific actions live audio pros can use to ensure maximum safety and minimal risk while engaged in load in tasks, we will use the details to model and explain broader safety topics like engaging in risk assessment by determining the factors that contribute to risk, and the way that details can be used to determine the weight to give each factor and to choose solutions tailored to minimize the most risk for each particular task.

Material Handling: Know Your Risks

Anyone who lifts, carries or holds, and sets down objects on a regular basis is engaging in a *material handling* task and has an increased risk of suffering from a shoulder, neck, or back strain or injury compared to someone who doesn't do this type of task as often. Though some enhanced risk is unavoidable when material handling, learning how to approach each task strategically to minimize risks and to make use of proper *body mechanics*, maintaining an ideal posture and following our bodies' natural range of motion to do tasks, will enable you to reduce this risk considerably.

This is essential as even a minor strain of a back muscle or joint ligament can knock you out of commission for anywhere from a week to as long as a month before it will be healed enough to allow you to return to work. Anything beyond a minor muscle strain, like a significant stretching or tearing of a ligament or of one of the core lumbar muscles, or a disc injury, which are all commonly known to occur when people lift weight without knowing how to do it properly, is no trivial matter.

Workplace Injuries: A Look at the Data

While it's difficult to track injury data for the live sound industry in particular (being so full of part-time workers and independent contractors), a look at US government statistics indicates that without some reason to account for why live audio techs are an exception to the rule then it's fair to assume they should experience injury rates proportional to the rates of back injury recorded in other professions where material handling is part of the job description. A look at the Bureau of Labor Statistics report titled "Nonfatal Occupational Injuries and Illnesses Requiring Days Away From Work, 2011" shows that across all industries and sectors:

- *Musculoskeletal disorders* (MSDs), injuries to the muscles and skeletal frame and joints, also commonly known as *ergonomic injuries* (avoidable injuries or accidents that arise from body motion or position), counted for 33 percent of all workplace injuries resulting in lost work days in 2011.
- Of these MSD cases *42 percent were back injuries*, with *overexertion due to lifting* being the most common cause of the injury, followed by falls, slips, and trips as the next most likely cause. Lifting overexertion is reached one of three ways: lifting too much weight in a single move, lifting a smaller amount of weight at too rapid a rate and/or for too

extended a duration, or lifting any amount of weight at any point while in an unsafe posture or moving in a risky fashion.
- When looking at the professions reporting these injuries the most often, the most affected are those workers who regularly engage in heavy lifting, with the top professions being laborers and freight, stock, and *material movers* (who load and unload trucks and pull or put away stock), as well as nurses, orderlies, and hospital aides (who must lift and reposition sick patients in order to attend to them).

According to the Office of Compliance, the body tasked with seeing that federal legislative branch jobs comply with legal standards:

> *Material handling can result in overexertion, which is the most costly injury in the United States, in terms of workers' compensation. Most overexertion injuries involve the trunk and back ... These strains may be "one-time" events; however, repeated stresses from handling materials can weaken joints, pinch nerves, inflame tissues, damage muscles, and result in chronic illnesses.*

A back injury can cost you only a few days off work, or just as easily can result in months of lost pay. A back injury might only leave you feeling sore for a week or two, but if you tweak yourself badly it could take a year or more of physical therapy to heal properly. And in either case if you don't let the injury heal completely or take care to avoid aggravating or reinjuring it, it could cause you chronic pain for years to come. Learn to lift safely; don't take risks with your health that you may regret later.

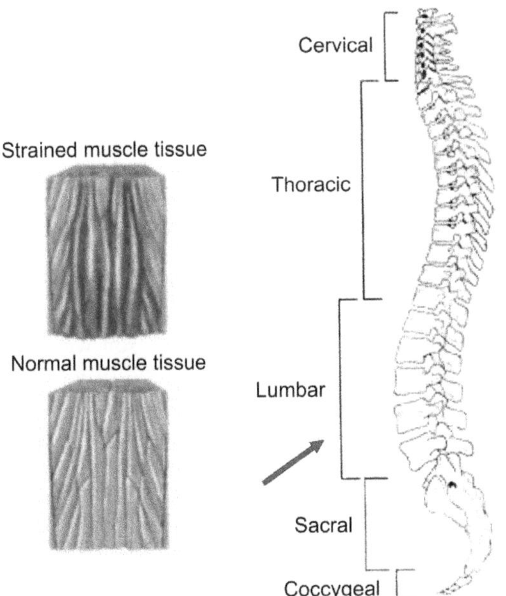

Figure 17.1. Back injuries are no joke.

Considering the data, it's fair to extrapolate that any audio tech or stagehand who engages in regular heavy lifting will carry some degree of greater risk for these types of injuries than someone who doesn't practice the activity as regularly. Even factoring in the fewer total weekly hours spent in high-risk material-handling tasks compared to industry sectors where the greater part of every work week is involved, the risk is still high enough to make material-handling risks and best practices topics that new live audio pros should learn about sooner rather than later. There are also factors that increase the gear-handling risks of audio techs as well. According to the National Institute for Occupational Safety and Health (NIOSH): "Disabling injuries are strongly correlated with job experience: new employees, regardless of age, experience a high and disproportionate number of injuries."

Torn Lumbar Muscle: A Personal Account

When Penny was 19, she injured her lower back while working as an assistant receiving clerk at a large record store in L.A., way back when people had no other option but go to brick and mortar stores to buy their music. As part of the job, she lifted and carried 30–40 lb boxes of books and CDs around half the day each day and despite receiving no training or even any information about methods for safer lifting she never felt particularly fatigued or considered the job hard. Since at least half the time she lifted while bending at the waist, there's a good chance she was accumulating strain without noticing it, but she never felt any—until the day she got injured.

She tore the muscle when she went to lift a 30 lb box while bent at the waist and twisted to one side, only to find that it was unexpectedly almost twice as heavy as she'd thought it was. Had she dropped it then she probably wouldn't have gotten injured. Instead, when she saw she wasn't going to be able to lift it slowly, she heaved hard to get it in place and felt her back go with a sensation that initially felt like tight rubber bands snapping in a row, which quickly resolved into sharp stabbing pains in her lower back with every movement and growing more painful by the minute.

By that evening she couldn't walk and had to crawl for three days. For the first five days nothing she tried was able to reduce the amount of pain she was in, though she could easily increase the amount of pain by staying in the same position for more than 10 minutes; the pain was so intense she got almost no sleep during that time. Fortunately, after the fifth day the pain slowly began to fade and pain pills combined with ibuprofen made it possible to handle the remaining pain well enough to return to work with restricted activity after only being out for five workdays, though she wasn't healed to the point of being pain free for a couple weeks after that. Never before or since has she been in as much continuous, unrelenting, and intense pain as that first week after tearing a muscle in her lower back, but it wasn't until long after leaving that job that she began to get an idea of the long-term effects she would experience as a result of the injury.

She can attest to the fact that they will persist; even if in the first ten years they only bother you once or twice, they will manifest more often as you age. To minimize the injury's long-term impact, it's important to be very careful to observe proper lifting procedures after an injury heals so as not to aggravate the injury, because in most cases you will always be susceptible to re-injuring that muscle. Scar tissue in the healed area of the muscle means it will never be as strong or flexible as it was before the injury, and it will take much less strain to cause re-injury than it did to cause the original injury.

Furthermore, as you age, even if you are careful or lucky enough not to re-injure your back so often that it becomes a constant low-level aggravation in the form of chronic back pain, there is still a good chance the old injury will come back in a smaller version to drop in to say howdy and make you miserable for a week or two (she has aggravated her injury from simply coughing too hard during a bad cold). Such an injury will also leave you more prone to occasional backaches, and will continue to do so for the rest of your life.

Direct and Secondary Consequences of Injury

The *direct consequences* of injury are primarily the direct effects to your body and your ability to work; an injury will cause you pain, and impair your ability function. If you are lucky, the pain and impairment will be limited and amount to nothing more than a visit to the doctor and a day off. As should be obvious by now, if you aren't lucky the pain could be extensive and the impairment could last a lot longer. Another direct consequence is the immediate costs of the injury to you and your employer, including the cost of medical treatment, the cost to you of lost income from days off, and to your employer for hiring a temp to replace you and lost revenue from the loss of productivity that happens when a worker must take days off. These immediate costs add up to billions of dollars a year when all worker injuries are tallied, making on-the-job injuries one of the most expensive health issues the United States deals with.

While these consequences costs are significant, they can draw focus from the fact that they are often only the tip of the iceberg in terms of overall consequences. The secondary consequences of injury are far more varied and many times hidden even from the person who suffers them. *Secondary/indirect consequences* are those consequences or outcomes of an event that aren't directly resulting from the event, but nevertheless can trace back to that event as being the first cause.

For example a back injury won't make you fail a class you are taking, but it will make sitting painful, making it harder to study long enough, and it may interrupt your regular sleep patterns, which could end up affecting your memory. If you were to fail your final and with it the class, it would be a direct consequence of sleep deprivation and inadequate test preparation. but since these conditions would not exist if not for the back injury anything that happens through them can be considered secondary consequences of that injury. Almost any injury is going to have secondary consequences, even if they are something as simple as the grass not getting mowed that week. In many cases, there are more secondary consequences than there are direct ones, since often consequences are like dominoes, as when a disability leads to a job loss, job loss leads to depression, which leads to both a breakup and a bankruptcy. Of course most injuries don't have secondary consequences nearly so dire, but it's far from unheard of.

Then there are the ones hard to predict and even harder to calculate. These losses are never reported, and often not even known by the person who suffers them. Some actually come in the form of what could have been but never will be. Others won't take effect until a decade has passed, for example early arthritis in an injured joint. While it's hard to quantify costs like an opportunity lost or future backaches, these less tangible consequences can sometimes hurt more than even the worst direct consequences and obvious secondary consequences. Whether minor annoyance or major catastrophe, secondary consequences can be roughly grouped as professional, personal, or financial.

Professional

Sometimes an opportunity only occurs once, so even a simple muscle strain or joint injury at the wrong time could alter the trajectory of your career if it knocks you off a touring crew, or ruins your chance to nab one of the top internships offered to your graduating class. In these cases the losses are not only six weeks employment, but also all the opportunities and connections that may have been made but won't be. While other opportunities will come along, they may not replace the ones that were lost.

Also, while live engineering isn't necessarily a strenuous profession requiring the ability to do heavy lifting in the long term, this is only due to front loading the sweat portion of the blood, sweat, and tears it takes to be a successful live audio pro. During the first half of the endeavor of becoming a live audio engineer, a not insignificant amount of gear humping and busting butt is standard, so a back injury that requires extended physical therapy before you are healed enough to work could set you back, and an injury that prevented you from doing so indefinitely could really limit your options. This type of injury would even make opening your own business difficult; if you can't carry your own P.A., you'd have to hire someone to do it for you. For a new business, it can take a while to start making enough money to pay yourself, much less afford to hire someone right away.

Personal

Even if such an injury somehow didn't impact your career, other secondary consequences that can stem from injuries are those that affect your personal life. Many of us have hobbies and other activities that are as much a part of who we are as our professions. An audio tech may be able to make up for being unable to lift for a few months by arranging a task exchange with their boss and coworkers, and one with their own P.A. rental business may need to pinch pennies to be able to hire someone to carry the load for them, but at least it's possible to hire a proxy for these types of activities. But you can't hire a proxy to go to the amusement park, snowboard, or take your vacation for you, to name a few activities that will have to be put aside for a while after an injury, or permanently, depending on the severity of the injury.

It is not uncommon for people who do permanent damage to their bodies through an on-the-job injury to find their personality affected and/or their personal relationships disrupted. This can be nothing more than being a little crankier now and then and feeling more frustrated with loved ones from time to time, but there are many examples of those who experienced a major depression and broke up with their significant others as well.

These are only a couple examples of how an injury can affect your life. Many people find the small personal consequences that stem from injuries far harder to bear than major direct ones. Whether it's short-term consequences like the aggravation of missing a vacation, or an entire ski season, or the long-term inconvenience and pain that come with a chronic condition, there's no question that workplace injuries can have an impact on your quality of life.

Financial

Finally, any injury is going to have both direct and secondary financial consequences. Many of these will be direct like the pay lost for each work day lost. Others will be secondary, like the lost tuition that would be part of the scenario of failing a class. When you work freelance you have to provide your own insurance; if you lack insurance, getting injured is especially difficult, and may create debt you will carry long after the damage is healed and the pain is gone. Even the insured incur extra medical expenses anytime they get injured. The cost for a visit to a general practitioner, to cover X-rays or an MRI, and to pay for ibuprofen and pain pills, if you have no deductible and a reasonable co-pay could be $80; but if you have a deductible or a high co-pay, or if tests and medicines aren't covered, or all of these conditions, then your bill will be closer to $500. And that's just the first day; at the least you'll make a follow-up visit, and very often you'll need to visit a specialist and could need regular rehab with a physical therapist.

That's bad enough, but freelancers also don't get paid sick days, unemployment insurance, or disability insurance either, so each day off will cost you too. Bankruptcy is a long-term secondary consequence of injury and illness that many have endured. Ironically, that leads right back to career impact when so many business owners run credit checks on potential employees.

Live Audio Risk Awareness

The odds are that if this data were tracked, for any audio tech just entering the field the chances of experiencing a shoulder or back injury is as great as experiencing hearing loss, another, much more widely discussed job risk for those in live audio. This is not to minimize hearing damage, but to express the importance of raising awareness about other injuries that are just as likely and only a little less undesirable.

Unfortunately there is currently not much discussion of the risks associated with general physical labor in texts about live audio, whether in print or online, though the custom of paying dues as stage hands or A3s for the newest live audio workers, and the gear humping this entails, is widely known and discussed often enough. However, the industry has only just begun widely championing hearing protection and making the consequences of hearing loss known by those who haven't experienced it firsthand in the hopes the lessons can be learned without needing to learn them the hard way. Material handling is not the only safety topic to get less attention than it deserves in live audio, but our hopes are high that other topics will follow the pattern seen with the topic of noise induced hearing loss, becoming more widely understood and discussed over time.

Weighing the Risk

Risk factors are the fundamental and situational characteristics of any activity that can contribute to the odds of someone sustaining an injury or some other undesired consequence while engaged in that activity. Some discussions of risk describe risk in terms of a *greater or fewer* numbers of risk factors: the more risk factors that exist for a particular activity, the greater the risk (i.e., the likelihood that something will go wrong and someone will end up with an injury).

We think that way of defining risk is an oversimplification, however, because risk is hard to quantify that simply. For example, in terms of the risks of injury associated with the task of gear humping, the fundamental risk factors we use below to frame our risk assessment for the injury risks of material handling tasks can be boiled down to object, posture, force, pathway, and time. These five factors could easily be further subdivided for factors (pathway split into distance and terrain), or condensed (object and force together could be called load). No matter how you break them down, these *fundamental risk factors* are always present to some degree for lift and carry tasks no matter who is lifting or carrying what; they are fundamental to the definition of the task (objects of variable shape and weight are picked up using a variable amount of force, taken along a pathway and set down somewhere else, and the person doing this will assume various postures and take time to accomplish this).

To weigh the risk for any activity, it is not how many risk factors that matters, but how much each of them contributes to the overall risk. The degree to which each increases or reduces the likelihood of injury not only depends on the details of each factor, but also how they interact with one another and with those situational factors unique to a particular event. When determining the risk inherent in any particular activity, the devil is in the details.

Some general risk factors common to material handling tasks include:

- *Weight and shape of objects to be moved*: A compact 20 lb box will be easier to carry than an oddly shaped item twice as large and 10 lb heavier. Ideally, the person who packed the items considered this and mitigated some of the risk posed by this factor in advance by choosing sensibly shaped containers with easy to hold *handles* and *evenly distributing the weight* in individual containers, as well as dividing the weight of the overall load between enough containers so none are too heavy.
- *Force required to move objects*: Items positioned all the way down on the ground or up above your head require more effort and force (energy, torque, etc.) to pick up, and in turn apply more force (mechanical stress like compression and sheer) as they are lifted to muscles and bones than items that are staged on a table that brings them into your body's ideal *power zone*, which is any level falling between your knees and shoulders, and represents the ideal range of height for items to be positioned in order to pick them up with minimal risk. Making sure that items can be *safely stacked* helps minimize this, as do making smart staging choices.
- *Twisting, jerking or sudden motions required to position the load*: Ever turned your head too quickly and strained your neck so that turning your head in that direction at all became painful and difficult? The biggest part of this factor is the sudden muscle contraction that any sudden jerking motion will create—usually in a muscle or group of muscles that are not warmed up or prepared in any way. This alone can create small tears in the surrounding muscle tissue. But when this kind of sharp muscle contraction is created in a muscle or muscles already bearing a load and/or then being extended and stretched with the twisting motion as well, then the muscle tissue has even less room to give (tolerance) and will pull or tear with the added strain of a tight contraction.
- *Contortion required while picking up or moving the item*: This is related to the factor above but covers any

other awkward positions which may or may not create the conditions for a strain. Any job requiring you to get into awkward positions in order to reach and pick up the objects to be moved, especially any requiring the back to be arched while stretching to grasp and lift an object, is more risky. This is what makes getting items off high shelves problematic; not only do objects above our heads require more force to pick up and position for carrying, but often they are just a hair's breadth out of reach and require us to contort our back and shoulders to get a grasp on them first, and both factors increase the risk of injury faced while performing that task.

- *Frequency and duration of handling and holding*: Is lifting and carrying the only task you do all day or one of several alternating activities? The less variety of motion used to complete your job and the more repetitive it is, the riskier it becomes. Even four hours of straight lifting is riskier than separating the same work into two-hour shifts with lunchtime or other work in between. In addition, not taking appropriate breaks or not knowing where to put an item so you are forced to hold it while you consider where to put it are both actions that increase risk by increasing the duration your muscles are under a load.
- *Condition of pathway between where an object is lifted and set down*: Will the ground be level concrete inside a well-lit and dry warehouse or have you been moving over wet cobblestones for 10 feet, wet gravel for 5 feet, five wet cement steps, and 5 feet of wet tiles, before finally transitioning out of the rain and onto a polished, now wet, wood ballroom floor? Obviously the second path is much more difficult as it isn't one you can use a cart on, requires you to transition over several different kinds of terrain, and in the wet most of the surfaces you need to walk over will be very slick. Since the path details and wood ballroom give away the second venue as being the type commonly used for wedding receptions and the size of the P.A. typically used at these isn't as large or heavy as one for a club or outdoor concert, this might be a situation where it would be appropriate to trade out the steel-toe work boots for a pair of super grippy sneakers—with the short path and a light P.A. weighed against an uneven and wet terrain, slipping becomes a bigger risk than dropping a piece of gear on your toes.

In addition to the core kinds of factors like those above, there will be additional *situational risk factors* that have nothing to do with the fundamental task, but can very much alter the amount of overall risk faced on the job. For example, the job poses different risks to workers doing it at 50°F and wearing bulky jackets than to those doing it in 100°F weather and wearing T-shirts. Some situational risks are conditions you and your coworkers bring with you to work, such as each person's overall physical fitness, or even daily physical preparedness. Given the exact same on-the-job-risk factors, you will face much less risk on work days that you arrive at work well rested and full with a good breakfast than work days you arrive hung-over, with only a few hours of sleep and a cup of coffee separating you from a zombie. Even the attitudes of those doing the tasks can serve as risk factors that can add to the risk of everyone doing the task with them. Ego/pride, competitiveness, or a belief that risk taking is a defining feature of masculinity are all attitudes that can increase the risks faced at a particular job.

Knowledge is Power: Risk Mediation

The risk factors above only describe the time you spend humping gear to and fro. Other on-the-job activities you may engage in, rigging, mixing, and even customer service, will each have a different set of risk factors with different degrees of relative danger and even completely different consequences associated with the risk. Nor will the overall risk be the same each time you carry gear; even though the factors used to calculate risk will never change, some or all of the details may change from job to job, or even from day to day. It is important to remember to *assess the risk factors for each new situation* (every new day and new set of tasks) rather than assume the risk is the same each day just because work is done at the same job. In this way, you can make choices enabling you to adapt to changing degrees of risk that accompany changing circumstances.

Sometimes safety requires you to weigh one risk against another and choose to mitigate the one that poses the greater risk. In the example given for considering the pathway you must follow a few sections above, the slippery terrain is more likely to injure you than dropping anything on your toes. In this situation, you are safest to reject the common wisdom of wearing safety boots to protect your toes, so long as you replace them with a pair of extra grippy sneakers that will help keep you stable on slippery wet surfaces.

The point is not to eliminate risk; that's impossible. You risk injury and death every day just by getting out of bed in the morning and putting on your pants. At the same time, on any given day you would actually still be risking injury and death if instead you decided just to stay in bed and lounge around that day without pants—ask any actuary and they'll tell you. You could avoid ladders your whole life for fear of injury and hit your bean on the floor falling out of a chair; every year people of all ages die and seriously injure themselves falling from chairs when they just wanted to sit there (and staying in bed that day wouldn't have helped them either: three times as many perish from falling out of bed each year as from chairs according to Nationmaster.com).

The point in calculating your risks is not to worry about how risky life is; the point is *risk mitigation*—to reduce the frequency of risk and how much damage it can do. If you take 30 seconds or so each day to scan your surroundings and weigh all the risk factors, you then can apply what you know to reduce the risk as much as you reasonably can in any situation, but to do that you need to be aware of the risks first. There is always risk, and you don't want to get trapped into endless worry about what could happen due to risks you can do nothing about. But just because there will still be some risk that you will never be able to control, you shouldn't conclude it's not worth the bother to look for risks you can do something about. If anything it's because there will always be risks you can't control that it's so important to look for those that you can.

Even though you can only manage some of the risk, it's usually enough to make all the difference. If your original chance of injury each day is 15–20 percent and through a couple of simple practices based on the particular risks unique to each day's situation, you bring that down to 3–5 percent, then you have shifted the odds so far in your favor that you might as well be at 0 percent. Sure a freak accident could happen, but it's more likely that you will buy another day. Take care to manage your risks each day and they can add up to many healthy, accident-free years on the job, many more than you can count on just leaving it up to fate and not taking responsibility for your own safety. The catch is that you need to practice safety practices consistently to reap the benefits; one careless day without taking steps to reduce your risk can negate all your carful safety practices up to that point, if you end up with an avoidable injury anyway.

Your brain is your primary piece of safety equipment. Though it can only make about two conscious calculations per second, don't underestimate its power; conscious calculations only account for about 1 percent of our thinking so a more accurate estimate of our brain's ability is roughly 100 trillion calculations per second. An immense amount of processing goes into the split-second choices that we make every day; choices that can make all the difference between someone getting hurt, or another good day on the job. Take a minute to evaluate each new situation, remain aware of your surroundings as you work and you feed it raw data that can maximize its ability to make accurate calculations and help you make the smart choices. But if you use it to let you move through the day on autopilot, you are wasting its power: when you make standby its default state, you and those around you lose many of the advantages it can confer.

Safety Is Mutually Beneficial

An added benefit to taking the proper precautions to watch out for your own safety is that in doing so you also *automatically lower the risk for those around you.* Your choice to take the time to secure your ladder before you carefully climb it means you are less likely to slip and fall from it, but it also means your friends below are less likely to have you take them out on your way down, and your friend across the room is safe from being hit on the head from behind by a falling ladder. When your choice to pay attention to your surroundings leads you to notice that the floor must have recently been polished, that helps your crewmates too, not only because you can then tell them to be careful in the hallway where it's slippery, but also because your awareness will lead you to step carefully and avoid falling, dramatically lowering the odds that the person walking behind you will suddenly be faced with the choice of stepping on you or falling themselves, or have no choice but to trip over you.

Safe Load In

Now that you know the consequences of unsafe practices, and have seen an example of how we assess risk by defining risk factors and looking at the details, you can use this knowledge to assess risk in any situation. Once you know all the risk factors and assess each situation's overall risk you can take the steps that will be most effective in that situation to reduce or eliminate some of the risk.

There are steps you can take to mitigate some of the risks of over-exertion injuries and keep yourself healthy when humping the P.A. to and from the venue.

Take Your Time: Arrive Early

Give yourself plenty of time for load in and set up on the day of the gig. The key to proper loading procedure is to

Calculator for analyzing lifting operations

Company | Evaluator

Job | Date

1 **Enter the weight of the object lifted.**

Weight Lifted

lbs.

2 **Circle the number on a rectangle below that corresponds to the position of the person's hands when they begin to lift or lower the objects.**

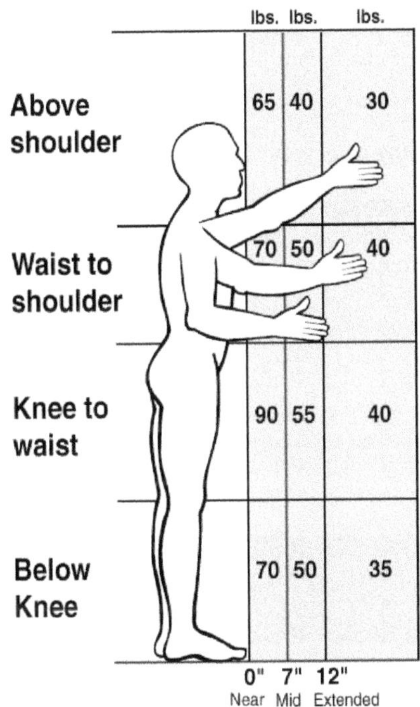

Department of LABOR AND INDUSTRIES

3 **Circle the number that corresponds to the times the person lifts per minute and the total number of hours per day spent lifting.**

Note: For lifting done less than once every five minutes, use **1.0**

How many lifts per minute?	How many hours per day?		
	1 hr or less	1 hr to 2 hrs	2 hrs or more
1 lift every 2-5 min	1.0	0.95	0.85
1 lift every min	0.95	0.9	0.75
2-3 lifts every min	0.9	0.85	0.65
4-5 lifts every min	0.85	0.7	0.45
6-7 lifts every min	0.75	0.5	0.25
8-9 lifts every min	0.6	0.35	0.15
10+ lifts every min	**0.3**	**0.2**	**0.0**

4 **Circle 0.85 if the person twists more than 45 degrees while lifting.** 0.85

Otherwise circle 1.0

5 **Copy below the numbers you have circled in steps 2, 3, and 4.**

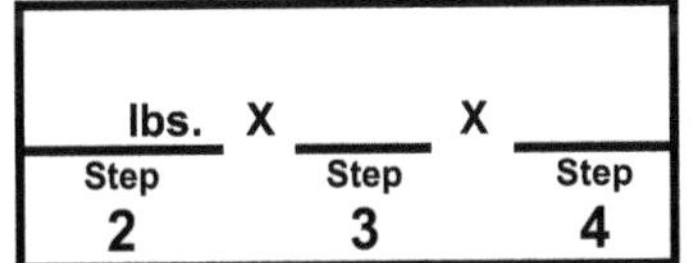

= **Lifting Limit** ____ **lbs.**

6 **Is the Weight Lifted (1) less than the Lifting Limit (5)**

Yes – OK
No – HAZARD

Note: If the job involves lifts of objects with a number of different weights and/or from a number of different locations, use Steps 1 through 5 above to:
1. Analyze the 2 worst case lifts—the heaviest object lifted and the lift done in the most awkward posture.
2. Analyze the most commonly performed lift. In Step 3, use the frequency and duration for <u>all</u> the lifting done in a typical workday.

Figure 17.2. Don't forget to use available resources. There are many free government resources available like this one from the Department of Labor to help you evaluate risks and calculate load limits. Check the online directory for the link to this document and other safety resources.

make your plan ahead of time and think strategically and also to always allow for more time than you think you will need; quality work is difficult to achieve in a rush, as is adhering to all best safety practices. Not to mention that work is also much more pleasant when you can take a break now and then.

Remember that your timely arrival will be pointless if you don't also make sure to arrange for somebody to be there early to let you in and switch the lights on. The same goes for double-checking on any stagehands that were promised by the venue to help with load in. Being shorthanded is the equivalent to being in a rush—both situations make it difficult to provide high-quality work. You can do your part by allowing plenty of time to set up properly, but to ensure your time isn't wasted, always ask for the support you need if it hasn't been offered or arranged for otherwise, and make sure you're specific about all details. Then, directly in advance double-check to be sure everyone remembers their promises and nothing has happened in the meantime to prevent promises from being kept.

Evaluate and Plan: Checking the Lay of the Land

Before you get started unloading the truck, head inside with your load-in crew and get to know your load-in path in advance. This allows you to check for any potential obstacles, difficult corners, narrow doorways, and trip hazards before anyone approaches them with heavy gear in tow. Even if there weren't any when you did your first inspection of the venue, you can't assume there won't be any on the day you need a clear path. Always check to be sure. A quick look around also allows the load-in crew, who may have never seen the venue, to look around and choose a load-in path and an out-flow path that don't intersect, or that intersect at the smallest number of choke points possible. Taking just a moment to do this will save you a lot of traffic jams and jostling as those going out can more easily keep out of the pathway of the traffic flowing in if they agree on the flow of traffic in advance. Finally, a quick look inside allows you to point out where you want different items staged so they are as close to where you will be setting them up as possible, and easy to access, but out of the way until they can be put in place and set up.

Make Sure Everyone's Ready!

Before carrying anything into a new venue, be sure everyone on your team is aware of the proper safety procedures and practices they are expected to adhere

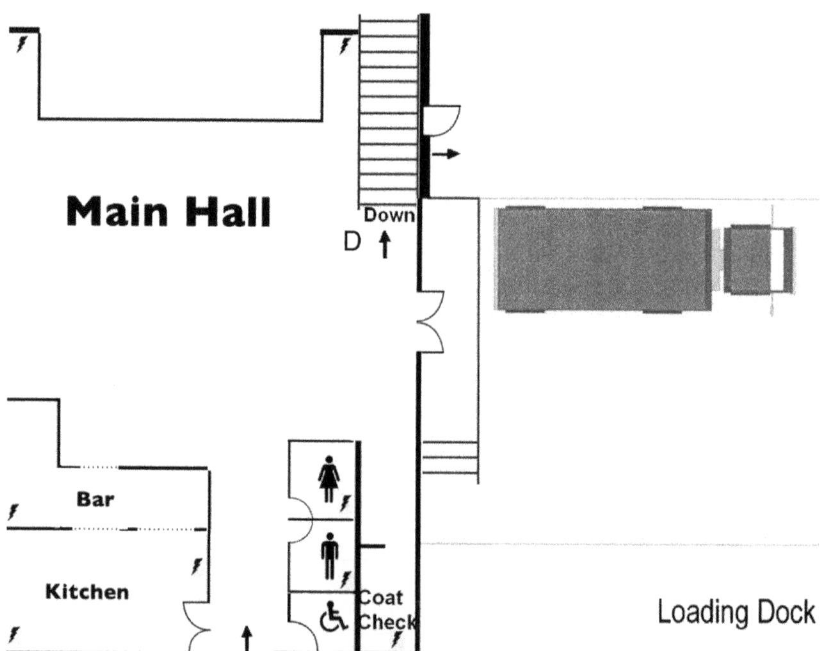

Figure 17.3. Map the route first before you come through with thousands of pounds of equipment.

to. This doesn't need to be gone over every gig, but a run through of the basics with anyone new is essential, and it doesn't hurt to periodically refresh everyone's memory if it's been a while since the last review. A method that can be used in lieu of going over it as a team is to just bring a copy of safety info to every load in and post it somewhere accessible so it can be referred to when needed.

Figures 17.4a and 17.4b. Safety requires you always use the proper tools and wear the proper attire, for the task.

Make sure everyone is wearing proper safety apparel for the task at hand. When loading and unloading that means at a minimum wearing appropriate gloves and boots. You can't mix very well if you smash or scrape your fingers between a flight case and doorjamb. Neither will you be able to set up efficiently if you're limping because gear slipped your grip and crushed your unprotected toes. Protect your hands with no slip-padded work gloves, and wear steel-toed boots and metatarsal guards to protect your feet and toes during loading and unloading.

Lighten the Load

The first rule of hardware handling is to start by minimizing your load. The second is to dress for the occasion. The best way to carry heavy gear is to make sure you can arrange to only have to carry most of it a few feet, and that you have arranged optimal conditions for carrying the rest. Ideally, all large gear you will be unloading will be packed in cases that are on casters. When this is not the case, a wheel board or hand-truck can be brought along to serve the same purpose (use bungee cords to secure items to device if needed). Also check inside the venue; many will have a cart or dolly available for use during loading and unloading.

Use Ergonomics

Body mechanics is the study of the way the body naturally moves and the limits to movement that should be observed to avoid injury. It is a branch of *ergonomics*, which is the study of how to match movement to tasks and vice versa so that tasks are accomplished with most efficiency and the least bodily strain. Practicing ergonomic principles can protect you from injury and make your job easier by reminding you to use your brain first to plan your movements and avoid actions known to cause injury, then by showing you how to best use your body's natural range of movement to utilize your body weight and leverage so your muscles don't have to do all the work.

Don't Force It

The amount of force placed on your back under certain conditions can be surprising. Think of your back as a lever. With the fulcrum in the center of the lever, it only takes 10 lb of pressure to lift a 10 lb object. However, if you shift the fulcrum to one side, it takes

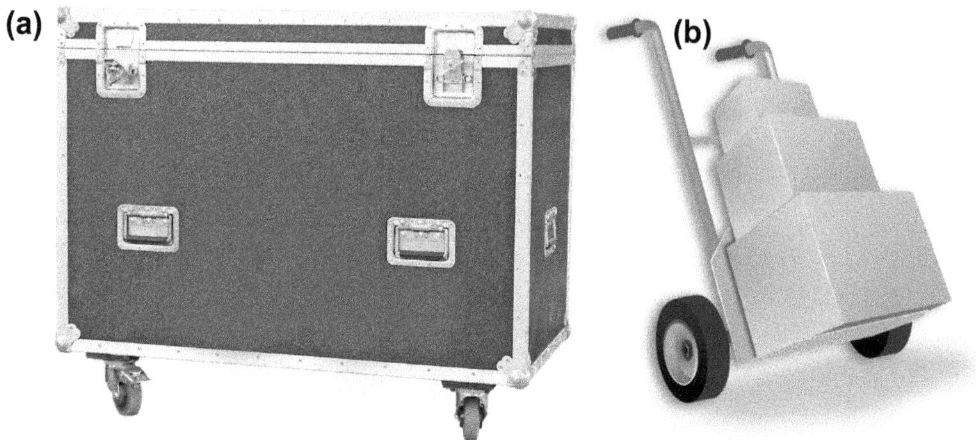

Figures 17.5a and 17.5b. Sometimes there's nothing wrong with taking the easy way out. Putting wheels on your gear is smarter than carrying it.

much more force to lift the same object: now it takes 100 lb to lift the same 10 lb.

Your waist actually acts like the fulcrum in a lever system, but it is not centered; in fact, it operates on a 10:1 ratio too. So lifting a 10 lb object actually puts 100 lbs of pressure on your lower back. That's not all though! After all, when you bend over to pick up 10 lb, there's more weight your back needs to carry than just what's in your hands. When you add in the 105 lbs of the average human upper torso, you see that lifting a 10 lb object actually puts 1,150 lb of pressure on the lower back. That's why anytime you bend or lean over to pick something up without bending your knees, you put tremendous pressure on your lower back and why even leaning forward while sitting at a desk or table can eventually lead to back-related problems.

How to Lift

Squat Lift

- Move close to the object; never reach or pick up items with arms fully extended.
- Maintain a wide base of support by planting your feet shoulder-width apart. This will help reduce the possibility of slipping.
- Squat down bending at the knees and hips with your back straight (its normal shape).
- Keep the lower back straight and the head and neck erect. This further reduces the chance of strain by safely distributing the forces experienced when lifting across the whole upper and lower back.
- Firmly grasp the load on the sides or bottom (below its center of gravity) with both hands and draw the object as close to your torso as you can. If the item is small enough to "hug" you can do so, but be sure not to arch your back; keep your spine straight.
- With your weight centered and your back straight, slowly straighten your knees and waist, lifting with the large muscles of the legs. Using the large muscle group of the legs to bear the weight reduces the forces on the low back.
- Exhale, and tighten abdominal muscles as you lift.

Two-person Lift

The standard two-person lift uses the squat lift. While squatting down, the lead person should first verify that both people have firmly grasped the object at the base and are ready before giving a 1–2–3 count. Upon the lead person's count, both people lift on the 3 and slowly straighten legs. As you move, allow the partner who is moving backward to set the pace.

The Half-kneel Lift

This is for those who can't comfortably squat. Stand close to the object to be lifted, and move down onto one or both knees. Keep the back straight and pick up the object and slide up onto thighs and then hug the object to the body. If on both knees, move onto one knee. Tighten stomach muscles and slowly stand, using the leg muscles.

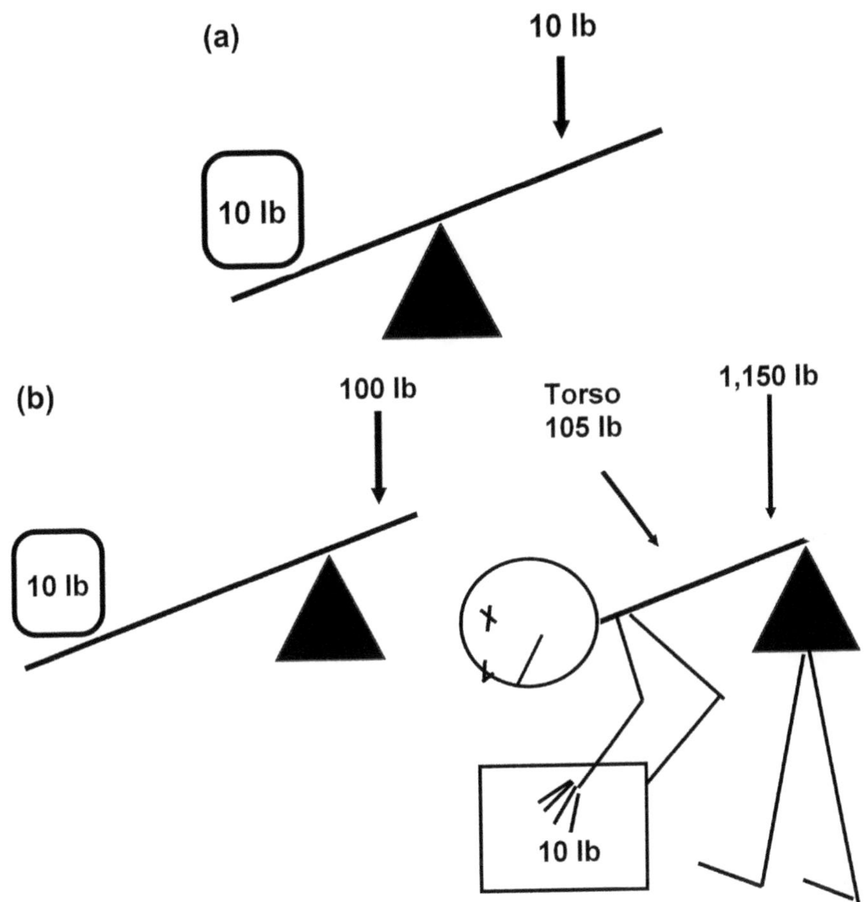

Figures 17.6a and 17.6b. Bending at the waist to lift puts too much pressure on the lower back.

Safe Material-handling Dos

The best way to prevent back injuries is to develop habits that reduce the strain placed on the back. There are some basic things you can do to reduce the risk of injury. To ensure you move gear as safely as possible, observe the following principles:

- Make sure the path to your item's destination is unobstructed by walking it without gear first. Move obstacles and people out of the way to minimize risk of collision or tripping and falling.
- Test the load first. Before you lift, make sure you can lift safely by checking the weight (carefully lift one side or slide the item to get an idea of how heavy it is).
- Ask for help or use a cart if an item is too heavy, too large, or is shaped awkwardly.
- Think before you lift. Have a plan for how you will lift the object, the path you will follow, and where you are going to put it, before you pick it up.
- Hold the load so it will not interfere with your legs while walking.
- Keep your back in its natural curve as you carry.
- Hold objects as close to you as possible. This reduces stress on the back.
- Keep your knuckles above waist level while pushing, pulling, and carrying.
- Push rather than pull. It is easier to utilize your weight when pushing.
- Use your feet (not your waist) to pivot or move through a turn.
- Watch where you're going when moving heavy gear, and move slowly. Even if other people have

Figure 17.7. You can be too punk rock for pants and still think proper lifting is a smart choice, notice the straight back and erect head on each of the three postures.

DON'T TRY TO LIFT MORE THAN YOU ARE ABLE

Figure 17.8. If you can't lift and lower an object into position slowly (but have to heave it into place), or if you can't hold it without leaning back and arching your spine or without your arms shaking...*then it's too heavy!*

obviously taken big chunks out of the wall, you are liable for the cost of repairs if you do any damage that was not there before. Besides it is generally not good for gear, even well-packed gear, to be slammed against door jams if this can be avoided.

- Take regular and periodic breaks to keep from getting fatigued and to avoid subjecting your hands and body to too much strain.
- Place objects up off the floor. If you can set something down on a table or other elevated surface instead of on the floor, do it so you won't have to reach down to pick it up again.
- Maintain good communication with the people you are working with. Let others know you are there if you approach them from behind, and that you are about to pass them before passing. Good communication prevents collisions.
- When two people are carrying an item pay attention to your partner so they don't have to work for your focus. Good communication is required for good timing, which reduces the likelihood of sudden, unexpected, or jerky movements as you lift and carry.

Safe Material-handling Don'ts

When lifting or carrying, be sure not to:

- twist or turn the waist or torso while lifting or holding a heavy load;
- lift or carry objects with awkward or odd shapes or that are too large or too heavy;
- obstruct your forward vision with a load;
- make any sudden jerking movements while lifting or carrying as this could cause a muscle spasm, or upset the load balance;
- heave to pick up an item and get it into a position where you can use leverage to carry—if you have to do this you are lifting something that is too heavy;
- hold your breath during exertion.

Strengthen Your Core

Your stomach muscles provide a lot of the support needed by your back. If you have weak abdominal

Figure 17.9. Be sure to pay attention to your partner! Let the person walking backward set the pace. Do not push as you move or you could push them over. Let them lead you with their tug.

muscles, your back may not get all the support it needs, especially when you're lifting or carrying heavy objects. Good physical condition in general is important for preventing strains, sprains, and other injuries.

Practice Good Posture

When your mother told you to sit and stand up straight, she was giving you good advice. Poor posture is a contributing factor for injuries of all types. Stay aware of your body position at all times. It is best to try to maintain the back in its natural "S" shaped curve. You want to avoid leaning forward (unsupported) when you sit, or hunching over while you're standing.

Stay Limber

Warm up before any strenuous activity; even a brisk walk around the block beforehand is better than starting cold. Gentle stretching is fine if you want to stretch, but

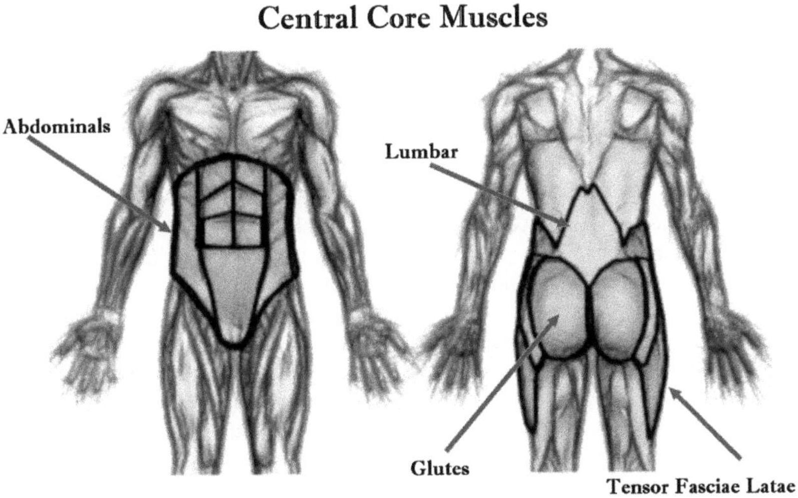

Central Core Muscles

Abdominals

Lumbar

Glutes

Tensor Fasciae Latae

Figure 17.10. The shaded muscles are called core muscles; these are the muscles you need to keep strong for lifting and carrying.

always warm up before stretching; stretching cold can weaken muscles and increase likelihood of injury.

Loosen Up

Try not to carry heavy loads and your baggage at the same time. You can't gnaw on your worries and stay aware and in the moment at the same time. Tense muscles are more susceptible to strains and spasms. Be good to yourself on and off the job and be sure to manage stress to stay safe.

Fuel Up to Stay Fit

It's important to know your body's limitations and strengths and to take care of yourself. You wouldn't start a trip on an empty tank, so don't start your day without fuel either. Get at least seven hours of sleep before doing physical labor, and always eat something for breakfast and lunch; calories aren't just to keep your body going—they are even more important for keeping your mind sharp.

Hardware Placement

Being smart about where you place the gear as you load in is important for several reasons in addition to avoiding injury as you set them down. First, it is necessary to ensure the safety of the people working at the venue, and to avoid fire-code violations. Next it is essential to ensure no damage occurs to the gear or the venue. Finally, it will save you a lot of time during setup if items are placed strategically to be close to where they will be needed, or placed centrally if they will be needed in multiple areas of the venue.

- Never block walkways or exits with equipment.
- Never cover the building's ventilation panels with any equipment.
- Put heavier objects on shelves at waist level, lighter objects on lower or higher shelves.
- Stack equipment against the wall where is not likely to topple over.
- Take the bins with microphones, mic stands, and related items onto the stage.
- Take the stage monitors and bin with snake and monitor cords onto the stage.
- Take the amp rack, P.A. speakers, and tub with related cords and stands and place them on the stage (right or left), or against the wall immediately to the left or right of the stage.
- Place the mixer, rack, and rest of items against side wall of main hall.

The Final Step: Reassess and Regroup

Before moving on to the next stage, and unpacking anything, use natural transition points like this to take a break and sketch a brief plan for the next leg of the campaign. Don't just let everyone start setting up without re-grouping; it may feel efficient but it'll actually cost you time in the long run. Folks should be tired after unloading the truck, and there are a few crucial things you need to do before setting up. Someone needs to park the truck if they haven't already; it is good practice to avoid being parked at the loading dock when your truck's empty as someone may come that you didn't expect and you'll be in the way.

While that gets done, you will want to hunt down the site coordinator to let them know you've cleared out of the loading dock and find out if they have any schedule updates. To explain for any who may be scrolling through the list of point people mentioned so far and wondering who this new character is: at any event or production, there will be someone who will be on site from beginning to end to coordinate the activities of the many people coming and going so that everyone has space to do their jobs and don't get in each other's way. The site coordinator is the one who knows everyone's schedule and plans so people like you and the lighting vendor don't have to know each other's plans; you both report to the site coordinator and they make sure you both have setup time without crowding each other. During the event, the site coordinator is the person who you talk to when you need to talk to someone but don't know who exactly that should be, and only have a minute to spare, or when you need to talk to a buck holder or their proxy.

It's often not anyone new or hired specifically for that task, and in fact it may even be a post manned in shifts, with the stage manager or venue owner serving as site coordinator for the first half of the event (i.e., setup), and the producer or production manager for the event itself. Or it could be the production manager every time; as you know by now, there is some variation in how different music business/audio types like to do things. At tour events it's always the road or tour manager. In any case, they like to know what's going on, so find them, let them know you're done loading and are about to commence setting up, and they will inform you of any schedule changes, for example the lighting vendor got stuck in traffic and needs access to the stage an hour later than planned. You will want to know specifically if the band load in times is the same as last time you checked so you can be ready.

After checking in you can figure out the order you want things done in and assign tasks so everyone who has finished taking a break can get to work. But before you start setting up remember to take a break yourself.

Chapter **18**

The Step-by-Step Guide to a Safe Setup

Once the P.A. is in the venue, you could have anywhere from eight hours to two hours to get everything done that needs getting done. A large complex show requires the whole day to set up for, while a small club show shouldn't take more than an hour and half. But whether it's a huge festival gig, or just some local bands on a Friday night, the day of the show is when all the aspects of a live engineer's job converge: science, art, and communication. To juggle them, the live sound engineer will need all their skills and knowledge to converge as well.

While we will discuss all the tasks the live sound pro may oversee or participate in on the day of a show, whether that pro is the engineer hired to oversee the entire operation, or the stagehand or tech reporting to him/her, they will have a lot to do before the show, and they won't be able to do everything by themselves. In light of this, the first thing any sound engineer must be able to do in order to be ready in time for the show is coordinate and/or delegate tasks, and manage time wisely to ensure all tasks get finished, without any being done twice.

We've already mentioned the importance of bringing along at least two copies of all lists and paperwork, and paper to make new ones that may be needed, and we were not solely thinking of having them "just in case" (though that's good too), but so you can use them to make your life easier. Tape up a copy of the channel lists and stage plan so those helping with setup can refer to them as they work (channel list at the mixer, stage plan and channel list at the stage). This way the stage can be set up and cables plugged in without requiring anyone to track you down and consult you over every move.

In fact, it's worth saying a few more things about checklists now, since we recommend cultivating the habit of tracking tasks and progress. Checklists are vital for an activity like setting up where there is so much information that needs to be juggled with each decision made, and so many different tasks to keep track of at any average concert production, that before we even reach the point where we can discuss what you do with the P.A. once it's powered up, we still have three chapters to go through to describe enough of what you must know to just set it up.

Don't Assume It's Done: Check It Off and Know It Is

Over time you will learn the procedures for setup by memory, from best practices for load in, principles of hardware troubleshooting, efficient cabling practices, and safety guidelines, and you need to, if only so you can reach the point where you can set up efficiently without stopping to consult your notes at every step. But even engineers and techs who can set up any system, in any venue, under any circumstances, will often use checklists to double-check their steps. While most develop an order for setup that works best for them, circumstances often dictate the flow of setup as much as planning does. Luckily, there are several different organizing principles that can be used for efficient setup, but having things thrown out of order does make it easy to overlook items that our memories often tend to jog free in terms of an order of operations.

In addition, except for the smallest portable P.A. setups, most P.A. setup situations involve more than just one person. This makes it hard to keep track of everything during setup. Small details can be overlooked when several team members mistakenly think another team member has already handled an issue. Even if you are the sort to make sure all essential tasks are yours exclusively, setup is a busy time, and it's easy to get called away from a task to help with something that can't wait.

To make sure nothing was overlooked, make up a checklist you can use to tick off tasks as you go, and to guide a final walk through to tick off the rest. At least include a basic final safety check to make sure there are no trip hazards, that fire escapes are all clear, that stacked speakers are secured or speaker stands stable and safe from jostling by the crowd. After this, over time, fine tune your list to add those things that seem to get overlooked till last minute. You might add a double-check of the stage to ensure sound absorption is in place, a double-check to make sure all storage tubs are tucked away safely, and a check that all tools have been gathered back up and returned to tool boxes.

Nor are checklists only for catching errors after the fact. Taking time to streamline your list to reflect an ideal workflow as well as all component tasks, and the checklist can be referred to as you go to help you manage your workflow efficiently, or can allow you to delegate tasks and provide some structure to assist the person as they learn to do the task. Such lists serve an important secondary function as well—by filling them in afterwards you can make them do double duty and have documentation of your activities. For example, if an accident were to happen to an employee on a job you were supervising, having a start-of-day safety checklist filled in (since you will be doing one either way right?) can provide evidence of due diligence and protect against charges of negligence. In a similar vein, having a load out checklist verifying all P.A. items were visually inspected before being packed in their cases can back up your claim if a week or two later a customer does damage and claims the damage was there when the items arrived. To serve this function they need to be signed and dated and kept for a while; nowadays, however, this is easy, since you can photograph it and store it in your smart phone.

Hardware Placement

Placing hardware first allows you to change things without tripping over wires; however, run cables first if equipment will obstruct cable access points later. We examine proper placement of transducers and console placement in the chapters following this one, but in general you should position all equipment and cables before plugging anything in. When lugging and positioning hardware, remember:

- Don't let speaker stands project into traffic flows. If you must put a stand near a flow of foot traffic, use caution tape around the bottom of stand and at eye level to make it harder to miss, or block it off with venue furnishings.
- Never obstruct walkways, stairwells, doorways, or emergency exits with equipment or create a trip hazard.
- It's just as important to keep fire doors closed as it is not to obstruct them. Fire doors keep fires from spreading. Never prop open a fire door for longer than is needed to move what you need to move through it. Never leave a propped-open fire door unattended.
- Speaker stands must be on level ground. Be careful of uneven or sloping surfaces, and use sandbags to secure speaker stand bases.
- Never stack speakers on items that are not designed to be stable and strong enough to carry their weight.
- Make sure all speaker stands are capable of bearing the loads you place on them. Don't raise speakers on unstable or weak supports.
- Never cover the floor vents with gear, or block the air vents on any equipment.
- Never place beverages where they could spill on gear, even if someone is present to attend to them.
- Don't forget to secure loose equipment that could fall on and injure anyone. If you stack it, don't forget to secure it as well. Thread a *ratchet strap* through handles if possible and across the top of the stack and tighten well to prevent boxes from sliding off a stack and onto someone's head. The life you save could be your own. Use stable stands or tables for other equipment as well (mixer, loose effect units).
- Locate power amplifiers as close to the loudspeaker loads as is possible to keep line losses negligible. If this is not possible, choose a wire gauge that is a large enough diameter for the length of the run.
- Place the reverb unit before the limiter, and place the limiter at the end of your effects chain. The limiter will control the absolute volume of your amplifier and prevent it from "clipping," an unpleasant noise that happens when you have exceeded maximum volume.
- Keep tape minimal until you know your setup works. Serious taping shouldn't happen until you've at least done a quick check to make sure all signals are going to where they are supposed to.

Curb Cable Chaos

The exact position of P.A. components is only an important variable under conditions where they will be connected to the rest of the P.A. The system components of the P.A. would forever be lonely islands if not for the sound equipment that ties it all together, the cable.

Cables get a pretty bad rap as being at the root of a lot of unexpected P.A. problems, and it is true, nine out of ten times when a P.A. stops working right, troubleshooting will lead right to a faulty cable. But this could be because no class of equipment is more mistreated, thrown about, yanked, tossed, and trampled upon than the humble cable. This lack of care probably stems from the fact that cables are inexpensive enough to replace, since the same person who might think nothing of tossing a cable 10 feet to land on a concrete floor would never even think of doing that to any other P.A. components. This doesn't make a whole lot of economic sense though. While among the cheapest pieces of audio gear it is possible to buy, the aggregate expense of cables over time will probably equal or surpass the amount spent on the purchase of other pricier piece of gear. It does not have to be this way though.

If you do right by your cables, they will do right by you. With just a little more TLC, your cables will last a lot longer. This could end up saving you hundreds of dollars over time— and that's not even calculating the savings in stress and wasted troubleshooting time from fewer unexpected last-minute failures. In addition, for every 5 minutes spent neatly labeling and wrapping

Figure 18.1. Use caution tape to call attention to trip hazards.

cables now, you save 15 minutes of time on searching, and general untangling later.

For general cable care:

- Mark the ends of cables with colored tape to label them by type (mic cables, speaker cables, instrument cables, etc.).
- Never wrap cables around your arm (over palm and elbow). This stresses and twists the cable, leading to failure. Coil in a figure 8 loop instead (look online for links to videos demonstrating proper coiling). If the welfare of your cables isn't reason enough to stop this bad habit, know that nothing screams audio newbie louder than letting people catch you wrapping cable between your palm and elbow.
- Use cable ties around cables in storage, plug XLR cable ends together once coiled to prevent tangles, and use a spool to keep your longest cables neat.
- Don't put a cable suspected of being faulty back in the box with the rest until it has been checked and repaired. A cable that has failed once will not "heal" or fix itself; if a cable proves faulty, don't use it until it has been fully checked out, even if it seems OK again later—otherwise it may fail again during use.
- Store cables with the gear it plugs into when possible. Cables that don't fit in gear cases, and spare cables should be protected by being stored together, no more than two or three layers deep, in small flat stackable tubs (not shoved in soft bags, or piled on top of each other in crates or larger tubs where the bottom layers get crushed).

Cable Handling

Keep your cables tidy as you set up. Secure all loose cables to avoid trips and wires being pulled out of the sockets. When running cables and connecting cables to gear:

Figure 18.2. Spools will neatly store your long cords and cables.

- Secure cables being run overhead with cable ties or with gaffer tape on the floor.
- Don't run cables on the floor across walkways or doorways. If there is absolutely no alternative then cover cable with a purpose-made cable strip and secure both to the floor with gaffer tape, or secure to the floor and cover completely with several layers of gaffer tape, use a spade to split the turf and bury the cables if on dirt or grass.
- Never run cables on the floor across a fire exit, always run over the door.
- Check that you have power where you need it before you tape power cables down, check that your speakers are working before you tape speaker cables down, and line-test each source quickly before you tape signal cables down.
- Leave a few feet of spare cable (neatly coiled) next to stage gear in case anything needs to be moved.
- Wind the speaker cables loosely around their stands and secure with cable ties to keep them neat and prevent accidents.
- Coil excess cable length about 2 feet from the stage box to keep it easy to access for quick setup changes between bands. Use a cable tie around each coil near the stage box to prevent them getting jumbled.
- As a better alternative to cable ties use reusable Velcro ties on-stage to keep excess cable tidy.
- Don't pull up the cable from the floor with gaffer tape still attached to the cable unless you want to spend hours trying to pick it off later. Instead, hold the adjacent untaped length of cable down while pulling the tape from the cable.

Cable Practices for a Clean Signal and Gear Protection

- Use the minimum number of connectors you can get by with to reduce the likelihood of hums, pops, and buzzes in the audio channel.
- Prevent damage to your speakers. Do not make audio connections while your sound system is on.
- Use the same polarity for all P.A. equipment and stage equipment.
- Use balanced audio lines wherever possible.
- Keep all cabling (i.e., excess lengths of power cables) well clear of transformer power supply units to avoid magnetic hum, (route all 3-core power cables directly away from power supply units).
- When possible, use the same length of speaker cable for both left and right speakers regardless of their location.

Electrical Safety

Risks include:

- Electrically powered audio, lighting, or other equipment which is damaged, faulty, or poorly installed can cause electric shock, burns, or fire.
- Electric shock has a variety of effects on the human body. Mild shocks cause an unpleasant tingling sensation; more severe ones cause muscle contractions, interfere with breathing, and can upset the heartbeat; and severe shocks cause extensive burns and are usually fatal.

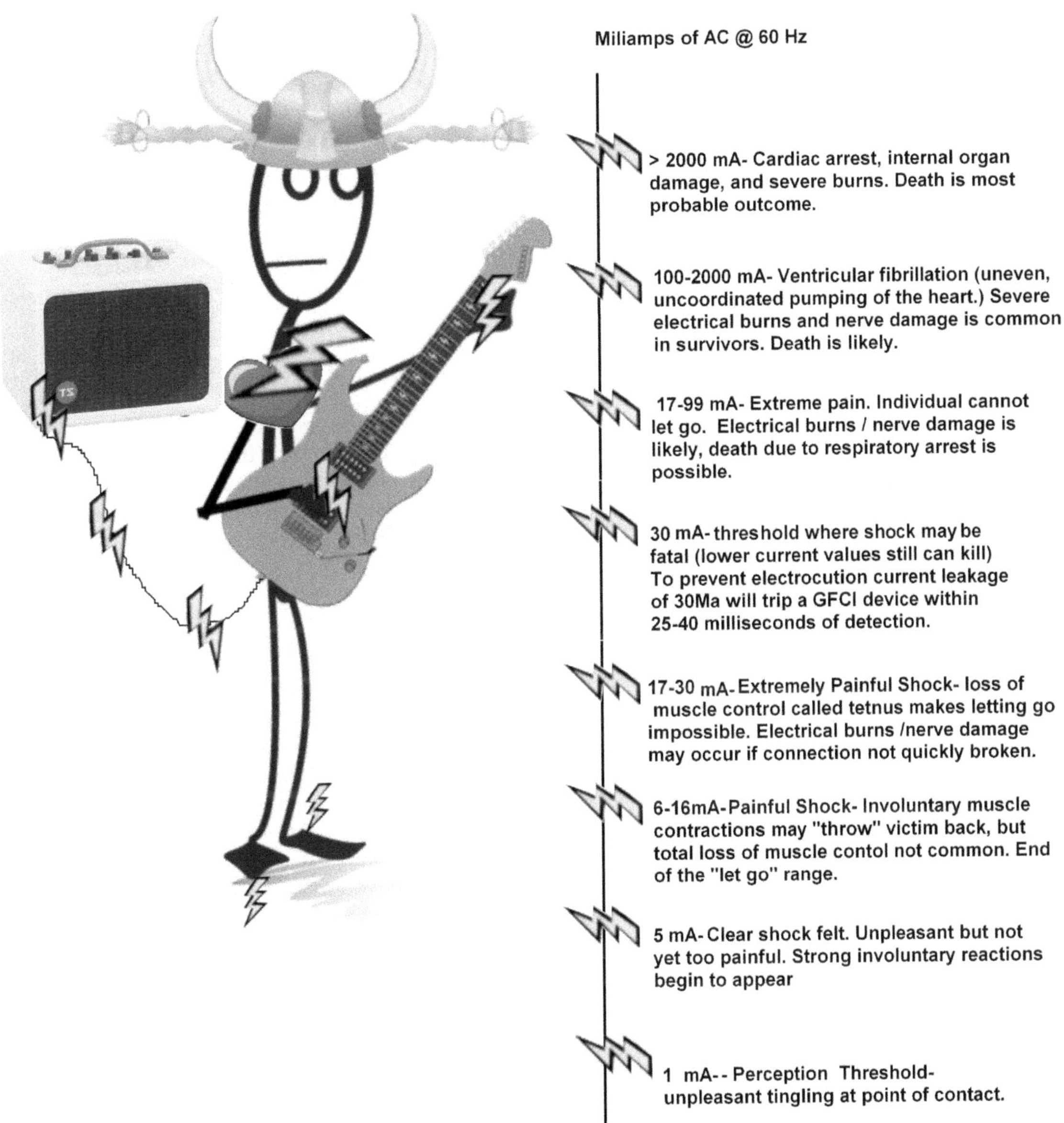

Miliamps of AC @ 60 Hz
> 2000 mA- Cardiac arrest, internal organ damage, and severe burns. Death is most probable outcome.
100-2000 mA- Ventricular fibrillation (uneven, uncoordinated pumping of the heart.) Severe electrical burns and nerve damage is common in survivors. Death is likely.
17-99 mA- Extreme pain. Individual cannot let go. Electrical burns / nerve damage is likely, death due to respiratory arrest is possible.
30 mA- threshold where shock may be fatal (lower current values still can kill) To prevent electrocution current leakage of 30Ma will trip a GFCI device within 25-40 milliseconds of detection.
17-30 mA- Extremely Painful Shock- loss of muscle control called tetnus makes letting go impossible. Electrical burns /nerve damage may occur if connection not quickly broken.
6-16mA-Painful Shock- Involuntary muscle contractions may "throw" victim back, but total loss of muscle contol not common. End of the "let go" range.
5 mA- Clear shock felt. Unpleasant but not yet too painful. Strong involuntary reactions begin to appear
1 mA-- Perception Threshold- unpleasant tingling at point of contact.

Figure 18.3. Electric-shock effects on the human body.

- The Bureau of Labor Statistics reported that 1,100 people suffered burn injuries and 1,480 people suffered electrical shock injuries due to electrical hazards in 2007.
- The National Institute for Occupational Safety and Health (NIOSH) reports that electrocution is the third leading cause of worker fatalities in the United States.

Even if you don't have access to a battery of calibrated tests, that doesn't mean you should forget about power safety. In fact electrical safety should be a habitual practice that occurs every time you plug something in, starting with a quick glance at the cord and plug to check it is sound before plugging it in. More problems are found by simple inspection than by testing. Visual inspection doesn't require special tools, beyond your eyes and common sense; if you can see something wrong with it don't use it! Always keep an eye out for unsafe equipment, and don't be afraid to speak up if it isn't yours. No matter who the gear belongs to, if it finds its way into your signal chain, or is pulling power too close by, its condition is your business.

To make sure you keep things safe:

- Always use a safety cutoff circuit, such as a ground fault circuit interrupter on extension cords running from power that you can't verify is grounded.
- Use a power conditioner when possible, or at least use power strips that offer surge protection.
- Fully extend each cable and check for breaks or kinks in the wire before using them.
- Check power cables frequently for signs of damage, and remove them from service for repair or replacement when damage is found.
- Use proper professional grade extension cords, of the right gauge for the application.
- Always use appropriate gauge speaker cable to plug into your speakers.
- Remove cords from receptacles by pulling on the plugs, not the cords.
- Use double-insulated tools and ground all exposed metal parts of equipment.

Ground Fault Circuit Interrupters

GFCIs are products designed to prevent serious injury or death from electrical shock by detecting ground faults at very low levels. If a GFCI senses minimal current leakage to ground in an electrical circuit, it assumes a ground fault has occurred and interrupts power fast enough to prevent serious injury from electrical shocks. GFCIs are now required by code in certain areas of the home, including kitchens, bathrooms, bedrooms, and anywhere that water may come in contact with electrical products. GFCIs are readily available, inexpensive, fairly simple to install, and are designed as wall receptacles, circuit breakers, and portable plug-ins. *Remember to test your GFCI units monthly and after every major electrical storm.*

How to Test Your GFCIs

Be sure to test your GFCIs once a month to make sure they are working properly:

1. Push the "reset" button on the GFCI to prepare the outlet for testing.
2. Plug an ordinary nightlight into the GFCI and turn in on. The light should now be on.
3. Push the "test" button of the GFCI. The right light should turn off.
4. Push the "reset" button again. The nightlight should now go on again.

If the nightlight does not turn off when the "test" button is pushed, then the DFCI is not working properly and may not protect you from shock or electrocution. Contact a licensed, qualified electrician to check the GFCI and correct the problem or replace the GFCI if necessary.

Common visible indications of trouble and practices to avoid are:

- Do not use equipment with missing screws and/or damaged casings, guards, or grills. These design elements are intended to prohibit contact between salty, conductive human skin and electronic components with enough live current to harm or even outright fry the owner of that skin. If you see an item like this, don't use it, and don't allow anyone else to use it.
- Do not use cables with plug casings that are cracked or damaged, cables and wires where the inner insulation or a bare wire is visible, or cables that appear crushed, badly twisted, or stretched. These should be removed from use.
- Do not use cables that have the kinds of fixes commonly used as "emergency" repairs (such as any that use tape or glue to fix cracked cable insulation); these are not safe and should not be used. Equipment with such fixes should not be stored with working equipment.

Figure 18.4. Power conditioners are ideal, but at least make sure all power strips have surge protection switches.

Keep the Prong

When you cut off the ground prong or use a cheater plug so you can plug a three-prong appliance into a two-prong outlet, you disable an important safety feature that protects you from electric shock if a hot wire comes loose inside an appliance you use. If that appliance has a metal case and the loose wire touches it, then the metal case is now hot, and anyone who touches it will get a potentially fatal shock. When the case is grounded, the electricity from the hot wire flows straight into the ground, and this trips the fuse in the fuse box. The appliance won't work then, but luckily your heart will.

- Do not use equipment in which fuses have been bypassed, or where items like screws, wire, or fuses of a different rating are being used as fuses. Plugs that have fuse ratings should use corresponding fuses. The same applies for any appliance-based fuse-points.
- Ground blades on plugs (third blade on three-prong plugs) should never be removed. If you must use a two-prong outlet as a last resort, use a cheater plug adapter, but refer to the test box in the right column first to see why that's still not recommended.
- Cables at entry or exit points in equipment casings should be securely clamped over the outer insulation. A plug's insulation should not be cracked or peeling. All cable conductors should be secure: screws should be checked fairly often.
- Don't use extension cords or power strips "daisy chained," or interconnected to reach greater distances or power more than the approved number of outlets. The resistance increase could cause them to become overloaded, creating a fire hazard.
- Never work in or near water with powered equipment or open the service panel while standing in water.
- Never plug in while hands are wet or sweaty.
- Cut unsafe cords to ensure no one accidentally uses them.
- Ground wires for equipment should never be disconnected regardless of noise.

There are a number of effective ways for dealing with ground hum, but removing a power ground is not one of them.

Shocking Facts You Should Know

A few more things you should know about electricity are related to shocks, or how to not get them:

- The best protection against shock from a live circuit is resistance.
- Resistance safety gear like leather gloves and rubber-soled boots help prevent the wearer from completing an electrical circuit to the ground.
- Resistance is additive, so boots and gloves together is better than either one alone; because a current must pass through the boot and the body and the glove your resistance will be the combined total of your resistant gear between the current and the ground.
- Conversely the easiest way to decrease your resistance is to get your skin wet. Therefore, don't touch or plug in electrical devices with wet or sweaty hands or wet feet.
- Metal jewelry is conductive, so remove it when working around electric circuits. Metallic items like rings or watchbands make excellent electrical conduits to your body, and can conduct enough current themselves to burn you even with low voltages.
- Be *especially* careful if you are sweaty. Sweat is essentially salt water, which is as close to a perfect conductor of electricity as you will ever have on your hands.
- Current will affect any muscles in its path. The path a current takes through the body makes a difference as to danger—hand-to-hand shock paths cross the chest making them one of the most dangerous.
- To guard against hand-to-hand shock use only one hand to plug into a live circuit and keep the other

hand tucked into a pocket so as to not accidently touch anything.

Two Must-buy Test Items for around $20

Any new live sound pro who regularly sets up and/or mixes at different venues should buy these two items. They could save you from frying yourself, your guitarist, or your gear, from hooking up to unfamiliar gear, or plugging into a badly wired circuit that's not well maintained. While you will eventually be learning to use more sophisticated test gear, you'll still want to keep these because when all you need is to test for one thing they are the easiest way to do it. Besides, together they won't put you back more than 20 bucks and they are idiot proof, so just plug in or touch and note what the light/s do.

- Invest in some form of socket tester such as the Buzz-It Check Plug from Martindale Electric. This simple device will check for the presence of ground, make sure the wiring is not backward (another common and potentially dangerous fault), and depending on the tester may test for other fault conditions. For example the Gardner Bender 120 VAC GFCI Outlet Tester; 1/clam, 5 clams/master will test for seven fault conditions (ground fault interruption, open ground, open neutral, open hot, hot/ground reverse, hot/neutral reverse, and correct wiring) and is $8. If you work with a band who is gigging regularly, it's a good investment.
- Another device worth investing in is a simple LED voltage tester, like the Gardner Bender 12-600 VAC Adjustable Circuit Alert Non-Contact Voltage Tester, a little over $12 at most hardware stores. These usually look like a fat pen and contain a high value resistor and a LED bulb—just touch the pointy end to a metal object (microphone, guitar strings, amplifier case), and the other end with a finger and if there is a dangerous voltage present, the bulb will light up. Some don't even need to touch the object, just come within an inch or two.

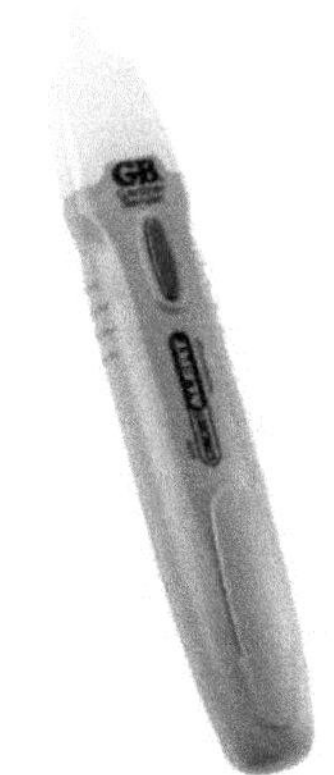

Figure 18.5. A voltage tester is a must for checking amplifiers and mics, the two biggest shockers in audio.

In an Emergency

Despite multiple layers of electrical safety rules injuries still occur. Most of the time, these result from people not following proper safety procedures. However they may happen, they still do happen, and anyone working around electrical systems and people should know a little about how to care for a victim of electrical shock, at least long enough for the paramedics to arrive.

If you see someone lying unconscious or "frozen" on the circuit:

- *Don't touch them* unless you are sure they are not touching anything hot, and check with the back of your hand first if you get frozen to them you won't get to be a hero—they'll just need an extra body bag.
- *Option 1*: Shut off the power by opening the appropriate disconnect switch or circuit breaker at the main distro box. (You should always know where that is, and check to be sure it's not locked). However, if a shock victim's breathing and heartbeat are paralyzed by current, or there is enough of it to "cook" them in 30 seconds, their window of survival may be short; so if you are far from the box, don't know where it is, or happen to have a power cord or broomstick right at hand and can do option 2 faster, then do that.
- *Option 2*: Use a dry board or broomstick to knock them away from the current, or use an insulated extension cord like a rope to throw over them and pull them away. Be prepared to put some muscle into it if they are "frozen" because they'll be gripping it with all their power. If what they are gripping with all their power is a microphone or guitar neck, don't try and pull it live from their hands, as it could bounce back at them after you free it and zap them again; instead yank the plug out (in this instance you may yank). Only do this If you can do this more quickly than the first option; if there isn't something nonconductive handy to use, don't waste time looking for one—go turn off the power.

Once the victim has been safely disconnected from the source of electric power:

- Check for breathing and a pulse. If weak or none call 911 immediately and give cardiopulmonary resuscitation (CPR) as necessary to keep the victim alive, or yell out to find if someone else present knows CPR. The main rule with CPR is, once you start, keep going until you have been relieved by qualified personnel. (If you don't know CPR, go learn; it costs no more than 50 bucks and a day of your time, and I can promise you won't regret the time and expense it took to learn it).

- If the victim is conscious, check with them to see if they have any burns. If yes, treat them for burns (check the directory for more information). If not, have them lie still until qualified emergency response personnel arrive on the scene. They could go into shock and should be kept as warm and comfortable as possible.
- An electrical shock insufficient to cause immediate interruption of the heartbeat may be strong enough to cause heart irregularities or a heart attack up to several hours later, so the victim should pay close attention to their own condition after the incident, and remain under supervision for the first 4 hours.

Thankfully the above scenario will probably never happen to you. However, because anyone working with gear is at some risk, both you and the bands you work with need to know power safety. If you at least consider the possibility as the above section hopefully forced you to do, the idea is that in the unlikely event it does occur you will be more likely to recognize what you are seeing and not waste precious seconds processing it first.

Now that we've discussed how to safely carry in gear and safely plug it in, we will look at precisely where to put the most critical pieces and how to connect everything up, then move on to Part 5 to talk about what to do with the P.A. after it's fired up and running.

Chapter 19

Good Sound Starts Here

Microphone Setup for Live Production

Whether you are house sound simply making minor adjustments, or a visiting engineer locally or on the road, any time you set up or adjust your P.A. system, there are some choices you have to make. As you will see when we discuss *gain structure*, getting the best system sound and the most from your P.A. starts before a single cable or plug is connected. Proper placement of the major elements of the *signal chain* should always be done strategically with the following goals as a road map to follow on the road to optimizing your sound:

1. Get as much direct sound to the audience as possible.
2. Get the level and tone as constant as possible throughout the audience.
3. Keep reflected sound to a minimum.
4. Keep unintentional direct and reflected sound away from microphones (*mic isolation*).

Microphones

When it comes to micing, there are many ways to skin a cat. The right technique to use is the one that produces the best sound for the situation. Some mic positions are more obvious choices than others, but in many cases there is no one ideal. As with so many choices the sound engineer makes, it's about settling on the tradeoff that's best for the sound and the situation. Any tips for specific placement are just guidelines to help you start making positioning choices. With time and experience, you will develop your own unique setup tricks that may

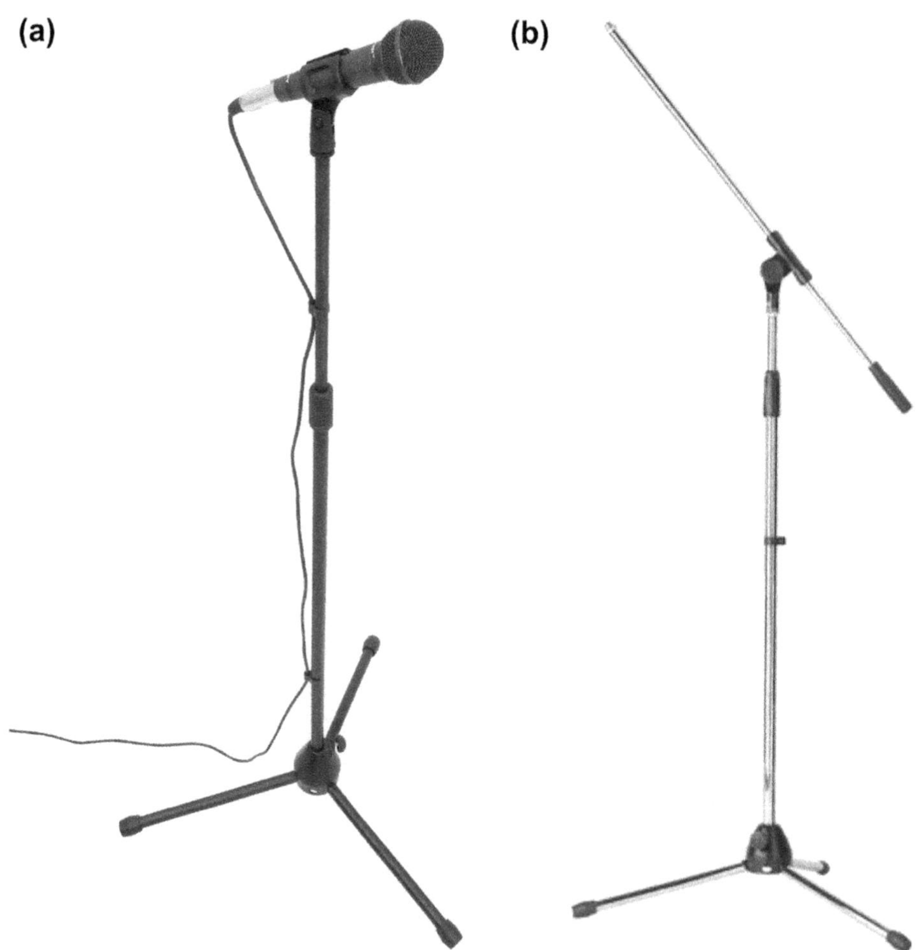

Figures 19.1a and 19.1b. Mic stand and boom stand.

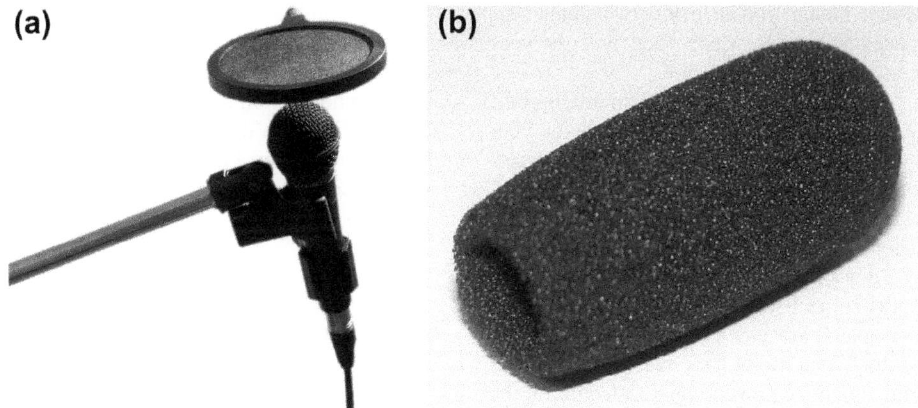

Figures 19.2a and 19.2b. Pop filter and wind screen.

include some of the ones included here, but will also include those among the variety of possible options that are most suited to you. First, some straightforward information of use to anyone who may need to mic a source, regardless of the method they choose.

Stands, Mounts, etc.

The microphone stand comes in three categories: the straight vertical stand, the boom stand, and the tabletop stand. Most vocalist mics are vertical stands, but the rest of the time a stand is used, it's going to be a boom stand. Proper use of a boom stand includes:

- For maximum stability, always extend the boom so its position runs directly over one of the stand's legs.
- Don't wrap the cord around the stand more than once around base and boom arm. Coil excess neatly at base, or even better at the other end.
- Don't over-tighten clamps. Tighten until firm, but don't use force.

Clamps offer even more versatility. Clamps can piggy back onto general-purpose stands and come in an array of designs that allow them to attach to almost anything else. Facts to consider when choosing between clamps or stands for a particular task include:

- Clamps free up floor space.
- Clamps are easier to transport.
- Clamps allow access to positions and angles not accessible otherwise.
- Clamps are more difficult to set up and move under time pressure.
- Clamps are more prone to vibration noise bleeding into mix.

Mics, etc.

- Pop filters are typically circular open cell with material placed between the vocal source and the mic. Used to combat sudden explosions of air pressure called "plosives" created by certain consonants like "P" and "D." These can cause pops or thumps in sensitive microphones. A pop filter will stop some of these pops or thumps. Well-trained vocalists also have learned to move the mic away or sing plosives to the side.
- Wind screens are usually foam coverings that go over the entire capsule and attempt to filter out wind blowing across the diaphragm of a mic, causing extremely loud unwanted pressure variations. These suppress breath pops better than a metal grille screen and also serve to insulate the guitarist from shocks.
- Most microphone manufacturers now offer miniature condenser mics often with excellent sound quality. They allow for close micing of instruments better than any other option and thanks to Moore's Law they are increasingly affordable.

Micing the Stage

When micing the performers for a live application, there are a number of choices to make and considerations to keep in mind. In most cases you will have to deal with certain limitations and also be sure that your choices are appropriate to the situation. Some of these limitations and considerations are:

- *Number and type of mics available.* Much as you may like to mic every drum and stereo mic every other instrument, often you will have a limited number of mics to work with. You will also be limited in the number of each type; if you have only two condensers you will need to decide where they will do your sound the most good.

- *Number of mixer channels available.* In small clubs especially, even if you have enough mics for your preferred setup, you may not have the enough channels, especially if you are monitor mixing from FOH.
- *The room.* It's always important to make your choices with the acoustics and requirements of the venue in mind. If you don't have enough speakers and amps to make sure the entire listening area is covered by left and right speakers, are you sure you need to stereo mic everything? You might have reason to, but then again it might not be reason enough to account for the trouble. And if the room only seats 50, do you really need to mic every drum? Venues may not be ideal for every setup; there are times when you have to put aside what you want, in order to provide what the room and the audience needs.
- *The band and genre.* If the band has indicated a preference in mic selection or placement, you shouldn't override their preference for your own without consulting them first, unless you must make a choice based on what's best for the room and audience or have another very good reason (beyond preference). If you think a configuration is not feasible in that venue because it would create feedback, do what you must; if you think your way would sound cooler, only change the setup requested in the contract riders if you can convince the band to give you the OK. Not all bands expect a specific setup; check the tech rider, channel list, and stage plot. Many bands will suggest a setup, but will indicate somewhere on the rider which are must haves, and which they don't mind being changed.

Choosing Mic Number

- Unless you have a reason to be micing for stereo sound, it is generally best to use one microphone or pickup for each sound source. Common exceptions are noted in discussion of specific miking techniques.
- When using more than one microphone in close proximity you need to be careful of polarity cancellation. Due to the way sound waves interfere with each other, problems can occur when the same sound source is picked up from different mics placed at slightly different distances. If a signal sounds weaker through two mics than through either mic individually, it is likely due to the signals being opposite polarities. (This has the same effect as when in-phase and 180-degree out-of-phase versions of the same sound waves combine and cancel each other.) Reversing the polarity of one mic signal should fix the trouble. The polarity-reverse switch can be found either on the microphone or on the mixer channel strip.
- Every time you double the number of mics on stage, the system loses 3 dB of potential acoustic gain (i.e., for every extra mic added, the volume level of the system must be reduced in order to prevent feedback).

Choosing Mic Distance

- The Inverse Square Law holds true for mics as well as speakers. So each time you reduce the distance between the microphone and the sound source by half, you increase the sound pressure level of the microphone (and the system) by 6 dB.
- To minimize interference from other potential sound sources (called spill or bleed) and increase *potential acoustic gain* (meaning the system can produce more gain before feedback), the microphone should be located close to the sound source.
- *Proximity effect* occurs when a voice or instrument is close to the mic (within less than a couple inches or so) and results in a boomy, bass-heavy sound.
- When using multiple microphones each microphone should be three times further away from other microphones as from the sound source. (A microphone that is 1 foot from a sound source, should be 3 feet away from the next closest microphone). This provides a margin of about 10 dB between adjacent sound sources which provides adequate isolation as well as reducing the likelihood of comb filtering effects. This is known as the *three to one rule.*

Choosing Tonal Quality

- The angle of address shapes tone. In general you'll get the brightest tone by aiming the mic directly at the sound source. Angling the mic in relation to the source produces a softer, mellower tone. Be careful though! Recall that some mics are top address, others are side address, so always know your gear by making sure to read the manual.
- Instruments that create sound over a large area don't usually sound as natural when close miced with a single mic. With these instruments a close placement can color the sound source's tone quality or timbre by picking up only one part of the instrument. If you want a sound that is more natural you can move the mic further away or add another mic. (Either

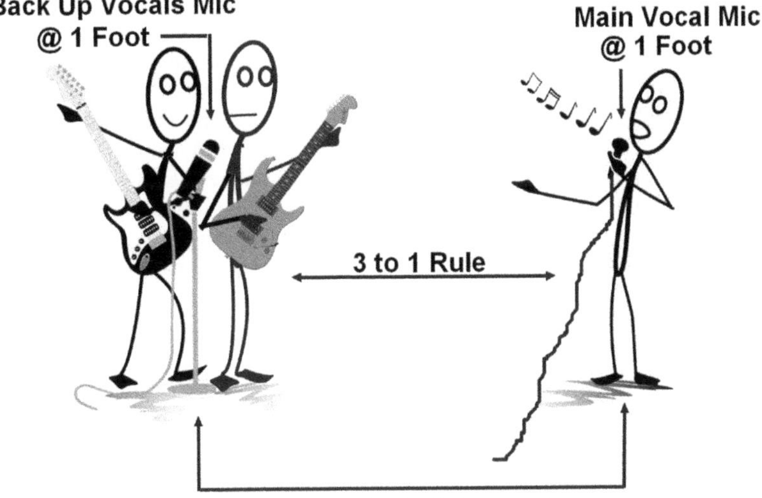

Figures 19.3 and 19.4. Two applications of the 3 to 1 rule.

choice has consequences—refer to the sections on distance and number above to weigh your options).

General Considerations

- Experiment with a variety of microphone and positions until you like what you hear. You don't always have to conform to standard ways of doing things. As long as you're not placing a microphone in danger there's no reason not to use them in unusual positions.
- Use a microphone with a frequency response that is limited to the frequency range of the instrument, or filter out frequencies below the lowest fundamental frequency of the instrument.
- Only use as many microphones as necessary to get a good sound. For each additional mic used the

amount of noise in the signal increases, along with the likelihood of interference effects such as comb filtering.
- If you can't get the sound you want, try different microphones and microphone positions before reaching for the EQ.
- With acoustic instruments it is generally preferable to use only one input method (pickup or mic).
- Aim unidirectional mics on-axis—toward the sound source and away from undesired sources (180 degrees off-axis for cardioid, 126 degrees off-axis for supercardioid).
- Place mics as close as possible to the sound source and as far as possible from unwanted sound (loudspeakers or other instruments).
- Large-diaphragm dynamic mics are a favorite for micing drums, bass drums in particular, because of their superior low-end response and good sound isolation.
- Dynamic mics tend to be less sensitive to off-axis pickup.
- Condensers are the rule-of-thumb for high-frequency detail on acoustic instruments.
- Dynamic mics handle high sound-pressure levels (SPLs in excess of 150 dB), plosives (the blasts of air associated with vocalizing the letters p, b or t), and wind better than condensers.

Choosing Mic Placement

There are many authoritative sites online that cover pretty much every type of method that can be used to mic each instrument, as well as just about every type of instrument you are likely to encounter. We could not hope to cover the same information here, so we will stick to the instruments you are most likely to encounter in a live sound environment, and the methods most likely to reduce feedback issues while providing solid sound. While we indicate the mics most often used for each application, in reality most musicians start buying gear to suit their sound before or very soon after they start gigging and will have their own mics for you to use. On occasion you will need to choose a mic for an unfamiliar application, and on these occasions, as well as if you should ever choose to buy mics to increase your on hand selection, turn to the link directory for a list of very reputable and helpful websites to aid in your research.

Vocal Mics

- *Lead vocals*: Ideally, handheld vocal mics are best kept around 5 inches from the mouth. In a noisy live sound environment, this distance can shrink to only an inch or so to ensure the performer's voice can be heard in the mix. At this distance, however, some mics produce low-end distortion due to the proximity effect. Work with each singer to determine the mic technique and mic selection that works the best for him or her (dynamic or condenser, cardioid).
- *Harmony vocals*: Same as for individual mics. If two singers are on the same mic their mouths should be slightly farther away and at roughly a 45 degree angle to the mic (dynamic or condenser, cardioid).

Instrument Mics

- *Acoustic guitar*: Aim directly at the bridge, not the sound hole and get as close as possible. Place microphone 6–12 inches from where fingerboard joins the body and aim toward sound hole (condenser, cardioid).
- *Guitar or bass amp*: Place the mic about an inch away and at a slight angle to the grill, aiming between the edge and the center of the speaker cone (dynamic, cardioid). Alternately use a miniature mic draped over the front of the amp targeting the same location.
- *Brass*: Place microphone 8–24 inches away, on an axis with the bell of the instrument (dynamic or condenser).
- *Saxophone*: Place the mic on a stand, roughly 3–5 inches above the bell and angled downward. Note that micing close to the finger holes will pick up key noise (dynamic or condenser).
- *Acoustic bass (upright)*: 4–6 inches, aiming halfway between the bridge and one of the F-holes (condenser, cardioid).
- *Grand piano*: Two mics placed inside, one pointing down, one at the middle of the bass string section above the hammers, and the other over the high strings (condenser, cardioid).
- *Accordion*: Use two mics, one for the left-hand and another for the right-hand keys.
- *Percussion*: Use two mics. Experiment—move the mics around and see what sounds best to you.
- *Keyboard amp*: Mike speaker same as guitar/bass amp (dynamic, cardioid). Alternately, run line from "line level out" on keyboard amp directly to a line input on the mixer.
- *Synths, drum machines, etc.*: Run a balanced line to mixer via a DI box.

Drum Mics

- Don't cage the drummer: Be careful when micing drums that the drummer can drum freely without hitting the mic.

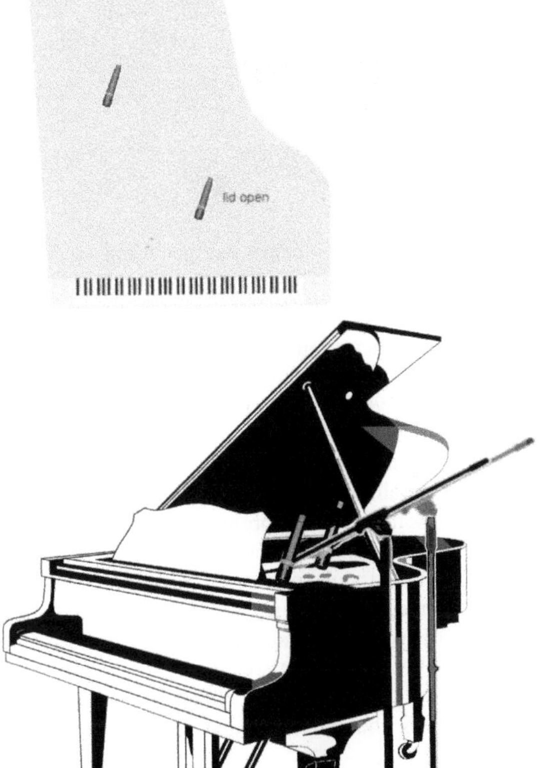

Figure 19.5. Grand piano micing.

- Don't strangle the beast: Even if you had the time, the spare mics, and the line inputs available to follow the lead of those who close mic every piece of kit, some twice, and then compress the hell out of everything to try and remove rattle, we advise the following setup. Drums rattle and roar: that is their nature. Set them free—we promise your sound won't suffer, and you'll have fewer headaches in the process.
- Unless the drumheads are new, look for the wear pattern on the head. This shows you where the drummer tends to hit and the best spot to place your mic.
- If using a second mic for the underside of a drum (i.e., snare), flip polarity relationship to the top mic.
- *Overhead mic*: Small diaphragm condenser placed somewhere above and slightly forward of the drummer's head, pointing downward. The easiest approach is to place the stand rear of kit, nearest one of the toms.
- *The kick*: Place mic inside the drum about 1 foot in front of the strike head and aimed slightly off center if there is no front head (dynamic, cardioid). Alternately, for front head with hole, place mic inside the hole so that the capsule is just inside the interior of the drum, and angled toward the spot where the beater makes contact to catch the best attack sound.
- *The snare*: Angle mic downward over the edge of the drum and aim it between the edge of the skin and the middle, slightly toward middle (ideally—in practice there may only be one available angle of approach out of line of the drummer's strike, in which case get as close to skin without touching (dynamic)).
- *Toms*: Position one mic on each tom using the same method as with the snare (keeping clear of drum sticks' path and wobbling cymbals). The best mic here is dynamic with a solid rejection of rear sound to keep too much sound from cymbals bleeding in. When possible, avoid placing tom mics directly underneath cymbals (keep cymbals or other drums oriented toward the rear of the mic, 180 degrees off-axis).
- *The hi-hat*: If you feel the overheard isn't catching it well enough, angle a condenser mic downward to just past the edge where the hats meet (angle so that the rear rejection keeps out bleed from the cymbals).
- *Bass drum*: close to the front drum head or preferably inside the drum (dynamic, cardioid).

Two-channel Drum Mic

- One kick mic, one overhead.
- Lay a Shure SM-57 (or equivalent) on the (carpeted) floor, under the snare and on the beater head side of the bass drum. Add one overhead. Experiment—move the mics around and see what sounds best to you.

Stereo Micing

Stereo sound was original developed for studio and broadcast recordings, though there are live applications for many of the techniques. Stereo recordings allow the listener to hear a sound image that corresponds with the

Figure 19.6. Two-channel drum micing.

location of the instruments. They provide an audio picture of the spatial qualities of the performance.

Stereo micing is more problematic in live applications because when you are micing for a live performance, the choices you make may be dictated more by the restrictions of the venue than that of obtaining the desired sound quality. If you want to mix in stereo, you are often better off close micing each instrument with one mic and creating the spatial soundscape using your mixer pan controls. However, there are certain live audio situations where stereo micing is useful, or even preferable. One application is when recording the live performance. Live recording is notoriously difficult, and is an advanced topic; we suggest you master getting the best sound for the show before you worry about trying to get the best recording. However, some sound sources may call for you to apply stereo micing techniques, usually those sound sources that create sound over a larger area

Figure 19.7. Coincident pair guitar micing.

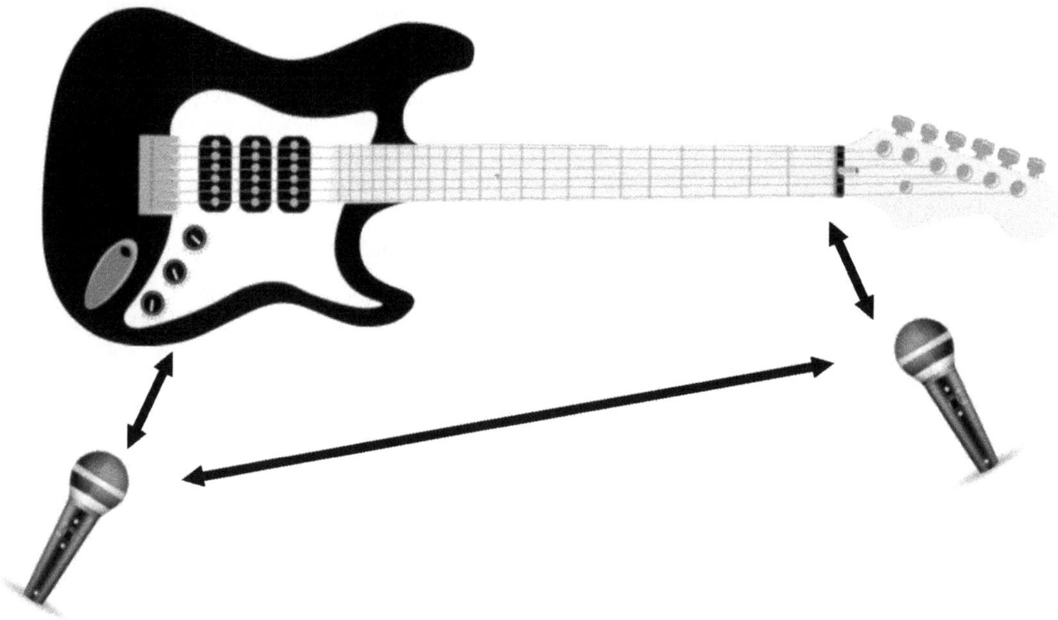
Figure 19.8. Spaced pair guitar micing.

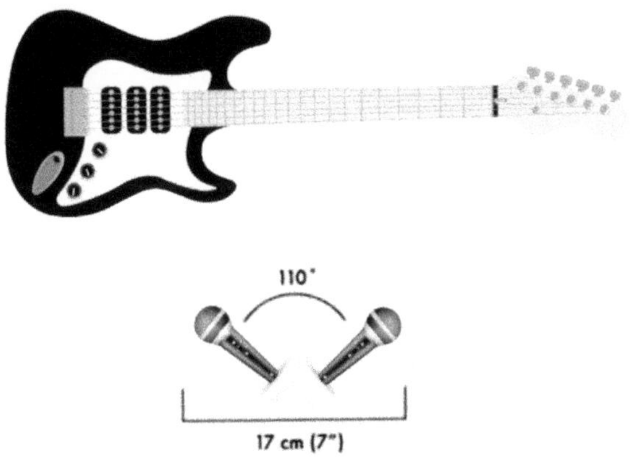

Figure 19.9. Near-coincident pair guitar micing.

than can be captured using a single close mic. Some of these include:

- groups of singers;
- piano;
- drum-set cymbals overhead;
- percussion instruments;
- string and horn sections.

Stereo Microphone Techniques

There are many factors that go into creating a stereo sound image/stereo field, including time of arrival, dynamics or volume differences, and frequency response. Microphone placement affects each of these parameters and plays the biggest part in determining the stereo image that is created. Using proven techniques will help you minimize phase issues which can reduce the intelligibility of any sound source. Phase problems are caused by the different times of a sound's arrival to each microphone.

Figure 19.10. Mid/side micing.

- *X–Y (coincident-pair)*: Each microphone is 90 degrees from the other so that they form the bottom side of an X while their patterns pick up a stereo field.
- *Spaced pair*: Two mics facing the same direction at the same distance from a sound source. The distance between the mics should not be farther apart than three times the distance to the sound source.
- *ORTF (near-coincident pair)*: Both mics are facing out from each other where the tails of the mics are close to touching the rears together and the front of each mic is 120 degrees separated.
- *M/S (mid/side)*: This technique is used in theater more often than in live sound reinforcement. Using both a cardioid mic and figure of eight pattern microphones you point the cardioid mic at the source while the figure of eight pattern is facing to the sides of the object being recorded. This provides a track that includes the environment sound of the object which is being recorded.

Chapter 20

From Signal to Sound

Speaker Placement for Live Production

So much of proper speaker placement depends on the speaker's properties and on the variables introduced by the venue acoustics. The reason why it's important for you to know something about acoustics is that no single setup configuration will work for every occasion. When the standard setups you're used to fail you, you may need to use the rule-of-thumb advice given here, along with your knowledge of acoustics, to tweak your system to suit the situation as best you can. *Always use your ears as your guide, and don't be afraid to make changes if needed.*

Small changes in the angle or placement of speakers can make a big difference in the sound. Taking 10 minutes to experiment with the precise position is not a waste of time if it saves you twice that time trying to accomplish the same sound improvement with your graphic EQ, and especially if it saves you from a night of feedback headaches.

In order for you to benefit from taking time to tweak and experiment with your speaker setup, you need to always keep in mind what you know about sound science and acoustics, and use it to guide your experiments; otherwise you're no more likely to solve sound problems than someone who uses a pair of dice to figure out which angle to try to shift their speakers next. To help you out, what follows is a summary of the key facts from Chapters 4 and 5 that any audio engineer needs to keep in mind when positioning and setting up the P.A.

Acoustics Review

- High-frequency, short wavelength sound, tends to be directional, and low frequency; long wavelength sound tends to be omnidirectional.
- Because of its short wavelengths, high-frequency sound has its power more easily attenuated by being blocked and absorbed by objects in the environment, while low-frequency sound is difficult to absorb and attenuate and travels much farther.
- Reflective surfaces tend to be smooth and rigid, while absorptive ones are flexible, soft, and porous.
- Concave surfaces focus sound and concentrate it at a single-point; convex surfaces scatter sound and help distribute it.
- Early reflections retain most of the signal's original power (arriving in under 30 ms after source sound) and reinforce sound—keep to short acoustic paths of under 30 feet. Reflections occurring later than that hurt intelligibility.
- Critical distance is the distance from the sound source where the direct and reverberant sound energy of a room is equal. When reverberant sound reaches a level that is +12 dB or more than the direct sound, the direct sound is masked and unintelligible.
- Critical distance is independent of decibels. Any attempt to increase the distance of intelligibility beyond the boundary of the critical field by increasing the decibels will fail, as the volume increase will only be canceled out by a matching increase in the volume of the reverberant field.
- Low-frequency sources will produce an additional 3 dB of output for each boundary they are placed against.
- Sound from two sources that reach listeners at different times will be out of phase when they converge and produce interference.
- The inverse square law predicts that 6 dB are lost each time the distance from the point source sound is doubled (3 dB for line source sounds).
- Wavelength = Velocity/Frequency = 1,130 (ft./sec)/ Frequency(cyc/sec).

Speaker Placement Rules of Thumb

Based on the fact that the science of sound is stable, even if the sound and venue are not, there are a few rules that always apply and rules of thumb that work best the majority of the time when setting up speakers. Often, a perfect placement will be impossible, and the end the result will be a matter of making the best compromises you can to suit the situation. However

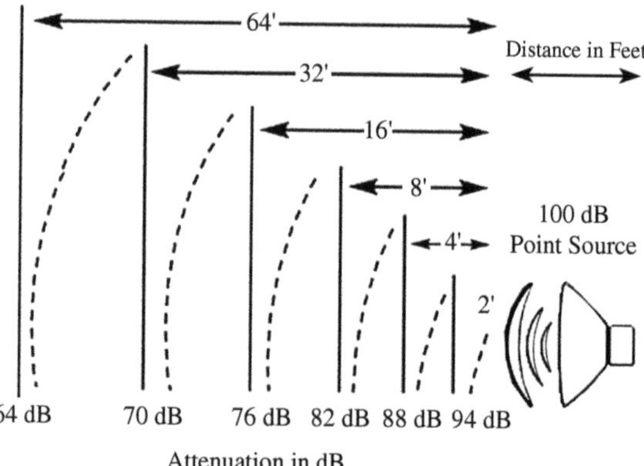

Figure 20.1. Inverse square law.

Frequency	Wavelength (feet)	Wavelength (meters)
20 Hz	55.8	17
50 Hz	22.3	6.8
80 Hz	13.9	4.3
120 Hz	9.3	2.8
200 Hz	5.6	1.7
1000 Hz	1.1	0.34
20000 Hz	0.0558 (= 0.67 inches)	0.017

Figure 20.2. Chart of frequency and wavelength.

bringing a tape measure and *observing these rules of thumb* can give your sound a fighting chance. As you read them, see if you can figure out what acoustic principles from the list above each rule is based on:

P.A. speakers should always be placed in front of the band, pointing towards the audience. This is to ensure that all microphones are out of the speaker's coverage area, because placing speakers close to the performers is a recipe for feedback. Ideally, P.A. speakers should be at least 5 feet in front of any stage mics. If this isn't possible due to space limitations, then anyone using a mic should be able to see the rear panels of any P.A. speakers. If this is not possible, then at the very least, no one using a mic should be able to see any part of the front of the speakers. Follow these rules even in a club without a proper stage, where the performance area on the floor.

Placing cabinets on the front edge of the stage is common in club venues, and is considered OK as long as the speakers are at the very edge and out of the line of the stage mics. Stage placement is considered desirable because of the added height it affords the speakers, and because it keeps them safely removed from the audience. Depending on the size and construction of the stage, this can end up a tradeoff that may prove troublesome. This is because most club stages are essentially big empty boxes (or in other words, resonators). Mic stands resting on the same surface as the house speakers are prone to pick up these resonating vibrations, creating feedback issues, even if they are out of the speakers' coverage area. Be aware of this and try to move the performers a little further up stage to put some more distance between them and the P.A., if forced to set speakers on stage.

P.A. speakers should always be placed so that everyone in the audience has a direct sight line to the mid and high speakers. Otherwise, audience members beyond the first several rows will hear only the low-frequency sounds (which travel through and around obstacles) and very few of the mid- and high-frequency sounds (which are required for intelligibility). If you have ever had the displeasure to be standing anywhere beyond four or five rows back at a show where this rule is ignored, we can assure you it sucks! Just as bad, as we can also attest to, is being on the side of one of the front rows where the mids are aimed directly at your ears, often cranked up beyond reason in a form of magical thinking where the FOH engineer figures extra dB SPL can somehow cancel out high-frequency attenuation. Therefore the politest course, even in very small venues where reinforcement not amplification is the key, is to raise the speakers above the audience to avoid beaming sound directly into anyone's ears.
If they are tall enough to keep the mids and highs at least above ear level of the front rows, the simplest solution is to put your speakers on top of the subwoofer cabinets. A more ideal solution is to raise your speakers on pole-type *speaker stands.* The best solution is to raise them on stands with sufficient reach to make sure the horns are at least 7 to 8 feet above floor level. If you have very high stands, and the stage is very low (more like a step), or if your stands are low but circumstances force you to place them on the edge of the stage, you can encounter the problem where the speakers are now perfectly placed so as to bounce sound right off the rear wall and back at the band, making it difficult for them to keep stage sound down. In this case, angling the speakers slightly down is advised, and often is a good idea even if reflected sound isn't an issue, since this will provide more throw and more even coverage overall.

Remember safety is primary, so *never raise speakers on tables, chairs or beer crates.* Always use proper speaker stands and mounting hardware and never exceed load limits. Also stands should never be tilted. Unless you have stands designed to allow control over the vertical angle, make sure to use a *tilt adapter.* Look online for links to relatively inexpensive tilt adapters (and *height extenders* as well, if you find yourself using stands that could use some extra reach). When using stands of any kind, be sure the legs are splayed properly, and use sandbags or weights to stabilize the base and keep stands from tipping over.

Unfortunately, bass speakers lose a great deal of their power when raised, so if you use full-range speakers you will have to deal with this—it is still less of a problem than almost no mids or highs for most of the audience. As usual it's a matter of picking the tradeoff that's best for the overall sound. If you have subs it's not as much of an issue; just make sure to keep these on the floor.

In general, loudspeakers should always be angled so they point toward the front-center of the intended coverage area. When speakers are pointed at an inward angle, this is referred to as *toe-in.* There are a number of advantages to having some amount of toe-in as opposed to facing speakers straight out. Doing so will cut down reflections from side walls, and help keep excessive reverberation in check. In addition you won't end up with the large gap in coverage over the front rows, and will avoid the situation where an even larger chunk of the audience is covered only by one speaker; situations that both occur when speakers are not set up with any toe-in.

The rule-of-thumb for toe-in is that the center-line of the speakers should preferably cross front of center to a point about one third of the room's length. In practice, the optimal angle depends on the speakers' coverage and the room, so take

Figure 20.3. Always keep P.A. loudspeakers in front of the band, with mids and highs raised to ensure audience line of sight. Toe in speakers for even coverage and to minimize reflections, and don't forget to secure stacks for safety.

a second to experiment till you find the best angle for a difficult situation. And of course the inward angle should never be enough that any stage mics are brought into the area of speaker coverage. Nevertheless, some inward angle will just about always be preferable to having none.

Ideally, all elements of the original sound should arrive at the audience's ears at roughly the same time. If the mid/high-frequency enclosures are on stage and the bass enclosures are downstage on the floor in front of the stage, then the arrival time of sounds emanating from each will be different. If the boxes can't be placed for proper *time-alignment*, then you need to correct the alignment by setting a *delay* in the system. Most active crossovers and controllers incorporate a separate delay for each output and digital boards usually allow for the user to set delays just about anywhere in the system. If neither of these options is available, then there are outboard delay units to do the same thing. *The rule of thumb is 1 ms of delay per foot; delays in small to mid-size venues will typically be around 1 ms to 3 ms.* (Percussive elements of sound often sound tighter if the "attack" arrives before lower-frequency fundamentals; be aware before setting a delay that it may be best if the bass signal arrives slightly later rather than earlier.)

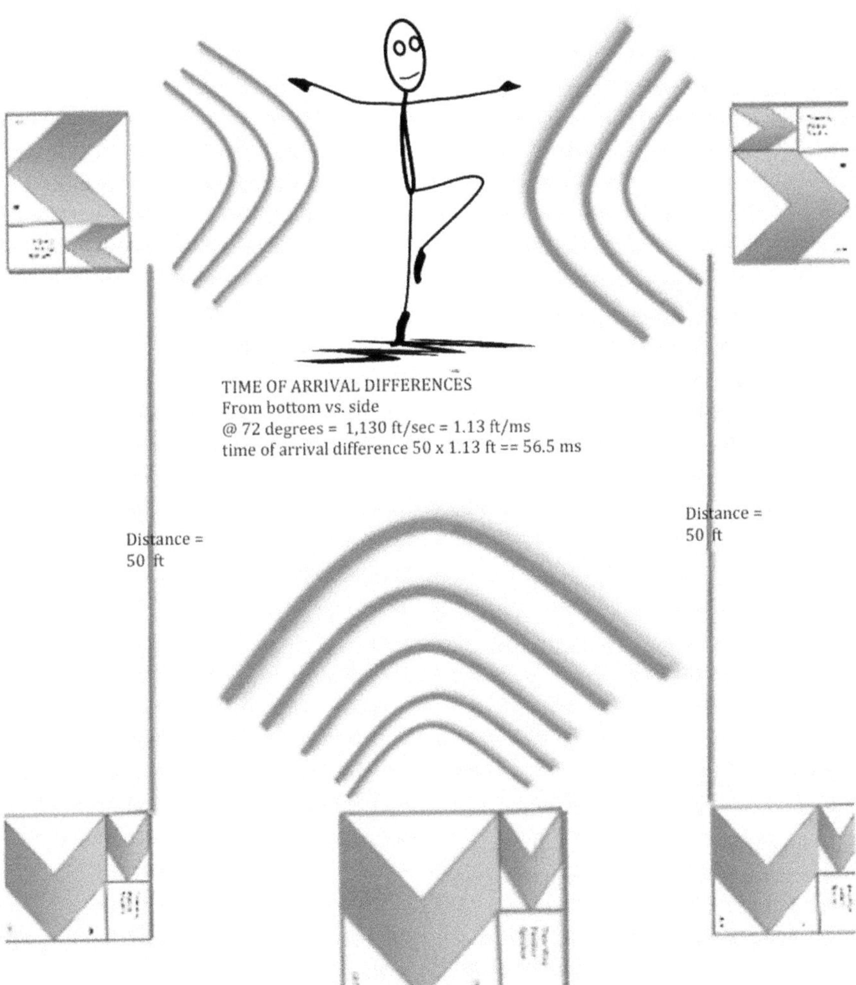

Figure 20.4. In venues that don't allow for speakers to be physically placed together, as well as in venues that don't allow full coverage from the front, set delay to ensure all sounds are in time.

Other considerations that may help you determine the best possible speaker placement are:

- Each speaker is designed with a particular "throw" in mind. That's the distance to which it can be considered to be reliable, especially in the high frequencies, since that's the most essential range for intelligibility. A *distributed speaker system* (using delayed cabinets part of the way down the room) will allow you to "throw" sound farther, or cover the needed space without the volume at front being as loud. When nothing seems to be enough to get even coverage or control those nasty reverberations, try using delay stacks to allow you to cover more space but turn down the overall volume. Remember, because high-frequency sounds attenuate much faster than low-frequency sounds you might need to boost the highs on your delayed speakers just a little bit.
- Room resonance never has positive effects on live sound—don't place the bass speakers against the walls or on or under the stage.
- Peaks and notches in level at various frequencies will be found in at least some positions in the listening area, and is not too problematic if kept to a minimum—remember this is somewhat offset by (reasonable amounts of) reflected sound. In troublesome venues, however, this problem can be exaggerated when both speaker stacks are mixed in mono.
- In most venues, especially when resonance is or seems a likely problem, the best plan is to place your subs symmetrically at a quarter of the room width from each side wall (for the smallest venues), or an eighth of the room width from each side wall (for venues 20–25 feet in width or wider). If this doesn't help with low-frequency issues, try positioning them centrally.
- Speakers with frequencies below about 250 Hz are omnidirectional for most practical purposes. If sensitivity to feedback proves more problematic than usual, place bass-sensitive microphones and pickups as far from bass speakers as possible.
- Don't stack cabinets beneath any kind of arched, domed, or similar concave surface—all this can cause is reflections that will interfere with your speakers' dispersion pattern.
- It is essentially impossible to provide quality sound coverage for listeners not positioned between the speaker stacks. If you need to cover listeners located at the far sides of the stage, you may have to set up an additional number of small speakers, called *fills*, to cover any listening areas outside the coverage of the main speakers.
- Venues with wide stages (40 + feet) can leave the front center rows without enough coverage. In these cases a front-fill may be necessary. Listen from the fill area at sound-check to determine how loud the fills need to be—just loud enough to allow vocals to be heard but not loud enough to disturb the main house mix.
- It is important not to have the speaker stacks too far from each other. If the distance between the stacks is too large (more than 2 to 4 meters for 40–80 Hz) dead spots will appear in the audience. With rooms that are too wide, rather than spread your stacks too far, if centering your subs isn't an option, you may need more than one full-range cabinet per side to ensure quality sound even if you don't need to play the P.A. loud.
- Dispersion angles define the speaker's "coverage" of the audience listening area. We want wide horizontal coverage, to spread sound across the audience, and narrow vertical coverage, to avoid bouncing sound off the ceiling, so speakers usually have a wider horizontal dispersion pattern than vertical. Be aware, before deciding to orient your speakers in a way other than intended by the manufacturer, that high-frequency dispersion will be affected if you place a box on its side that wasn't made for it. Be sure you don't just cause more trouble for your sound by bouncing sound off the ceiling or limiting your horizontal dispersion when you can't afford it.
- Try to keep the bar out of the main sound field. If sound interferes with the bar's ability to turn a profit there will be complaints no matter how reasonable the overall sound levels.
- Some venues are just bound to be a headache; be ready to face defeat and find an out-of-the ordinary solution rather than waste time trying to make the usual setup work. Even in venues where a bulldozer is the only proper solution, do what you need to do to keep from making the problem worse. Reduce stage monitoring and stage micing to a minimum, use fills for wide venues and time delays for long ones, be ready to use single stacks in very small or odd-shaped venues (store the second stack, or use them as time delays). *When your back's against the wall, make it happen—the show must go on.*

Stage Monitors

Monitor placement is critical, both so the artist can hear what they need to play well, but also to avoid feedback problems and keep stage sound from interfering with the house mix. Many problems can be avoided by placing the monitors appropriately relative to the mic

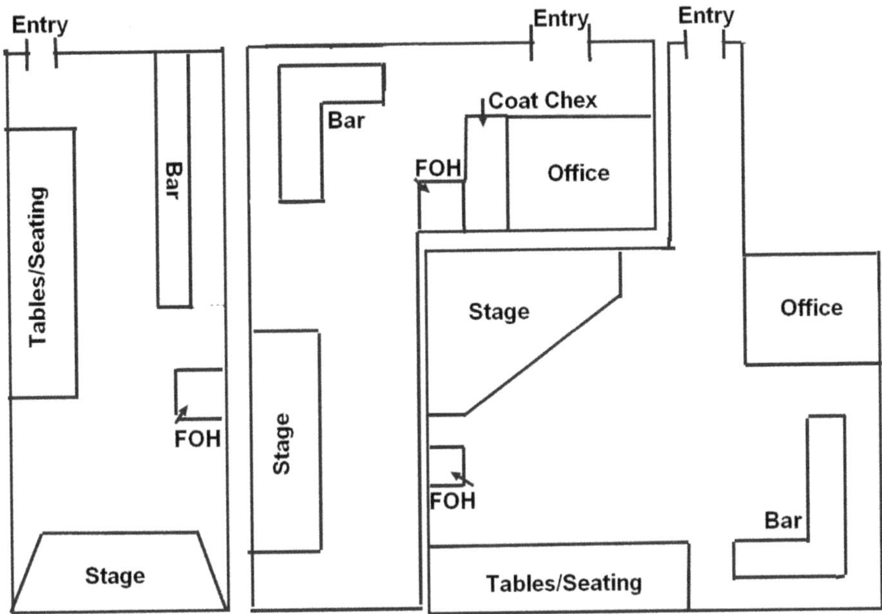

Figure 20.5. Some venues are shaped oddly and present extra difficulties. Depending on how reflective the finishes of these venues are, they could range from problematic to major headaches.

placement, and by also paying attention to providing musicians the maximum support so they can play in comfort. Much of the decision making as to the number and spread of monitors should ideally be made with the band's input ahead of time. Providing adequate monitoring while keeping stage noise controlled is a team effort. We will look more closely at noise control stategies in coming chapters. However, the methods in terms of set up are fairly straightforward:

- All vocal mics should face away from the monitors, amps, and drums.
- Monitors should preferably point away from the audience, unless they are required to help boost P.A. sound.
- Point the monitor's horn at the ears of the musician instead of their knees. Most monitors are built for musicians who are standing but are less ideal for those sitting—using a two by four to prop up the rear edge can be an effective way to adjust the monitor's angle so sound is directed properly. Also, if a standing player is too close or off axis, monitors will not be as effective.
- Super cardioid mics have a very narrow front-pickup pattern but achieve this by sacrificing some rejection directly behind the mic. If you point the back of a supercardioid mic directly at a monitor, feedback is more likely than if you offset the monitor by roughly 15 degrees (30 degrees for hypercardioid mics). With switchable pattern mics becoming so much more common, make sure to pay attention to this fact and educate your singers about where to stand in relation to their monitors if they want to switch back and forth between different polar patterns.
- Unless setting up paired wedges, try to place monitors at least 5 feet apart.
- Trumpets and trombones will spill into vocal mics if placed at the back of stage. Put horns at the front of stage at one side or the other if you can.

Unless the band has a monitor engineer you trust handling things, always double-check the monitor and backline position to make sure it's as agreed upon and no one has decided to move things around at the last minute. While you're there, make sure they've set up agreed-upon acoustic treatments or assign someone to set them up if the band ends up slacking.

Apply/Double-check Acoustic Stage Treatments

- Double-check to make sure that all agreed-upon setup arrangements for keeping stage nose managed are

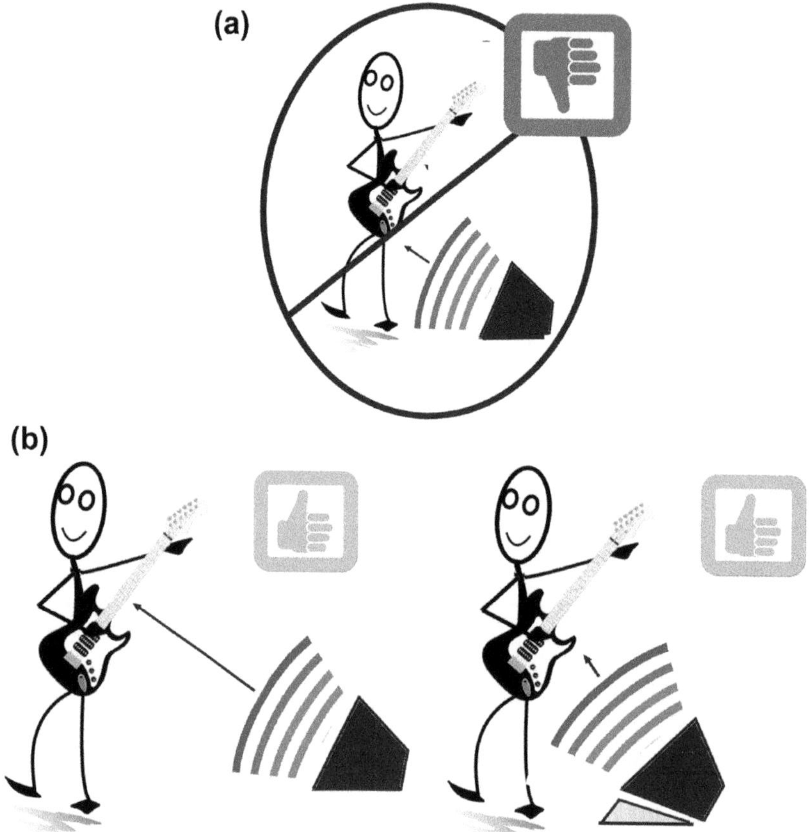

Figures 20.6a and 20.6b. Make sure monitors are pointed where they can be heard so that they can be used without having to turn them up too loud. Monitors aimed at knees can contribute to excess stage noise.

adhered to; a gentle reminder is usually all that will be needed if they have not. (We will discuss working with bands to reduce stage sound and apply acoustic fixes in upcoming chapters.)

- Stack two milk crates or bottle crates for each amplifier. This prevents blocking amplifier vibrations from spreading through the floor and to nearby microphone stands, creating a rattle and with it noise creeping into mic's signal. This also elevates the sound to the musicians' ears and placing them at a convenient height for users to make adjustments without stooping down.
- Cover highly reflective stage surfaces. Reflected stage sound also adds to spill and adds to feedback issues. Heavy curtains or other absorbent material can make a noticeable difference. A cheap option is to hang a curtain or blanket across the back wall.
- Use sound curtains/moving blankets to cover side stage walls; remember to drape the fabric.
- Check drums for carpet. The band should have one, but keep a spare handy in case. Off-cuts large enough for this cost next to nothing.
- Peek in the kick and make sure it's dampened if you and the band agreed on that.
- Look at amps and make sure they aren't at 11.
- If the venue is problematic, put up your acoustic ceiling pillows.

Mixing Board

Ideally, mixers are placed centered to the stage, far enough back to be out of the way of the audience, but still in the "sweet spot". Very few gigs allow for the ideal in all things. There are practical reasons to choose a different position for the mixer than the traditional ideal. One is when positions further back or to the side provide a better picture of what most audience members are hearing. Another is when you simply don't have a choice, since many venues have a specified FOH setup and not a lot of flexibility allowing for easy repositioning. When you do have the option to choose where to position the FOH mixer, these are the things you should take into consideration:

- *Connection to the stage*: Whatever mixer position you choose, you need to be able to set up the snake so it doesn't present a tripping danger right through the main audience position. This most often means running the cable along a wall. Note that you will need to run the snake above any fire exits, and preferably above any high traffic entryways. This will require secure anchor points (think eye bolts). Gaffer tape will not cut it for securing cable that is heavy over doorways. This will also add to your setup time.
- *Crowd control*: If you are too front and center, patrons who don't know better will likely mistake you for "hospitality", and try to ask you for everything from directions to the restroom to the club lineup for the following month. Patrons who do know better but are too drunk to worry overmuch about perfect etiquette will stop to make suggestions, first among them to "turn it up" (regardless of current levels or time). These annoyances only matter as considerations if you have a booth where only one patron at a time can stand at the entry and shout or gesticulate wildly. If you do not, then your concerns are far more serious—in this case your worry will be how to keep drinks from spilling on the rack and mixers and manage to focus on the task of mixing. If so, then find a more remote location.
- *Usable sound*: While front and center is not necessarily most representative, and there are many less "perfect" spots that will still allow you to hear well enough to provide a good mix, do your best to avoid spots where reflections or shadowed sound will make mixing well much more difficult. This includes: directly against a side or rear wall, under a balcony or behind obstructions, or in a fully enclosed booth or other separate enclosure. If you find yourself in any of these spots, you will absolutely need a monitor or headphones for the main mix in addition to whatever monitoring set up you use to track your monitor mixes. During sound check you will need to confirm that what's being heard in headphones is representative of what the audience is hearing.
- *Representative sound*: If you find yourself rear center or rear offset from center, even in the optimal FOH position, you need to keep the big picture in focus. Be aware that at the center you will be hearing more low end than is representative, while if offset you will be hearing slightly less than is representative. Make sure you know the overall sound conditions so wherever you are, you can make a mental "mix adjustment" to account for any such deviations and provide low end that is consistent for as much of the room as possible.

Part 5

Show Time and Beyond

Whether mixing a show is as comfortable to the FOH engineer as slipping into a favorite coat, or whether the engineer is mixing one of their first shows and wound so tight with anticipation and anxiety they'd vibrate if anyone thought to test the theory and pluck them, the day of the show encompasses all the technical knowledge and skills that go into being a sound engineer, including but not limited to:

- the scientific understanding needed to know how to account for the ways people hear and deal with venue acoustics;
- the practical knowledge of how to prep for a show: how to keep organized, how to load and unload safely, how to keep track of priorities and processe;
- the social skills to deal with everyone from production crews, venue owners, grips, fello live sound engineers, and of course the performers you are mixing;
- the technical knowledge of gear, frequency, power, etc. that you need to set up a P.A.system, test and EQ for best sound, and finally run the system;
- the artistic skills—from developing your ear to identify frequency, instruments, and notes, to the understanding of genre and music, to the basic feel for what sounds good;
- the knack for making it sound just a little bit better, achieved from countless hours listening and practicing at adjusting with the EQ.

These make up the aspects most focused on and no wonder—they are the skills you can't do without in order to simply get the job done; your overall objectives, what it is that you are looking to achieve. This aspect is no less essential to making good sound, even if the fact that it isn't needed to simply mix makes it easier to lose track of.

Chapter 21

Setup

Connecting and Tuning the Chain

Matching Amps to Speakers

Before you can finish up, you will need to determine how many speakers you want to wire to each amplifier channel, what wiring configuration you want to use, and finally how thick and long that wire should be to handle the current according to how everything is configured. Speakers have two features to consider when choosing how to wire them to a power amp: the amount of power they can handle, expressed in watts; and their impedance, expressed in ohms, typically 4, 8, or 16 ohms. Not coincidentally amps are rated as outputting a set amount of power, expressed in watts, against a specific impedance load, typically 2, 4, or 8 ohms.

As already mentioned in Chapter 12, it is generally agreed the amp output power should be somewhat greater than loudspeaker power handling rating. Though there is some debate as to how much more, most suggestions fall in a range of between 50 and 200 percent more powerful (note that double the power is only 3 dB more; there are plenty of situations where a conscientious user may even want four times the power, or 6 dB headroom). The fact that amplifier power output is based on tests that use sine waves, which draw more power than audio waves (and so make amplifier ratings overstated), combined with the fact that loudspeaker rating can be based on any number of tests that are also often overstated, makes it difficult to be more precise. Add to this that factors like crest factor, frequency, and a number of harder to calculate details also contribute to determine power transfer/power handling, and that speakers can be driven to thermal or mechanical failure by distortion due to an underpowered amp almost as easily as they can by being overdriven by an amp that is too powerful, and it becomes clear why there is less to lose by having an amp with a bit too much power than one without enough. Ultimately it is going to take a user who observes best practices, pays attention to their system, and refrains from clipping to ensure system safety either way; anyone who does not is as likely to do damage no matter how well their amp power and speaker capability is matched.

Since it is impossible to be entirely precise in matching power, it becomes essential to be extra careful to observe practices where there are fewer unknowns, and always be sure to match speaker and amp impedance according to accepted guidelines. Fortunately it is not too difficult a chore; the ability to wire the final stage of your system together in several ways creates avenues for finding an acceptable work around in almost any situation. In many situations there is no need to wire more than one speaker enclosure per amp channel, as it is not uncommon that these wiring configurations will already exist in the system within your multiway speaker enclosures, all you'll need to do is plug in. Because of this, however, regardless of any advice you find here or elsewhere, always consult the user manuals for your speakers and follow their recommended best practices first. Many enclosures contain more than one speaker, and you don't want to wire enclosures together as if each were one speaker unless you've consulted the manual to check if this is advised and that you are including every speaker in your calculations. That said, there may be times when having the option to wire speakers/enclosures in parallel or in serial will provide the exact solution required to solve a particular configuration quandary, so it's a tool every live sound tech should know how to implement.

Impedance Matching

When plugging in one speaker/enclosure per amp channel the rule is that in order to utilize the full-rated power of the amplifier the speaker and amp should have the same rated impedance; 8 ohm amp with 8 ohm speakers, 4 with 4, etc. This is so simple an equation that it is clear that what many of the people asking the question "How do I match my speakers and amps?" are really trying to find out is "How much can I mismatch my speakers and amps and still get away with it?" The answer to this is that *it is never advisable to use an amp to power a speaker that has lower impedance than the amplifier, but it is perfectly safe to power speakers that have higher impedance.*

This is because as impedance goes up, current draw decreases. So when matched exactly power transfer from amp to speaker is the expected 100 percent as rated. For each step up in speaker impedance, power transfer is 50 percent lower. So a 4 ohm amp rated at 500 W will deliver 100 percent of its rated 500 W power to 4 ohm speakers, but will deliver only 50 percent of that power, 250 W to 8 ohm speakers and only 25 percent, or 125 W, to 16 ohm speakers. This is why it is always safe to use speakers with higher impedance than the amp's rated impedance, and in some cases may actually be preferable to an exact match, as the amplifier is sure to run cool and work reliably. Conversely, it is also why it is not safe to use speakers with lower impedance than the amp is rated for, because the amplifier from the examples above will attempt to deliver twice its rated power or

1000 W to the speaker. If this is not enough to fry the speaker, it is more than enough to fry the amp!

There are two caveats to this that we must mention. The first is that even though some amps now claim to be able to deal with 2 ohm impedance, and even 1 ohm, there are some experts who advise against ever powering a load with nominal or combined impedance below 4 ohms. We have not used this technology so our only advice is that if you have a 2 ohm speaker and 2 ohm amp, go ahead and use them together if the manufacturers indicate it, but be aware that there are some who feel the jury is still out on how hot an amplifier will run under these circumstances, so to be safe keep an eye out and preferably evaluate it within the exchange period; if you notice trouble, bring your speaker impedance up to 4 ohms. The second caveat is that the ability to drive higher impedances than their rating demands is a feature of modern solid-state (transistor) amplifiers only, and is not shared by older tube amplifiers. If you are running a system with antique pro or personal gear you should be aware that this is not recommended for tube amplifiers, and driving speakers of different impedances than a tube amp could damage the amplifier.

Wiring the Speaker to the Amp

When one needs to solve a configuration problem or bring speaker impedance in line with available amplifier channels, it may become necessary to wire multiple speakers/enclosures to a single amp channel. There are three standard configurations for doing this (if you have dual coil speakers check the chapter directory for info about additional options), wiring in parallel, serial, or a combination of both. Before you connect speaker and amps, even if "wiring" only amounts to "plug in speak on connector," be sure to disconnect the power supply to the amplifier rather than just turn it off. Always connect to power last to be as safe as possible.

- Parallel circuits use dual wires to connect positive to positive and negative to negative terminals along the circuit.
- Series circuits use single wires to connect positive to negative along the circuit.
- To combine series and parallel wiring, connect the speakers on one channel together in series, then wire the channels in parallel to achieve an acceptable resistance level.
- When two or more speakers are placed in series, the impedances add together to produce higher impedance. R = resistance (the ohm rating of your loudspeaker): R = R1 + R2 + R3 + R4.

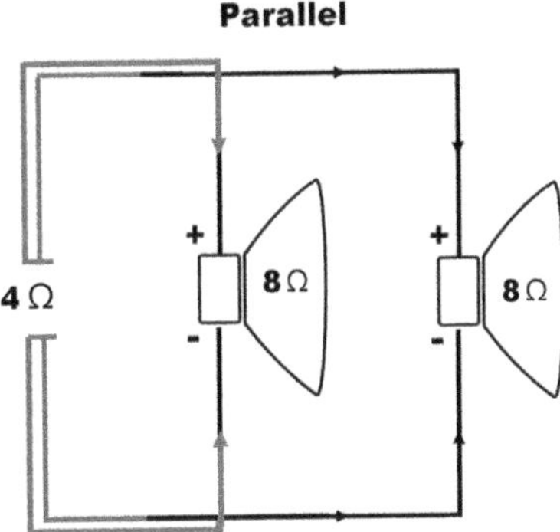

Figure 21.1. Note the two wires vs. one wire in this close-up of each wiring configuration.

- When two or more speakers are placed in parallel, the impedances combine in a different way to produce lower impedance. Multiply the resistance of all speakers and then divide the product by the sum of the resistance of all speakers. For two resistors, this simplifies to: R = (R1 x R2) / (R1 + R2), or if the resistors (speakers) have exactly the same value: R = R1/2.
- By combining the two methods, the opposite effects on impedance cancel out; a series-parallel circuit can be given the same impedance while allowing the use of many speakers.
- Using identical speakers; using different types of speaker on the same amplifier channel is not advised.
- Be aware that in most systems it is difficult to put speaker enclosures in series—a special wiring harness or adaptors are required, so plan ahead if you are thinking this may be required.
- Be sure to use the wire of a gauge that is optimal for the impedance, power, distance of the circuit.

It is also important to use large enough gauge speaker wire so that all the power running from the amp to the speaker makes it there without being lost along the line. The lower the impedance of the amps and speakers being used and the longer the length of the speaker wire connecting them, the larger the wire gauge should be to keep power loss and heat buildup minimal; it is a standard rule of thumb that you want to always keep power loss in the wire to 0.5 dB or less. Figures 21.4a, 21.4b, and 21.4c show the effects of wire length on power transfer, and provide suggested wire gauges to use for different situations.

Unless you will be buying amps of your own, or helping someone else do it, you shouldn't need to worry about matching speakers and amps immediately: earlier on in a sound engineer's career, the sound system

(a)

# of Speakers	X Impedance	Combined impedance
2	16Ω	8Ω
2	8 Ω	4Ω
2	4 Ω	2Ω
3	16Ω	5.33Ω
3	8 Ω	2.67Ω
3	4 Ω	1.3Ω
4	16Ω	4Ω
4	8 Ω	2Ω
4	4 Ω	1Ω

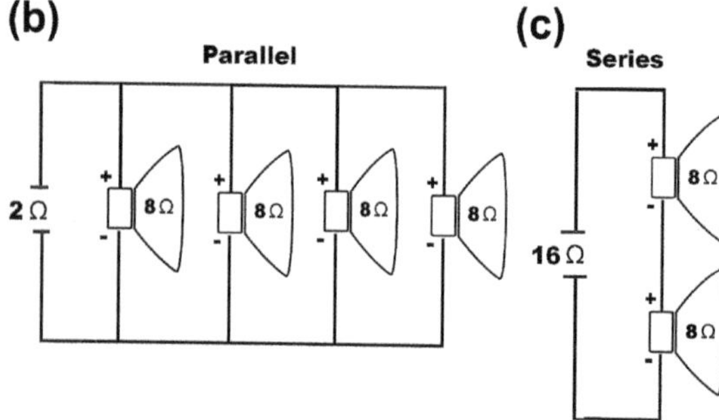

Figures 21.2a, 21.2b and 21.2c. Parallel and series examples and common parallel loads.

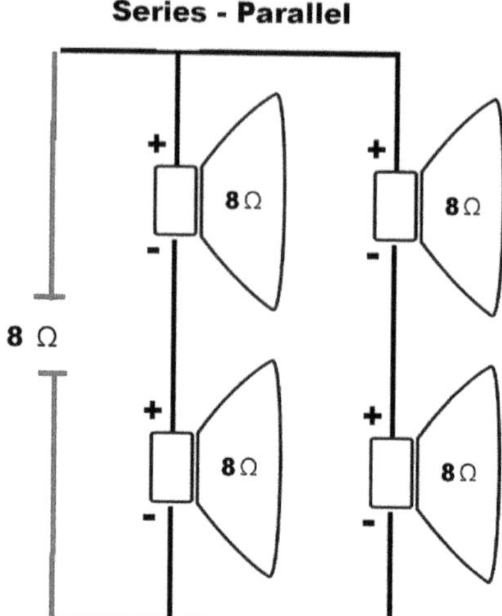

Figure 21.3. Example of combined series parallel wiring.

will be in place or already purchased or rented and at most will need to be hooked up during setup. Should you work at a sound company, you will assist at many events before being lead on one and designing the rental package, so will have plenty of instruction in how to do it right before you will need to worry about doing it on your own. However, for anyone who is planning on purchasing amps soon or is just curious about the calculations used to figure this out, there are links in the directory leading to some great basic tutorials. For any who want even further help, there are also links in the directory leading to calculators that are designed to make matching amps to speakers a cinch.

Connecting the Dots

Once everything is in place, it needs to be connected. Be sure before plugging it that you lay out all your cords first, making sure they are long enough and leaving a little excess in case something must be moved a little. To keep everything oriented when running XLR to XLR cables, remember that connector pins "point" in the same direction as the signal flow (female connectors end at the "source", male connectors at the "destination"). Be sure to turn down gains at each device that has gain controls before you connect it. As you read the general steps for connecting a P.A. system, be aware that this list assumes you will be following all the cable best practices already mentioned, and will be taping and leaving cables neat. Also be aware that these steps are a general guide. No two systems are exactly alike, and it is expected before you set up your own system you will become acquainted with its unique configuration.

Turn Isolated Components to a Signal Chain in 11 Easy Steps

1. Connect the female XLR ends directly into the on-stage mic signals.
2. Connect male XLR ends into stage box inputs labeled channels 1–9, using channel list to connect correct mic signals to each corresponding snake channel.
3. Plug instruments to DI boxes for each instrument (stereo for keys); plug male outs from each DI box into female XLR ends.
4. Connect male XLR ends leading from DI boxes into corresponding stage box inputs labeled channels 10–13.
5. Connect all sound sources to mixer channels 1–13 using the respective male XLR ends at fan end of snake multicore cable. Label all audio

(a)

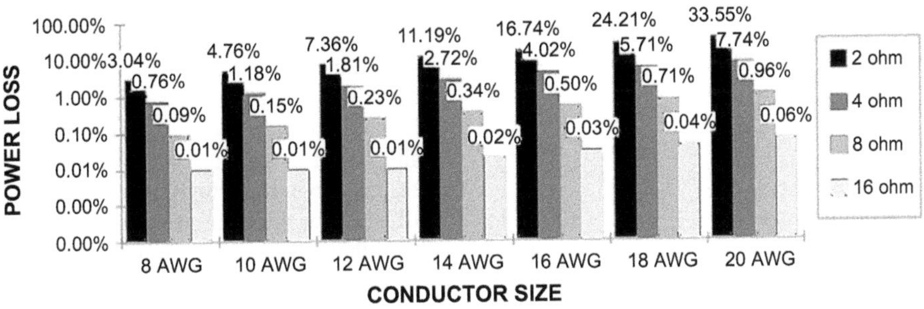

(b)

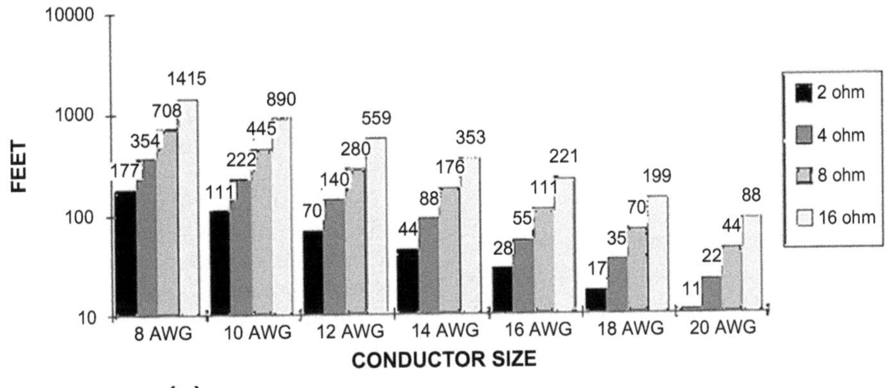

(c)

Load in Ohms	1-100 Ft	Over 100 Ft.
16 Ω	16 gauge	14 gauge
8 Ω	14 gauge	12 gauge
4 Ω	12 gauge	10 gauge
2 Ω	12 gauge	10 gauge

Figures 21.4a, 21.4b, and 21.4c. Wire gage and power transfer 21.4a & 21.4b Source: Image Courtesy of ProCo Cables.

input channels by applying a small strip of masking tape on each input channel and writing the name of that audio input source.

6. Connect the audio inputs of outboard effect units, such as reverb, to the aux send channels or inserts on your mixer. The exact configuration depends on your mixer and is detailed in your mixer's manual. If your mixer contains built-in effects, you may skip this step.
7. Connect line-level signals to outboard processing units such as EQ and compression, using channel inserts (and either mix inserts or aux sends).
8. Connect line-level returns for speaker and monitor amps from mixer to dedicated snake return channels.
9. Connect the male-to-male connector from the designated return channel at the stage box to the corresponding powered speaker/monitor.
10. Connect the male-to-male connector from the designated return channel at the stage box to the corresponding powered speaker/monitor or to

the corresponding amplifiers for unpowered passive crossover systems, or to the corresponding amps via the active crossovers.

11. Using the speaker cable only, connect the audio outputs on your amplifiers to the audio inputs on your speakers.

Essential Concepts

Before we talk about turning on the system, we need to examine a few more concepts that are crucial to understand before you can safely run the P.A. system. To make sure you don't overload any elements in your system, you need to maintain a proper gain structure. This is what ensures you always have enough headroom to handle any signal you may encounter in the system. While many aspects we've covered already, like speaker and mic placement, are essential to allowing for optimal gain structure, the factors below must also be taken into consideration.

Metering

As you set your gain structure (and indeed any time you use the mixer) you will have to pay attention to both your ears and the metering (in the form of indicator lights or needle gauges) furnished by your mixer. All boards handle signal metering differently, but no matter what form a board's meters take, they all serve the same function: to ensure that signal levels do not exceed your mixer's capacity or the capacity of components in your system further along the signal chain. The meters allow you to adjust each signal to the optimum level for processing as it enters and ensure it is at an appropriate level to exits. You will need to read the manual to be sure of the scale and configuration of any board's meters. Providing even a basic description of metering that applies for all cases is impossible in light of the different accepted scales and configurations that can be (and are) used. In general, however, the majority of meters will be variations on a similar theme.

Output/Main Signal Meters

Most mixers include at least one pair of bar meters comprising LED lights (one strip for each stereo out). Some mixers will have another set for groups and monitor out; those that do not will allow for the main output meters to be assigned to either the stereo outs or the groups/monitors. LED meters comprise green and red lights at each end of the meter bars, and amber lights in the center. Regardless of scale used, most of them usually work something like this. As the first weak signal is sent through the board, the first green light will flash and as the level of the output signal (and correlated volume) increases, the next light in line will light up, with subsequent lights following as signal strength increases, and so on. When the red lights begin to light up, you have either reached or are just past your maximum output signal (most often it will be either the first red light, or the light immediately above that indicates maximum output).

Needle gauge VU meters work in much the same way, with the needle moving only slightly at the weakest signals, and swinging further right as a signal level is increased. The area to the furthest right is usually colored red. As with LED indicators, at maximum you will want to see the needle bounce into the red only during peaks, not remain there continuously. No matter what meter you read to help you determine maximum output, this level is the point beyond which you can't raise the signal further and expect maximum performance from your system (to leave headroom so you don't clip). You will need to read your manual, or carefully experiment at a pinch, to be sure which specific light represents your mixer's safe upper limit.

Note that maximum is not the same as overload, since to be a useful level it must be far enough below clipping to allow for minimal headroom (at least enough so that you can find it on the meter and dial back without automatically clipping). That said, neither is it a level you will want to run automatically. You do want to become familiar with it so you know the outer limit of your system, not necessarily to run at those outer limits. Many live sound applications only require the sound to be boosted slightly and evenly distributed. Running your sound too loud is no better than not having enough sound—your job is to ensure optimum sound levels, which are not always maximum sound levels.

Look into which light indicates your mixers max ahead of time, but on the day of a show take a few minutes after setting your input gain and adjusting your amplifier gain, and check your meters and main fader to get a feel for how far you can safely increase your output volume. For the ideal maximum volume you will want to set your volume so that this light does not remain lit continuously, but instead flickers on and off during your maximum volume spikes. When normal peaks cause the lights immediately below it to flash most of the time and the light indicating your max output flashes only occasionally (at the most extreme peaks), your output signal is set exactly at your board's safe maximum level. Note where your knob or fader is before you back off on the volume again. After this check you will know clearly how loud you can go, and most importantly where to stop increasing volume levels to avoid distortion.

If your board also has a single "clip" LED, make sure to keep an eye on it too. If the clip light is flickering, it means your board is reducing the signal to keep from clipping (a safety feature in some boards). You don't want to depend on this, though, so if you see this light flicker, dial back until even peaks don't engage it. You want to avoid clipping at all costs; aside from the damage it does to your sound, too much clipping anywhere in your system is likely to damage your speakers. If it seems to contradict your main meter, you can look into making sure your meters are calibrated and matching later, but if it flashes when you are running close to max levels, err on the side of safety and dial back your volume a bit.

Input Meters

Most mixers, even quality mixers, often furnish minimal channel metering due to space and cost considerations. Often, channel meters are nothing more than a green LED indicating the presence of a signal and a red LED which flashes to indicate a signal overload. Depending on the mixer, the red LED may flash to indicate imminent overload and only remain lit after reaching overload, or may only flash at overload. Since "close to distortion" and "distortion" are experienced quite differently, make sure to read the manual to be sure. For more precise channel metering, which is handy for adjusting your signals accurately, most mixers come with a pre-fader listen (PFL) pad. When pushed, PFL assigns the mixer's main meter to display the pre-fader signal level of that channel. Note that some variations on PFL, like "solo", will require you to set your channel fader to unity (marked on most faders) before you can assign the main meter to a channel.

Volume Down

While powering up seems like it should be a simple process, you will want to turn things on in the right order to save your gear from the worst "pops" that potentially could damage your mid-range horns and are just as potentially bad for the ears of all present. Before you start, make sure that the master output levels on the amplifiers and the mixer (both monitors and main) are turned all the way down. If you don't turn down these levels, Murphy's Law could mean your lack of attention happens to occur the very day one of these level controls is bumped to max in transit, resulting in instant and terrible feedback the second you flip the on switch. Feedback like that can easily blow a speaker, so err on the side of safety and always check the position of your knobs before powering up.

Zero the Board

You will want to make it a habit to always normalize or zero out your board both before and after you use it. As you check that the output level controls are lowered, take a minute or two to normalize the mixer. To do this, make sure to:

- Lower all volume faders (channels, monitors/subs, and main mix).
- Switch off all phase and phantom power pads.
- Set all gain controls, such as at channel inputs and aux sends, at unity.
- Flatten/center all EQ controls, and switch out any channel high pass pads and switchable EQ controls.
- Center all pan controls.
- Switch all routing/group buttons out (not selected).

The last is a common practice because it ensures that in the rush to set your system, get through sound check, etc. you don't overlook a setting from a previous session (and assume you just selected it) that may not be best for your current mix. By starting with all controls in a neutral position, you can be assured that any settings were actively selected to enhance the current mix. Also making it a habit to check the status of every knob, switch, and fader before using a board ensures you don't get a nasty surprise from unexpected errant signals interfering with your gain structure and overall sound quality. Whether or not you make a habit of doing it for yourself before using a board, you will absolutely want to do it after using any board that others will be mixing on after you. It is considered bad etiquette to neglect to do so, and any time you guest on another engineer's board, or will be having someone guest mix on yours, they will expect to find all controls normalized.

Power Up

Once you've ensured everything is at safe levels, turn the system on in this order: backline, rack, mixer, crossovers/speakers, and finally amp/s (see below for procedure of turning on amps). The most important element for you to remember is to always turn on amps last, and turn amps off first when powering down (full power-down procedure simply reverses the power-up procedure). After powering up, whenever anything connected to the P.A. system (microphones, keyboards, guitars) needs to be turned on/off, make sure that the "level" slider for the channel it is plugged into is pulled all the way down. Switching active mics on/off is especially likely to produce loud pops unless care is taken to mute the channel or lower the volume first.

Turning on Amplifiers

First plug in a CD or MP3 player to a free channel to provide some music, adjust the gain for that channel, and turn up your main mixer volume to unity or two-thirds of the way up its fader track (set your EQ gains at unity also, for any units that lie in the signal chain before your amplifiers). Go to your amps, but before leaving your mixer, if there are too many people about, make sure to leave someone to watch your board so no one turns the main fader while you have your ears right next to your system's speakers.

Turn amplifiers (or powered speakers) on one at a time, leaving a few seconds between each one to allow power supply capacitors to charge (allow the same time between each when powering down as well). When they are powered up, raise the volume/gain on a single amp channel, and check the amp's channel level sound from the connected speaker(s). Turn the amp level back down. Repeat for each amplifier channel until you have checked all channels for each amplifier and speaker. At the end power the subs up one at a time, this time keeping the volume of the first raised. This serves as a basic phase test: if the two subs together are louder than the first alone was, they are in-phase.

Finally, raise the gain levels on the first amp channel till you reach the desired speaker levels, and then do the same with the remaining channel. Make sure that amp gain is set so that the same level is being output by both sides of all paired speakers. Repeat with any remaining amps/channels. Don't forget that other speakers in the system, back line, and audience sound will raise the final overall sound level in the venue. Also make sure you are familiar with your amp and speakers, as with any of your equipment, before the first day you will be operating it.

Line Level

Proper gain structure starts with making sure the various components are appropriately matched. Before you can worry about setting "gain" or "trim", you need to properly match the line levels of your devices. In addition to the three standard signal levels that are used in professional mixers (mic, line, and speaker), many mixers will have pad controls for each channel that allow you to choose between the two common types of line level you will find on gear that can be plugged into your mixer channels, which are 10 dBV and +4 dBu. When a mixer allows for you to match your channel line input to the line output level of the gear you are interfacing with that channel, you will see better gain structure if you properly match the items.

The difference between the two is the original signal strength generally found in personal audio gear (−10 dBV) and that used in professional audio gear (+4 dBu). Pro equipment (especially for live applications) must produce higher output levels, allow for more headroom, and have robust components enabling it to withstand rigorous use. Home-based equipment is used differently and requires less signal strength to achieve adequate sound quality for the purpose. So as a cost-saving measure, equipment intended for home use interfaces using a lower signal level than pro equipment. Historically there were few occasions when there was reason to mix the two, but the home recording/digital revolution has created more demand for the ability to do so and still maintain optimal gain, so most new mixers allow for the operator to select levels to match channel inputs to gear outputs.

Basically the voltage difference between the two is 11.79 dBV, not 14 dB, because the two values have slightly different reference values (dBV = 1 V, while dBu = 0.775 V). Your mixer and most of the P.A. gear you may use will usually have ins and outs rated at +4 dBu, meaning:

- +4 on output = more voltage, stronger output,
- +4 on input = less sensitive, ability to handle greater voltages means it can work with higher signal levels before distorting (i.e., more headroom).

You also have personal gear you may have reason to interface with the P.A. (e.g., an iPod for intro music, personal headphones). Much of this type of gear will have outputs rated at −10 dBV, meaning:

- −10 on output = lower voltage, lower output,
- −10 on input = more sensitive input, potential overload if high voltage input (+4 dBu) produces peak values greater than relatively lower headroom found in personal gear.

Because of this difference, if you plug in low-voltage gear into your pro system without matching the levels, the signal sent to your mixer will be a power that the mixer is not expecting. Because of that you have to amplify the signal much more than the rest, so you introduce more potential unwanted noise into the signal flow.

If your mixer or other gear allows for matching line levels, make sure everything matches. If you are unsure if a piece of gear matches, usually your ears will tell you, as the signal will have a different strength than the signals of equipment known to match. If you are dealing with an older mixer that doesn't have this feature, you will sacrifice some sound quality when plugging in unmatched gear, but both values are safe to plug into the other's line inputs (unlike mic, line, and speaker levels which are too different to safely use interchangeably without using a DI box to convert line signals to mic). You will simply need to bring the level up or down to the appropriate level using your gain controls.

While you're at it, make sure to check if your mixer channels automatically adjust according to mic or line inputs or if they include pads requiring you to select which input is to be used, pay attention to the outputs of your wireless receivers (some deliver mic level, others line level), and use direct (DI) boxes (not line inputs) for keyboards or other instrument-level devices such as samplers. By matching line levels before setting gain, you can make sure to start with the maximum amount of headroom and the best initial gain structure.

Ringing out the Monitors

Equalizing monitors require more time and focus than equalizing the main P.A. because their proximity to the microphones makes them more prone to feedback. Make equalizing the monitors your first priority so the biggest source of feedback gets equalized even if your setup time is cut short. If needed, the mains are a lot easier than monitors to equalize once the band has started. Also, ringing out the stage monitors requires raising the volume level until it borders on feeding back, so make sure no one is on or near the stage to ensure that no noise enters the mics during the process—otherwise all kinds of damage (in the form of blown-out ear drums and speakers) could result. To ring out the monitors, follow these steps:

- Flatten the graphic EQ's sliders a bit below center to leave room for moving these upward as you set your EQ, while still leaving a little room to adjust them downward as well.
- Gradually slide the master monitor volume control up until you can detect faint feedback. Don't let feedback get out of control, but keep volume right on the edge of feedback. If it threatens to become real feedback, pull the volume back just enough to end any continuous feedback.
- Slowly slide up the highest frequency (far right) slider on your EQ until you begin to hear faint high-frequency feedback coming from the speakers. This is the "ringing" referred to by the procedure's name. Once you hear it, don't move the EQ up any farther. Instead, tweak the slider down slightly until the feedback subsides.
- Move over to the next slider and follow the same procedure. Continue to work your way from high to low repeating the same steps with each of the remaining sliders on the EQ.
- Return any sliders that move up all the way and don't feed back, and adjust them by ear until they sound right, then move on and keep ringing out the rest.
- If a lot of them will move up without feeding back, slide those that do feed back down even farther than where they started and begin again.
- The frequencies that were rung out in first pass will allow for enough increased volume before feedback to allow for some of those frequencies that didn't ring on the first pass to respond and be "rung" out as well.
- Return any sliders that again move up all the way without feeding back, and this time adjust them by ear until they sound right.
- Keep ringing out the rest until finished.
- Turn the master monitor volume back down to the level you will be using during the performance.
- Check a microphone and see how the results sound. If all sounds good, move on to the mains.
- If it doesn't sound the way you want it to yet, adjust the equalizer's sliders by ear a little more, but make your modifications to any sliders that did cause feedback by reducing frequency levels rather than raising them. The others can be adjusted up or down without creating any feedback problems in the monitors so you will want to concentrate your efforts there.

As your ear gets better at identifying frequencies, you will get to a point where you can identify the approximate frequency you are hearing feedback and locate the corresponding slider on the EQ needing adjustment without needing to tweak them one by one. This will make equalizing the monitors easier, because you will be able to simply raise the master monitor volume and identify by ear which frequency feeds back and move its slider down. Basically, you will be able to do the process in reverse—making feedback happen with the master volume and then reducing the problem frequency.

Equalizing the Mains

If you have a second graphic EQ, you can use it to equalize the main P.A. If the venue is small enough and problematic enough to benefit, you can repeat the same procedure and "ring" out the main speakers the same way you did the monitors. Because the P.A. is farther from the mics and not as prone to feedback, this is usually not needed. Instead, you will get the most benefit from sparing (less is more when it comes to equalizing) and strategic adjustment by ear. If your sound check promises to be leisurely enough, you can adjust the FOH graphic EQ then.

You usually don't have much time to even deal with the board during sound check, so most often you will use a CD to set your EQ. Make sure to use music that is of the same genre and sound as the bands you will be mixing, and also is something you know well enough to know how it should sound. As you equalize the room, make sure to follow the best equalizing practices found in the following chapter, to have a reason for all adjustments (the slider has not yet been tweaked is not a good enough reason), and to check your adjustments by moving to selected spots in the venue a few times to assess the sound (and make adjustments to the sound) from locations throughout the venue. When you are finished, take a break and relax for what ever time remains until sound check. Or, if you can't resist, pull out a list and double-check all your settings.

Chapter 22

All in the Mix

Fundamentals, Mixing Monitors, and Sound Check

Setting Gain Structure

You must get the gain settings correct at each mixer channel in order to deliver a clean signal to the rest of the chain. You can boost the fader up as high as you want, but if the trim is off you'll get nothing but noise. Conversely, if you have the trim way up and the fader way down, the chances for distortion are much higher. In addition, it is as important to concern ourselves with setting our gain structure so that it is balanced across all signals as it is to make sure individual channel signals are proportioned well.

Most mixers have plenty of headroom, typically +20 dBu to +24 dBu, occasionally even more. However, the system's overall capacity will be limited by the point in the signal chain with the least headroom. This is why you need to adjust each channel gain so that signal levels and headroom are the same in each channel, since system headroom will be determined by the channel with the least amount anyway, any extra headroom in other channels is wasted, and variable levels only make it more difficult to estimate your overall headroom. Also, individual channel levels combine to produce a higher overall level. If the peak level in each channel is +20 dB, each individual channel may be (just) below clipping, but the sum of all the channels will be considerably above +20 dB, causing the main mix to clip. If you are using many channels, using even a fraction of the headroom in each channel may be enough to overload the mixer's summing amplifiers.

Nab band members one at a time and have them create their respective signals into their mics for the short time it will take you to set the line levels. With each channel, select the PFL for that channel and gradually adjust the gain until the loudest sounds from that signal indicate 0 VU on the meters. Refer to the procedure for determining max output, only instead of the red LED indicating max output (which should be somewhere around +9 VU), in this case the green LED representing the value 0 VU should flicker only at the highest signal peaks.

If the dynamic range of the signal is too great to make this level practical, you can use a compressor to compress that signal, or otherwise you should keep any extra gain applied to a minimum.

Aim to set each channel's fader near the unity position if possible (this should be marked, though if not go for somewhere two-thirds of the way up). If trim is all the way down and the PFL signal is still far above "0", try using the pad switch—it will lower pre-amp sensitivity enough to allow for a lower signal less likely to cause distortion. Either way some of the channel faders should be at or near the "0" mark. If too many faders are very low or very high, then something is wrong with the gain structure. Keep in mind that other junctures along the signal's path, like the main out controls, should also be set at or near "0." If setting the master fader at unity makes the volume in the room too loud, turn down the level controls on the power amps. If you need to bring the master fader far beyond unity to get adequate volume in, then your amp gain is too low, or your system is underpowered.

Monitor Mixing

The monitor mixer uses a dedicated monitor console located at the side of the stage and makes independent mixes for all the band members, tailoring the mix composition and signal levels to each performer's needs so they all can hear themselves and the music.

The monitor mix is one of the most important elements of the show. When musicians can't hear what they are doing, they make more mistakes than when they enjoy a perfectly balanced stage sound. The monitor mixer is in charge of helping the band hear themselves, so the monitor mix is the single biggest factor influencing the performance of the stage talent. A good-quality monitor mix will make performers more comfortable so they can perform their best. A weak or distorted monitor mix will leave the performers feeling a great deal of frustration and anxiety. There is nothing worse for a performer than to be left without a clear reference to their own performance or to wonder if they are in sync with the other performers on stage. The kind of anxiety this creates for performers guarantees that they will not have their best show. Mixing monitors can be really confusing and becomes even harder in smaller venues when you are tasked with mixing the monitors from the FOH position, and dealing with both the monitor and FOH sound. This can get pretty intense as you juggle many essential tasks at once so expect to make the normal mistakes as you learn bad monitor mixes are one of the first reasons cited for a bad show by most bands.

There are some controls on the main FOH mixer that must be located and set up and are imperative to have ready before sound check: aux sends are the only outputs that will enable a full mix to be output at each of the aux sends used. You will want to prepare one monitor mix per aux send as well as putting each aux send monitor mix to pre-fader send, because you don't want the stage mix levels to change when you adjust the mix being sent to the P.A. It's important to figure out how many different monitor mixes you need for your performers and to make sure you have the aux sends prepared in advance so you are free to focus on the mix instead of set up tasks.

It is also important to figure out in advance how you will "check" each of the monitor mixes you are sending. You must to be able to check them during the show without having to leave the FOH position. Headphones-based monitoring of the aux sends is usually sufficient; however, headphones can't detect sound issues that may occur along the signal chain after the signal leaves the mixer, so the FOH engineer with reason think to that the output signal could be powerful enough to drive the monitor wedges to distortion will want to use a monitor speaker at the console to check monitor mixes—ideally one similar or identical to what's being used by the lead vocalist.

Before preparing your mixes, you need to set the gain (or trim) at each send, as well as at the master aux. With the whole band playing during sound check, after mapping your aux sends to pre-fader send, turn your master aux gain control to 5 (or 12 o'clock). Then increase the gain of each performer's signal at their assigned monitor in their monitors (also to 5 or 12 o'clock). There should be an audible signal at each performer's monitor at this point; if not while tweaking the aux send gain of any signal/mix that's too loud or soft compared to the rest, continue increasing gain at the master aux send till you are satisfied with the overall sound. Finally, using the chart provided and consulting each of the performer's preference, start adding in other elements to each mix and adjusting the relative gain of each until you've finished building an initial mix for each performer.

The charts to the right suggest preferred sound levels for various mix elements according to each performer's own instrument. The left-hand chart column lists performer instrument which set the baseline, the center list selectable channels for each performer's mix, while the instructions to the far right suggest the best output level for each mix element relative to the baseline output of the performer's instrument (i.e., stronger = higher then 5, same = near 5, lower = less then 5).

Stage Sound

Because the stage is the only place in the club where noise sources, microphones, and speakers are tightly grouped together, it represents one of the biggest potential points of failure and sources of feedback, and usually will take the most effort to keep under control. As FOH engineer, much of your careful setup and planning focused on achieving the best sound for the audience will hinge on your ability as monitor mixer to

Background Vox	Drums	Much lower
	Percussion	Much lower
	Bass	Lower
	Keys	Lower
	Horns	Much lower
	Guitar	Lower

Guitar	Drums	Slightly lower
	Percussion	Lower
	Bass	Same
	Keys	Lower
	Horns	Much lower
	BG VOX	Lower
	VOX	Same (touch lower)
Keyboard	Drums	Lower
	Percussion	Much lower
	Bass	Slightly Lower
	Guitar	Slightly lower
	Horns	Much lower
	BG Vox	Lower
	Vox	Slightly lower

Horns	Drums	Same
	Percussion	Lower
	Bass	Same
	Guitar	Lower
	Keyboard	Lower
	BG VOX	Same
	VOX	Same

Drummer	Percussion	Same
	Bass	Higher
	Guitar	Low
	Keys	Much lower
	Horns	Much Lower
	BG VOX	Much Lower
	VOX	Much Lower
Percussionist	Drums	Higher
	Bass	Same
	Guitar	Lower
	Keys	Same
	Horns	Lower
	BG VOX	Lower
	VOX	Lower

Figures 22.1, 22.2, 22.3 and 22.4.

work with the band to maintain the balance of sound on stage. As monitor engineer, your main concern is with feeding enough sound to the stage to please the performers; as FOH your concern is with not letting stage monitor sound build up enough to spill off the stage into the listening areas, or worse, to enter the system via one of the stage mics. Working with guitarists to keep their stage volume down is of major importance, because their amps tend to get the loudest, and are most likely to start the volume escalation over the course of the show, upsetting the delicate balance between the sound on stage and the main mix. Hopefully, you began dialogue with the band earlier than the day of the show, and reached out diplomatically to begin to establish trust, as that gives you the best chance to secure their cooperation, but whether you did or not, you need to go over the factors that will provide the best odds for keeping stage sound low enough to keep it from interfering with the FOH sound.

- Talk to the band one last time before or during sound check about the importance of keeping stage sound under control while reassuring them you know how essential it is that they can hear the monitor mix and are committed to helping them hear themselves and making them sound as good as possible to their audience. If they trust you as an ally and understand the issues created for their overall sound when they crank up their monitors and stage amps, they will be more open to working with you.
- Make sure they note the position in which they need to generally remain in relation to their monitors in order to keep the sound aimed at their ears. If they don't like the distance, reposition them, or put a wedge under their monitors, whatever they find most comfortable.
- Ask them to start playing at lower levels and ramp up slowly so stage sound doesn't get too loud too early.
- Make sure to only feed the channels that each band member really needs to hear in each mix.
- Mix channels should be no louder than what the respective performers really need; otherwise it may mask other sounds in the mix and muddy the overall sound quality.
- See if there are any other stage tweaks you can make together to help them keep sound down (e.g., dampening drums, raising or re-angling amps, turning off an amp and plugging in directly with a DI box).
- Once their monitor mix is exactly what they want it to be, just leave it alone, checking now and then in case it must be adjusted, but making no changes unless by request.
- Cut frequencies not in the audible output range of frequencies of the instruments on stage.
- Avoid using compression in the monitors, because this actually encourages really bad mic technique for most inexperienced vocalists. Compressing vocals is only crucial in an in-ear monitor mix, but is not needed in wedges.
- Note that vocals and guitars generally compete for gain most often, and therefore are most prone to creating feed back.

If encountering resistance from the performers, assure them that you will deliver them exactly what they ask for on monitors but that they should consider meeting you halfway as it's in their best interest to work with you to keep stage sound manageable. Let them know that if stage sound gets so loud that you are forced to choose between thinning out their sound yourself in the main mix, or turning up the master volume to distorted or unsafe levels, you will always choose to thin out their sound, even if you would prefer not to.

Sound Check

- Keep all main P.A. faders down to begin with.
- Set all aux sends to the monitors to pre-fader mode.
- Prep each channel that needs high-pass filters and phantom power.
- Add compression and gates to channels that may need it.
- Depending on the amount of time you have for sound check, have the band move through each instrument and on each channel's pre-amp and re-check the gain to make sure it is still right after having equalized the room.
- Turn up monitors and set each one to beginning settings outlined in monitor mixing in the next chapter (at this point the monitors on stage should begin to sound out, because your aux sends are in pre-fader). Before turning up the main faders ask the talent if there are any changes they desire to their monitor mixes.
- Arrange for a few clear hand signals you and the band can use to communicate basic ideas during the show.
- Assuming you were provided with the set list and cue sheet ahead of time, double-check that no changes have been made so you can be sure you know when events like solos, etc. will occur. Have the band play a number, and after starting with all your faders down bring up all instruments and keep your main output as close to 0 VU as possible.

- Apply general equalizing to each instrument where needed (following the guidelines given in the next chapter).
- Walk around the venue (leaving the FOH position) and listen to how it sounds everywhere around the venue (FOH position affects tonal balance).
- Return back to the FOH position and adjust the mix accordingly.
- Record the settings of the board either by using a scratch sheet, or by taking pictures of the board, eight channels at a time.
- Move on to the next band.

Dynamic Processes

In live sound there are three dynamic processes that you will need to use in order to mix a professional sounding show: compressing (single channel and buss compression), gating (or using a noise gate), and limiting.

Setting Your Compressor

Hook a compressor into any instrument or vocal channel where the signal range varies between too loud and too soft. To set a compressor, start by setting the ratio to one of these suggested ratios:

- vocals 3:1;
- bass 2–4:1;
- drums 4, 5:1;
- guitar (solo) 3–4:1.

Then, with the attack and release set to somewhat fast times (about 9 o'clock for attack, and 12 o'clock for the release), bring the threshold from its highest setting down until you begin to see some gain reduction on display meters. Depending on your need, which will almost always be a slight compression, turn down the threshold until you have 3–4 dB of gain reduction. If there is a need to preserve the signal's transient, slowly increase the attack until you can still hear it. Finally add back in the gain lost by using the makeup gain parameter, and hit by-pass till you need to apply that compression during the show.

With standard uses of compressing a single channel some common uses of compression include the bass guitar and on lead vocals to control variable dynamics of the singer. When a compressor is used to compress many channels routed to a sub-buss, it is referred to as buss compression. A common example is applying buss compression to the drums set, by routing all drum channels to a sub-buss before it goes out to the main channel.

Gating

The processing type used in live sound is focused on an automated way to control noise on channels which have gotten noisy from power getting on the cables or more commonly noise coming from the performers. The gate, sometimes called the noise gate, is a set-and-forget way to keep unwanted noise getting into the mix from effects. You set a noise gate to be closed (no sound gets through) till the instrument is played. This will raise that channel's amplitude to a level that will push open the gate and all sound on that track will be heard—until that instrument goes quiet and the noise will then return, unless the gate is set properly, and will shut again after the level drops below the threshold point and the track again is kept silent.

Setting Your Limiter

Every installed audio system has a brick-wall limiter as the last process in the audio chain before the signal reaches the power-amps then the speakers, if the club owners or audio installers know their trade. This is the only way to protect their system speakers from getting cooked by an overloaded signal leaving the mixer or amplifier. If in any doubt, where don't know for sure whether a limiter is or is not present in the signal path use your own limiter to make sure you don't damage the club system.

When setting a limiter, first turn up program(music)-based material till your system reaches a level of amplitude you determine as the loudest levels you want to allow. Set the limiter's attack infinitely small (1–3 ms) and set the release somewhat long (1 sec). While watching the limiter gain reduction meter, turn down the threshold of the limiter till the meter just begins to show the smallest amount of reduction. If your system is installed, make sure it is out of reach of anyone who might alter its settings. If your system is a mobile P.A., you will have to reset the limiter's threshold at each new location.

Setting Effects

The signal leaving the mixer will pass through the master fader first and then out to the processors that affect the entire mix. Without enough experience any experimental processing in an attempt to be "creative" could be disastrous, so *until you have had a chance to practice hearing and applying creative effects, don't stray from the standard processing tools that are nearly always used in the signal path after mixer.* Any time-based processing (reverb and delay) will only hurt the clarity of the mix and should not be used here, especially not by an inexperienced engineer in a small-venue setting. That said, for those times you get to experiment, or if a band insists on a touch of reverb in their monitors, you will want to know how to set up effects units.

Properly adjusting the levels for the effects is similar to adjusting the input signal levels of your mixer channels. An effects unit usually has some kind of clip light or gauge indicating the strength of the input signal. For best performance you want the gauge to be right before the upper edge of the allowable levels. The gauge is usually marked clearly to indicate the maximum input level it can handle before it begins to distort, but when in doubt read the manual.

To adjust this input signal level:

- Turn all the output signal controls for the effects signal path to around halfway; you will want to actually hear the effects while you set them.
- If the effects unit has a "bypass" button, make sure it isn't in bypass mode.
- Turn the "master" effects send knob on the soundboard to about halfway.
- Turn the "input" knob on the effects unit up to around halfway.
- As the microphone is receiving a signal, gradually turn up the effects or aux knob on that channel until you see the gauge on the effects unit reach the proper level (or until the clip indicator lights up, then dial back). You will start hearing the effects on the channel as you turn them up. Ideally, you would want the channel effects knob to end up at around halfway when the signal reaches its proper level in the effects unit. This is so you will have room to adjust it either up or down if you need to later.
- If you find that you have to turn it all the way (or almost all the way) up to get enough signal, back it off a bit and turn up either the "master" effects on the board or the "input" on the effects unit.
- If you find that you get a strong signal when barely turning it up at the channel, try turning down either the "master" effects on the board or the "input" on the effects unit.
- Once you have the balance between those three knobs set to an acceptable level, set any other channels using only the effects/aux knob on each individual channel (as the master mixer level and input on the effects unit is already set).
- Once all the channels are set, turn the "master" effects knob on the board or the input level on the effects unit down just a hair so that you can have a little headroom to adjust the individual channels up some later if it becomes necessary.

Once the input levels are set, you can set your effects as needed:

- Adjust the effects volume to the main signal by using the "effects level" or "return" knob on the mixer.
- The "mix" knob on the effects unit adjusts the ratio of original signal to effects in the signal leaving the effects unit. Turned all the way one way, you get only the effects (i.e., only the echo, not the original sound); turned the other way, you get nothing but the original signal.
- Some mixers have a "monitor effects" knob so you can route the return signal of the effects to your monitors.

A few more details about effects that you should know include:

- You can loop your effects into a spare channel in your mixer instead of to the "effects return" jack. Your level adjustment will be at the channel fader and your monitor effects adjustment will be on the "mon" control for that channel.
- You can also equalize your effects by using the EQ knobs assigned to that channel. *Just make sure to keep the effects knob on that channel turned all the way down* (otherwise an internal feedback loop can occur if it is left up, potentially damaging your speakers from the horrible sounds that will result).
- You can place effects in line with the compressor between the mixer and the amplifier. The downside of this method is that you give up control over individual signals and adjustments affect everything in the mix equally.
- You can hook up effects in line between an instrument/microphone and a single mixer channel using the sends and returns assigned to that channel.
- You can chain effects in sequence, one after the other, but pay careful attention to each signal meter in the line and make adjustments to the input levels to avoid overloading the signal.

Be aware of problems that arise from effects in general:

- If set too loud, effects can contribute to feedback issues; if feedback becomes a problem, check to make sure your effects aren't too loud.
- Don't fall into the trap of trying to "cover up" bad musicianship or vocals by turning your effects up. Effects are meant to enhance good sounds, but they don't do anything pretty to ones that don't already sound good raw.

Essential Mixing Concepts

The balance inside any kind of live show is fundamentally based on the stereo image, and how to organize that stereo field is completely dependent on what is being reinforced. When you're reinforcing a bunch of electric instruments, that type of mixing is just fitting everything into the sonic space. Whereas with an acoustic instrument show it's much more important to put the acoustic instruments in the mix originating from their approximate stage location. If this is not done it will ruin the intimate feel of an acoustic show.

Panning

The main instrument or the instrument of focus should be slightly louder than everything else and somewhat supported by the other instruments on stage. Panning is the main way to separate each instrument in the mix. Panning full left and panning full right are the edges of your sound field. Everywhere in between those edges is space to separate tracks to ensure no sonic in-fighting occurs between instruments. When getting instruments' sounds separated so they don't "fight" for sonic space, it is more reasonable and more effective to pan the instruments away from each other in the stereo field than it is to try to separate them by equalizing.

Equalization

When using equalization in a live situation you're not going to achieve precise control of frequencies; more often than not your frequency paintbrush is actually wide and general as there's a limit to the sort of sound issues that you can truly fix in a live show format. Graphic EQ and pre-set EQ controls are the norm for live sound mixing boards because very narrow band equalization (parametric) boost/cutting changes are barely perceptible in a live venue.

Simple and common examples of using equalization to fix an instrument in a live setting include:

- Lowering the higher frequencies 5 to 7 K for a violin.
- Taking the screaming, tearing sound out of a brass type horn (i.e., trombone).
- Removing certain frequencies out of the guitar distortion so it keeps the warmth without overpowering the musical notes.

Even with this difference in what an equalizer can achieve in a live show, it still holds true that when you use the EQ, it is always better to cut frequency bands before you boost any frequency. Always think what should be removed before considering a boost. Remember there is no frequency issue that can't be solved by turning the instrument or mic level down at the fader.

Dynamics

Controlling dynamics in your mix is exactly that: you're controlling the distance between the louder and softer portions of every instrument or for groups of instruments as the situation may call for. When controlling dynamics, first start with the proper setting of the gain or pre-amp at the beginning part of every mixer (usually the top most knob on every channel on a mixer); during sound check you found the proper amount of gain for each input to have each input at roughly the same VU level. If that was done properly then hands-on dynamic control of the show can be done by literally keeping your hands on every channel fader and adjusting each channel in relation to the overall sonic feel of the entire mix. Realize your first line of defense against clipping and overdriving the system is to keep your finger on the fader to adjust from moment to moment.

Overall Volume Concerns

An underlying problem in many live sound (installed or mobile) setups is lack of attention to the gain structure of the system as a whole, which can be especially sonically damaging when the issue is malformed gain structure in the FOH mixer heading output to the speakers.

The most common newbie mistake that live sound engineers make is failing to understand how a limiter does its job. If it is not understood that a limiter is there for speaker protection and not for sonic quality control, assuming the limiter can be regularly engaged to "catch" any dangerous excess without any consequences to their sound, the newbie will turn up the main output level keeping the limiter fully activated. In doing so, the newbie sound engineer will figuratively push the sound in-between a rock and a hard place. A limiter does its job by lopping off the sound wave at the top and the bottom of the original complex wave. The more they turn up the main the more the limiter cuts the wave into a square. Once the signal level is activating the limiter, the sonic clarity only degrades, and amazingly quickly your amazing mix is nothing more than a square wave of ear-tearing horribleness! Always remember, first trust your ears, and then confirm your observation by checking all VU displays to ensure proper gain structure and, in the end, the best-sounding show possible.

Chapter 23

Mixing the Live Show

From Curtains Up to Last Call

15 Minutes to Showtime

Even after you've gotten some time and experience under your belt, you will want to try and be in place 15 minutes before curtains up, but it's especially important not to rush in and take your place at the mixer at the last minute when you are new and still a little out of your comfort zone. Being at the mixer a little ahead of time gives you a chance to center yourself, and go over everything you want to remember. Check to make sure everything you need is in place; though last-minute checks at this point are more a matter of knowing what you will be doing without, since if anything more complex than your flashlight or a trip to the bathroom got overlooked until this point, 15 minutes isn't going to be enough time to produce it. However, it is going to be enough time to discover that you missed something so you are not surprised by it mid-show, and have time to get a head-start on figuring out how to work around it. It also does ensure you can attend to a few essential minor details. If you haven't already you can use this time to:

- use the rest room, and get a drink/refill;
- add fresh batteries to your handheld recorder, flashlight, or SPL meter as needed;
- have your monitors/headphones and talkback mic plugged in and within reach;
- have any track sheets, cue sheets, a timepiece, etc. on hand;
- have your camera, notebook and pen, phone, gum, tissues, aspirin tucked away but easy to access as needed;
- throw up your "Do not disturb," "This is not the concierge's desk," or "Man/Woman at work" sign;
- breathe and chill out.

As the band is about to take the stage, the FOH board should have some initial conditions that were set up during sound check already in place:

- all channels clearly labeled with board safe tape and sharpie type pen (the importance of quick identification cannot be understated, making sure you don't waste time searching the board for channels or controls is crucial);
- every input coming into the live mixer has had its preamp gain set to levels that are similar in VU levels across every channel with headroom to spare;
- much of the planned outboard processing (compression, sends for delay, sends for reverb) are patched into the signal path where appropriate (all but the noise gates and limiter should be bypassed even if set);
- individual channels are routed to sub-mixes, group channels, and, if available, a voltage-controlled amplifier (VCA), in a logical and useful way;
- initial monitor mixes that the talent has agreed work for them are set up as rough mixes going to the stage wedges;
- if there wasn't time to set up the initial EQ setting during sound check, keeping the EQs in neutral position is fine.

Curtains Up

As the band takes the stage, turn down the intermission CD by half-volume, but keep it playing to give them time to get in place, plug anything else in, sip water, and give you the nod that they are ready to start the show. Once you have the go-ahead, bring down the CD volume completely and pause or stop the device playing music or just mute the channel and take care of it later when you get a chance (*remember to always mute any channel that is not supposed to be in one of your mixes and has sound/noise on it, i.e., a device plugged in and powered on*). Bring up your faders for the lead instrument and singer first, and follow with the rest section by section, remembering to bring up any sub-group/VCA faders too. Because it's the beginning of the show, effects will be ready to go but still bypassed so you shouldn't need to take care of that now; however, note that you need to *always be sure and bypass all of the time-based effects on the lead singer's mic before they begin to address the crowd between songs and at the start and finish of each set.*

As the band begins to play, *take a few moments to listen before you make adjustments*. Things will sound

Figure 23.1. Make sure to label your channels.

different than they did at sound check because people are squishy and absorptive, so your highs will sound muted in comparison to sound check. Also the drums may sound louder simply because no matter how much you ask them to play at the same level as show time, drummers hit harder when the curtains go up. After you've taken a second to assess the sound do a quick fader adjustment to make sure your levels are balanced as they should be, usually you will need to make sure to raise the vocals so they pop and lower the drum faders so they are still solid but sit in the mix instead of above it. You will also want to make sure the bass is audible in the mix and to bring the guitar forward without it interfering with the intelligibility of the vocal.

It is *very important to keep your overall main output below* 0 VU especially early in the show. This provides enough space for peaks and dynamic changes in the performance, and for the inevitable sound escalation that will take place over the course of the show. Once your volume faders are somewhat set, begin to apply the compressors and other dynamic processes that you set at sound check, as needed (except for noise gates and limiter which should already be in place as the faders and curtains go up). The final step in this initial flurry of action would be to apply equalization if it is obviously needed. Note that *as needed* is the operative instruction that should be guiding your actions here. All this should be accomplished quickly, within the first few minutes of the show, so you are taking care of any glaring problems, but shouldn't be fine tuning yet. Only change levels if an element is obviously not sitting right in the mix; only apply processes if you can hear they are needed already, and only apply a little equalization to account for the bodies in the room.

Naturally, the exact details of how to start your mix will vary according to the sound of the band and the conventions of the genre/genres they play. It is the job of the FOH mixer to slowly and delicately change the balance of the mix to highlight the lead instruments for each section of every song so it's assumed you did what you needed to do to become familiar with the act, or if you did not have the lead time or cooperation from the band to enable that, you should at least know the genre.

During the Show

Now, take a breath, take a step back, and listen to the mix. From here on out through the show, it is your job as the FOH engineer to keep your ears tuned to the needs of the mix as well to any audible problems, so you can act immediately to make adjustments. For the rest of the show, the FOH engineer will make subtle and very slow changes that should not be noticed by the crowd. What follows is a set of guidelines and principles that can guide your FOH choices throughout any show:

Take Care of Business

- Don't make any sudden or drastic changes. Most adjustments should be subtle. Occasional major equalization adjustments should be made smoothly.
- Cutting EQ bands is always better than boosting; whenever possible make choices based on this. For example, a sound can be made warmer by either increasing the energy between approximately 200 and 600 Hz or by reducing the energy in the 3–7 kHz region.
- Close your eyes every now and then to make sure your eyes aren't influencing your mix more than your ears.
- Remember to make choices strategically even as you keep making small adjustments. You will be making adjustments regularly, but don't twiddle knobs just because they are there. If it sounds good, sit back and listen and use the time to think about the mix. There will be reason to adjust soon enough.
- If you can't make out the lead vocals then nothing else matters. If it's already very loud instead of trying to make vocals intelligible by increasing the vocal level, try carefully lowering the level of everything else, a little bit at a time, until you can hear them. If the backline instruments are loud enough without the P.A., take them out of the mix altogether.
- Don't run out of headroom. Pay attention to those overload VU meters on board channels and outboard gear. Remember, boosting frequencies will cut into headroom.
- Keep an eye on all incoming signals and look for spikes in your LEDs to pick up on signal peaks, mismatches of output levels and especially any hint of gear overload.
- Get an SPL meter and keep an eye on it, — 110 dB SPL is very loud — if it is hitting the venue's limiter the whole mix can degenerate into an indistinct mush of noise.
- Remember, your first line of defense against clipping and overdriving the system is to keep your finger on the fader to adjust from moment to moment.
- Don't lose control—keep attentive to what is going on with your sound so you catch potential problems early. If you *do* miss a change don't try to catch up to what you missed, as it is jarring and just makes things

sound much worse; deal with the sound in the moment, not the sound from 3 minutes ago.

Take Care of the Band

Keep an eye on the band on stage, and watch for cues that can fill you in on what they need. A bit of empathy is a good skill to use throughout the show. Are they fiddling around with the equipment around them more than is usual? Do they look uncomfortable? If they are looking uncomfortable or upset it often will have something to do with equipment not working, or a monitor mix being too low or distorted; do your best to try and feel what might be off for them, check their monitor mix via headphones at the FOH now and then—you usually should be able to fix issues before they get bad enough for the band to signal you. Stay attentive, keep an ear tuned to signs of trouble with the sound, keep your eyes tuned to the band's cues, and be around if they need to get your attention.

- Keep a close eye on the performers. If the vocals suddenly drop out of the mix, check the stage while reaching for the fader; it may be something the lead did on stage. If you check there as well you'll see them reaching to fix the mic they accidentally knocked and now is pointed at the floor, instead of hunting phantom problems at the mixer.
- If any of the performers are staring at you, it may mean something's not working, it may mean they are out of water, but it also may mean they had an awkward eye-contact moment with a freaky fan one too many times and need a safe focal point. Try and figure out what may be going on as well as you can but don't sweat it when you can't; if you prepared right they know how to signal you. Keep an eye out for gestures, otherwise just smile and keep doing your job until they clue you in on what they need.
- Keep your ears open. Throughout the show, listen out for booming, ringing, or any untoward sounds. You should be the first one to notice if a speaker cone dies or other major P.A. malfunction.
- Additionally you should know before anyone asks you that the bass-player's strings are causing that rattling noise, and not your mixer, power amps, or loudspeakers.
- Watch the band for signs they need you to act on, or for changes of instrument (you may need to mute the guitar player's channels while he changes to another guitar for another song and un-mute when he is jacked back in; otherwise you will get a very loud pop throughout the P.A. system during such a changeover).

Take Care of Issues

- If you are losing control of the mix, don't freak out. Keep cool and stay calm, and take a moment to think about the problem before you try to correct it. Remember, if it hasn't hit feedback levels of badness, eight out of ten audience members probably won't notice it immediately if at all. Of those that do notice, half will look to verify they are right before they'll be sure. Don't be caught scowling and looking like there's a problem—just nod, move to the music, and keep looking at the console as you correct the problem.
- If you feel the mix is totally falling apart, it's still not time to freak out. Slowly flatten settings for the channel strips and start over if you need to; focus on what needs to be done and you'll be back on track in no time.
- If you experience technical issues, follow the advice in the next chapter, but as with problems with the mix, stay calm, stay focused, and stay positive.

Take Care of Yourself

- When problems occur, as they inevitably do, take care of the problem, and then let it go. If you are busy beating yourself up you can't be attentive to catching the next issue that arises. If it's something important, note it and deal with it later; your job now is to stay with the show, and in the moment.
- Stuff happens, realize you are not superman and don't claim responsibility for more than your share of the blame, if you are like the rest of us you have enough to deal with just by dealing with what *is* your fair share.
- If after later thought, you decide you made a mistake or caused an issue due to your own fault don't waste the opportunity but evaluate your mistake honestly and figure out how you can make adjustments so you do a little better next time. Then avoid the trap of wasting time with calling yourself stupid or focusing on your failing, treat yourself with the same compassion you'd show someone else, and drop it. If your tendency with other people would not be compassion but to harp on mistakes, refer to the bonus online Chapter 29 and work on that.

When things are going well, keep attentive for issues that may arise, but also give yourself room to enjoy yourself. Sure you are at work, but it's a pretty kick-ass job; if you don't let yourself have some fun why the hell are you there? So if it sounds fine don't fiddle with controls, but rock out, take some pictures for your portfolio (the band will appreciate good photos too, so be sure to pass on the best ones). Stay attentive and when the time to turn back to business arises,

Sub Bass (Growl, Rumble, Power) 20Hz to 80Hz	Range of kick and bass. These frequencies felt more than heard. Due ti eqaul loudness principle, difficult to hear at low levels. Produces a sense of power and intensity in an instrument or mix. Cut all below 40 Hz as instruments don't create audible frequencies below that. Boost single instruments as appropriate, but not the entire mix at this range, or it will sound mushy at high levels.
Bass (Boom, Thump, Mud) 80Hz to 250Hz	Produces a sense of thickness in an instrument or mix. In excess, results in muddiness, boom, and loss of clarity.
Low Mid-Range (Fullness, Body) 250Hz - 500Hz	Much of the fullness or richness of sounds comes together here. Range of low order harmonics. Excess here can muffle higher range instruments.
Mid-Range (Horn, Cheap) 500Hz to 2kHz	Range of most fundamental harmonics, this range creates the foundation for a solid mix. Careful cuts of instruments dominant in this range (Guitar, Piano,Vocals) produces a cleaner sound. Determines prominence of elements in a mix. In excess makes sounds tinny, and contributes to listener fatigue. Cut at 500-1k to reduce Horn like quality of sounds, and at 1-2k for tinny sounds.
Upper Mid-Range (Prominence, Horn) 2kHz to 4kHz	Range essential to clarity, and responsible for attack in percussive and rhythm instruments Boosts can add presence to instruments. In excess this range adds a harsh tone to sounds and contribute to listener fatigue.
Highs/Presence (Clear, Bright) 4kHz to 7kHz	The range where s sounds and the attack and definition of most vocal consonants are defined. Boosts can increase closeness of sounds. Cuts can increase the distance transparency of sounds
Brilliance/Treble (Air, Sparkle) 7kHz to 20kHz	Sounds in this range consist exclusively of harmonics. Vocal sibilance and "metallic" attack of drums at 7kHz. High end "sizzle" of cymbals and vocal breathiness at 15kHz

(a) 40 - 80Hz	• Boost to add thump to Kicks & Toms. • Boost to add fullness to low frequency instruments • Cut to decrease the "boom" of the bass and to increase overtones and the recognition of bass line in the mix. • Cut vocals in this range, these frequencies not produced vocally
100Hz	• Boost to add a harder bass sound to lowest frequency instruments. • Boost to add fullness to guitars, snare. • Boost to add warmth to piano and horns. • Cut to remove boom on guitars & increase clarity (Helps separate bass and guitar in the mix
200Hz	• Boost to add fullness to vocals- can be used to compensate for thinness created by high frequency boosts. • Boost to add fullness to snare and guitar (harder sound). • Cut to decrease muddiness of vocals, mid-range instruments, and bass.. • Cut to decrease gong resonance of cymbals
250 - 350 Hz	• Cut to thin out vocals • Cut to remove boom from piano
(b) 400 - 500 Hz	• Cut to get rid of the cardboard box/hollow sound of kicks and Toms • Cut to decrease ambiance on cymbals. • Cut to decrease bark from piano
600 - 700Hz	• Cut to reduce rattle of snares. • Boost to add clarity to bass lines especially when speakers are at low volume.
800Hz	• Boost for clarity and "punch" of bass. • Cut to remove "cheap" sound of guitars. • Cut to reduce nasal quality of vocals • Cut to reduce tinny quality of snares.
1.5KHz	• Boost for "clarity" and "pluck" of bass. • Cut to remove dullness of guitars.
2k - 3 kHz	• Boost for more attack on low piano parts. • Boost for more clarity / hardness on voice. • Cut to disguise out-of-tune vocals / guitars
3.5k - 4.5kHz	• Boost for more "pluck" of bass. • Boost for more attack of electric /acoustic guitar. • Cut to increase breathy, soft sound on background vocals. • Boost for vocal presence.
(c) 5KHz	• Boost to make kicks/ toms cut through the mix • Boost for more "finger sound" on bass. • Boost attack of piano, acoustic guitar and brightness on guitars • Cut to make background parts more distant. • Cut to soften a thin, hard edge in guitar.
7KHz	• Boost to add attack on low frequency drums (more metallic sound). • Boost to add attack to percussion instruments. • Boost on dull singer. • Boost for more "finger sound" on acoustic bass. • Cut to decrease "s" sound on singers. • Boost to add sharpness to synthesizers, rock guitars, acoustic guitar and piano.
10KHz	• Boost to brighten vocals. • Boost for "light brightness" in acoustic guitar and piano. • Boost for hardness on cymbals. • Cut to decrease "s" sound on singers.
15KHz	• Boost to brighten vocals (breath sound). • Boost to brighten cymbals, string instruments and flutes. • Boost to make sampled synthesizer sound more real. • Take care with boosts to avoid highlighting hiss and noise

Figures 23.2, 23.3a, 23.3b, and 23.3c. Equalization and Hz notes.

you'll know it, and you'll do it, and odds are you'll do it that much better for having taken a few minutes to enjoy it too! If you can't enjoy working FOH now and then you might as well take up dentistry, or anything else—sure you won't enjoy that either but it'll pay better, and there will be room in live sound for someone else who appreciates it.

Equalization

Remember to always keep in mind what you know about the human ear and the nature of sound and use these to guide your choices. As you mix, don't be afraid to jot down a few notes to fill out later, and if you try a mixing tip or equalization adjustment you read about and it doesn't sound right with your mix, undo it and try something else—just because it's written down, doesn't mean you should take it as gospel. Always let your ears be your guide. Figures 23.2, 23.3a, 23.3b, and 23c show a frequency spectrum outline of general frequency adjustments according to the changes that frequency will accomplish. Remember, though, the point is to use these to listen and to learn what works where by what your ears say, not to memorize formulas and become stuck with them. We provide these because everyone wants them to start with and you could easily go online and find the same thing, but our hope is you use them in pursuit of never needing to look at them, and not because you memorize them, but because you develop your own understanding and make them irrelevant to you.

Finally, remember that while you should learn as much as you can each time you get a chance to mix, this is not your practice time but other people's show. To do some odd processing in an attempt to be "creative" at too soon a stage could be disastrous; until you gain experience it is not recommended to stray from the standard processing usually used in the signal path once the signal has left the FOH mixer.

Figure 23.4. Let your ears be your guide, stay in the moment, and enjoy yourself.

Get the Next Gig

Some problems you want to avoid at all costs. If you expect a call for the next gig, keep these to a minimum:

1. *Feedback*: always completely unacceptable. If the system is set up correctly and the monitors have been properly rung out, this should not be an issue. However, you cannot control the talent and if the lead singer wants to point his/her mic at a monitor you can't control that. You *can* cut his/her mic as soon as anything like that occurs. Be ready and able to do so.

   Common places that feedback can come from:
   - un-gated floor toms;
   - vocal mics with no high-pass filters on;
   - mics that have been moved by accident;
   - overly bright reverb returns.
2. *Too much subwoofer*: We all like the feel of the walls shaking with tons of subwoofer energy but you don't always need that much subwoofer. In small venues it will overrun every part of the sound field all too quickly. Apply high-pass filters to all channels that don't need the lowest frequencies in their output sound.
3. *Reverberating speech*: Don't forget to turn off all effects on the main microphone between songs (reverb and delay will truly annoy your crowd while the band talks in between songs).
4. *Mixing too loud*: If you are in a club that has a decent system it will have a limiter on the main outputs. If you overload the mixer to the point of smashing signal into the limiter, all that will be left is a horribly muddy, ear-fatiguing ball of dung. Give a hoot and don't noise pollute.

When it comes to what to do at your board, know the rules that are absolute and use them, but the smartest way to approach rules of thumb are as guidelines to use as a starting point. There is no magic recipe or perfect mix that suits all occasions, and even some technique that has worked in the past may not work this time. Ultimately you need to use your ears and your overall understanding of all the aspects of sound on your way to the end point—the clean and clear sound reinforcement that provides the best experience for the audience.

Chapter 24

Beat Trouble to the Draw

Tactical Troubleshooting for Today's Audio Outlaw

Tactic #1: Prevention

The best kind of trouble shooting is shooting trouble down before it becomes your problem. While sound companies who live or die by the efficiency and profitability of their inventory make it a point to regularly maintain their equipment (much of it done, with oversight, by young audio techs new to the field), most audio equipment does not get serviced nearly as often as it should. If you decide to buy and rent your own P.A., or if you end up working house sound at a small venue or monitors for a touring act, it'll be up to you in most cases to see that gear gets regularly maintained (at least once, preferably twice a year). It is worth the time to take care of this, or to pay to have it done as it will keep your gear from slowly degrading in sound quality over time and help it last longer. Equipment that lasts until the next upgrade is equipment that doesn't break mid-show.

However, be aware, that while carefully and safely performing maintenance tasks on equipment is recommended, and will not damage sound equipment in the vast majority of cases, there is always the chance that existing problems may be revealed immediately that would have taken longer to manifest otherwise, especially in gear being maintained after a long time of neglected maintenance. Do not maintain gear that must be used the same day, without a backup (especially if gear has gone too long without maintenance). In addition, always refer to your owner's manual and observe their recommended maintenance practices first. Only perform any other maintenance procedures if they do not directly contradict your gear manufacturer's maintenance best practice. Note that, if under warranty, check to make sure that maintaining an item yourself doesn't void the warranty, and if warranty requires a specific number of professional service visits in a set timeframe in order to remain in effect. Finally, only perform maintenance tasks on unplugged, unpowered gear.

Microphone Maintenance

Among classes of gear, microphones are second only to cables in the amount of abuse they receive, especially during live shows. In the course of your average weekend round of bands, your typical club mic gets molested by singers who practically eat the microphone, held out to audience members who shower it and shout into it in equal measure, and gets smeared with lipstick at least once or twice. As the mic gets drooled on and the windscreen starts to reek like smoke and corn nuts, the little crannies in the outer grill start to accumulate crud. Some folks try to wipe this off from time to time, but wiping the basket just because it's hidden behind foam doesn't mean it's not lingering under there! This makes regular cleaning of your system's microphones a must if you want to maintain good hygiene and optimal performance.

Removable Grill Microphones

If the mic has a removable grill, you will want to remove it first. Most microphones have grills that can be removed simply by unscrewing them, but if yours does not unscrew easily, that does not necessarily mean it is not removable; however, you don't want to apply a lot of force or yank on it to find out, as excessive force could damage the mic even if the grill is removable. First, look at the point where the mic grill meets the body of the mic. Some mics' grills will be secured to the microphone body at the base of the grill, with a small screw on each side. Just take care not to lose the screws, put something underneath before unscrewing them to catch them if you lose hold of them. If you can't find any screws, and can't easily unscrew or pull off the grill, you can try checking the manual to see if it indeed is a mic with a non-removable grill, or you can simply try jiggling the grill back and forth gently while applying light but firm and steady pressure to pull it away from the cartridge. As long as you only apply firm steady pressure, and don't use excessive force or torque, after a few minutes one of two things will happen: either you will feel it begin to gradually loosen, in which case keep at it until you have removed the grill, or nothing will happen and you won't be able to budge the grill, in which case refer to the instructions for mics with non-removable grills.

After removing the grill, find a safe dry place to store the mic where it will not accidentally get knocked over or wet while in this naked vulnerable state. If the foam windscreen can be removed, do so; if not, examine it to see if it is sound, or if it is flaking and disintegrating, in which case just tear it off. Clean with cleaner like Microphome (see directory for link); otherwise, using a soft bristled toothbrush, water, and a gentle soap-like dish soap, brush the grill until the gunk is gone; if gunk remains in crevices wooden toothpicks can be used to safely scrape out stubborn deposits (if windscreen still attached rub soap in gently with thumb; if not wash separately). Rinse thoroughly. Dry with a lint-free cloth or paper towel and tap excess water out, then leave to dry overnight, or aim a fan at grill and dry for two or three hours. *Do not put the grill back on microphone body until completely dry; even a small amount of moisture can do damage to condenser microphones.*

Non-removable Grill Microphones

Hold the microphone upside down (you will want to keep it upside down for the duration, so might want to rig a harness to hold the mic in place upside down, or enlist a helper). Using a *dry* soft-bristled toothbrush, loosen the gunk on the grill (taking care not to press the bristles any farther into the grill than is needed to reach gunk in the crevices). Brush out loosened gunk with a very soft clean artist's paint brush (or a clean makeup brush). Repeat until all gunk is removed; if gunk remains in crevices wooden toothpicks can be used to safely scrape out stubborn deposits.

With unpainted grills only, if required, the end of a toothpick can be dampened (not wet) in *isopropyl alcohol* first to loosen stubborn crud (99 percent only if cleaning a condenser mic, 91 percent OK if dynamic). Once all crud has been removed, wipe and polish outside of grill with lint-free cloth.

Headphone/Earbud Maintenance

For *earbuds*, remove silicone or foam cover and wash separately in mild soap and warm water. Using a dry soft-bristled toothbrush, lightly brush off any loose and dried earwax, taking care not to push it into the grill. Use a toothpick or pin to carefully clean out the dried wax remaining in any crevices, holes or in grill. Dampen a Q-tip in isopropyl alcohol and clean the surface of the earbud without driving liquid into the body of the earbud, allow to air dry for an hour, and replace the fully dried and clean foam or silicone cover. For *headphones*, wipe the entire surface down with lint-free cloth, damp, not wet, with isopropyl alcohol. Follow up with leather or pleather treatment on padded soft materials or the alcohol will dry out the material. Clean plug with isopropyl alcohol, or *contact cleaner*, such as Kontak.

Amplifier Outboard Gear Maintenance

Use a vacuum cleaner extension and low- to mid-powered vacuum cleaner and clear dust from between units and from unit vents. Dampen lint-free cloth or paper tower with isopropyl alcohol and use to clean oils from front surfaces and knobs. Dampen a Q-tip with isopropyl alcohol, or Kontak contact cleaner, and clean all jacks. When spraying contact cleaner into equipment wrap the surrounding area and control cavity with a paper towel to catch and absorb any overspray.

Speaker and Flight Rack Maintenance

Use vacuum cleaner extension and vacuum outside carpet on any carpeted speaker enclosures and flight cases. Then vacuum inside of flight cases. Dampen lint free cloth or paper tower with isopropyl alcohol and use to clean oils from all metal and resin surfaces. Grab a screwdriver and inspect and tighten all loose bolts, screws, or connectors on outside of speaker enclosures and cabinets (tighten until snug, but do not over tighten, care should also be taken to avoid stripping screws) Dampen Q-tip with isopropyl alcohol, or Kontak contact cleaner, and clean all jacks, terminals, posts. When spraying contact cleaner into equipment wrap the surrounding area and control cavity with paper towel to catch and absorb any overspray.

Mixer Maintenance

Always keep the mixer covered or cased when not in use. At least monthly and preferably every two weeks move all pots and faders through two full excursions (move each control through its entire range of motion twice). This will keep all controls lubed and prevent rarely used channels and controls from accumulating enough dust to become scratchy. Use a broad, soft-bristled paint brush to remove dust from top of mixer, and loosen dust caught in crevices. Use a vacuum cleaner extension and suction dust from faders and air vents. Dampen a lint-free cloth or paper towel with isopropyl alcohol and use to clean oils from front surfaces and knobs. Dampen a Q-tip with isopropyl alcohol, or Kontak contact cleaner, and clean all jacks, terminals, and posts.

When spraying contact cleaner into equipment wrap the surrounding area and control cavity with a paper towel to catch and absorb any overspray.

Cable Maintenance

Test all cables with a cable tester and remove any faulty cables from service. Clean plug ends with isopropyl alcohol, or contact cleaner. Follow cable best practices already covered in previous chapters.

Troubleshooting Tactic #2: Detection Tools

Avoiding issues is not always possible, but if you can catch them before the show and fix them, that's almost the same as avoiding them all together!

Cable Testers

Cable testers can let you know which cables are faulty before you hook them up to your system, though many models do much more than that. For example the *Ebtech Swizz Army 6 in 1 cable tester* doesn't just tell you which cables don't work; it also allows you to wiggle a cable and know if you have an intermittent connection, so you know which cables are about to go bad as well. In addition it generates test tones at two different frequencies and can tell you how cables are wired among other functions.

SPL Meters

Aside from the fact that you need one to be sure you are not doing the audience harm, you also can use it to keep track of your system output so you know everything is working. Your meters only tell you if the signal in the mixer is too large to handle or within system limits, but you can't be sure how loud it is without an SPL meter. Luckily, since of the two tools mentioned this one is the only *essential* tool, there are models like the Galaxy CM-130, which are available for 50 to 60 bucks, a price even an audio tech can afford.

Figure 24.1. SPL meter.

Tactic #3: Start with a Game Plan

Murphy's Audio Law #5: "The probability of catastrophic sound system failure is inversely proportional to the amount of time left before the show starts."

Sometimes issues occur so late in a sound check, or even during the show, that the engineer must fix the issue on the fly and under pressure. Since by the time you are running the P.A. we can assume you will have a firm grasp on the basics, success or failure depends more on your attitude and composure allowing you to access and use what you know or driving you to approach the problem erratically like a decapitated chicken, testing and inquiring in a random manner, skipping and not keeping track of steps. While you may stumble across the problem this way, you are more likely to get stuck going in circles, retesting the same connections while missing other obvious potential culprits each time—this reaction only wastes valuable time, creates even more pressure, and creates an imminent engineer meltdown. To master the state of mind required for troubleshooting follow these rules:

Troubleshooting with the Heat Turned Up

Don't panic or let your emotions take over. Remember, you have all you need and you're covered. Take a breath and do your thing and let go of distractions. This means don't get mad at a person or a piece of gear as well as not panicking. If there's reason to be mad later, be mad later, but right now the only thing that's going to solve the issue is calm, focused, and methodical action. Displays of emotion or gear smacking will not solve your problems (except for gear that needs it—sometimes a bad mixer needs nothing more than a good firm smack on one side to behave again). But like a good parent, never hit your gear in anger; only as a last resort and for a good reason. Know your gear, use your judgment instead of your emotions to make that choice, and of course scan the room subtly to time your snack for when most folks aren't looking. Note that smacking people is never an option, even if like the mixer in the first example, they need it, and even if you check to verify you can do it when no one is looking.

Focus on the problem and nothing else. If the problem is not one that can be fixed immediately and is influencing the sound or has even stopped the show, give a nod to the band to let them know you are on it first, and then get to it. If you are lucky and have a pro band they may help you out by filling in with amended program content to divert attention from the problem until you can fix it, whether that's a base and cowbell jam with some improvised yips and yahoos from the vocals to cover a dead guitar signal, or a charismatic front man turned comedian shouting out jokes to keep the energy alive while you reanimate the P.A.

If you are unlucky the band will stand fidgeting and murmuring to each other, much like venue patrons—staring at you intently while you troubleshoot as if their focus on you could help boost your tech power. If you are sublimely unlucky they could join with the mildly inebriated, pony-tailed audience member in a Doobie Brothers shirt and shout out conflicting and rapid-fire instructions and tips on what your current/next step should be. But as far as you're concerned, it shouldn't matter. You need to be able to tune out all three situations equally.

Keep your data baggage free. Tune out irrelevant outside noise and keep focused on the problem till you fix it—the key word to outside distraction being relevance. If a knowledgeable member of the band steps forward and appropriately offers you advice as one pro to another rather than random advice shouted rudely, then this advice should be integrated and used as you make decisions. Remember this advice will be based on his/her insight from having literally approached the problem from a different location and facing a different direction, so it may include important information you don't have from your FOH perspective. The point is, when it comes to getting the job done, you don't want to reject valuable data out of hand based on ego. Don't be blinded to clues that lead you to the amp you don't want to admit broke because you chose it over the suggested one. Don't ignore a solution because you don't like the tech or band member who hit on it first. It should go without saying, but since it's a trap even the brightest people fall into at times, to be safe don't go with any knee-jerk reaction without evaluating why, and only reject data or advice after considering it first.

Be methodical and think first, act second. This means move in a logical and efficient order. Remember your steps need to lead away from the symptom back to the cause—only at the point in the chain that is a potential cause of symptoms can any fixes possibly produce improvements. When there is more than one potential cause, test for them efficiently using common sense instead of randomly. First, verify the symptom, then work back and make a mental list of potential causes and work your way along the signal chain between the symptom and each cause (either starting with the most likely cause for the situation, or if that is not clear, starting with the easiest to test) until you find your particular cause, Then make a mental list of known solutions to fix this type of cause to your problem, and apply possible solutions one at a time till the correct one is reached (again starting with most likely to work, and/or easiest to try).

Always use all your knowledge. This means always have a rationale for any action and base each decision both on the evidence provided by the symptoms you perceive in the sound and on your overall knowledge about audio and P.A. systems. For example, even though connections are often a point of failure, don't start checking connections first without a reason to, but first think about how your plan fits with the symptom too. (If feedback is your problem, you aren't going to find anything at your connections.) If you have a reason to suspect connections are the cause (for example no guitar sound is reaching the speaker indicates there may be a connectivity problem) remember to use your brain to verify if it is the most likely cause (it could also be the cables or a signal not going where it should) and if so which fix to try first. Taking a few seconds to formulate a plan for the most efficient way to assess and address your issues is always faster than randomly trying all possibilities in any order. Sure you will probably stumble across your answer sooner or later, but your aim is sooner.

So in the above scenario you wouldn't want to immediately go to check the wire from speaker to amp, since you are already at the mixer start there; first, you would check your board to see if the problem is at the beginning of the chain by checking to see if a signal is present in the mixer by listening to the channel in PFL, and if that's good you would check mute and routing to make sure it's not accidentally misrouted and is leaving the board by the right route. Then you would check its path through the processors to see if everything is OK there. Only then can you walk to the speaker confident you are not about to waste a bunch of time checking the signal connection at the speaker when a few checks at the mixer would have fixed the problem or told you the problem was with the signal connections at the other end of your signal chain (like at the stage box and DI box connections). But before you go, remember your procedures and best practices, so turn down your faders or mute if you plan on unplugging anything without turning your amp off; otherwise you could end up exchanging a simple connection issue for an exponentially worse problem consisting of a blown amp and angry people with blown ear drums.

If brain freeze ever makes you unable to access all the possible causes of a familiar symptom (it happens) then use a block diagram of your system to keep track of your process.

Maintain a Professional Demeanor

Remember that even in the worst-case event of total P.A. failure, patrons and band usually will take their cues from you, and if you have their support (or support from most of them, ponytail guys and similar sorts are bound to pop up in any event, but don't forget these types aren't relevant) you will feel less pressure, have an easier time maintaining focus, and getting from A to Z will happen more quickly. The same goes for the band (at either sound check or mid-set), and especially if they don't know you well enough to have supreme confidence in your skills. *To be treated like a pro, act like those present expect a pro to act.* This means a calm intent facial expression, purposeful movement, and steady progress through your steps. If they see this, they will know a pro is on it, and wait patiently(ish) as the pro takes care of the problem.

Because one very common ego defense people (myself included) use when they feel put on the spot and all eyes are on them is to revert to being class clown, I should also add that what they *don't* want to see is the FOH engineer joking around, acting like they are too cool to care, and they may see it as wasting time or moving slowly on purpose. Considering clowning is so common because normally, as an ego defense meant to deflect pressure, it's a pretty sure-fire tactic, I have to warn that short of freezing in panic and standing there not even trying, playing around is the best way to ensure folks start heckling as soon as they are convinced you will fail, which will be barely more than five minutes for a clown. So avoid acting too jovial unless it is upon finding the problem and while in the process of applying the speedy fix. If they've been waiting around for ten minutes and you are still in discovery mode, they will have little patience for the wasted seconds spent clowning. Remember it may be just a day at the office for an audio pro, but for the audience, it's their Saturday night out. All too soon it will be over and they will no longer be drinking beer and rocking out, but doing house chores or back in the weekday grind. It's not that you can't smile—a confident smile will go a long way to allay their anxiety—but they expect you to take their party time seriously.

On a similar note, if you go through your troubleshooting steps while scowling at each component/connection malevolently and muttering curse words like you'd like nothing more than to take an axe and turn that piece of crap P.A. into kindling, they will get antsy sooner and you will lose their support. First, this approach will make you come off as inherently deranged, which is not what people look for in a pro. Second, your apparent lack of confidence and respect for the system will convince them that even if maniacs made good sound engineers, the maniac charge of this P.A. clearly thinks this system is obviously beyond hope. At five minutes they will start grumbling and becoming convinced that you and your piece of crap system just murdered their weekend leisure time.

So when the whole room is watching with fingers crossed, keep a straight face and be matter of fact as you do your job if you want them to keep rooting for you. You'll find that when they can tell you are serious about fixing the problem, they will be incredibly patient while you get things fixed, and will be more likely to assume it was gear and bad luck if it can't be fixed.

Tactic #4: Turn to Known Troublemakers

Ground Loops

Many people know that the United States runs on 120 V from the power outlet, but what you may not know is that because it runs on AC (alternating current) power, there is an inherent noise that comes from alternating current, which alternates at 60 cycles per second, and we should know that 60 CPS is 60 Hz. The reason we don't hear the AC usually is because when we plug in a lamp it is changing the power into light, but we change power into sound-pressure waves for people to enjoy, which means we can hear the hum!

Alternating current alternates between positive and negative, with the third prong in a 3-prong plug being our ground reference. If the ground reference is not properly grounded, then the ground voltage may change slightly on each alteration, or the ground reference on different plugs may not be the same ground reference. If we have a wavering ground reference, it will "waver" at the same frequency that the current alternates. We know that waver to be 60 Hz. So, basically, a ground loop is an auditory representation of our AC current. That current can originate from a device that has a wavering ground reference, or (more likely) the ground loop can be caused by an unshielded power source or cable in which some of the electromotive force escapes and gets into our cables.

Fortunately there is an easy way to fix device-based ground loops—use a power conditioner. This often solves ground-loop problems because many device-based loops come from badly or poorly wired power plugs, a break in the wire, loosened wire contacts, or wiring defaults at the power source. Power conditioners do two main things:

- They take the power from the wall and put it into short-term storage (basically a huge inductor/capacitor circuit).
- They then use that "stored" power to provide a steady stream of power to all devices plugged into it.

As there are very minor power fluctuations that happen on our power grid all the time, the power conditioner conditions the power to always be at a steady stream of power. In addition, a power conditioner doesn't let the ground reference wander in the middle of any power cycle. Instead, it takes a ground reference and "nails" the ground reference down within itself. This means the audio devices plugged into a power conditioner have a steady ground reference and the AC fed to them will alternate cleanly from + to −. In addition to power conditioners, DI boxes can serve as temporary fixes in some situations, their transformers isolating the problem and preventing it from moving to the mixer.

Ground loops are also caused by power (electricity) getting out of one cable and getting on to another; when we consider the fact that audio devices have both a power cable and so many audio cables coming from the same set of circuits, it's amazing we don't have ground loops in every device. Proper cabling practices are required if we want to be sure to avoid ground loops. The simplest way to avoid getting any stray power source onto any audio cable is:

- Use balanced cables wherever possible, and never use unbalanced cables over long cable runs.
- Minimize the surface contact area between any power line and any audio line.
- Always try to have power and audio cable lines apart from each other, but when this is not possible try to have their intersections at right angles to each other therefore minimizing the contacting surface areas.

Dealing with Distortion

Distortion occurs anytime our signal exceeds a component's maximum dynamic capabilities. When it can be heard the first step toward correcting it is identifying exactly where the problem is, because distortion can occur at almost any point in the signal path, from the microphone to the speaker.

Generally speaking, to avoid distorting the audio signal you should keep the level below about 0 dBu at every point in the pathway. This would be easier to do if you could measure the signal level at every point it is subject to changing, using a VU (volume unit) meter or similar device.

As you usually can't measure the signal level outside the mixer, you have to do some detective work. Since when you are looking for something, the best place to start is the beginning, follow the signal through the system from where it enters to track down the point where the level gets out of control.

- A distorted mic signal is created by a very loud noise being too close to the mic. Move the mic further from the source and see if that helps.
- Though only the mixer output has a meter, other mixer stages and gear does come with "clip" displays. Look for the first place a clip light shows up and the problem will have occurred after the last dark clip light and before the one that keeps lighting up.
- Are any volume or gain controls in your system turned up higher than normal? Are there any obvious points where you could drop the level?
- Are your speakers being driven too hard? If you have an amplifier which is pushing the speakers beyond their design limits, then be careful to see if turning your amplifier gain down will improve the sound.
- Does the distortion match peaks? Try a compressor.
- Is the problem really distortion? There are many noises which could sound like distortion; for example a dirty volume knob, or even bad room acoustics.

When Venues Attack: Learn How to Fight Back

Every aspect of a venue impacts the sound made there, and when sound is changed by the room, it is rarely

Need More Gain Before Feedback? Here's How to Get Some:
(In order of effectiveness)

- Bring Microphone and sound source closer
- Move Microphone and loudspeakers further apart
- Bring loudspeakers and listeners closer
- Reduce number of open microphones
- Use more directional microphones and loudspeakers
- Control stage reflections (use accoustic treatments)
- Control room reverberation (use accoustic treatments)
- Use Graphic Equalizer to lower gain at frequencies that are creating feedback

Figure 24.2. Gain before feedback.

improved by it. On many occasions, there isn't much that can be done about it. Just as frequently, however, those actions that could be taken to improve room-based issues are not taken. While taking care with speaker placement and micing is essential, especially in problem venues, there are often other small treatments that could improve venue sound which are ignored.

While small-venue sound pros often all take a hit for the bad choices of a few, it's our observation that far from being purveyors of bad sound due to lack of talent or laziness, they are often working like demons to coax lackluster sound from a system that, if not for their efforts, would serve up nothing but atrocious sound. So when we attend small venues where the FOH engineer has not made even a small move to making acoustical improvements, we wonder why.

Since we find it hard to believe that in all these cases it is ignorance of acoustic principles, we think it might be that too many people believe the myth that acoustical treatments are not worth bothering with unless you are able to sink thousands of dollars into a full renovation performed by licensed installers.

After a brief look online I found there are still some professional installers who are happy to perpetuate this myth in blog and forum posts that claim only a full professional treatment is able to create *any* improvement in a room's sound quality. There even appears to be a small dedicated contingent who visit each thread started by a DIY newbie asking about affordable DIY acoustic treatments to heckle them for being so stupid or cheap to dare to hope.

But those who say there is no realistic DIY option for those unwilling to pay them thousands of dollars for the top-of-the-line results are wrong. You will never end up with results as good as hiring a pro, but saying that this means you should not make what improvements you can is like saying that those who can't afford a top chef should just starve rather than cook themselves some Top Ramen.

Fortunately, a closer look reveals there are plenty of pro installers who provide a more nuanced and honest view which is that no \$200–\$500 treatment is going to sound as good as a professional treatment costing ten times more, but when the sum is spent wisely on strategically targeted treatments, a budget, do-it-yourself solution *can* produce a noticeable improvement in sound quality. The first blows against this myth were made by new home producers after digital technology enabled the home recording revolution. A look at the buzz online makes it clear this idea is finally starting to lose its grip, as there is evidence that more and more small-venue sound pros are finally waking up to the small affordable changes that can improve their sound.

A Little Acoustic Treatment Goes a Long Way

No one should expect their DIY treatments will make Mel's Bar sound like The Met, but when there are often small treatments that can be made with only a little effort and cash, and the rewards in terms of reduced struggle with feedback alone will make it more than worth the investment—all the more so if this breathing room leaves you free to attend to the sound and make it better instead—the sound engineer begins developing a reputation as a pro who takes their work seriously at any level.

In such cases, you may want to figure out where you can apply your limited resources to create the most improvement with the least amount of additional investment and energy. The first step that's required for you to improve a room with some DIY acoustical treatment is for you to know what you have to work with, and what options are available. To that purpose, you need to use all your resources and do your research first. There are links in the directory to help plan small DIY projects as well as links to calculators that will allow you to estimate both the room's reverberation time and its critical distance. As you go online to begin your research, you will also find links to articles that discuss the benefits and limitations of DIY acoustical treatments for the thrifty FOH engineer. In the same directory there will be links to schematics for how to build your own simple acoustical panel absorbers and acoustical hangars.

When DIY Beats Doing Nothing

For most venues the best focus will likely be to create a stage with enough reflective surfaces to have a live musical environment with lots of early reflection, but not one so reflective that excessive reverberation contributes to the need to raise the noise level on stage in order to hear. If the stage only needs a little work we may turn our focus to dealing with low-frequency resonance, if needed, or if the room is too live to get a good sound, then getting some absorption into those spaces is what you need. This should include some bass trapping, but things like thick curtains can also be very effective.

Any long-term treatments will only make sense for those who are house engineers and some of the solutions will be ones that can be employed even by guest mixers at those venues where even simply being able to cover a large reflective brick wall will make a difference in the sound. Eventually, even engineers who don't currently have acoustical sound issues bad enough to make treatment an immediate priority will want a few things that are suggested in the articles. Most sound engineers will at least want a rug or patch of carpet to put under the drummer onstage, a pillow or wool blanket to dampen drums for those drummers who forget to bring their own, as well as a few moving blankets which can be used for many purposes, but are also one of the best of the cheap sound absorbers an engineer can find (and they're fire rated too). Small treatments can improve some of the worst problems that can be found in small venues, but they are less effective at optimizing venues that are already well designed acoustically, so if this is your situation, stick with the smaller acoustical accessories that we mentioned might be useful to any small-venue engineer.

Acoustic Treatments

All materials can be assigned an absorption number along a scale that varies between 0 and 1. The 0 value indicates lack of absorption (in other words the surface is totally reflectant), the value 1 indicates total absorption (lack of reflection). Sound absorbers are generally porous material.

Porous absorbers: Common, lightweight porous absorbers include carpet, draperies, fibrous mineral wool and open-cell foam. The amount of sound energy dissipated depends on the thickness of the material, its density (which determines the amount of difficulty that the sound encounters traveling through) and its resiliency (flexibility with the ability to spring back to its original shape).

Sound attenuation blankets are manufactured with higher density than thermal insulating blankets and should be mounted without being stretched taut, otherwise their resiliency is compromised, and may render them ineffective in dissipating standing waves.

The further curtain fabrics are placed away from walls, the better the absorption is at lower frequencies. This is because of the quarter wavelength rule.

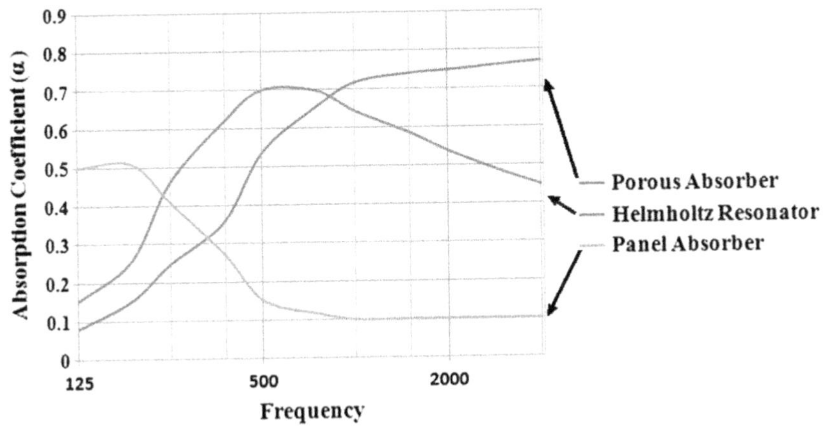

Figure 24.3. Absorption by frequency.

Sound	Frequency	Wavelength
Lowest audible C	16.4 Hz	21.03 m
Lowest C on piano	32.7 Hz	10.51 m
Middle C on piano	262 Hz	1.29 m
Violin A string	440 Hz	0.76 m
Four octaves above middle C	4,186 Hz	8.25 cm
Highest audible tone	20,000 Hz	1.7 cm

Figure 24.4. Wavelength matters.

The quarter wavelength rule: Acoustical absorbent material must be placed away from the walls and ceiling at a distance of a quarter wavelength of the lowest frequency to be absorbed. This will include all higher frequencies if the absorbent material is soft furnishing or fiberglass, but also enable some of the lower mid frequencies to be absorbed.

While fabric treatments are best for higher frequencies two ways to control low frequencies that are almost as simple include acoustic hangers and panel absorbers.

Acoustic hangers: basically fiber board panels wrapped with insulation and hung freely using wire or rope. These can catch mid to low frequencies and are only a little more work than simply hanging curtains.

Panel absorbers: non-rigid, non-porous materials which are placed over an airspace that vibrates. Common panel (membrane) absorbers are usually most efficient at absorbing low frequencies. According to the SAE Audio Reference Material (www.sae.edu/reference_material/), a panel of plywood or particle board is placed over an air cavity with insulation glued to the back of the panel. The panel has a resonant frequency and when it resonates the insulation absorbs the energy.

The great advantage of panel absorbers is that *you can apply front panels with angled or curved elements so when mounted on a wall or the ceiling they stop parallel wall interference and prevent standing waves creating diffusion.*

One of our aims throughout this text has been to make sure the new live sound pros reading are aware of what options may be available to help them make the transition from apprentice to pro as smoothly as possible and more rapidly than might happen without being shown some of the ways that out-of-the-box thinking can empower them and benefit their sound. Some of you will benefit more than others. However, if you find yourself in a competitive market and working in a venue more like an echo chamber than a sound venue, then doing every little thing you can to improve your sound and your resumé while you are in a position to have the opportunity could help you make the most of it. Just remember that any big project requires thought and planning, because you will only get out of it what you put in. When cash is not something that you can put in much of, then DIY provides an opportunity for you to make that up with careful thought, planning, and effort. Since such a project can be a resumé builder, if you bother taking it on, then take the time to document your process and the final results.

Part 6

Becoming a Live Sound Professional

In the previous chapters, we provided a simple and easy-to-read P.A. guide and access to more audio information in terms of gear and equalization and mixing tips than you could read in the next year. And then we added a few of the lessons that most live sound books don't include. These chapters include more topics you may not find in most other Audio textbooks but will really need to know.

This is the stuff that you either start learning early, or, once you've figured it out way down the road, you wish someone had just told you so that you could have learned it sooner and avoided a lot of headaches. Or because you overlook these topics completely you never arrive, which would be too bad, though hardly uncommon. The reality is that live sound is a very competitive field. That's not meant to scare you. You can make it as a live sound pro. But you won't make it far up the ladder if you are a substandard sound engineer. And you don't get to be a good sound engineer from reading a P.A. guide, even if you visit every link in the directory and collect a million mix tips. But you *can* get very good, if you are motivated and serious about putting in the effort. Because there are a lot more people interested in working in audio than there are sweet gigs, and because everyone else is studying their P.A. guides too, the first leg of the road from here to there is all the "extra" stuff you know.

The problem in live sound is that to become a good sound engineer, you need to gain some experience, get some practice, and survive long enough to make it up the ladder. That's going to be more difficult if all you worry about is gear, setup, signal chain, and mixing theory. If all you having going for you is studiously learning about gear and signal paths, well, it's you and tens of thousands of other rank amateurs competing to get a leg up. So, to turn theory into practice, you need to hustle to get into a position where you have a mixer to play with. Oh, and a band, or some other such thing worth amplifying so that you can practice, practice, practice. This part is important, and is why you can't just get a small mixer and practice at home, and why it's not easy to make the transfer from live sound amateur to live sound pro. It's all about becoming a professional, which means being part of a community and knowing about your entire field and not just the niche you settle in.

Chapter 25

Getting Your Foot in the Door

Making the Transition from Out in the Cold to in the Live Audio Fold

---

Training Options: Three Paths to a Career in Audio

If you are reading this as part of your coursework at a public or private school, or vocational program, you are beyond needing these initial sections, but may find information of interest here anyway, or may choose to rejoin us in the section on internships. If you have not yet decided on which track you want to take to get the required training you need, in this section we provide you with the guidance you need to evaluate schools, compare programs, or develop your own self-guided training, ideally in conjunction with working part or full time at a local sound provider.

There is a real diversity of opinion when it comes to whether or not the aspiring live sound engineer gains any benefit from formal training, and if so, whether this must take the form of a traditional four-year B.S. or B.A. degree, a technical two-year degree, or a shorter program which may or may not offer some sort of certificate of completion (which may mean something positive to a potential employer if they know about and have a good impression of the school, but most likely will mean nothing at all to them, or in the worst case may be detrimental, if the person hiring has a bad impression of that specific program, or short-term training programs in general).

There are some who insist that you can't make it to the top without a four-year degree, and many music departments now offer audio engineering degrees that cover everything from studio, live, and broadcasting audio, to audio science, music theory, and optional business courses. Others feel that some school is beneficial, but don't feel training needs to amount to a four-year degree. These people argue that a two-year degree in electronics, and a one to two months training program in audio engineering is more than sufficient or even ideal. Still others will argue that you will start at the bottom regardless of a degree, and are better off learning on the job, since anything but on-the-job learning is theoretical anyway, and even the best schools can only offer limited amounts of real-world practice, if they offer any at all. Often, these views will match the view holder's own experience or choices, but that makes the arguments for each position no less valid.

All three viewpoints have some evidence to support them and some evidence to contradict them. The truth is it is possible to find examples of successful audio engineers who have followed each route and the same goes for people who have failed to thrive in the field and who eventually chose to move on. In the directory for this chapter we include links to articles whose authors argue in defense of each of these views so you can see for yourself the arguments for each option. Before choosing any option, we encourage you to read these opinions as well as our own below, and to also seek out the opinions of local sound company professionals to consider, as you decide your own path. If you don't know any, call around and tell them you want to prepare for a career in live sound and want to know which applicants they prefer to hire and what training they chose for themselves. You may need to follow up as most are busy folks, but if you are friendly and persistent, enough will get back to you with an answer to make the time you spent asking worthwhile. Be sure to ask the same of the sound engineers working at any local venues as well, especially if work of this type is especially interesting to you.

Then take all this information, plus what you know of the demands/benefits of the field, and combine it with what you know of your own options (a four-year degree is much more attractive to those with college accounts, or eligibility for scholarships or grants, than to those whose only option is to pay completely via loans), aptitudes (some very smart people are not good students which reflects their particular way of learning and not their aptitude or lack thereof), and circumstances (some people have familial obligations that make moving to pursue training currently impossible even if eventually they might be able to move to pursue more job opportunities), and make the best choice you can based on your circumstances. An informed choice won't always turn out to be the right choice, but the odds of it are better than the 50 percent afforded to a random choice, and usually a lot better than an ill-informed choice which at best has no more than a random chance of getting it right, and usually is much less.

Pros and Cons of Each Training Path

Four-year Degree Pros

- The greater breadth of a four-year degree usually makes graduates eligible for entry-level positions in all four main audio niches: studio audio, live audio, broadcast audio, or sound design.
- In a tight employment market even "entry-level" work traditionally available to those without schooling now often goes to degree holders first.
- Degree holders gain status advantage from the cultural view of those who have a four-year degree as being more "educated" and often are also mistakenly identified as more "intelligent" or "capable" than adults without a four-year degree. This explains in part why the four-year degree is considered more

transferable to other fields than most other kinds of training, which provides more chances to find well-paid jobs in fields outside the degree holder's major.

- With only two to four years of additional training/experience the degree holder is eligible to teach audio at the high school or college level.
- The long-term lifetime income advantages still make college the best investment you can make even if you are on the low end of return on investment (ROI). For example, if after paying for the cost of the degree, all you see is $5,000 more a year in pay, what is the return? For a $50,000 degree (state resident at a public state college, all costs included including room and board) your lifetime returns will be $200,000 more in income than without a degree (and that's a conservative estimate). The graduates with an average ROI make $500,000 more over their lifetime than those with only a high school degree.
- Even after eliminating the influence of income disparity as much as possible, degree holders still rate higher for many significant quality-of-life factors that benefit not only them, but their families and communities as well. They are more likely to vote, be aware of current affairs, read for pleasure and read to their kids, eat their veggies, do volunteer work, etc., but most importantly they rate themselves as happier, even those at the bottom of the income scale.

Four-year Degree Cons

- If a graduate of a four-year college has trouble getting the kind of work they had hoped for and many do, they are out more in time and expense than those who chose other training options and work outside major is less likely to bring full career benefits to those seeking it.
- A $50,000 investment hardly represents the true cost for the average student. If the entire amount is paid for in subsidized undergraduate loans with 3.4 percent fixed interest and a ten-year payback schedule, the true cost adds $10,000 in interest. And this assumes the full cost is available as a government loan; many people are not eligible for more than half that amount in subsidized loans or even eligible for these loans at all. This is why private loans often bridge the gap, or are the only option, covering from 50 to 100 percent of the cost of the degree. However, these variable rate loans are typically three times the interest, or more, so depending on the loan that extra $10,000 in interest could be more like an extra $30,000+.
- In a field with high competition for jobs and a starting income ranging between $22,000 to $50,000 a year for the first five years, ability to begin payment before the six-month grace period after graduating is over is difficult to be sure of, but one sure thing is that a monthly payment of $492 (for the lowest interest loan described above) will be hard to swing on $22,000 annual pay if food is one of your vices. This amount effectively reduces your yearly income to around $16,000 and that's without taking out for social security and taxes. (This is possible even if it would be a great hardship, depending on the cost of living. But the $700 or $1000 monthly payment if some or all the debt is through private loan would leave just enough for nine to 12 months rent and nothing else.)
- A four-year degree can be a great investment, if you can pay for it, but if it makes you take on debt you can't afford to pay back it's a trap that can cost more than it's worth. Considering the questioning of the same old way of doing things in education occurring over recent years, and that in 2012 over half of new graduates are unemployed or underemployed, it's not unreasonable for those whose only option would be a great deal of debt to keep college in the playbook as an option, but wait until there are jobs for graduates opening up again before borrowing for school.
- While those who can hang in there will likely be promoted to top positions faster than someone who must spend time at each tier learning the trade, graduates will usually still be required to start their careers in the same entry-level jobs as those with no training, and will be expected to be just as patient and humble while waiting for promotion as anyone. Though not as bad for all of them, after four years and much expense spent becoming audio experts this is a burden and a trial so outsized for others their difficulty with it is powerful proof that it is utterly essential practice.
- Not all people have the disposition for four years of full-time classes.

On-the-job Learning Pros

- For those who aren't really sure if live audio is for them, but are interested and motivated enough to take a look, working for a year in the field before deciding to invest a lot of time and money in training is smart. If the experience moves them to look into live audio training, the time spent will leave them in a much better position to make an informed choice for how to proceed with their education, Conversely, if it leaves them sure of nothing but that live audio is not a good match in the long term, they saved four years and $50,000 by taking the time to field research first.

Figure 25.1. Whether a four-year degree or learning on the job, a good career starts with getting the right training.

- Those who do choose to learn everything on the job, supplemented by a self-guided program of learning and a selection of audio seminars or a well-chosen short-term audio "boot camp"-style program in live audio, can save a great deal of money and achieve the same levels of career success in only a few extra years, or even in a similar time frame as someone with a degree, if they are sharp.
- Hands-on learning for those whose learning style favors this.
- Does not preclude pursuing any other training options later or concurrently, and can support other options by reducing some of the risk/problem factors inherent in full-time formal training alternatives but which are not as much of an issue for those who attend them part time.
- Employer looking to keep good workers in long-term positions are likely to be more interested in helping a worker choose the best training program to pursue concurrent with work.

On-the-job Learning Cons

- There is no guarantee that you will find full-time employment right away or that once you do an employer will be able to keep you for the years it will take to work up to being an advanced audio tech, much less levels beyond that; if you have to move on to another employer before you've learned a significant amount of transferable skills, you will lose ground. For example, four years of experience at one employer will get you far enough that a change in employer will typically allow you to continue on at the same level after an initial adjustment or trial period, but four years job experience working for three different employers will only get you half as far along in terms of real skills or perceived value, since instead of steady progress and learning, you probably will only end up going over the same ground as a first-year worker three times, at three different companies.
- No matter where your on-the-job learning takes place, until you can claim on a resumé that you worked for someone as a fully-fledged live audio engineer and not only as a tech, you have no proof but the word of an employer or two that you are qualified to do any of the things you have learned on the job. You will want to be sure to leave on good terms and hope that they are diligent enough to take the time to provide a detailed letter of recommendation outlining the skills you learned under them. If you act badly on your way out or if their ethical business values place golf with business contacts over loyalty for ex-employees, you could find it hard to pin them down long enough to provide a good detailed reference, and find yourself looking for a new job not too far ahead of square one. Odds are another sound company owner or venue employer can take your years on the job and quiz you to get a general idea of where you were at, so you may have the opportunity to continue your training elsewhere, but again, there's no guarantee that this will happen right away, or at all in a small enough market.
- In difficult economic times you may well find yourself spending a few years in the same position as any young person with limited work experience and only a high-school degree, which unfortunately is not an advantageous position to be in and can usually be summed up by one word—unemployed (and no short-term vocational training program will count for much in this case, nor will seminars or a handful of classes).
- Not all employers are equipped to be teachers even if they care to be. In these cases, on-the-job learning will have to be arranged by you.

Two-year Degree Pros

- The middle road is perhaps the wisest option and if not the wisest then arguably the safest option of the three, but as with all safe options and many middle paths, it definitely lacks in the sex appeal of the other available options. This is to pursue a two-year associate's degree at a public community college (also called junior college, or J.C.). However, while it only nets about half the monetary benefits of a four-year degree, it can be had at one third the cost, making the ROI actually higher than a four-year degree, according to the proverbial bang-for-buck way of accounting these things. It also confers many of the same non-monetary, quality-of-life rewards shown for those who graduate from four-year schools to only a slightly lesser extent.

All the professor at Half the Price

If you live anywhere near a four-year school and you want to be sure to have the most qualified instruction possible while attending a junior school, go to the four-year college's website, pull a list of all the faculty, and compare the list of names against the course schedules of the surrounding J.C.s to find the one or two that university professors prefer to pick up hours from. When Penny attended community college, many of the instructors at her first J.C. were tenured faculty from California Polytechnical University, Pomona. When she moved closer to work, she transferred to a community college where many classes were taught by instructors from UCLA and USC.

- While any associate degree is an advantage over none, certain majors are extra good options. An associate degree in electronics is related to the audio field and will give a definite advantage in climbing the live audio career ladder, but also opens up a wide range of opportunities in other fields related and unrelated (making it as close to recession proof as any training gets, audio or otherwise).
- As long as you pursue the option within approximately a decade of your Associate of Arts (AA) classes, with only two additional years at a four-year college, you can have your bachelor's degree. In fact attending J.C. for the first two years of college is a good option for anyone who wants to minimize the cost of a four-year degree.
- If you are unlikely to be accepted at a public college because you got Cs in high school, don't test well, or never took your SATs, etc., instead of paying more than the cost of a state degree to attend a private college you can attend J.C as a path to a four-year school and save money instead, as two years of J.C. with a 3.0 grade point average (G.P.A.—equivalent to a B or 85 percent) will make you eligible to apply to any state college. Nontraditional students also often use J.C. as a path to a four-year degree. For example, older students returning after some years in the workforce, or students without a high-school diploma who have a General Education Diploma (GED) instead, will find they need to do at least a year at J.C. in order to be accepted at a four-year college, after which they can transfer to their four-year college as sophomores or juniors. Penny was one of these nontraditional students; Penny took the California High School Proficiency Exam when she was 16 (the legal equivalent of a high-school degree) and started community college, where she got most of the freshman and sophomore credits for her B.A. degree.
- Though J.C.s are thought to offer a lower-quality education, Penny can attest to the fact that this is not the case at all of them. In fact, she'd argue that many actually offer an education as good as or even better than attending the same classes at an average state college. Whether you agree with this not, it is a fact that even if you must endure the notorious "overcrowding" of some J.C. classes, where you might have 33 desks and 35 students, this is nowhere near the "horde" of 300 classmates you can get lost in while attending similar general education courses offered in auditorium "lecture" formats at most four-year colleges. Also, community college instructors may not always have a Ph.D., but you will always be taught by the postgraduate degree holder listed as the course instructor, and not a single class will be taught by a graduate student instead—unlike many classes at four-year colleges. Penny's two years of community-college classes prepared her well enough at least to go on to receive a B.A. degree in English, complete a two-year High School Teaching Certification program, and eventually earn an M.A. degree in English, all with a 3.6 grade point average or above.
- Yes, it's true all you need to be accepted at a two-year is some form of legal identification, a pulse, and tuition, but students who are unprepared for college-level coursework typically must take one or more remedial classes to prepare, before they can register for college track courses. Also, for a variety of reasons J.C.s tend not to be as subject to grade inflation as other higher-education programs, so it's easier to flunk and to actually do well it takes the same effort and ability as it does to do well in first-

and second-year classes at an average four-year school. Having attended both two-year and four-year schools Penny never found the quality of class discussion better or worse at either one, except perhaps in a way that favors two-year colleges, since the chances of finding a general education class with good class discussions is higher with classes of 35 students than with classes of 300 (which is why these are called "lecture" courses, since that's how 95 percent of class time is spent).

- If you supplement an electronics associate degree with an audio boot-camp-type training program or several good seminars, you will be well positioned to be hired on an entry-level basis in almost any live audio position, and will have a backup degree that with only a little extra work will prepare you to seek a half dozen certifications in everything from computer networking, to repairs, to installation, or that could be followed up by apprenticeship leading to an electrician's license or two to three additional years of schooling leading to an engineering degree. In fact if you pursue a "backup" by using your electronics associate degree to prepare to take a certification test in electronic repair, or audiovisual installation, not only will you have more options for employment in hard times, and more options to supplement what you make seasonally in audio with repair work which is year round, but you will have skills many audio engineers would kill for (and that those who hire them will notice too).

Two-year Degree Cons

- The difficulty lies in the fact that this path is not free, though it should cost under half what a four-year degree costs, and is a far better investment than any audio vocational training alone (though such training makes a great addition to this associate degree).
- While it doesn't require nearly the same amount of math as a four-year degree, and the math requirements are achievable by anyone, the math phobic will need to get over their issues to pursue this path, and some tutoring may be required to get up to speed.
- Finally, it may take persistence to get the classes you need if your local J.C. is overburdened and underfunded. You may need an extra semester or two to graduate if you have to wait on courses to open up. If you feel your local J.C. is hopeless in terms of class availability, there are correspondence schools and online options for earning this type of degree as well; just look into them well before spending money on any of them.

The Best of Both Worlds: Apprenticeships

If you are looking for a way to pay for classes, one path that combines elements of school and hands on learning is an apprenticeship. *Registered apprenticeships* combine structured on-the-job training with classroom instruction. When apprentices complete a registered program, they receive a nationally recognized certificate from the U.S. Department of Labor or an approved state agency. Because employers develop the training plans, training keeps up with the needs of the industry, so these certificates are accepted by employers nationwide as proof of an apprentice's training and qualifications. Apprenticeship also can be combined with other kinds of training. Classroom instruction credit often can be used to count toward licenses, certifications, and college degrees.

Learning a skilled occupation takes time. How much time depends on the occupation and training program (audio installation apprenticeships in California appear to be two to three years or 4,000 to 6,000 hours on the job). All apprenticeship programs require at least 2,000 hours of work experience to complete, which equates to approximately a year of work. Most programs require about four years, or 8,000 hours, on the job to complete. In addition most require at least 150 hours of formal classroom instruction, usually taken concurrently with on-the-job training. Employee benefits also vary. Some programs offer new apprentices full health, dental, and retirement benefits immediately; others do not offer benefits at all. A few programs pay apprentices for the time they spend in class. However, at a minimum, all apprenticeships offer a fair hourly wage, which begins at a rate only slightly below entry level and increases at intervals as skill and experience goes up.

When apprentices are accepted into registered programs, the sponsors and the apprentices sign an agreement which details the specifics of the apprenticeship program including: the skills apprentices will learn on the job, the related instruction they will receive, the wages they will earn, and the time the program will take. Employers agree to train apprentices and make a good faith effort to employ them for the duration of the apprenticeship (though if a company goes bankrupt or shuts down, understandably they can't keep employing apprentices, even if they usually will do all they can to help them transfer into other programs) and also typically agree to pay for all or most of an apprentice's tuition and fees for classes. In turn apprentices agree to attend and pass all required classes, and make a good-faith effort to stay with the employer until they have completed their apprenticeships. Apprentices may also be required to pay for their own books and lab fees, as well as purchase their own tools and safety equipment.

Any occupation can be registered as apprenticeable if it meets four criteria:

- it has clearly defined tasks and duties;
- it is customarily learned on the job;
- it requires manual, mechanical, or technical skill;
- it requires at least 2,000 hours of work experience and at least 144 hours of related instruction.

Currently, 858 occupations are recognized as meeting these standards and being eligible for registering apprenticeships. Sound technicians and workers who install and maintain sound equipment are included among these recognized occupations. Note that if you look at some of the sound provider companies given as examples in the first part of this book, a number offer system design and installations as well as rentals, so even though this kind of training would still require you to get additional training and/or practice in operating sound systems, training in installation can get you in the door.

Not every occupation has apprenticeships available at all times, and they may not all be available in all places. In some areas competition for apprenticeships is greater than in others. To find every opportunity, apprenticeship seekers need to check several sources. A good place to start is simply by Googling your state along with the term "registered apprenticeships." This is how we found the two main sources of technical apprenticeships in California, which between them listed half a dozen sound technician and sound installation apprenticeships (check the directory for links to these two sources). Another way to look for apprenticeships is to check with your state's office of the U.S. Department of Labor and/or State Bureau of Apprenticeship (check link directory for portals to state and government agencies). These agencies list current programs, and some will help people contact businesses that might want to start new programs.

Also try career counseling offices at any schools you are attending or have attended that give access to recently graduated alumni. Many apprenticeship sponsors publicize openings at career centers and local high schools and community colleges, and career counselors usually know about the programs in their community. Trade unions and professional associations usually have information too and are good places to inquire, so check the links to professional associations included in the web directory.

Before choosing a program, consider whether it is registered with the U.S. Department of Labor. Many employers have greater trust in the training offered by registered programs than in the training offered by unregistered ones, and because only registered programs confer *journey worker status* on graduates, graduates of these programs have more job choices. Also, be sure to check to see if a program matches exactly what you want to be trained in. Some sound technician apprenticeships are specific to sound installations; others are much broader, involving sound and several other types of system installations which may not be of interest. Just because an internship is listed under sound technician doesn't mean it will be right for your career path.

Evaluating Schools and Programs

While many private or vocational colleges, correspondence/*online schools*, and short-term, fast-track training programs like *training boot camps* are reputable and teach the skills necessary to get a good job, others may not be as trustworthy. Their main objective may be to increase profits by increasing enrollment. They do this by promising more than they can deliver. For example, they may mislead prospective students about the salary potential of certain jobs or the availability of jobs in certain fields. They also may overstate the extent of their job-training programs, the qualifications of their teachers, the nature of their facilities and equipment, and their connections to certain businesses and industries. It's not always easy to spot the false claims that some schools may make, but there are steps you can take to make sure that the school or program you enroll in is reputable and trustworthy.

Certification of Completion

Note that there are no official certificates for studio or live audio engineering, and any certificates offered by schools only show that you have completed that school's training program, *not* that you are an accredited audio engineer, a trained audio engineer, or anything near the equivalent. If a training school indicates the certificate they offer is *anything* more than a certificate of completion (i.e., claim it is universally accepted as proof of skills qualification, or makes you a "certified audio engineer") then tread carefully, as this is a dishonest claim and may indicate that other claims they make are also questionable.

Do Some Homework

Before enrolling in any program, do some homework. Here's how:

- Consider whether you need additional training or education to get the job you want.
- Investigate training alternatives, like community colleges. The tuition may be less than at private schools. Also, some businesses offer education programs through apprenticeships or on-the-job training.
- Compare programs. Study the information from various schools to learn what is required to graduate. Ask what you'll get when you graduate—an internship placement or eligibility for a clinical or other externship?
- Are licensing credits you earn at the school transferable? If you decide to pursue additional training and education, find out whether two- or four-year colleges accept credits from any vocational or correspondence school you're considering. If reputable schools and colleges say they don't, it may be a sign that the vocational school is not well regarded.
- Find out as much as you can about the school's facilities. Ask about the types of equipment—computers and tools, for example—that students use for training and supplies and tools that you, as a student, must provide. Visit the school; ask to see the classrooms and workshops.
- Ask about the instructors' qualifications and the size of classes. Sit in on a class. Are the students engaged? Is the teacher interesting?
- Get some idea of the program's success rate. Ask what percentage of students complete the program. A high dropout rate could mean that students don't like the program. How many graduates find jobs in their chosen field? What is the average starting salary?
- Ask for a list of recent graduates. Ask some of them about their experiences with the school.
- Find out how much the program is going to cost. Are books, equipment, uniforms, and lab fees included in the overall fee or are they extra? Be wary of boot camps that cost more than a year's tuition at a private college for a couple weeks of instruction. Always compare the number of hours of instruction and cost of programs. Also ask if any instruction will be hands on. Be aware that online degree programs that charge as much as regular degree programs aren't able to offer any hands-on instruction for that price.
- If you need financial assistance, find out whether the school provides it, and if so, what it offers.
- Ask for the names and phone numbers of the school's licensing and accrediting organizations. Check with these organizations to learn whether the school is up-to-date on its license and accreditation. *Licensing* is handled by state agencies. In many states, private vocational schools are licensed through the state Department of Education. College *accreditation* is usually through a private education agency or association, which has evaluated the school and verified that it meets certain requirements. Accreditation can be an important clue to a school's ability to provide appropriate training and education—if the accrediting body is reputable.
- Check with the Attorney General's office and the Better Business Bureau in the state where you live and in the state where the school is based, and with your county or state consumer protection agency to see whether complaints have been filed against the school. Note that a record of complaints may indicate questionable practices, but a lack of complaints doesn't necessarily mean that the school is without problems. Unscrupulous businesses or business people often change names and locations to hide complaint histories. This is why knowing a school's history is important; if a school is brand new and can't show a history of good practices, or seems to have a history indicating a name change every few years, tread carefully.

Society of Broadcast Engineers Certification Options

If broadcast audio is an area of interest, a four-year degree from an accredited college can be used to supply four of the five years of experience required to qualify to take the exams offered by the society of broadcast engineers, if you want to pursue accreditation as a Certified Audio Engineer® (CEA®) or Certified Broadcast Radio Engineer® (CBRE®), though a four-year degree is not the only route to these certifications. Five years on the job is just as good, and an A.A. degree or training program can also be used to count for one to two years of the required five. There are also less intensive certifications offered without the same level of eligibility requirements. Check the directory for a link to more information about this group of certifications, which are one of the two limited certification options currently available in the field of audio.

Taking Care of Business

Once you decide on a school, review the materials the school gives you, including the contract. Avoid signing up until you've read the documents carefully. Check the contract to see whether you can cancel within a few days of signing up and, if so, how to go about it. If the school refuses to give you documents to review beforehand, take your business to another school. Its refusal may be a sign that the school isn't trustworthy. *If a school official tells you something other than what is in their documents, ask the school to put it in writing. If the promises aren't in writing, the school can deny ever having made them.*

To finance your vocational training program, you may apply for financial aid through the school's financial aid program. If you take out a loan, be sure you read the agreement and understand the terms of repayment before you sign. Know when repayment begins and how much each payment will be. Also, realize that you're responsible for paying off the loan whether or not you complete the training program. If you don't pay off the loan, you may run into some serious problems.

Self-guided Learning

In many cases, no matter how much you may be able to learn in your formal training, there will be essential topics that aren't included in the training, or that won't be included as soon as you will need them. For example, an apprenticeship will most likely teach you a great deal about the technology, and more about electrical skills than any other training, but even if you get training in how to tune a room and how consoles function, most installation apprenticeships are unlikely to teach you much about operating a console to mix a show, much less provide ear training. Even a four-year degree in audio may leave you without enough electrical knowledge to satisfy you, or will be lacking in business training you will need at least a bit of if you want to go into business for yourself. You will need to be able to learn topics on your own in order to ensure you are fully trained. Even once you have established a career, you will need to be able to keep on learning to remain up to date on industry practices and stay competitive. These tools will enable you to guide your own learning for specific topics.

Seminars and workshops: Seminars may last anywhere from a few hours to an entire 8–12-hour day and include detailed instruction and eye-opening demonstrations. Seminars often allow for the chance at hands-on practice and are a great place to learn about solving problems concerning sound system setup and operation. Workshops may be one-day events similar to seminars, or may take place from over several days up to two weeks in length, sometimes called *"intensives,"* which are more like mini-classes that allow for more in-depth coverage or hands-on experience. Many will offer certificates of completion indicating the topics covered.

Seminars by local contractors may put on a one-day event held at their own place of business or at a local hotel convention center. Training professionals travel the country offering seminars at convention centers, or local schools or studios. Manufacturers of hardware and software also offer seminars held throughout the country each year, and software that offer certification programs often offer regular classes and testing contracted through local training centers. Conventions like the yearly Audio Engineering Society (AES) conventions and National Association of Music Merchants (NAMM) shows offer seminars and workshops that convention goers can sign up for in advance. Community colleges also offer weekend workshops and two-week classes on general topics like soldering, electrical safety, CPR, and incorporating your small business.

Seminar schedules are usually released at the beginning of the year or at six-month or quarterly intervals, so they usually can be signed up for well in advance. To find seminars, look for them advertised in industry publications, announced on the web pages of large local sound companies, on well-trafficked industry discussion boards, and through industry organizations. Also look for them through your local community-college adult-education programs. In the link directory we include some links to help you get started finding these.

Some seminars are offered for free, but most charge fees ranging from $50 to several hundred dollars. Two-week workshops may only cost a hundred dollars at your local community college and meet four times, or may involve 40 hours of onsite instruction per week and cost a few thousand. As with any training, look at the number of hours and what you will be getting when deciding if the cost is fair. As with all products, look into the better business ratings and online reviews of any training seminars or seminar providers and ask around to find the most recommended and reputable ones before signing up for any.

Audio reference books: Audio books may provide a broad overview of many audio topics or an entire field of audio, like this one, or can focus on providing detailed information on a single subject or limited range of topics. These are good resources and you should expect to invest in a number of them, especially if you don't attend a course of formal classroom training. Even if you pursue a four-year degree, however, expect that once you have a nice library going from books bought for your classes, you will still find use for more books on topics not covered in as much detail in school. Look for directory links leading to some good book lists and sources for reputable audio texts.

Internet: The World Wide Web is a fast-growing resource for ideas, operating hints, and new products to help solve problems. Many manufacturers have good websites with training on their equipment. There are also sites providing education and sound system principles, operating tips, and advice. Even video sites usually thought of as places to find entertaining viral videos are filled with educational tutorials by live audio industry luminaries like Dave Ratt and many others. See the Links page under Resources for good web pages for learning about every audio topic under sun, as well as other topics of interest like how to handle paying taxes as a freelancer and how to market your new audio business.

Videos: There are several educational video series available on live sound topics. Some of these are packaged as full training packages with 50 or more video tutorials created by well-known engineers, and others cover single topics. Some are so pricey, costing more than many hands-on training courses, while others are much more reasonable and are well recommended. Many of these provide valuable and accurate information on almost any audio topic imaginable, and are useful for learning processes that books have a harder time conveying. Again, check out the Internet link directory for links to some reputable sources of video tutorials and video training, both free and for a charge. Again even if something comes recommended by one trusted source, it's up to you to look for further information to verify the worth of any product.

Developing Your Own Internship

Internships are often an essential step to getting a shot at finding employment because many jobs ask for applicants to have some experience first. Internships are a way to get experience that can help you get your foot in the door. Many companies that offer internships

Is the Position Secretly Seasonal or the Internship Intentionally Irrelevant?

Not too long ago you could have been confident that, barring layoffs, your aptitude and effort would ensure extended employment, but this is no longer always the case. While contract and temporary work has always been around, it has become much more common than before, and some employers find it easier to let employees go when business slows and hire more when it picks up. This is not a problem in most cases; unfortunately not all employers who engage in this practice always clearly label advertised positions as temporary or seasonal.

Similarly, there are internships which accept interns more for the cheap or free labor than a desire to shape the next generation, who don't teach their interns much. Though these employers and internships are the minority, there are enough in every industry nowadays that it's worth asking a tactful but tactical question to try to asses if your suspicions are founded. In a competitive field, you don't want to pass up an opportunity at a position with potential for growth for a job that's not likely to get you anywhere, or an internship that won't teach you anything or enable you to make any worthwhile contacts.

Unfortunately, asking about a position's potential for longevity or advancement is unlikely to give you a clue since these folks will probably give the same standard reply as most employers—that they can't guarantee anything, but for the right employee there is of course potential. It's easy to affirm the likelihood of a hypothetical positive outcome for a hypothetical ideal employee or intern, and easy to deliver an answer to a question asked by 90 percent of interviewees.

A follow-up question to test the real chances for growth at a position would be to ask about their promotion policies in terms of the last entry-level new-hire who successfully transitioned to long-term employment. How long ago was it and why were they chosen? What did their timeline look like and are they still with the company? For an internship, ask the interviewer to describe specific projects or tasks assigned to recent interns. Also ask when they last hired an intern, and what qualities that intern had that influenced the decision. These are questions they won't have much, if any, practice answering. Because the answer must refer to real events and not hypothetical possibilities, if they are fudging, it will show. If they can't provide a straight answer, or if they can, but it turns out they haven't put any low-level employees on a promotion track or hired any interns in almost a decade, odds are not good they will do so for you. If they provide an answer that shows they are on the up and up, it's still not a sure thing you will be a good match for them, but at least you'll know there's a fair chance.

even hire their best interns, so a good internship can often lead right into a good job. Many companies have well-established strong internship programs. Others may have assorted positions available as the need arises. There are many different ways to find a great internship experience and we link to several sources for finding internships online.

Often, good schools or training programs will help you find good internships and even may be able to place top students in internships. However, there seem to be more studio and broadcast than live audio internships offered, and competition for internships often far outstrips available positions making it hard to find the position you're looking for. As you look for internships, don't just rely on what you find listed. If there is a particular local soundco you are especially interested in, it is in your best interest to contact that company directly and ask. Give them a call, send an email, or write a letter. Ask if they hire interns, have hired interns in the past, or if they might be thinking about hiring an intern. They may not be advertising until next month, or their internship positions are currently filled but an opportunity may open up at a later date.

If you have finished training and find your job hunt isn't getting you anywhere right away, and internships are just as hard to find, try researching and developing your own internship proposal to bring to sound providers who don't have an internship program of their own. It takes time and effort but creating an internship proposal that a company can use to establish their own internship program even after your internship can get you a foot in the door or get you hired, but also gives back to the industry right from the get go by helping develop a new internship program that benefits the company and students well into the future. It shows an incredible amount of initiative and confidence if you can pull it off, and in addition to getting you an internship can be listed as project that will look good on your resume on its own.

Define Your Goals

It is important to keep your goal in mind while developing your internship, defining what you are looking for. To legally qualify as an internship a position must have the following features:

- on-site work experience that is directly related to a career goal or field of interest;
- duration agreed upon ahead of time and is short term (typically a summer, a semester, or a year);
- emphasis on learning and professional development, which may involve guidance of a mentor figure;
- can be paid or non-paid and have full-time or part-time hours;
- may be for academic credit (but not required).

What Are the Advantages of Interning?

- gaining valuable work experience and competitive edge for your job hunt and graduate school applications;
- developing professional and marketable skills;
- "test driving" a chosen field—do you like the type of work, atmosphere, hours, coworkers, etc.?
- practical learning environment—using theories learned in class;
- networking—developing business contacts, securing letters of recommendation, and connecting to employers who may offer you full-time work in the future.

Develop a Strategy

In order to write the contract and proposal you will need to:

- *Use online resources:* We include links on the book's companion website to internship information to get you started there, and you can also look up the websites of sound providers in your area. Also go on industry discussion boards with questions; if you let folks know what you are doing, you can get lots of good advice there.
- *Network:* Talk to people who are connected to the live audio field or who know about the internship process (professors, family, mentors, career counselors, friends, etc.).
- *Use your local library:* Tell your local librarian you are trying to find information on internships, including how to register one and what legally binding internship contracts look like, and they might steer you toward some resources you couldn't find online.

This process should involve a fair amount of research to find out what successful internship programs in live audio look like. One of the best ways to gather information about internships in a given field, for a specific position, or at a particular company is to talk to people directly. Through informational interviewing you can gather information and identify opportunities that you can't get from reading a company brochure. People working in the field can provide the most up-to-date information on how people get started, what employers are looking for, career paths, corporate culture, and industry trends. Begin by networking with everyone you know: family, friends, professors, career counselors…everyone!

Let them know that you are researching the live audio field in order to develop an internship proposal and ask them if they have any information or know anyone who might. Attend career panel presentations, and contact companies and schools. Your goal is to get the names of people who might be willing to meet with you and talk about the field.

You will want to contact local schools and ask what they require of companies in order to partner with them for internships and find out what qualities an internship must have to be "official," and what the process is for registering an internship. On top of using this information to develop the most legitimate internship proposal possible, you might even be able to foster a partnership between a sound company and one or more local schools.

Reach out to companies that already offer internships and ask if anyone might meet with you or even email you to answer questions such as does offering internship positions require additional insurance or affect their liability insurance costs, or do paid internships require any different handling by payroll or are they no different than regular paychecks. Even an open-ended request for any information that might help or a simple question like what details/advice about offering internships would they have for a company like themselves considering it.

Just because a company does not have a formal internship program or any current internship openings does not mean that you can't approach them and inquire about setting up an internship. Reach out to local sound companies to ask why they currently don't have an internship program, and inquire if someone did the work of putting together a proposal for an official internship with their company specific needs and capabilities in mind, would they be interested in offering internship in the future. You can be honest and let them know you finished your training but found too few internships and entry-level jobs available and while you continue your search one way to keep busy and be proactive is to develop your own internship proposal and make it good enough that the company who accepts it can use it to base future internships on. You might do this by putting in the effort to track down the person responsible for hiring or by sending your initial inquiry to one of the owners via the mail.

The Proposal

As you gather information you should start to get a picture in your head of what your internship should look like, once you have enough information you should be able to put together a contract and details into an internship proposal. Most employers are too busy for people who will "do anything" so be as specific as possible. Based on the information you gathered during your research, prepare a proposal for the internship that details both the labor an intern could provide and specific lessons a sound company could teach the intern in exchange. If you haven't gotten any companies to bite already (you might get a company on board with you early and develop an internship specifically for them and with their input) you will still want to contact each of them again with your finished proposal. If they hadn't considered using an intern earlier, your initial inquiries might have planted the idea in their minds, making them ready to consider a proposal they weren't ready for earlier.

Take your proposal to local schools and see if you can get them to endorse you or point you toward opportunities you hadn't considered. Don't be afraid to approach employers with your proposal—what you are offering them is a highly valuable commodity. Be willing to adapt your original proposal to work better with a specific company and let companies know this is an option. At the same time, do be prepared for companies not to be interested in what you have to offer. It may be that they simply don't have a need for your services at this time, or have had a bad experience with previous interns. Don't let this discourage you and don't take it personally. Just put together the best proposal you can and keep searching until the right opportunity presents itself. Also remember that you can keep developing your resumé and pursuing regular employment too.

What to include

Make sure you take the time to prepare your materials carefully so that it's clear that you are offering something of value. There is no prescribed format for your proposal, but a letter written specifically for each employer is a good place to start. Explain that rather than wait a year for the limited available internships to open up again, you have put together an internship proposal for yourself to bring to companies in your field, but furthermore to offer something of value right from the start have done the research to make sure the internship proposal includes all the details and documents required to sign up as an official internship so any company who offers you an internship that wishes to start a long-term internship program can use the documents you prepared as a template to develop their own internship program. If you have arranged for this add that you have already contacted local schools who partner with local businesses and who they might work with.

After your letter should come the body of the proposal which includes a clear and concise description of the internship you seek, and either copies of all documents such as a sample contract, or a description and partial example of the additional documents that will be included in a full proposal if they get back to you indicating interest. Be sure to indicate that you encourage the company to tailor the internship details to reflect their unique needs and what they can offer and make a counterproposal if your proposal is not perfect for them.

Finally, include a cover letter and resume as you would to apply for any internship—the letter should highlight why you are the right person to intern for the company, indicate that you are equally open to a paid or non-paid position, and include your dates of availability and should be followed by a resumé that illustrates your strengths, academic experiences, and marketable skills.

Where to Send It

Be sure that your proposal gets to the person who has the power to hire you. It may take some detective work to find the people who might have a better idea of the company's staffing needs and most importantly who can actually offer you a position. Websites and company literature can help you find the name of these people (look in the "about us" or "contact us" section), or ask your networking contacts if they have any suggestions. Telephone receptionists and department secretaries can help too. It is better to invest the time into a few phone calls than to send a letter to the "Director of Personnel" or "To Whom It May Concern."

Follow-up

Follow up with a phone call or an email within a few days to a week of the receipt of your proposal. Students sometimes feel it's pushy to make such a call or that they are bothering the employer. Don't be pushy, but a polite follow-up is expected and reinforces your interest and commitment. Everyone in the working world understands the need for persistence.

Don't Give Up

As was mentioned at the beginning, there are many different ways to find a great internship experience. Developing your own internship takes time and effort. Don't be discouraged if you can't find a position that is readily available; keep at it. Persistence is key! An internship experience that you've created will be far more rewarding than a position you settled for because it was all you could find.

Your Job Search

There comes a point when your only options are to find a job as an employee, find gigs as a freelancer, start an audio company of your own, or move on to another career option. Note that taking a temporary sales job at a music store to pay the rent does not count as the fourth option until you stop putting any energy into one of the first three. It may take some time working in related fields or working another job part time while continuing to put the rest of your energy into the first, second, or third option before you will achieve your goal. Even then it will often take being willing to learn related skills like booking bands to supplement your audio work if you are going to make it work.

We include all kinds of links to helpful information on job hunting, freelancing, and starting your own business in the web directory, and a lot of the information in the final chapters will be helpful to you no matter which path you choose. However, since these topics are all covered in great detail in the sources we link to, by experts with more experience in helping job seekers than we have, we have only one piece of advice for your job hunt, since it's something you will want to begin working on while you are still studying audio, and is important to be aware of before applying for internships as well—be careful about how you use social networking sites.

Your Online Presence and Finding Work

Social networking sites are a great way to connect with your friends, but they can also spell trouble for your job search or potential business partnerships. If you have a profile, photos, or videos online, take the time now to review them with your job search in mind. Are you comfortable with how your online persona might come across to a potential employer? You need to be: more and more employers (and companies hired to run background checks on potential employees) look to the web for details on candidates, so your online presence could be a factor in whether you get the interview or if your resumé gets tossed. Freelancers and small business owners need to be just as concerned, since potential clients and partners are getting more savvy as well and, in addition to looking up a company, are starting to look up company owners as well before making choices.

Don't let an embarrassing photo or stray comments undermine your job search or tarnish your freelance or business reputation. Understand that anything out there could be fair game. Don't assume employers or clients won't look at or find your profile. (It's a good idea to limit who can access your profile, but don't simply rely on this, privacy settings can get reset with policy changes behind your back.) Remove any questionable photos, videos, or posts. You likely have the good sense not to brag in an interview that you won the naked beer-drinking contest at your college three years in a row, but if one of your online profiles broadcasts it, a potential employer may find out anyway. The rule of thumb is, if it might cast you in a bad light in the eyes of an employer or client, remove it. You want to be perceived as professional. Also check to see what others are posting about you and try running a search on yourself in several search engines: you might be surprised by what comes up.

If you find any unpleasant surprises or old secrets broadcast before you were thinking about things like your business reputation refer to the web link in this chapter's web directory that leads to an excellent article about removing or burying information from the web.

Chapter 26

Navigating the Audio-technology Marketplace

To complete our introduction to the individual components that together form the signal chain, we need to discuss the audio-technology marketplace, because developments there have a big influence on the entire audio industry, including live audio reinforcement. Live audio engineers benefit from keeping up with developments in the audio technology marketplace for a number of reasons, the most obvious one being because they typically work with a much wider array of gear than any other audio engineering niche. While broadcast and studio engineers use many different electronic devices to do their jobs, they work with the same gear and setup every day. Because they mix sound at so many different events, live audio engineers tend to work with new equipment on a regular basis. Whether a live audio pro works a different tour each season, or a different event every night, over the course of their career they will use many different models of equipment from all sorts of different manufacturers.

There is certainly a lot of equipment available that they might run into; at any one time there are hundreds of different kinds of microphones, mixing boards, EQ and effects units, and speakers in use at live entertainment venues and available for rent through sound companies. When you consider that these components can each be mixed and matched in different configurations in any P.A., it becomes clear that there is almost an infinite variety of potential P.A.s that an engineer might encounter on any job. The factors responsible for the sheer number and variety of components used in live sound and available for sale in the audio marketplace are the same ones that have been driving forces of the digital revolution; working together these have fueled the progress occurring throughout nearly every area of society including the audio industry—capitalism and computer chips.

Over 30 Years of Digital Audio: How Silicon Changed Sound

For most of the history of pro and consumer audio technology, technological advancement progressed at a steady but moderate pace. A good stereo would last the consumer a decade before new technology would be far enough advanced to make it worth upgrading, and the same was pretty much true of pro audio P.A.s as well. When digital audio technology was introduced to the marketplace in the early 1980s, with the advent of the MIDI protocol and Yamaha's release of the first FM digital synthesizer in 1980 (which didn't gain much traction at $16,000), followed by the release in 1983 of the first Yamaha digital synthesizer to be widely adopted (priced so even musicians could afford it at $2000), audio technology became tied to computer technology. Since then the rate of advance for digital audio has been subject to the rule of thumb used to describe the advance of computer technology, called *Moore's law*.

This rule of thumb is based on a prediction made in 1965 by the co-founder of Intel, Gordon Moore, where he stated that the number of transistors that can fit on an *integrated circuit* (also known as a computer chip) would double approximately every two years. Though the prediction simply took the trend that had been observed since the silicon chip was invented in 1958 and projected it forward, his guess would prove more accurate than he ever imagined. He was far less accurate when, just ten years later, he stated that by 1980 the exponential growth described by Moore's law would reach the limits of what could be achieved within the laws imposed by physics; he and the many since who have predicted it would soon be impossible to continue increasing computational power at the same exponential rate have all been wrong and Moore's law has remained in full effect for 50 years, and currently continues unabated.

In practical terms this means that for five decades we have doubled our computing power approximately every 18 months, more or less doubling capabilities such as memory size, energy efficiency, and processing speed right along with it. (If it seems that this can't be true because our computers don't seem to run that much faster that's because the features and hardware requirements of our software has increased at a similar rate.) Furthermore, this pace is likely to continue for the near and foreseeable future. The latest forecast for when the breakneck speed of progress will wind down has been made by theoretical physicist Michio Kaku, who predicts it will gradually slow down but will continue for ten more years, noting in several interviews in 2011 and 2012 that we are actually finally approaching the limits of what physics will allow us to achieve with silicone, though he also adds that before that time researchers will likely have developed a new combination of methods and materials that will ensure that the progress of Moore's law doesn't come to a complete halt, but is replaced by a new law describing a rate of progress suited to that new evolution in our technical progression.

Moore's law has been the factor underpinning every aspect of the technological revolution that has unfolded over the last 50 years, including the introduction and growth of digital audio, and promises to usher in even more big changes over the coming decade or two before the transition is complete. (Much like the change from an agrarian to an industrial society, the digital revolution, which we are right in the middle of, promises to bring changes to how we live at almost every level by the time it's over.) As a result of our growing computing power the functionality of all our electronics have been increasing at an exponential rate as well, from the number of songs we can store on our MP3 players, to the number of features in our cell phones, to the number of megapixels in our cameras. While computer technology has by now found its way into every analog electronic niche and influenced the functionality of their product offerings, few electronics industries have been wedded to digital technology for as long, or seen its own technological progress influenced more by advances in computational power, than audio technology has over more than 30 years of steady development of digital audio equipment.

On top of this, few industries in general have experienced as much shake-up or undergone as many structural changes in how their business is conducted because of digital technological advances as the combined audio and music industry has. While that is beginning to change, and other media niches are beginning to experience the same restructuring as audio/music production has, the audio industry definitely served as the canary in the coal mine and the issues it began wrestling with almost a decade before other industries has foreshadowed the changes that all media industries are now trying to come to terms with. The audio/music industry is still figuring out a way to restructure that will be viable for all players over the long term, and won't be able to completely settle into a final form until our entire culture is done transitioning anyway, but the many adjustments made so far indicate those professionals in live sound will be able to meet any challenges to come with the same flexibility and ability to laugh at absurdity that it has shown in the past.

When Going Toe to Toe Makes Both Players Grow: How Capitalism, Competition, and Computers have Teamed Up to Create Change

Another factor that has combined with the explosion in computational power predicted to create change is competition for market share (both in business, but equally in the marketplace of ideas). The first way this occurs is a bit more subtle but is required to drive Moore's law to predictably provide more and better digital power to tools that directly and indirectly fuel societal change; this first requires the sort of conditions

Figure 26.1. Computer technology and audio technology paired up earlier than similar partnerships, spending over 30 years together means digital tech has contributed to many developments in audio technology.

needed for a prediction like Moore's law to be possible. After all, Moore's law is not law of physics, and in fact doesn't predict or describe anything remarkable about physical processes or materials other than requiring the unspoken implication that such doubling is not prohibited by physical laws. Instead the prediction is about human capacity—specifically it predicted what a group of electrical engineers can do with a little sand if someone would just herd them into one department, provide them with material support, and apply the proper motivation. Of motivators that have fueled Silicon Valley, after novelty and need stopped being factors, competition (more often for bragging rights than profit it seems) has been the most important.

The end result of the above is simply that across almost all areas of human activity, the greater organizational capabilities, access to data, and tools for real-time communication created by new computer and communication technologies fueled by Moore's law have increased the ability of smaller entities to compete against the status quo, and have allowed organizations large and small to increase productivity, both situations that are by themselves proven engines of progress. This is why analog audio technology has been developing at a more rapid pace as well, not directly tied to gains in computing power as with digital audio, but because the rise in computing power as described by Moore's law has resulted in changes like broader and more convenient access to existing research, greater ease and speed of communication, and better and faster scientific tools, from handheld test devices to simulation and modeling software.

Developments such as these have advanced just about every area of inquiry and empowered scientists and engineers in research and development (R&D) departments and universities everywhere to make discoveries and invent new technologies much faster. Some of these can then make their way into the analog audio technology that is released in the marketplace every quarter.

The exponential increase in computing power has also allowed manufacturers to make equipment smaller and smaller without a reduction in power, creating a whole new slate of products where upgrades are not for new functions or power, but for fitting current function and power into ever smaller versions. The technological revolution has also created increased efficiency of manufacturing paired with decreasing costs, and access to resources and markets all over the world through the internet, which have had the effect of opening up the marketplace to inventors and entrepreneurs and giving them the means to compete with the big corporate companies.

All this increased activity and competition has combined to lead us to the current state of the audio marketplace, where new audio equipment is available each season, and the lifespan of a piece of equipment is rarely longer than five years before it becomes obsolete, much less a decade like it once was. Gear manufacturers develop and release new audio equipment at a dependably rapid pace, making sure the available offerings cover the entire range of functionality and price points. Any sort of equipment that may be of interest to their customers is developed, produced, marketed, and sold, then upgraded several times before being retired and beginning the cycle anew with another model.

These developments and other industry trends are relevant to live engineers because they inevitably will at some point inherit any general duties that relate to audio technology. Live engineers should just expect this will happen as a matter of course, though many don't think much about the tasks they may be expected to do outside of setting up and running the P.A. To most people's way of looking at it, the understanding that live engineers are able to run any type of P.A. presupposes that live engineers come equipped with a fair amount of knowledge about live audio technology in general, making them the resident audio-tech experts. This makes the live engineer the natural choice to handle tasks related to evaluation and selection of audio equipment, and they will end up needing to if they work in the field long enough whether they are prepared to act in that capacity or not.

Because we tend to think it's smarter to prepare in advance, in this chapter we will examine some of the tasks live audio pros may be expected to handle in their capacity as general live audio-technology experts. Taking steps to learn about the broader trends and practices of the audio-technology industry is especially important for the newest generation of live engineers. The end result of the influence that competition and computational power has had on audio technology means that today's engineers must become fluent with a depth and breadth of available audio equipment that is unprecedented, and will need to confidently navigate in a marketplace that is larger and more complex than ever before. In this chapter we suggest some simple strategies audio students can use to begin developing this expertise right away.

The Live Audio Pro: Audio-technology Expert

The main focus of your training is on the specific knowledge required to set up and run a typical live P.A., and how to mix audio in a variety of live settings, but while these activities represent the primary function of most audio engineers, they aren't the only tasks live engineers will be expected to do. Most audio engineers are required to handle a variety of other duties in addition to setting up and running the P.A. Additional duties will vary depending on whether the engineer works for a venue sound crew, a sound company, or is working a seasonal contract for a touring act. Details may differ, but each type of position can involve tasks that may require that the engineer handling them knows how to research, evaluate, and purchase gear. The live audio engineer who already has a general understanding of live audio technology and is up to date with current developments in the marketplace is more equipped to do these tasks confidently and with a minimal amount of extra work and stress.

The examples given below are for illustrative purposes only and not intended to represent what your tasks may look like, just to provide a few examples of what could be requested of you.

With regard to renting gear, the audio pro should expect they may be asked to handle or help with the following tasks, either acting as proxy or consultant:

- assessing gear requirements for specific events;
- choosing the best soundco from available options;
- selecting an appropriate rental package and negotiating the best price;
- negotiating last-minute rider substitutions regarding gear and price on behalf of client or soundco.

With regard to maintaining P.A. equipment belonging to the client or employer or assisting with purchasing new or replacement items, the audio pro should expect they may be asked to handle or help with the following tasks, either acting as proxy or consultant, either from a dedicated budget or requisitioning funds for each occasion they will be needed:

- replace faulty cables, broken microphone stands, and loose connectors;
- purchase essential supplies such as batteries, amplifier fuses, sharpies, gaffer tape, first-aid kit, storage bins, etc.;
- select a repair shop and arrange and pay for equipment repairs (and know when the estimate is higher than the replacement cost so you can consult with the gear owner before agreeing to repairs in this range);
- make purchasing decisions on behalf of an employer or client or consult them in purchasing replacements or upgrades of gear based on a set of criteria or your own judgment.

While you are guaranteed to find yourself needing to research equipment even if you never asked to purchase it, the mere possibility you may be asked to assist with any purchasing tasks is reason enough for learning a little about navigating these situations without getting ripped off. The responsibility to be prepared to make safe and conscientious choices when entrusted to handle another person's money is important even when it's only a modest budget, and even more so if you will be entrusted with spending a large sum wisely.

In addition to being prepared to act as proxy for an employer or client, becoming an audio-technology expert and being familiar with the marketplace can prepare you for opportunities to act as a consultant for others who need help with tasks involving selection and purchase of gear. In these cases you won't need to make purchases yourself, but you will still need to know about selecting and purchasing gear so you can pass on the best advice clearly, as well as be able to articulate the reasons for any recommendations when asked.

Personal Purchases: Tools and Supplies

Finally, among all the other good reasons to develop a system for learning about gear, choosing between options and making good audio purchases is your own bottom line. The odds are good that if you work as an audio engineer for any length of time, you will buy some of your own gear to use at work. Cords, microphones, testing equipment, outboard units, and DI boxes are the sorts of equipment live engineers often choose to purchase to have on hand as spares, even if they use the ones provided by their employer or contracted party much of the time.

You may decide to buy your own meters and tools over time as well, even if they are provided at work. You will usually want to be able to take essential tools home with you; in a career like live audio, working full or part time for an employer doesn't mean you won't find freelance gigs from time to time that you won't want to pass up. When opportunities arise to work at events scheduled on your days off, you'll want to be free to take your tools too. (While some bosses are laid back and may be fine with you taking items like meters or mics home for your own use, we don't recommend it if you can avoid it; it is simply not worth the hassle in case something breaks.) It goes without saying that the above applies doubly for engineers who freelance a great deal—their need for their own set of test equipment would exist even if they were sure to be supplied tools at each job, since having to get used to using new meters (which like all gear will each have unique features and quirks) at every job would defeat the purpose of such tools, which is to make the engineer's job easier.

Personal Purchases: Your own P.A./P.A. Add-Ons

Even if you don't want to run a full service P.A. rental business, you may even choose to purchase a small P.A. of your own to supplement your income by renting out your P.A. and mixing services on your days off or between freelance gigs. Many live engineers do this once they have the knowhow to do it. Some slowly grow into larger rental businesses as they find themselves enjoying it, putting a lot of effort into growing the business, and eventually doing well enough that they don't need to work anywhere else—most audio rental companies, even the big ones, started with one or two engineers and a P.A. and grew from there.

Other live engineers never do more than pick up a few gigs as needed. They may or may not upgrade their gear a little over time, but either way they are happy to stay small, one engineer/one P.A. operations and pull in extra money in the off-season as free agents, while growing their freelance career, or position within a larger rental company, the rest of the time.

Finally, you may not want a full P.A., but will want one or two specific pieces to supplement the venue P.A. or rental equipment. The venue P.A. may only have one effect, or not enough compressors, and the venue owner isn't likely to provide funds to fill out the area that is lacking. The freelancer may feel it's worth bringing their own parametric EQ to each gig since a rack mounted parametric is not a sure thing but the ability to target the exact troublemaking frequency can make a big difference when mixing in difficult rooms.

Personal Purchases: Other/Personal Audio Gear

Buying gear is also likely because so many live engineers have overlapping interests in other audio niches, as this has become more common over the past decade and will be even more so in the foreseeable future. Some live audio pros who feel no draw to buy a band P.A. may still want their own D.J. rig, or a home studio set up for basic composing and production, or both of these and their own live P.A., depending on their goals and plans. Most of the information and tips in this chapter also can be applied to buying the right home studio set up or D.J. rig, and there are links in the directory that provide additional guidance for these particular purchases, as well as to guides for buying a karaoke P.A.

Getting to Know the Audio Marketplace

Knowing how to research and buy audio gear is part of the job, but what, when, and how much you choose to purchase personally will vary according to your needs and your budget. In any of the sections that discuss evaluating gear using terms that imply an intent to purchase, don't feel this in any way implies you must buy gear to work as a live audio pro—but don't be too quick to assume that there's no use in developing the ability to evaluate gear with an eye to selecting the best option and decide to simply skip over them just because you won't be buying for yourself. For those who want to buy essential items but not break the bank, don't worry—if done right, these purchases really don't have to be expensive.

However, no matter what you may intend to buy down the road, or even if you don't plan on buying anything, your first step is simply to get acquainted with the lay of the land and become familiar with the categories and classes of equipment instead of worrying about any individual item. The first three categories you need to understand in order to be an expert of audio technology are the layers or tiers that are sorted according to price point and features.

Market Tiers and Price Points: Budget, Mid-Range, High End

Price tiers are categories for separating any grouping of like items according to relative quality/and price. Many vendors will group certain items together according to shared price tiers. At the hardware store it'll be lawn mowers lined up by ascending price. In department store clothing sections each department is organized by racks hung with items in the same price range starting with clearance or bargain racks in the back, and with premium priced items in the front. Whether vendors arrange by price point or not, many shoppers do, as almost all products inherently fall in one of three categories (average price, bargain price, expensive price) whether that is acknowledged or not.

Product tiers are categories, usually the three indicated above, sometimes four with a low and high mid-range, that manufacturers use to indicate the target market and price point of different product models or model versions. Manufacturers who offer products by tier may produce several versions of the same model (often called a product series or line) where the base model name, product design, components, and feature types are almost identical across different tiers and the tier/cost of each item is determined by the number of each type of feature (this may be indicated by the number appended to the base model name). This is very common with consoles; almost all models come in at least three versions, and some models come in series of six or more versions. For example, a console manufacturer may sell "Model 8/4 Fex6," "Model 12/6 fex10," "Model 16/8 fex12," "Model 20/10 fex16," and "Model 24/14 fex20" at price points $200 apart, where the first number indicates the number of channels, the second number indicates the number of aux buses, and the "fex" number indicates the number of included board effects. Other manufacturers follow a nearly identical method; offering a tiered product line that shares the same base model name and essential qualities but is a tradeoff of the number of added features for cost. Instead of appending numbers to indicate relative tier/feature numbers, value weighted words are used: *light* or *basic* for budget items, *standard* or *pro* for mid-range items, and *deluxe* or *premium* for fully featured high-end items.

Other gear manufacturers will use different model names for each value tier, sometimes using a slightly coded model name to indicate tier and an additional number or value indictor to order each model within that tier. For example, a speaker manufacturer might call their under-$500 budget speakers the "Beer-bashers" and append watt power to further order them by power and price. The same pattern would repeat with the mid-range tier's models the "Bar-hoppers" and the high-end models the "Country-clubbers" (note the "value-weighted" model names indicating the tier from low to high by the "status" associated with the model name).

Finally, there are those manufacturers that create products for all tiers or target one specifically with all but a few products being offered in that tier; however, they don't distinguish between tiers using any obvious naming scheme and don't indicate they are arranging their products by tier. Only looking carefully at their features and pricing or how their products are positioned would reveal any patterns to indicate their products are tiered.

Most audio gear is tiered according to value and price by the consumer even when the vendor or manufacturer they are buying from doesn't, but many manufacturers do tier their products even if they don't call attention to it with a naming scheme like the ones described above. Fortunately, almost all modern audio gear is able to meet a certain baseline standard as to how it will sound. If it doesn't, most manufacturers won't even allow it out of development and into the production and manufacturing phase, much less let it get to market. Gear that meets this base standard and has features and functionality comparable to what is usual for that type of gear is usually going to be priced to fall in the mid-range. *Mid-range* equipment represents the industry "average" in terms of the capability and sound quality of the currently available audio technology and as such is the equipment price tier representing the market "standard" for comparison that determines what equipment qualifies as *high end* or *budget*.

As the group representing what is "typical" in audio technology, the difference between any two models within the mid-range category will usually be negligible in terms of sound quality and overall performance. Though models from different manufacturers and product lines will vary in feature sets and functionality according to their intended application and target market, and each will have a slightly different sound coloration based on their frequency response, all will still fall within the same baseline quality/performance range.

Because of the basic similarity in quality, the variety available due to differences in layout, available functions, and product design becomes an important means for marketing to set that product up as something unique amid a field of competitors with offerings of similar quality. Even equipment that is otherwise identical will have at least one or two features to distinguish it from similar models. Some will have shells made with thicker materials and boast sturdier construction; some will come with a greater number or range of features than comparable models; some will have a sleeker design and wider dispersion, while others will be made for portability, both lighter and more compact than comparable equipment but still reasonably sturdy. Those speakers that have no distinguishing feature may come with one of the best features out there, a good two-to-three-year warranty. This is why it's worth taking the time to research purchases to make sure to get the "best" equipment for you; even though most mid-range gear will be of similar quality, there will be models more suited to your needs than others.

All gear in this category will also fall in the same range in terms of *list price*, the retail price suggested by the manufacturer (also called the *MSRP* or *manufacturer's suggested retail price*), and variations in price within this range will usually reflect the particular combination of available features, power and size, and brand popularity. This price range is determined by a combination of factors. The low end of the range must cover all costs and provide a reasonable profit margin; the high end of the range is limited by what people are willing and able to pay (which in turn is influenced by factors like the availability and cost of alternative products, and the state of the economy).

It is important to remember that the quality and price of the baseline is determined by market forces, not by any committee overseeing things. As noted above, price depends on factors that vary over time so not only does it vary between different products and tiers, but will fluctuate over time for a single product. The standard for quality and performance of gear varies over time as well. This standard is mainly determined by what is already available (any new product that isn't at least as good as what's already available won't sell), and therefore the standard tends to go up over time. As companies compete with each other for profit and market share, the persistent drive to produce a better product steadily raises the bar as to what is considered standard over time.

Look at Figure 26.2 to see the current price points for a few types of typical mid-range equipment; despite the variation in how high it can go, notice that the overlap for all gear types is the range between $500 and $1000, though above this level speakers become high end while the mixers aren't high end until they go over $8,000. Now notice that the price points for mid-range are quite different for analog mixers than for

	Budget	Mid -Range	High End
Speakers	Behringer $100-400 100-400 W	JBL $400-900 500-800W	Peavey $1000+ 900+ W
Amps	Behringer $200-400 2 x 500W @ 2 Ohms 2 x 300W @ 4 Ohms	Peavey $500-800 2 x 900W @ 4 Ohms 2 x 550W @ 8 Ohms	Crown $900+ 1550W/ @ 2 Ohms 1000W/ @ 4 Ohms 600W/ @ 8 Ohms
Mixer (Ana)	Mackie ProFX12 $200-500 12-channel Compact Mixer with 3-band EQ per Channel, Built-in Effects, 4 Stereo Channels,	Yamaha MG206C $600-1,500 20-channel, 6-bus Live Mixer with Channel EQ and Compression,	Midas VeniceF32 $1,600-10,000 32-channel 4-bus Analog Mixer with FireWire, Four-band Swept EQ, Channel Inserts, and Direct Outs
Mixer (Dig)	Yamaha O1V96i $500-2, 500 24-Channel Stereo 24-bit/96kHz Digital Recording Console with USB audio streaming	PreSonus Studio Live $2,500-8,000 24-channel Digital Mixer and FireWire Audio Interface with 24 Microphone Preamplifiers, Effects, QMix, and Recording Software - Mac/PC	Yamaha 02R96VCM $9,000-30,000 56-channel 8-bus 24-bit/96kHz Digital Recording Console with Surround Support and 16 Pres

Figure 26.2. A rough estimate of current tiers and price points. Knowing where gear falls can help freelance engineers determine on the fly if rental prices are fair or if there's room to haggle and can help soundco engineers quickly pick the right gear to suggest.

digital ones; a $5,000 digital mixer is mid-range, but the same cost for an analog board would qualify as high end. At the same time, look at both budget mixers category. A budget digital mixer can go as high as $2,000, which would be mid-range for analog. But the cheapest budget digital overlaps with the most expensive budget analog mixer, since both can be had for $500.

One more thing for you to note in looking at this figure is the fact that one manufacturer, Yamaha, has mixers available at every tier quality. Even though many manufacturers will produce most of their products for a single tier, most companies will have products available in at least two tiers and many will cover the entire range. This is why even companies that might be identifiable as a little more expensive or a little more affordable, depending on where the majority of their products fall, rarely can be classified by a single tier. When companies do manufacture products for only one tier, it will usually be budget or high-end, and generally will be a company selling from its own site and not one of the well-known names found at all the big-box vendors.

You may have noticed that since the mid-range is set first the tiers for budget gear and for the high end simply become whatever products are priced below or above the mid-range. While this is often true, it isn't automatically the case, especially when it's priced right on the edge where tiers meet. Just because an item is priced to fall into a certain range does not necessarily mean it fits in that category. Though usually correlating with price, the number and kinds of features may indicate the situation is a little more complicated than simple black-and-white categories can account for. Another factor that can bump an item into a different tier is being on sale, so don't just look at the product's sale price, look at its list price. We will talk a bit more about the variations to the basic rules as we explain high and budget gear in a little more detail.

It's not just mid-range gear that has benefited from the high bar set by a baseline high-quality sound standard. Providing relatively good sound quality is no longer very difficult to do. In many cases, today's budget gear is nearly as good as mid-range gear. At the same time there are still numerous examples of budget gear that is flimsy and cheaply made. The quality of the gear depends on the source, so doing your research is especially important before purchasing a piece of budget equipment. Some tips you can use when considering budget gear are:

- It is possible to buy quality equipment at budget prices, so there is no good reason to sacrifice on quality.
- No matter how much you save on an item, buying poorly made budget gear rarely proves a wise investment in the long run, since the money saved is usually not enough to cover the added time and expense of buying a replacement item within the first year of purchasing the first (and as often as not, you may need to replace a shoddy item as early as one to three months after the original purchase).
- Although all companies make mistakes—bad production lines don't get noticed until it's too late, or quality control lets a shipment of faulty components pass—in general manufacturers with a good reputation usually won't be caught selling substandard gear, no matter what tier it's priced at.
- Don't buy budget gear if it looks flimsy or if it sounds noticeably different from gear at mid-range. Budget gear may have fewer features, less lovely design, a little less power, but the two areas you don't want to compromise on are sound quality or build quality. (A third area is just as important but harder to tell from a short demo—performance quality; for example does it overheat easily despite being of average efficiency).
- For any gear that you don't know much about, you'll always want to check for reviews. Some companies are pretty good at making a product look appealing, even if it's going to fall apart like a cheap cardboard toy in only a month or two. Luckily products like that almost always get torn up on review sites. But you won't know unless you look. You don't need to hit a million review sites to find out either; just Google the make and model, and if there are burned customers, don't worry, you won't miss them.

If you remember old-school prices then practically all the gear out there that's below high-end price range seems too good to be true. If you get suspicious of any gear that seems like too good a deal, don't assume something must be wrong with it any more than you should assume because it looks good on paper that you must snatch it up immediately. Instead look it over and ask questions to try and figure out why a piece of gear is at a certain price. To help you figure out if your worry is founded or you are just being suspicious of every good deal, here are some things to look for that might help explain why a good-quality product may be found "slumming" it in the budget bin:

- The product is on sale; normally it would qualify as mid-range but deep discount has transformed it into a budget bargain.
- The product is slightly less powerful than comparable items at mid-range; for example it only goes up to 105 SPL instead of 110+ SPL.
- The product was made with less expensive, but not necessarily cheap materials. A very sturdy particle-board is comparable with some kinds of wood, but cheaper.
- The product is made by a budget company that saves by not being branded, not having an R&D department, etc.
- The product has a slightly smaller feature set.
- The product is a late model (last year's leftovers).
- The product was a display item.
- The product saves cost and adds quality by using good original equipment manufacturer (OEM) parts.

What sets high-end gear apart is usually better materials and construction, making for greater sturdiness and a longer life before anything needs to be fixed. Another factor may be better design, so the box looks better and/or will have better sound dispersion at certain frequencies. More power and the most recently developed technologies/features are two additional qualities setting high-end gear apart from the rest.

Since the industry baseline for sound quality is pretty high already any sound improvements at average sound power are usually only going to involve fine differences that aren't going to be that audible to the majority of listeners, even sound engineers, such as a few degrees more perfect response at some frequency ranges. However, the added dynamic range and improvements in response at the extreme ends of gear capability (for example it can reach higher SPL without distortion) are often clearly audible. It's important to remember that the bulk of a typical program won't occur at the extreme ends of gear capability if you bring gear with adequate power to do the task, making this quality something that may not be worth paying twice as much for, at least not for most first time P.A. buyers.

There really is no reason to buy high-end gear (unless it is drastically reduced and sounds good to you) for your first P.A., and won't be noticeable as better than average for 95 percent of small-venue gigs. Your money is much better spent buying two or more power amps at mid-range prices than on one high-end speaker that gets a little more SPL before distortion.

One factor to be aware of when purchasing high-end gear is that while the low end of any price range is set by factors that can be verified like manufacturing cost, the high end is determined by the market assessment in a product's quality. This is not the same thing as true quality. If you encounter high-end gear where it is difficult to understand what the price is based on, it may be based on nothing but hype or brand prestige. High-end gear that invests in clever marketing tactics to be perceived as top quality can charge hundreds even thousands of dollars more than an identical sounding item, merely based on a public perception that it is a superior brand and a hard-working marketing team. When shopping high-end gear, mistakes are more costly, so it's even more vital than when shopping for budget gear that the precept "caveat emptor" or "let the buyer beware" should rule your purchases. No matter what the press says, or the ads, or what you think you should hear *never pay a premium for sound quality you can't actually hear!*

Figures 26.3 and 26.4. Photos of the digital Yamaha mixers from the figures above. Some companies like Yamaha make great products for all tiers. Top: The Yamaha O1V961 24 Channel, 24-bit Digital Recording Console. Bottom: The Yamaha O2R96VCM 56 Channel, 24-bit Digital Recording Console *Source: Photos courtesy of Yamaha.*

The Proper Use of Gear Specifications

Specifications are data sheets that report equipment performance based on test results of equipment capabilities and features. The testing that produces these results may be done by the component manufacturer, or by one of the independent organizations that set capability and testing standards for electronics and/or audio technology. Of course, tests by independent testing bodies will be more reliable than manufacturer tests. This is because such independent tests are typically standardized, so you know the testing methodology used for each test is the same and what those methods are, making it possible to compare the results for two different pieces of equipment reliably. Specifications are available for just about any audio equipment you are interested in, and are freely downloadable at audio manufacturer sites and most audio vendors.

Specifications are sometimes represented as containing all you need to know about that piece of gear. They can be very informative (if you understand what a rating is saying), but most of them still *don't reveal much about how a piece of equipment will sound*. This is why specifications shouldn't be the deciding factor used to choose between different pieces of gear; only the sound of the gear should be. Rather than the last information you should consider while evaluating or shopping for gear, specifications are most useful when used as the information you consult first. When you have several dozen different makes and models to choose from, specifications are one of the best tools to help you strategically narrow down your choices to a more manageable number.

For example, when speaker shopping, you should have already an idea of what your system will be used for and figured out how much sound power will be required for that application. You should also have an idea of the type of speakers you need, the number you want, and how they should be powered/matched to amplifiers (and have drawn up a block diagram to record this data). So if you're planning has helped you determine that you want two each of highs, lows, and subs, need them to be able to put out 115 dB SPL, and have an impedance of 8 Ω, then you can use the information available in speaker specifications sheets to weed out any models that don't meet these parameters. If the pool of choices is still too large to deal with, you can eliminate further according to your next priorities by considering factors such as weight, dispersion pattern, and efficiency. Then once you have narrowed down your options sufficiently, you can start researching the remaining choices in more depth by reading manufacturer marketing materials, gear reviews, and arranging to listen to them.

We recommend looking at specifications first even if you don't need them to whittle a large group of options down to size. Even if you are comparing a limited set of options, if read carefully specifications can give you a general idea of each component's comparative capability and can provide useful details to keep in mind as you read product reviews and other marketing materials. Specifications are a great way to initially get acquainted with each of your equipment options on the same standard set of features; because not only do specs provide a similar set of vital facts about each option, they are written to be simple data sets and will include few of the value weighted words that marketing material and reviews are known for.

Even when you aren't comparing options and are considering a specific model on its own merits, specifications are an ideal way to begin your research. You are less likely to be impressed by an over-the-top review, the expressive language of the product brochure, or the salesmanship of the regional rep after carefully reading the specifications and noting they are generally standard in comparison to similar products (as almost all products are). However, the reverse is not always true; if you don't read the specifications until after you've been sold on how wonderful the product is, it's harder to read the spec sheet with a critical eye. People who read specs after already deciding they like a product tend to interpret ratings more favorably and overlook missing information more easily than if they start out reading the specifications first.

To get the most out of reading specifications, it helps to have an idea of what to keep an eye out for as you read them, both in terms of what you are looking for in a piece of equipment, and what specification ratings for that type of equipment are based on and should look like. Knowing what you're looking for in a P.A. component can help you direct your focus on the specs that are most relevant to you, and save you time by allowing you to eliminate items that don't meet or exceed the range of values representing your requirements. Knowing what necessary mean and how they should appear is required in order to glean any useful information from them. In the previous chapters, definitions and details of some of the most essential gear specs are discussed for each type of gear, as well as what each rating should look like (including the parameter that should be referenced for each, if any). Also check the directory for this chapter; Rane audio has a very good guide to reading specifications, and we include a link to it and other good resources and guides to help you interpret specifications.

You should also be aware of some of the methods used by marketing departments to downplay negative information about their products and lead you to draw conclusions they want you to draw. These usually are found in the marketing materials describing specifications rather than the specifications themselves (which is why we recommend reading specs first and separately from other marketing materials); however, certain omissions can occur in the specification sheet as well. We are not talking about fraudulent marketing practices either, or any kind of blatant misrepresentation, which are actually quite rare and unlikely to occur in the marketing material of any established manufacturer, even though you should take care not to be fooled by those when they do occur with lesser-known or unscrupulous vendors.

What we are talking about instead is the corporate equivalent of what happens when people go on an interview or a first date. Not dishonesty, but the right to present oneself with the best foot far forward, and without being forced to highlight information immediately upon first introduction. All the information presented is true, but it is done in a way that highlights the best attributes; less impressive attributes are overshadowed and left up to the observer to pick up on by themselves. Sure it can be misleading, but if it's dishonest, then so is every woman wearing makeup and every man who does not reveal their occasional flatulence problems to all potential new house mates in the spirit of full disclosure We've all done it to one degree or another, and marketing departments are paid experts at creating the best impression for the products they are tasked with presenting.

You would be hard pressed to find any specification sheet with falsified data or marketing materials making any claims that could be described as technically untrue. Specification sheets are provided in the form they are and provide the information they do out of custom; however, they are still ultimately a courtesy provided for the customer's convenience and not because they are required by any industry or legal standards. Without such standards, as long as all the information required by law is available elsewhere, there is nothing outright dishonest about omitting a spec, or presenting a rating without including the reference used.

Any adult of average intelligence will not be misled by the contents of a specification if they do a little homework first to learn what each spec should include if it is to be meaningful and thoughtfully considers and sorts the data, discarding any that isn't complete enough to be useful, then ranking what's left according to how much meaningful information it provides, and how much weight if any it should be afforded in relationship to other information before all the useful data is used to arrive at overall assessments and conclusions.

Conversely any time someone reads a specification sheet and leaves with an impression of product performance and capability that is better than the audible or otherwise detectable real performance of the product they won't be doing so because they were presented with falsified data or statements that aren't factual, they will be doing so because they are careless enough to ascribe positive value judgments based on their own assumptions about what the data means instead of looking up established definitions to determine the objective meaning. If a quick fact check reveals any data that can't be objectively placed into context, it should be discarded, not used to formulate assessments or draw conclusions from.

Omissions, when they occur for any reason other than accident, are almost never going to be hiding significant quality defects, but at most be intended to downplay minor differences in ratings that put them at a disadvantage in a field where all the products are so good that choices might be made based on such insignificant inaudible differences. At the same time, when tactics are used to make a product seem much more exciting or useful than it is, it won't be intended to trick an unwitting customer into purchasing an inferior product, such hype will only be used to lead consumers into assessing a solid, good-quality product as a cut above the rest in a field of solid, good-quality products with nearly identical capabilities. The most common method to hype a product is to boast about high ratings that make no difference whatsoever, because to the uninformed consumer the high number will be assumed to mean better sound. For example, amplifier specs often boast damping factors of up to 1000, even though gains in damping factor between 100 and 200 are miniscule and gains over 200 are meaningless.

Will anybody be harmed by choosing a quality amplifier because of a hyped damping factor of 1000? Probably not, as most middle-range audio gear is of similar high quality. The person buying a good-quality amplifier out of many others with almost identical sound quality loses nothing, even if that high damping factor cannot be heard. This assumes, however, that the purchaser did not pay an extra $500 per amp for the "high quality" implied by the rating.

So even though it is mostly harmless, you still might not want to be so impressed by marketing hype that you don't catch it when you come across it, You could fall for hype created by good companies who are only trying to capture their share of the crowded market 100 times, and wouldn't be disappointed; they would use it only to grab your attention long enough to sell you a quality item at a fair price.

But not all marketing is as benign, and some companies will jack up their price once they see their hype is garnering interest. If you are on a budget, you don't want to ever find out you spent $500 more apiece on your P.A. speakers than you needed to. So don't believe the hype, always verify information you aren't sure how to interpret and once again be careful about making decisions based on information that you accept because "it seems obvious," without looking closely to see if you actually know what it means.

A few tricks to keep an eye out for include:

- Pumping up what doesn't matter. Some audio specs are very informative; others sound impressive but have literally no impact on the quality of that type of equipment.
- Sometimes an impression is created implying a score or rating is good because it is "higher," when higher scores for that particular feature are actually not preferred or don't usually matter.

- If a specification that generally appears is conveniently missing, look into it. You can be sure that companies don't neglect to include specifications that make their product look good.
- Be aware that some tests are not the best reflections of how a product will respond under normal conditions, even if everybody uses them. It's not a perfect world, but many products are only tested under perfect conditions.
- Don't be fooled by specifications without a reference. Sometimes ratios and percentages are included in product descriptions without stating what they are in relationship to. We are used to using 100-point scales, so seeing the term 99 percent can automatically give a good impression, but it really means nothing unless we know "of what" or "in comparison to what."
- Most audio tests should indicate conditions that they are true under. A single-number rating often requires the condition be stated to understand it. Typical conditions could be frequency (20 Hz–20 MHz), level (±3 dBU), ratio (3:1), or impedance (8 ohms).

Do Your "Presearch"

Everyone knows they need to do some research prior to buying gear, but a lot of people only start researching at the last moment, after they have the money to buy and have already made up their mind about the item they want based on hearsay and hype instead of research and careful consideration. The problem is that these folks then rush through research mainly seeking to confirm their choice, so they may be blinded to other options or ignore or minimize information about their choice that would get in the way of allowing them to do what they've already decided they want to do, which is to buy that product pronto.

While a purchase stemming from an impulsive choice hastily researched at the last minute is better than making a completely un-researched impulse buy, it's only marginal improvement and hardly ideal. To ensure you make the best buy you can, if you are thinking of getting your own live P.A., D.J. rig, or studio gear, you should start "presearching" long before you are ready to buy so when you enter the research process a few weeks in advance of going shopping you will have the advantage of doing so with an understanding of the marketplace as a whole. Just make sure to keep an open mind and take the time to fully check out all options.

If you choose to become a live audio pro, part of your job description is knowing the variety of audio equipment commonly used in your part of the field, and having a general knowledge of the audio marketplace anyway. To keep up with what's going on in the audio-technology industry you should make it a practice to see what's new on the market and to read about upcoming developments. Some habits to practice to ensure you are up to date include:

- Shop around online for an overview of what's on the market, and what current prices are like now and then, even when you're not shopping for anything in particular.
- Sign up for the quarterly catalog from a couple of the big online vendors and check out what's new whenever the catalogs arrive.
- Pick a couple of gear blogs from the web directory and check them weekly for developments.

In the meantime whether you're currently planning to purchase a P.A. or not there's plenty of presearch for you to do if you want to become acquainted with the audio marketplace and become a knowledgeable audio pro who knows how to navigate it.

Meet the Manufacturers

- Look in the web directory for the links to audio-gear manufacturers and start learning who's who, including who is related to whom (for example, Crown Audio, JBL, and Soundcraft are all brands that have name recognition in their own right, but are also all owned by the same well-known company, Harman).
- Go to the website of each one and see what they make; sometimes you may find a surprise—many manufacturers are known for making a single item like speakers or mixers but if you go to their websites, you can discover that they often make many more kinds of equipment than just the items they are best known for.
- As you visit each manufacturer site check out their product lines and compare their prices (if they don't show list price then go to a vendor that carries them). You're not shopping for your own items at this point, just getting to know the field. Part of that is knowing which manufacturers tend to be more expensive, which ones have a wide variety of budget items, and which ones have mostly items solidly in the mid-range.
- Other things to check out at each manufacturer's website are their resources pages. Many

Figures 26.5 and 26.6. The Line 6 stagescape M20d and Mackie's DL1608 are digital mixers that have some never-before-seen features, including one which allows you to mix remotely with your iPad when away from the main console. As an expert in live-audio technology, you want to know when companies release mixers with innovative new features.

manufacturers provide archived articles, calculators, videos, test tones, forums, and other resources that can be very useful. If you find tools and resources you think will be useful to you, make sure to grab the web link.

- Look for each manufacturer's manuals download page and bookmark it.

Meet the Vendors

- Once you feel you have gotten to know the manufacturers go to the directory of vendor sites and start visiting the vendors. Look to see where each is located to find out if you have any big-box vendors within driving distance of where you live.
- To get an overview of items and prices for sale in each category go to the first page of that category and look for the options to sort items at the top of the listings page, select "low to high." Then look if they let you select the number of items per page and select the highest number or if possible select "view all." You will end with all items for sale in that category, listed from lowest price to highest, on a single page, or a few pages instead of a few dozen. This setup makes it much easier to see each manufacturer's items across all price tiers, and to get an idea of where the budget, mid-range, and high-end price tiers are.
- Browse the products and see if there are any manufacturers you overlooked. If you haven't already, compare manufacturer prices with the prices listed at the big-box stores.
- Compare prices between the stores you like the best. Note which items tend to be the best bargain at which store. Pay attention to price ranges for each item type to note the price range where each manufacture tends to fall most. You should still be trying to learn which manufacturers are known to be on the expensive side and which have a name for extensive lines of budget gear.
- Just like the manufacturer sites, many online stores often have great resource pages, so find them and link to them. Make sure to look for those vendors with gear manual download pages and link to them.
- If you have an audio-gear vendor in your geographical area, go check one out in person, spend some time listening to items they have out for customers to demo. Pay attention to price tiers that different items fall in and see if you notice any differences in the materials used to make them or in the sound you hear from them.

Something to Talk About

While new live sound learners and pros have plenty of reason to learn how to navigate the audio marketplace already, this one final reason resonates with some people most of all. Being able to talk with a bit of knowledge about industry technology and development because you've taken the time to learn the landscape and keep updated gives you a way to find common ground with some of your industry colleagues more easily and with less awkward false starts than if you couldn't engage on the topic with any authority. If all audio pros were live audio pros it would not be necessary, but not all are. Many you will work with may be so different from you that you will be grateful to have the safe and neutral option to talk to the awkward accordion player or church audio installer sitting at your table about a shared interest in speakers, instead of spending the same minutes in uncomfortable silence wracking your brains for something to say.

Visit your Resource Pages

In the chapter directory a variety of great resources are linked to. Take the time to check them out and get a feel for them. Gear forums are great places to ask questions. Just remember to follow the rules of forum etiquette in the online etiquette guide included in the chapter about working with others. Other resources include review sites and blogs and links to the gear buying guides put out from time to time by some of the biggest audio industry publications like *Mix* magazine and Pro Sound Web.

You'll Never Remember All That: Put It to Paper

Dedicate a notebook to your gear and industry research and don't do research without a pen and notebook at hand. The amount of gear out there is huge, and the number of features for each piece are extensive enough that it will be impossible to remember everything without notes. Take notes as you pull information from product documentation—if you have to look at a manual or brochure twice to fish out information, that's time wasted.

As you research your first buy, you are likely to come across info about other products you intend to eventually purchase too, so take notes on anything you learn that could help you later; you don't want to have to try and track the same information down again later. To keep information from getting lost, dedicate sections to specific topics instead of writing everything in an ongoing column. If taking notes makes it hard to do research because it slows you down too much, try using a digital voice recorder, or use the recorder on your computer or laptop. Use that for taking notes, so you can simply speak your notes, and then you can transcribe the information at your leisure later or even send it to an online transcription service; there are some very affordable ones to choose from.

In the next chapter is a quick-and-dirty buying guide with more specific shopping tips, but getting a headstart to knowing the marketplace is important and will let you enter the task of shopping already equipped with some knowledge. Make sure to check out bonus Chapter 28 on the companion website, as there will be useful tips and interesting information that will help you continue to learn about the audio-tech market-place whether you intend to buy anything soon or not.

Chapter 27

Working Like a Pro

Preserving Your Assets: Noise-induced Hearing Loss

Anyone, whether musician, sound engineer, or venue patron, who is exposed to harmful sound levels is vulnerable to the risk of noise-induced hearing loss. While many readers may have already been warned of this risk, it bears repeating, especially since many young engineers fail to heed the warnings and proactively engage in practices to help limit their exposure and avoid the consequences of hearing loss.

What is Noise-induced Hearing Loss?

Noise induced hearing loss (NIHL) is damage to the tiny hair structures of the inner ear that allow us to hear. It is cumulative, silent, and permanent—increasing gradually over months and years unnoticed until it begins to interfere with our ability to understand the speech of those around us, but unfortunately by then it is too late, and the 30+ decibels of hearing, depending on severity of loss and frequency, is gone forever and can't be replaced. It is easy to misunderstand the difficulty sufferers go through as loss of hearing begins to isolate them from the social interaction around them, as our eyes are so primary that we don't realize how much of our experience of life enters our minds by way of our ears. If this is hard for you to imagine, we'll address that later, but for now this should be easier—imagine not being able to hear the music you so love anymore, finally reaching the point where you have to give up the job you love so much and worked so hard to break into, because you can no longer hear frequency well enough to even tech a system much less control the mix. Even mild hearing loss can compromise the ability to understand speech in the presence of background noise or multiple speakers, leading to social isolation, depression, and poorer quality of life.

To make matters worse, as if this isn't a severe enough consequence, is the related condition that in many cases will accompany NIHL, or can occur as a result of noise exposure without including noticeable hearing loss, and is known as *tinnitus*. This condition has a few different "flavors" it can come in, but in most cases it involves a permanent sensation of a buzzing, whooshing, or high-frequency tone that is only audible to the sufferer, coming from within the damaged auditory system rather than the environment. Depending on the person this tone may be louder or softer, more easily "tuned out" or impossible to escape, but it is usually constant and while there are a variety of treatments (most dealing with teaching the sufferer to learn to tune it out or shut it down), none work reliably for all patients, and many patients don't ever learn to shut it down. Some learn to live with it with less regret, others are very distressed by it for a long time, and suffer from inability to sleep or concentrate, but eventually learn to cope, and on rare occasions a few are so driven to distraction and annoyance by it that it drives them to suicide.

Either of these symptoms of high sound level exposure can occur suddenly and temporarily, disappearing within hours or days (if you have ever had your ears ringing and/or hearing muffled for the day after a concert you have experienced this). When these effects abate after some time away from the loud sound and turn out to be temporary, the condition is described as a *temporary threshold shift*.

These symptoms can also occur suddenly and permanently after even one particularly bad exposure (imagine the days after a concert where you expect the ringing and muffle to go away but after waiting 24 hours, then 48, and then 72, when you realize with horror that you messed yourself up for good this time). The idea that symptoms of loud noise exposure are only either temporary, or occurring slowly over many exposures is a myth—*sudden and permanent NIHL and/or tinnitus is less common, but not unheard of.* If the above scenario ever happens to you, see your doctor after 48 hours pass with no improvement, and visit your emergency room if you can't get a swift enough appointment to see a doctor. A *steroid injection* within the first days of symptoms can cure or minimize symptoms and the sooner the better; if you wait a week to make an appointment, you may be too late.

How High do Sound Levels Need to be, to be Harmful?

It depends on the level and the amount of time you are exposed. Osha standards limit exposure to an allowable daily "noise dose" based on what we can tolerate before the risk of damage begins. The baseline is 85 decibels for eight hours, which must be followed by at least ten hours of "rest" at decibel levels of 70 or below. The formula conversion is that you need to cut the allowable time in half for every 3 db you add in levels. This means you should only be around 100 dB SPL environments for 15 minutes a day. As an engineer, you will be exposed to levels that high (and at times considerably higher) for hours at a stretch. Unless steps are taken to reduce exposure by making do with fewer decibels to achieve the same fat sound, and wearing hearing protection, you are risking falling prey to noise-induced hearing loss.

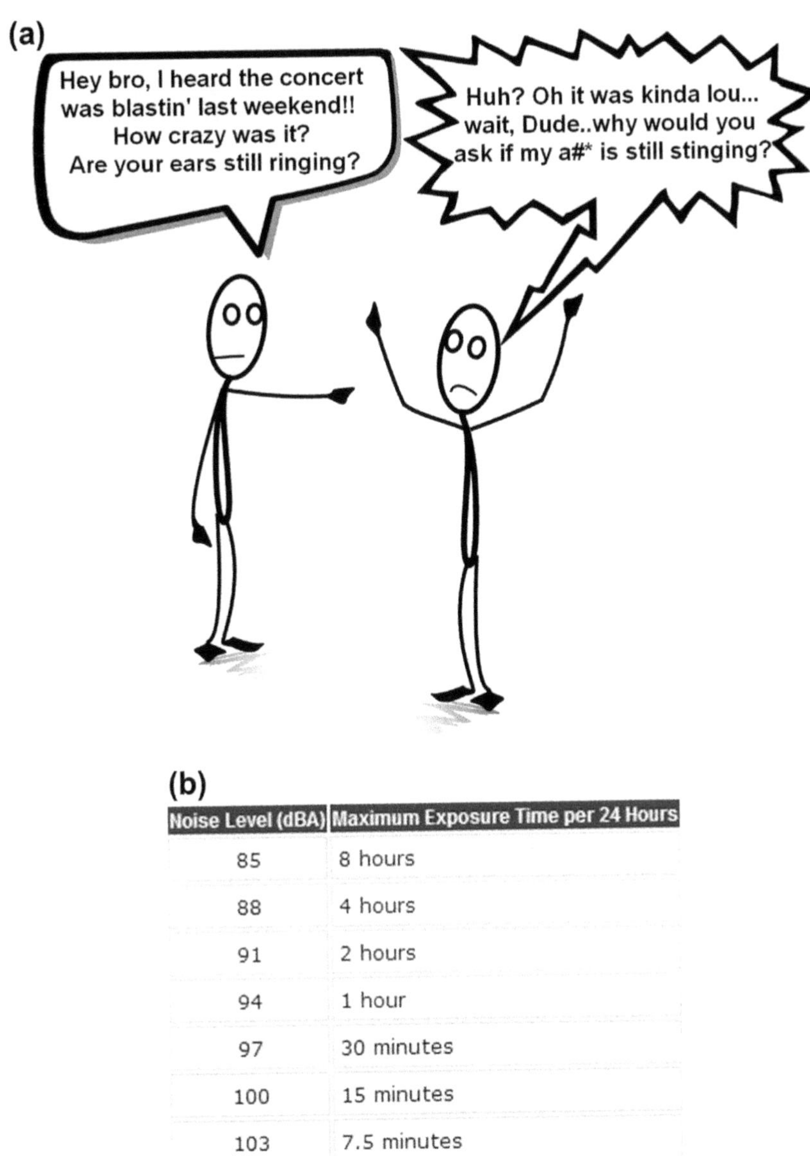

Noise Level (dBA)	Maximum Exposure Time per 24 Hours
85	8 hours
88	4 hours
91	2 hours
94	1 hour
97	30 minutes
100	15 minutes
103	7.5 minutes
106	3.7 minutes
109	112 seconds
112	56 seconds
115	28 seconds

Figures 27.1a and 27.1b. Temporary threshold shift, a taste of what could happen for good, maximum SPL exposure times.

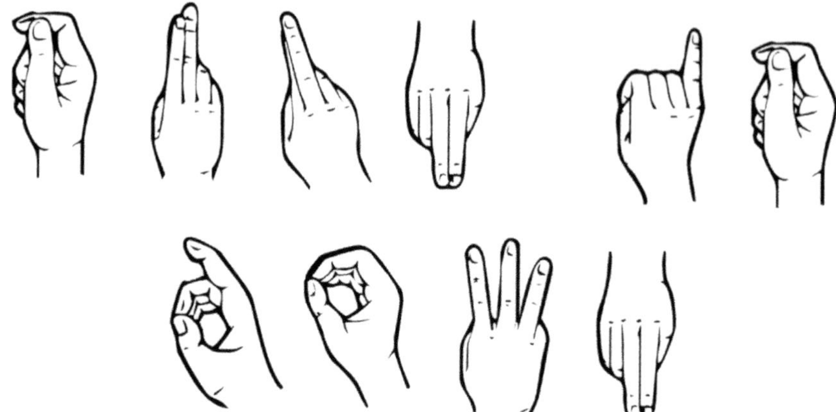

Figure 27.2. "Turn it down" sign.

What can I do to Protect Myself?

The good news is that studies show that musicians and engineers can minimize or even avoid consequences of harmful sound exposure with a few precautionary habits.

- See an audiologist for routine hearing tests at least once every six months if your livelihood depends on your ears being in good health. If you don't have health insurance (like a lot of musicians and freelance engineers), consider your options: a hearing screening can cost around $100, which is a small investment for a lifetime of hearing health. Also, check with your local universities and teaching hospitals. You might be lucky and have an audiology clinic nearby that can help you meet the needs of both your ears and your budget.
- Don't use sound levels higher than are needed to produce good sound. Learn the methods, such as creating dynamic interest by selectively varying levels according to what is going on with the music that can make a show seem louder without needing to be as loud.
- Use the lessons you learn about acoustics to help you keep sound levels down both on stage and in the main mix, such as proper speaker placement, acoustic treatments to minimize reflected sound on stage, and aiming monitors and amps at ears and not knees.
- Involve bands in controlling stage levels by explaining what they can do, why it helps their sound, how it works, and why they should do it to protect their hearing as well. Make up an info sheet to print and give them, and send them to some of the hearing loss links in the directory.
- Use a limiter on your system to ensure peaks don't hit the red zone for ears as well as speakers.
- Get an SPL meter and use it occasionally as you mix, ear fatigue and the build-up of sound gradually can combine to fool you into thinking it is not as loud as it really is.
- Buy flat-frequency earplugs (the non-molded kind are very affordable), and use them. Wear them for a little bit at a time over the week before you first intend to wear them at a gig. They will be more comfortable to mix with and you will be more likely to use them again if you make sure to get used to them before you try mixing in them. If occlusion is such a bother that you can't mix with both plugs in at all times, use in one ear and switch back and forth between ears at regular intervals, cutting your full exposure in half—some sound protection is better than none.
- Buy custom-fitted musician's earplugs with a filter to allow unattenuated high frequencies to pass through, presenting a much better listening experience. They run about $150 plus a visit to your audiologist, but you can purchase universal-fit versions for as little as $10–15.
- Avoid medications that can be dangerous to your hearing. Researchers have found that women who took ibuprofen or acetaminophen two or more days per week had an increased risk of hearing loss.
- Research in this area has found correlations between serum magnesium levels and noise-induced hearing loss. So buy a bottle of magnesium supplements at your local drug store and you'll be less likely to receive permanent ear damage.
- If you can afford them and can try them to make sure you can use them before shelling out, try mixing with sound attenuating headphones to monitor the mix

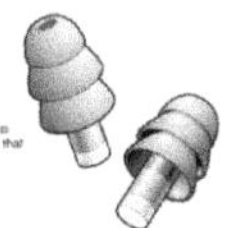

Figure 27.3. For only a few bucks a pop, flat-frequency earplugs can save your ears.

(and if the band can, encourage them to use personal monitoring also).

- Visit the directory to read the articles in the section on hearing loss, use the interactive tools to see the relative sound levels at the root of NIHL, and read the testimony of sufferers who were once willing to ignore warnings too and take their tales to heart.
- Don't just protect your ears at the mixer; bring your earplugs with you when you go out, and use them in noisy environments wherever you encounter them. Don't forget to protect your ears even from temporary harmful noise in otherwise safe environments (like reaching up and covering your ears for the seconds it takes for an ambulance with siren blasting to move past). Other places, things, times to protect your ears are: video arcades, fire crackers, discos, music concerts, shooting a gun, movie theaters, sporting events, motor boards, motorcycles, the trailers at the movies, amusement parks, mowing the lawn, etc.
- When it's essential to listen on closed headphones, use a pair with good insulation from the outside world so you can keep the volume down.
- If you can't motivate to care for your own ears, at least follow the best practices for controlling venue noise by remembering that you have the power in your hands to truly harm your patrons who don't know enough about the risks to protect themselves. Make sure to keep sound levels sane and only serve up your good vibrations. As more and more people are affected by the consequences of overly loud venues, regulations will control your mix if you don't anyway, and even if not regularly enforced yet it is the law in many communities and regulation is spreading, so control your own sound now to head off overly restrictive future regulations.

Finding Information Online

Searching is something we take for granted until you can't find what you are looking for and are forced to wade through pages of irrelevant results while searching for information you can't find anywhere. Use these tips and the tutorials and resources linked to in the directory to find the information you need when you need it.

(a)

Noise level in dB	Select a protector with an SNR of:
85-90	20 or less
90-95	20-30
95-100	25-35
100-105	30 or more

(b)

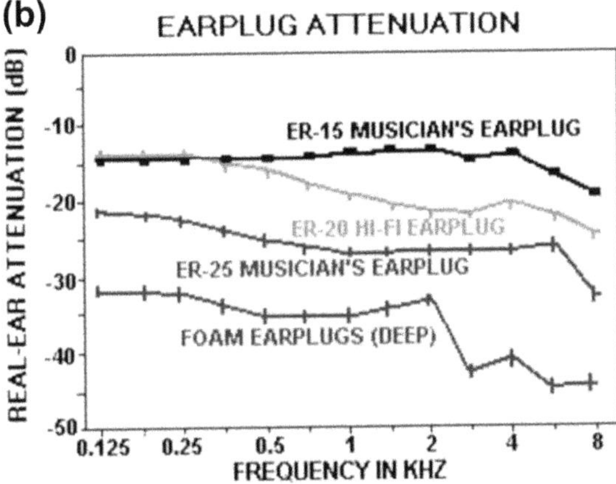

Figure 27.4a and 27.4b. Earplug info.

To narrow your results, stop searching only one-word terms in search engines such as Google. If you type in a term such as "audio," you get hundreds of thousands of pages, far too many pages to search. Because a search engine cannot read your mind to know your real intent you need to learn how to speak "search engine" so you can ask for what you need.

- When you need combinations of things or ideas use phrases in quotation marks, for example "live audio." This search finds these words in this exact order so you won't have to wade through pages about studio or broadcast audio unless they have informtion on live audio too.
- Define clearly what you are looking for. Ask yourself what way others would have looked for it: "live audio engineer", "live audio tech," etc.
- A good search may contain more than a couple of words. Don't be afraid to try something really specific that defines your problem: "live audio engineer salary," "live audio tech entry level salary," etc.
- Boolean basics: When looking for a document that contains several concepts use the word "+" between each term. The result will be only documents that contain all these terms.

- Boolean basics: When looking for documents that do not contain a term put a "–" before it to exclude it. Make sure there is a space before the minus sign so it isn't mistaken for a hyphenated term.
- "OR" lets you search for sites that contain either one word or another, i.e., "audio OR sound."
- To search within a specific website use your term followed by the word site and a colon and the site address, for example cables site:sweetwater.com
- Use the "*" to fill in the blanks if there are terms you want to keep open.
- "~" will include related terms for your search term.

Don't always assume what you are looking for will be on the first page.

If the term you've chosen isn't working, rethink your key terms and look for:

- synonyms: audio, sound;
- plurals: child, children;
- abbreviations: california, CA;
- alternate spellings: meter, metre;
- variations in terminology: earth wire, ground wire;
- alternative terms: audio tech, system tech.

The more you search, the better you get!

Your Local Scene

Clubs and bars are a great way to learn more about live sound; one way is to just go and watch what the sound engineer does. Most are also very open to talk about what they do, so long as approached with a little common sense and respect. When the sound guy is rushing by with a crazed look on his face is a bad time, so is mid-show when his eyes are locked on the board or the stage. Expect a few short conversations before the chance for a longer one pops up.

And if you run into one who just doesn't want to talk to you? Well, you can always give it one last shot where you ask if there is a time you could help them with any heavy lifting or with teardown/setup. Offering a fair exchange will work with some, but if not, find another club. Stalking the sound engineer who is too busy to or won't chat won't do you any good, but if you really want something don't give up either. Hit the next club.

Local Bands

Another foot into your local music scene is keeping your ear to the ground about local bands looking for a monitor mixer. The web directory has a few national band directories and classifieds, and craigslist is another place to look as well. Sometimes a local scene will have their own loop and if you only look online you won't find them. (This is not just a band thing, it's a local thing—but bands use them too, so whatever you are into just look for ones where people in your target demographic are likely to frequent.)

- Keep an eye out for random cork boards anywhere customers can access them, usually in the front of a store, but not always.
- Another version of this is small unobtrusive tables, shelves, windowsills, also easily accessible by customers, with lots of loose flyers and notes, stacked or spread on them.
- Also keep your eye out for free local classifieds at any places like this or similar.
- You don't need to run all over town; there may be none around you, and if there are any there will probably only be a few at most (if people advertise local parties on them, they will be in odd places—I have found them in diner and venue bathrooms). At the same time, it's not unheard of for people to hear about this and discover one right at nose level somewhere they've been dozens of times. So it's worth it to keep your eyes open.

Places to Keep an Eye Out for Local Ads/Network

- local music venues;
- record shops;
- skate shops;
- alternative clothing stores;
- coffee shops, diners or hang outs, especially all-nighters;
- tattoo/piercing shops;
- practice spaces/warehouses;
- guitar centers or equivalent (also small music shops);
- comic-book stores.

Time-management Skills

- Work overload—feeling you have too much to do—is a common cause of job stress. You may not be able to affect the amount of work you have, but you can use time management to help you be more efficient and feel less under the gun. Try these tips to improve your time-management skills and lower your stress level.

- *Set realistic goals.* Create realistic expectations and deadlines for yourself, and set regular progress reviews.
- *Make a priority list.* Prepare a list of tasks and rank them in order of priority. Throughout the day, scan your master list and work on tasks in priority order.
- *Protect your time.* For an especially important or difficult project, block time on your schedule when you can work on it without interruptions.

Chill Out: Manage Your Stress

In small doses, stress is good—such as when it helps you conquer a fear or gives extra endurance and motivation to get something done. But prolonged or excessive stress overwhelms your ability to cope and can take a severe psychological and physical toll. High stress levels have been linked to depression, anxiety, obesity, diabetes, high blood pressure, heart disease, musculoskeletal problems, impaired immune response and cancer. The good news is that you're not powerless. You can learn ways of coping with and managing stress.

- *Follow a healthy lifestyle.* Exercise at least three times a week to positively affect mood and reduce stress. Get plenty of sleep, and follow a healthy diet to fuel your mind and body.
- *Take a break.* Make the most of workday breaks. Even ten minutes of personal time can be refreshing. Similarly, take time off, whether it's a two-week vacation or just a long weekend.
- *Have an outlet.* All work and no play is a recipe for burnout. Make sure to spend time on activities you enjoy, such as reading, socializing, or pursuing a hobby.
- *Get other points of view.* Talk with colleagues or friends you trust about the issues you're facing at work. They may be able to provide insights or offer suggestions for coping. Just having someone to talk to can be a relief.
- *Surround yourself with positive people.* Make sure those in your life are positive, supportive people you can depend on to give helpful advice and feedback. Negative people may increase your stress level and make you doubt your ability to manage stress in healthy ways.
- *Use imagery.* Visualize a relaxing experience from your memory or your imagination. It can help you get a more balanced perspective, and calm your mind.
- *Be open to humor.* Give yourself permission to smile or laugh, especially during difficult times. Seek humor in everyday happenings. When you can laugh at life, you feel less stressed.

Five-minute Breathe

- Breathe deeply, from your diaphragm. Breathing from your chest won't relax you, so picture your breath coming up from your "gut."
- Breathe in deeply through your nose, and then out your mouth.
- Slowly repeat a calming word or phrase, such as "relax" or "take it easy." Keep repeating it to yourself while breathing deeply.

Much like the "check engine" light on your car's dashboard, if you neglect the alerts sent out by your body, you could have a major malfunction, so pay attention to the warning signs listed here. These are just some of the ways that your body is telling you it needs maintenance and extra care: headaches, muscle tension, neck or back pain; upset stomach; dry mouth; chest pains, rapid heartbeat; difficulty falling or staying asleep; fatigue; loss of appetite or overeating "comfort foods"; increased frequency of colds; lack of concentration or focus; memory problems or forgetfulness; jitters; irritability; short temper; anxiety.

Checking Lists Twice

Regardless of which of these lists are provided to you, or which you are expected to provide (depending on whether you represent a venue engineer, soundco engineer, or band engineer), you also have lists you need to use to keep yourself organized, both for each event and in general. (In fact, if you are up to date you should never have to make any of the above lists from scratch after the first time—just use your most recent copies and tweak them for the occasion.)

If you are the engineer for a few local bands, you should have current set lists, tech sheets, stage plots, etc. for each. As soon as you have to make the first set, all you need to do is save them, and then update them as needed (new gear to the respective tech sheet, new tunes to the respective set list). If you tech for a sound company, you most likely will have lists built up from larger inventory lists for each gig, but most will be based on a core list of tech must haves. If you are house sound, your venue info sheet will remain consistent, and your gear lists will only need alteration as items break or occasional new ones are added.

No matter what your focus, you should have checklists to help you keep track of your priorities, task order of operations, current projects, etc., as well as any reference notes you may need. For example, most small-venue house sound (whether FOH for gig bars, or house DJ's who run events from dance nights, to pub quiz nights, to karaoke nights) should have:

- *Venue gear lists*: To print fresh as needed. Used to check off as each item gets locked up or packed away at the end of the night. (Don't check it off till the lock clicks.)
- *Personal gear lists*: If the venue won't provide it, but it makes your job easier, and makes you more effective, you will often find it worthwhile to provide for yourself. All personal gear should go in a tool box or small lockable cart, and as with venue gear double-check you haven't forgotten something and left it lying on a ledge somewhere, by checking your list off as you pack up at the end of each shift.
- *Venue calendar*: If your venue books a month in advance, or six months, you should have a calendar keeping track of upcoming gigs and notes indicating how much prep is finished and still needed for each upcoming entry (and attached checklists/info for those immediately up the pipeline).
- *Contact lists*: If you are wise, when you meet cool talented people in your field and related ones, you make the effort to keep in contact the same as when you meet cool talented people in your personal life. If you are doing it right, you will also be on some contact lists too. You don't just keep your contact list handy to call in favors, since it'll be useless for that anyway if you don't also keep it handy to be cool to those you like—so remember to hook up the soundco contact with the band looking for a good company to rent a P.A. from to throw their first record release party, and then if you know people who would dig their sound, spread the word!
- *Procedure lists*: If pros who have been doing this stuff for years use checklists to keep track of their steps, don't be stupid and try and do it by memory. You should have lists to remind you both of when to do certain tasks, but also to keep track of them as you do. If you tune your system and check your gear on a regular schedule you should have a record of the last time you completed the task, and a date set for the next. Then as you go through each step of each procedure, you should have a checklist off procedural steps to check off as you go. When you're checking a few dozen cables that all look alike, it's easy to lose track of your progress. It would suck if after all the good effort to keep on top of your tasks, you fell short due to lack of organization and missed that crucial cable about to give up the ghost until the middle of a big show.
- *Reference notes*: Gear settings you don't want to look up again, cheat sheets and conversion tables, frequency references for that band next week indicating their lead singer plays accordion and spoons. Whatever, you should have it handy when you need it.
- *Time schedule*: Not only useful on show days; keep some blanks around to use on project days, and to fill ahead each week.

First-year Tips

Starting to work in an organization is a critically important time that requires you to use special strategies to be successful. Many recent graduates/new hires hang on to (childish) student attitudes and behaviors too long. There is breaking-in stage that lasts from the time you accept your job until about the end of the first year that can make or break the early part of your career. There's a special game being played during the first year, and most new hires don't know all the rules. Because you're the "new kid on the block," people will respond to you differently, work with you differently, and judge you differently. You, in response, have to approach them differently. It takes time to understand and earn the rights, responsibilities, and credibility of a full-fledged professional. It's by learning those rules that you can get the strong start your career needs. In that first year, you have to know how to establish yourself, learn the "way things are done," and figure out what you need to do to earn credibility and respect.

Much of your early-career opportunity and success will be charted by the impressions you make on the people you work with and the perceptions they develop of you in the early weeks and months on the job. Research suggests that how you approach your first year will have a major impact on your future salary, advancement, job satisfaction, and ability to move within the organization. Mess up your introductory months and you may find yourself labeled as "immature" and relegated to lesser assignments while your colleagues are busy impressing the boss with their professional maturity and success. This is not to say that an entire career is made or broken, but the simple fact is that it can take years to recover from a poor start.

Slow Down

The crux of the problem for some new hires lies in how they try to make a positive impression. Conventional wisdom says that you need to show your new organization how smart and talented you are by using what I call the "big splash" approach. The problem is that if you do this before you have earned acceptance and before you understand your new organization, you will most likely stick your foot in your mouth and embarrass yourself. What makes the most positive impression is not showing how much you know, but rather demonstrating the maturity to know how much you don't know. This means keep your eyes and ears open, and mouth shut, until you learn as much as you can about the company and the people in it. You need to learn the ropes, to understand the nuances of how things are done before you can hope to make intelligent suggestions for change or have new ideas accepted. Managers know that college has given you only part of what you need to be successful, so don't make the mistake of believing that you are ready when you first walk through the door.

Learn the Culture

Every company has its own culture. This, in turn, translates into unspoken and informal sets of rules and norms about how you should behave. Organizations want employees who "fit" the culture and enthusiastically embrace it. It is critically important that you take the time to understand the culture and politics of your new workplace. If you don't, you are almost assured of making many dumb mistakes.

Pay attention to "the way things are done around here." And for heaven's sake don't try to teach them "a better method"—at least not yet. Watch your colleagues, paying attention to the things they spend their time on. Learn what the norms and values of the organization are by watching how others behave. Find out the basic mission and philosophy of the organization. Understand what people expect of you, particularly the accepted work-ethic and social norms, and the limits. Pay attention to the political climate and how people communicate and work together. Find ways to "fit in." And remember, you can't change the culture until you are accepted into it.

Make a Good Impression

Impression management is important in your first year. Whenever you start any job, there are a lot of people watching to assess your ability to succeed. Even the smallest mistakes are magnified when you're new to the job. As you progress in your career and build a good professional reputation, your track record will give you a safety net against mistakes. But in the first year it's the impressions that count, since you won't have a rep yet. Since every organization is different, you will need to figure out what the "right" impression is. That's why the first element of a good impression is the ability to read the organizational environment. They are looking for someone who has good judgment and can build good relationships with colleagues. They want to see a readiness to change and learn, plus a healthy respect for the experience and expertise of older employees. They want to see that you have confidence in your potential, but humility about what you can do at first. They want to see that you have your expectations in check and are willing to work hard to learn how to make a contribution. Most of all, it's an attitude they're looking for—one that says you are realistic about your role as a new employee and are willing to do what it takes to earn your spot on the team.

Being New

Most new hires don't really enjoy being new. Like it or not, during this transition period you're going to be the "new kid" at work. Learn the art of being new. The better you become at acting like a new employee, the better off you'll be in the long run. This is contrary to traditional thinking, which says you need to stop acting like a new employee as quickly as possible, but leave that up to your employer—they will let you know when they want to see you take more initiative. Effective new employees understand the importance of accepting their role as the newcomer patiently, and attack the tasks of learning the organization and getting accepted.

Manage your Expectations

A major frustration of many new hires after being on the job a short time is that their expectations are not met. Frustration is nothing more than the difference between expectations and reality. If you work at keeping your expectations realistic, you won't be disappointed. That's right, welcome to the world—surely you've heard of lowered expectations? (Don't worry it's not forever.) The odds are that many things about your job won't be what you expect them to be. It's important to remember that the image painted of the company for applicants is always a bit rosy. The reality of your first job is that it probably won't be nearly as glamorous, or as important, as you'd hoped for. The way decisions are made won't be nearly as logical as you expected, often because of politics. People skills and teamwork will be much more important than you imagined. Most new hires won't find the reality much like what they had expected— there will be more pressure, extra hours, unexpected types of tasks to perform. But stay positive and don't complain—most employers are frustrated with the naive expectations of new workers, so you'll score lots of points if you keep yours realistic.

Become a Good Follower

The single most important person in your first year of work is your new boss/manager. You have to be sure that what you do supports your boss. Learn what your boss wants, needs, and expects—and then do it. Bring your boss solutions, not problems; make your boss look good, you will succeed. Most of all, remember that it takes skill to be a good subordinate; you can't become a good leader until you've learned to be a good follower. Remember, too, that a bad boss is not a legitimate excuse for a poor performance; it is your responsibility to make the transition a success. The good bosses will help, but it's your career. Also, you may not like conforming to your new employer's culture. In time you will be able to assert your individuality and find your own style. Get yourself accepted by the organization and respected by your colleagues, and become productive; only then will you have the right to assert yourself in the organization. That is the art of being new.

Forum Etiquette (adapted from the public domain usenet etiquette guide available at faq.org)

- Only post something after you've followed the group for a few weeks, after you have read the Frequently Asked Questions posting if the group has one, and if you still have a question or opinion that others will probably find interesting.
- Never forget that the person on the other side is human. Do not attack people if you cannot persuade them with your presentation of the facts. If you are upset at something or someone, wait until you have had a chance to calm down and think about it. A time out or a good night's sleep works wonders on your perspective. Angry words create more problems than they solve. Try not to say anything to others you would not say to them in person in a room full of people.
- Never assume that a person is speaking for their organization.
- Be careful what you say about others. Information posted on the net can come back to haunt you or the person you are talking about.
- Your postings reflect upon you—be proud of them. Minimize your spelling errors and make sure that your post is easy to read and understand.
- Be brief.
- Use descriptive titles.
- Be careful with humor and sarcasm. Without the voice inflections and body language of personal communications, it is easy for a remark meant to be funny to be misinterpreted. Subtle humor tends to get lost, so take steps to make sure that people realize you are trying to be funny.
- No double posting: only post a message once, and do not post the same message on many boards.
- Summarize what you are following up. Don't reply to a long post without indicating what part you are responding to.
- Cite appropriate references.
- Don't correct someone else's spelling and grammar.
- Don't use message boards as an advertising medium—keep your spam to yourself.
- Ignore flames, flame wars make everyone like idiots, no matter who is "right."

Potentially Awesome: Why Good Sound Matters

You won't always be able to achieve the sound you want. You need to be able to forgive yourself and others for the mistakes that happen that can interfere with sound, if only for your ability to move on and achieve good sound after the mistake has been fixed rather than get bogged down by past events. Not to mention the good being able to get over things does for one's blood pressure and sanity. As long as you prepare adequately, do the best you can with what you have available, and fix those mistakes you can when you can, then there are few mistakes bad enough to keep you back or ruin the whole show. The only deficiency bad enough to make it better that you just give up and quit is if you don't care enough to try to do your best. Good sound is never a guarantee, but you can be assured of less than stellar sound if the engineer doesn't much care about working hard to make the sound the best he/she can.

And if you remember why good sound matters and who good sound is for, it should make sense to you why you need to give each event your best effort. If you remember that people only expect P.A.s at events they hope will be Potentially Awesome, you should be able to remember why what you do is much more important than it seems on the surface. Whether you are making their anniversary or graduation party the night to remember everyone hopes for, helping mediate their spiritual expression, ensuring that the Friday nights they live for sound as good as you can make them, or simply remember that it may be just a business convention but it's the most exciting business event of the year for those attending, and make sure they can understand the keynote speaker—*people are the reason you do what you do.* No one is always 100 percent attentive or on task, and you won't always be either, but remember on those days you may be tempted to just dial it in or not bother to care that it may be just another day on the job for you, but for everyone else it's one of the types of events that make life sweet, or just make life bearable. Just as important, remember it when you are tempted to feel like what you do doesn't matter. *There is reason to be proud of your efforts, and you should not discount that without the sound engineer and their P.A., there is no Potentially Awesome, no matter how cool the band!*

Making Ethical Choices

Ethics is a framework for doing the right thing when faced with choices where a number of actions and outcomes are possible. No one can tell you what your ethics should be, as there is more than one framework one can use, and good people may come up with different choices when given the same dilemma, yet still all qualify as having made an ethical choice.

An *ethical choice* is a choice of one action out of all other reasonable actions we might have chosen in response to an *ethical problem,* a choice where other parties beyond oneself have a stake in the outcome and one or more *tradeoffs* are involved. Some of these choices are simply personal choices reflecting one's personal *values* (such as which charity to donate $50 to, or whether to keep it for bill payments)—in these cases there is no unethical choice. Other situations represent more weighty ethical choices.

Ethical problems, where no matter what the choice, one principle must be betrayed if the other is to be honored, are *ethical dilemmas.* There are minor ethical dilemmas and major ones, depending on the amount of harm and type of harm, and who the stakeholders are.

To make ethical choices, options need to be examined to assess whether they are in accord with the chooser's ethical principles, many of which will be in line with *universal ethical principles* (loyalty, honesty, fairness, responsibility, non-violence, etc.). It's also important to consider how the outcomes may benefit or harm stakeholders, which at minimum will include the decision maker and greater society, and often will include other parties. Then, depending on the decision maker's values and priorities, they will make the best choice they can using their reason and intuition. Often this process will take place in their head in a split second, other times it will take hours, even days of deliberation.

The Ethical Live Audio Engineer

So why should a live engineer worry about ethics? Well. because without it you can't always be sure to do the right thing, even if you're a good person, because what feels right can be the wrong choice. Most people have a guiding ethical principle that informs their choices even if it is vague and unarticulated. But making ethical choices is easier and more consistent for people who have taken some time to clarify their values and prepare for those dilemmas where it is difficult to decide on the right path. Those who want more info can check the chapter directory which includes links to great resources on ethics, but you need to develop your own ethical approach based on what's important to you if the process of articulating your ethical foundation is going to be effective.

Engineering Dilemmas

P.A. Reception or P.A. Deception?

Until you have had a year of pretty regular chances at mixes you will still be learning and not as skilled as you will eventually be. You've been charging student prices in line with your limited experience, and renting your services and P.A. to bands for club gigs, and to frat parties, but have not yet worked a big or high stake gig as a professional. However, say you get a call from someone who doesn't know you aren't a pro yet, offering to pay you full market prices to mix a large outdoor wedding. Do you refer the gig to someone with enough experience and ask to tag along so you can learn, do you hire someone to help you pull it off but keep up an appearance of doing it all on your own, or do you book the client and hope you won't mess up their reception sound too much?

Fix Her or Stay at the Mixer?

You are mixing at a private party at a large cabin/camp site outside of the city with a mix of P.A. bands and DJs all weekend. Right now a live DJ/band act is up and you are finding the mix very interesting, since there still aren't too many of these hybrid bands. However, you notice a girl who looks passed out right now near the wall vomit and have what might have been a little seizure, and you are not sure how OK she is. None of the party staff are around because it's a daytime set, and no one but you saw her, or is sober enough to get help for her. You realize that to go get help you'd have to walk halfway across the camp site. Do you go get someone just in case, at least to check if she is breathing, or do you put it out of your mind and keep mixing because she's probably fine and that's not what you're hired for anyway?

P.A. Principles or Deafening Decibels?

You are mixing at a festival and the sound at the mixer is already 115 dB 25 feet from the main speakers. While there was some tape around them keeping festival-goers from getting within 10 feet, several fans are now behind the tape and literally a few feet from the speakers and security doesn't seem to care. Just then the deaf-in-one-ear producer walks up and tells you to turn it up, and scowls but doesn't care when you point out the now growing cluster of fans already in danger of permanent damage right in front of the stacks. Do you turn it up knowing you are dangerously close to immediately deafening some festival-goers (as in it won't take years to show up, because 140 dB can knock out a nice chunk of hearing at once)? Do you say yes but fake it, turning down gain as you increase mains? Do you say sorry, but that's illegal and irresponsible? Do you play deaf?

Summer Stampede or Sound Guy Save?

You are mixing a show at an amphitheater with tiered step-ups in the center bowl where the audience is and as the very popular band took the stage 15 minutes ago you began to see the fan per square meter density get kind of sketchy, but now just saw four or five kids at the bottom of the first step-up fall down and not get up and the stream of fans continue to flow down into the bowl. For 90 seconds frantic-looking security have been trying to tell the band to stop, to allow them to be heard by the surging crowd but they are being ignored. You can not see security any more so hope they dragged the kids out from under foot, though you know they could just be blocked from view by the crowd and are still desperately trying to avert another trampling tragedy (see web directory). You might look stupid if you shut down the sound and say something, if they already have it handled, but if they do not you don't have long to decide before it's too late. Do you turn down and talk back?

Business Ethics Earns Trust

Business ethics also make good business sense. Many of your ethical dilemmas will be ones your clients, employer, audience, and/or bands have a stake in. Ethical professionals are more trusted and sought after by employers and clients. Ethical pros consider all stakeholders when making decisions, rather than only themselves, so doing business with ethical partners means all stakeholders can feel confident they have a chance of being treated fairly.

Ethics is messier than systems of imposed morality because you have to think about each choice, with no set of inflexible rules to turn to. There is a grey area even after choosing a framework and method to guide one's choices, but without ethics everybody would be left with doing what felt right. Most would either then turn to religious precepts, which may or may not apply fairly in a modern context, or will otherwise choose the path benefiting their own self interest. Ethics doesn't allow for unconsidered choices and requires a lot more work on the part of the user.

Our choices might not be the same but we could respect each other's choice as being deliberate and based on something agreed upon even if we didn't agree on the outcome. It's not a neat tidy system with no gray area, because it has to have a gray area to cover every possible situation. And it leaves room for human error but that's not as bad as leaving all choices up to personal whim.

Index

Photographs and diagrams are indexed in bold. Where figure (bold) and text locators are identical for a given entry, only the bold range is given.

M

Q

R

X

Z

Milton Keynes UK
Ingram Content Group UK Ltd.
UKHW052016071024
449327UK00027B/2292